Lecture Notes in Civil Engineering

Volume 47

Lecture Notes in Civil Engineering (LNCE) publishes the latest developments in Civil Engineering - quickly, informally and in top quality. Though original research reported in proceedings and post-proceedings represents the core of LNCE, edited volumes of exceptionally high quality and interest may also be considered for publication. Volumes published in LNCE embrace all aspects and subfields of, as well as new challenges in, Civil Engineering. Topics in the series include:

- Construction and Structural Mechanics
- Building Materials
- Concrete, Steel and Timber Structures
- Geotechnical Engineering
- Earthquake Engineering
- Coastal Engineering
- Hydraulics, Hydrology and Water Resources Engineering
- Environmental Engineering and Sustainability
- Structural Health and Monitoring
- Surveying and Geographical Information Systems
- Heating, Ventilation and Air Conditioning (HVAC)
- Transportation and Traffic
- Risk Analysis
- Safety and Security

To submit a proposal or request further information, please contact the appropriate Springer Editor:

- Mr. Pierpaolo Riva at pierpaolo.riva@springer.com (Europe and Americas);
- Ms. Swati Meherishi at swati.meherishi@springer.com (India);
- Ms. Li Shen at li.shen@springer.com (China);
- Dr. Loyola D'Silva at loyola.dsilva@springer.com (S-E Asia and Australia/NZ).

Indexed by Scopus

More information about this series at http://www.springer.com/series/15087

Zinoviy Blikharskyy · Piotr Koszelnik ·
Peter Mesaros
Editors

Proceedings of CEE 2019

Advances in Resource-saving Technologies and Materials in Civil and Environmental Engineering

Springer

Editors
Zinoviy Blikharskyy
Lviv Polytechnic National University
Lviv, Ukraine

Piotr Koszelnik
Rzeszow University of Technology
Rzeszow, Poland

Peter Mesaros
Technical University of Kosice
Kosice, Slovakia

ISSN 2366-2557 ISSN 2366-2565 (electronic)
Lecture Notes in Civil Engineering
ISBN 978-3-030-27010-0 ISBN 978-3-030-27011-7 (eBook)
https://doi.org/10.1007/978-3-030-27011-7

This Springer imprint is published by the registered company Springer Nature Switzerland AG
The registered company address is: Gewerbestrasse 11, 6330 Cham, Switzerland

Preface

Building constructions, construction technologies, renewable energy sources, heat and gas supply, ventilation and water supply of the buildings are the areas where extreme progress has been made in recent years.

After many successful international conferences in the field of construction and environmental engineering (CEE) «LVIV- KOŠICE - RZESZÓW» Current Issues of Civil and Environmental Engineering took place over the past 20 years in Rzeszow, Poland (1995, 2000, 2004, 2009, 2015), Košice, Slovakia (2002, 2005, 2011, 2017) and Lviv, Ukraine (2001, 2003, 2007, 2013). Lviv Polytechnic National University has organized and held the XVII International Scientific Conference «LVIV- KOŠICE - RZESZÓW» on 11–13 September 2019, in Lviv.

The Lecture Notes in Civil Engineering (LNCE) contains the latest advances, innovations and applications in the field effective methods of calculation, resource-saving technologies and progressive materials in civil and environmental engineering, as presented by leading international researchers and engineers at the XVII International Scientific Conference «LVIV- KOŠICE - RZESZÓW» Current Issues of Civil and Environmental Engineering «LVIV- KOŠICE - RZESZÓW» dedicated to the 130th anniversary of the renowned professor A.KURYLLO (1889–1980), held in Lviv, Ukraine, on 11–13 September 2019. It covers highly diverse topics, including structural shaping and optimization, aspects of structural behaviour and modelling of advanced methods of analysis, experimental tests and numerical simulations, design codes, in particular Eurocodes and other national and regional limit state codes, highway and bridges engineering, modern architectural and structural solutions, innovative materials and products, durability and maintenance, fabrication and erection, sustainability in construction, renewable energy sources, heat, gas and water supply, ventilation and air-conditioning, ecological and energy-saving technology, modern technology in water purification and treatment, and protection of water ecosystems. The contributions, which were selected by means of a rigorous international peer review process, highlight numerous exciting ideas that will spur novel research directions and foster multidisciplinary collaboration among different specialists.

In particular, this book introduces a number of new advances and includes interdisciplinary research, experimental and theoretical studies that advance achievements in civil engineering and environmental engineering. These advances have significant implications in a number of research areas such as sustainable construction and infrastructures, waste and wastewater management, enhanced sustainability, renewable and clean energy, sustainable materials and industrial ecology, and building automation.

Seventy-two papers have been carefully reviewed and selected from over 120. A total of over 200 researchers are represented in this book.

We express our sincere gratitude to the members of the Organizing Committee, to the members of the Scientific Committee, and in particular to all authors and participants for their substantial and valuable contributions.

We express our sincere gratitude to Mr. Pierpaolo Riva, Publishing Editor at Springer for his great support in preparing the final book.

We hope that the reader will find the book inspirational for further research and professional life.

Zinoviy Blikharskyy
Piotr Koszelnik
Peter Mesaros

Organization

Chairmen

Zinoviy Blikharskyy
Piotr Koszelnik
Peter Mesaros

Scientific Committee

Zinoviy Blikharskyy, Ukraine
Piotr Koszelnik, Poland
Peter Mesaros, Slovakia
Natalia Fialko, Ukraine
Khrystyna Sobol, Ukraine
Myroslav Sanytskyy, Ukraine
Serhiy Solodkyy, Ukraine
Yevhen Kharchenko, Ukraine
Volodymyr Chernyuk, Ukraine
Vasyl Zhelykh, Ukraine
Maosheng Zheng, China
Vincent Kvočák, Slovakia
Nadezda Stevulova, Slovakia
Rybka Adam, Poland
Dushan Katunsky, Slovakia
Zuzana Vranayova, Slovakia
Stanislav Kmeť, Slovakia
Renata Gruca-Rokosz, Poland
Bartosz Miller, Poland
Lucjan Ślęczka, Poland
Lech Lichołai, Poland
Barbara Tchórzewska-Cieślak, Poland
Dorota Papciak, Poland
Marek Gosztyła, Poland
Zbigniew Zuziak, Poland
Jacek Selejdak, Poland
Mária Kozlovská, Slovakia
Magdaléna Bálintová, Slovakia
Orest Vozniak, Ukraine
Petro Kholod, Ukraine
Todor Donchev, UK
Mariusz Adamski, Poland

Contents

About the Editors

Prof. Zinoviy Blikharskyy is Professor of Civil Engineering and Director of the Institute of Building and Environmental Engineering of the Lviv Polytechnic National University, Lviv, Ukraine. He has more than 35 years of teaching, research and administrative experience in higher education institutions and universities in Ukraine and abroad. He has received engineering and construction education in the field of industrial and civil engineering, as well as both Ph.D. and Dr. Sc. degrees in the field of building constructions. He is the winner of the State Prize for "Restoration of the Palace of Bandinelli", which is an award of the President of Ukraine. His scientific interests are researching of corrosion of reinforced concrete structures with taking into account load action; operational suitability, strengthening of reinforced concrete constructions under the action of loading; use of non-steel reinforcement for strengthening of reinforced concrete constructions; reconstruction and restoration of buildings and structures of historical and modern buildings. He has published more than 150 scientific papers, supervised ten candidates' theses and was a co-founder of 6 international conferences "Current Issues of Civil and Environmental Engineering and Architecture" (2007–2017 years). Also, he is Active Member of the Academy of Construction of Ukraine and Editor-in-Chief of the magazine "Theory and Practice of Construction" since 2007.

Prof. Piotr Koszelnik is Associate Professor of Environmental Engineering at the Faculty of Civil and Environmental Engineering, Rzeszów University of Technology, Poland, and since 2010 Head of the Department of Environmental Engineering and Chemistry. He serves as Dean of the Faculty and Member of the Committee on Environmental Engineering of the Polish Academy of Sciences. His main academic interests include environmental chemistry especially water chemistry, carbon and nitrogen cycling, stable isotopes, eutrophication, micropollutants in water and man-made lakes. After obtaining the habilitation, he conducted the project "Analysis of nitrate sources in the surface waters of the San River catchment based on nitrogen and oxygen isotopic ratios". These studies were thematically related to the implementation of the EU Nitrates Directive, under which it is necessary to identify sources and then to limit the runoff of nitrates from agricultural sources to waters. Moreover, he was one of the investigators for the research project "Emission of greenhouse gases from dam reservoirs". As part of the project, the study was conducted on the production of methane and carbon dioxide in the bottom sediments of six dam reservoirs located in south-eastern Poland and emission of these gases into the atmosphere. He was also the manager of the project "Study on resilience to degradation and the possibility of reclamation of small water retention reservoirs with the use of calcium preparations". He is the author of more than 100 scientific papers.

Prof. Peter Mesaros is Associate Professor at Institute of Construction Technology, Economics and Management and Dean at Faculty of Civil Engineering, Technical University of Košice. He obtained his PhD. degree in sectoral economics with specialization on construction industry at University of Economics in Bratislava. He has many years of experience in the development and implementation of information systems and technologies in construction practice. His research addresses the issue of knowledge and information management in construction industry, and he is Head of several research projects focused on research of knowledge-based and virtual technologies for intelligent designing and realization of building projects with emphasis on economic efficiency and sustainability, with specialization on identification and mapping the non-graphical data in building information modelling. He worked as Member of Working Group SOC-1 "Vocational training" in the European Construction Industry Federation (FIEC) in Brussels and Representative of the Association of Construction Entrepreneurs of Slovakia in the social dialogue FIEC-EFBWW within the European Union. He had close connection with industry while working at several managerial positions in applied research in construction and information technologies. He is the author and co-author of over 100 publications in the field of economics and management, information and communication technologies in construction industry.

Reinforced Concrete Rod Elements Stiffness Considering Concrete Nonlinear Properties

T. Azizov[1], D. Kochkarev[2], and T. Galinska[3(✉)]

[1] Pavlo Tychyna Uman State Pedagogical University, Uman, Ukraine
[2] National University of Water and Environment Engineering, Rivne, Ukraine
[3] Poltava National Technical Yuri Kondratyuk University, Poltava, Ukraine
galinska@i.ua

Abstract. In this paper results of stiffness iterative determination technique while torsion the reinforced concrete rod elements considering concrete nonlinear properties is given. The section of the element is divided into several rectangular elements. The shearing torsions at each iteration step are determined by the distribution of torsions, which magnitude is higher than the limit, over the elements where the torsions are in the elastic area. An algorithm for obtaining points on the "Torque moment is relative twisting angle" diagram of a reinforced concrete rod elements is given. Its advantage from a similar technique is that shear torsions are determined directly from the concrete shear diagram, and not according to the formulas of the elasticity theory. The above algorithm enables to calculate reinforced concrete rod elements using any concrete shear diagrams obtained both experimentally and theoretically, as well as elements of any cross-section. Limiting torsions are determined directly from the concrete shear diagram.

Keywords: Torsion · Reinforced concrete rod · Nonlinear properties · Diagram · Iteration

1 Problem Statement and Research Analysis

Quite a lot of work has been devoted to the calculation of reinforced concrete elements. However, the calculation of the strength and stiffness of reinforced concrete elements spatial sections and in the case of cracking, spatial torsion cracks are considered in most works (Krainskyi et al. 2018; Khmil et al. 2018; Ponamonov 2012; Karpenko 1976; Cowan 1972). Investigations concerning the calculation of reinforced concrete elements with normal cracks in torsion have been carried out in the works (Azizov 2009; Sribyanik 2009; Azizov 2007). Mainly the problem of determining the torsion state of a block separated by normal cracks, which torque moments are transmitted via part of the section is solved in these works.

An approximate solution of reinforced concrete rod elements torsion problem considering nonlinear properties was studied in (Kochkarev et al. 2018; Jurkowska 2018; Jurkowska 2019; Yaremenko and Shkola 2010; Karpenko 1976). Nonlinear properties of concrete were taken into consideration. However, in this paper,

Z. Blikharskyy et al. (Eds.): CEE 2019, LNCE 47, pp. 1–6, 2020.
https://doi.org/10.1007/978-3-030-27011-7_1

a significant error was made when the tangential torsions in the cross-section were determined by the theory of elasticity formulas. It cannot be fallen short of expectations.

Taking into account the above mentioned, the purpose of this article is to develop a method for determining the torsion stiffness of reinforced concrete rod element considering nonlinear concrete properties.

2 Main Material Presentation

Consider the typical section of the rod element subjected to torsion. For the clarification simplicity, it will be considered a section without reinforcement. Considering reinforcement is carried out like in the calculation of reinforced concrete elements for bending by multiplying the area of reinforcement elements by a modular ratio equal to the ratio of the reinforcement elastic modulus to the elastic modulus of concrete. Besides for clarification simplicity, it will be considered a rectangular section. Any other section elements can be calculated using the method described below.

The cross-section is divided into several rectangles (Fig. 1).

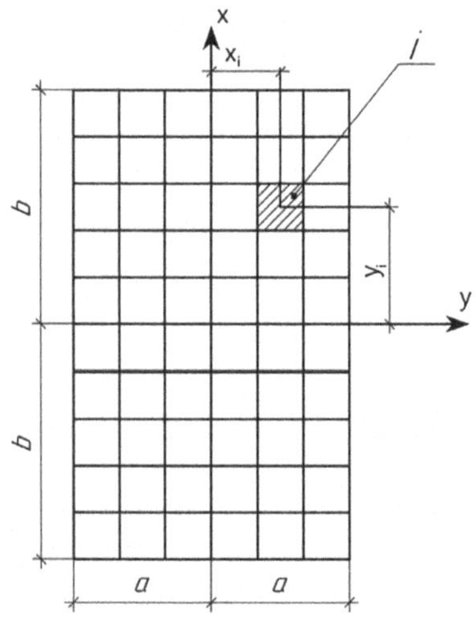

Fig. 1. Cross section dividing into separate elements diagram

Draw the X and Y axes from the element center of torsion, which in the general case may not coincide with the center of the section gravity (Harutyunyan 1963). The definition of a torsion center is not difficult. Each i element (shaded in Fig. 1) has coordinates Xi and Yi. Each element has tangential forces:

$$T_{yz,i} = \tau_{yz,i} A_i; \; T_{xz,i} = \tau_{xz,i} A_i. \tag{1}$$

If the distribution of tangential torsions over the cross section is known $\tau_{yz,i}$ and $\tau_{xz,i}$, it is not difficult to calculate the tangential forces $T_{xz,i}$ и $T_{yz,i}$, and, therefore, the torque in cross section:

$$M_t = \sum_{i=1}^{n} \left(T_{yz,i}X_i + T_{xz,i}Y_i\right). \tag{2}$$

Where n is the quantity of the total elements, the section is broken into.

From the theory of torsion, it is known (Harutyunyan 1963) that the torque moment Mt in cross section is associated with the relative angle of twist Θ dependence:

$$M_t = D \cdot \theta. \tag{3}$$

Where D is rod stiffness (in the section under consideration) in torsion.

It should be particularly noted that the dependence (Azizov 2010) is valid for any distribution of torsions in the element cross section. In other words, if the distribution of tensions in the section (and the torque moment in the section), as well as the relative twisting angle, is known, then the stiffness of section D is certainly determined from the expression (Azizov 2010). And vice versa, if stiffness D is known, then the relative twist angle is certainly determined from (Azizov 2010). This is a significant point to consider. It also helps to determine the stiffness of the element during torsion.

Consider the algorithm for determining the stiffness in the cross section of the rod element considering concrete nonlinear properties. In this case, first, the shear diagram will be taken in the form of the Prandtl diagram. Such a presentation of the diagram is based on experimental studies conducted in (Vildanova 2015).

1. Set the preliminary value of the twist relative angle $\theta 1$;
2. Assuming elastic rod work, determine the elastic moment $M_{t,e}$, which corresponds to a given twist angle:

$$M_{t,e} = GI \cdot \theta_1. \tag{4}$$

Where GJ is torsion stiffness of the rod assuming its elastic work;

3. According to the theory of elasticity formulas (for example, according to [9]), determine the tangential torsions $\tau_{yz,i}$ and $\tau_{xz,i}$ in the center of every i- element;
4. Determine the maximum stress value. $\tau_{yz,max}$, and $\tau_{xz,max}$
5. Determine the ratio

$$k_{yz,i} = \tau_{yz,i}/\tau_{yz,max}; k_{xz,i} = \tau_{xz,i}/\tau_{xz,max}. \tag{5}$$

6. Calculate all ratio sums. $k_{yz,i}$ and $k_{xz,i}$, which values are less than one:

$$k_{tot,x} = \sum_{j=1}^{m} k_{xz,j}; k_{tot,y} = \sum_{j=1}^{n} k_{yz,j}. \tag{6}$$

Where n, m is elements quantity where the stress exceeds the limit, respectively $\tau_{yz,i}$ and $\tau_{xz,i}$.

7. If in any elements, the shear torsions are greater than the limit shear torsions $\tau i > [\tau]$, then the stress in this element is taken equal to $[\tau]$. It is considered as a stress τ_{yz}, and as τ_{xz};
8. Calculate the value of the "extra" moments in each element, where the torsions exceed the limiting:

$$\varDelta M_{y,i} = \left(\tau_{yz,max} - \tau_{yz,i}\right)A_i \cdot X_i;\ \varDelta M_{x,i} = \left(\tau_{xz,max} - \tau_{xz,i}\right)A_i \cdot Y_i. \tag{7}$$

9. Calculate the total "extra" torque moments

$$M_{e,y} = \sum_{k=1}^{n} \varDelta M_{y,k};\ M_{e,x} = \sum_{k=1}^{m} \varDelta M_{x,k}. \tag{8}$$

Where m and n are the same as in point 6.

10. We distribute the "extra" moments between the elements where $k_{yz,i}$, and $k_{xz,i}$ are less than one, in proportion to the ratio:

$$M_{y,i} = \frac{M_{e,y}}{k_{tot,y}} k_{yz,i};\ M_{x,i} = \frac{M_{e,x}}{k_{tot,x}} k_{xz,i}. \tag{9}$$

11. Determine the new values of the tangential torsions in the elements where, at the first iteration, the torsions were less than the limiting ones (taking into account the added "extra" torsions:

$$\tau^r_{xz,i} = \tau_{xz,i} + \frac{M_{x,i}}{A_i Y_i};\ \tau^r_{yz,i} = \tau_{yz,i} + \frac{M_{y,i}}{A_i X_i}. \tag{10}$$

12. Further, the calculation is repeated, starting with point 6. In this case, the elements quantity m and n, where the torsions exceed the permitted ones, will already differ from the previous ones.

Note: at each iteration step, the position of the torsion center should be specified. Accordingly, Xi and Yi also are specified.

Thus, at the end of the iterative calculation, obtain the final distribution of tangential torsions.

13. Knowing the final torsion values $\tau_{xz,i}$ and $\tau_{yz,i}$ in all elements the cross section is divided into(see Fig. 1), using Eq. (1), calculate the tangential forces $T_{yz,i}$ and $T_{xz,i}$;
14. By the Eq. (2) determine the torque moment Mt from intensification $T_{yz,i}$ and $T_{xz,i}$. At that time, the torque moment Mt will be different from the first moment $M_{t,e}$, determined by Eq. (4). This is because some torsions exceeding the value $[\tau]$, will no longer contribute to the total torque moment.

So we get the torque value Mt, corresponding to the relative twist angle given in point 1 θ 1 or the first point on the diagram «Mt – θ».

15. Next, increase the value of the relative twist angle θ2 and, after performing the calculation using the above algorithm, obtain the second point in the diagram.

Continuing incrementally increasing the values of the twist angle, get the diagram «Mt – θ», it is easy to obtain the values of the twisting angle with the real torque moment acting on the real rod, as well as the value of the torsion stiffness D corresponding to the specified real torque moment, since knowing θ and Mt, by the Eq. (3) the stiffness D is easy to determine.

If at the initial stages of calculation, it turns out that all torsions are less than the limit, and then this part of the diagram «Mt – θ» is built on points considering elastic laws.

It should be mentioned as a significant remark. If the real diagram of the concrete shear is known (Vildanova 2015), then the torsions in the elements according to the point of the algorithm should be taken directly from the diagram. As mentioned above, expression (Azizov 2007) is valid for any distribution of tangential torsions in the section. So, to set the specific value of the twisting angle, then, using the concrete shear diagram, determine all torsions (exceeding elastic values) and by the formula (Azizov 2010) taking into account (Azizov 2009) determine the moment Mt. And get the point on the diagram «Mt – θ». Further, similarly to the above, we repeat the calculations with other values of the twist angle and get the diagram «Mt – θ».

Rod bearing capacity exhaustion will be the top point in the diagram.«Mt – θ», when the value of torque moment is reduced because in separate elements there will be deformations equal to the limit deformation in the concrete shear diagram.

According to the described algorithm, it is easy to get a diagram.«Mt – θ» for rods with a section differing from a rectangular section. The principle of calculation does not change for any section. There are solutions to the theory of elasticity by definition of tangential torsions.

To calculate a rod with a cross section which has no solutions to the theory of elasticity by definition of tangential torsions, the following technique can be used. Similar to (Azizov 2010), based on calculations using well-known programs like Ansys, Lira, and others, using finite bulk elements, obtain the laws of the distribution of tangential torsions over the cross-section. Then the calculation considering the concrete nonlinear properties is carried out according to the algorithm described above.

Longitudinal reinforcement are not marked in Fig. 1. The presence of reinforcement is easy to take into account (while calculating reinforced concrete elements by bending) by replacing its cross-section with an equivalent by multiplying its cross-sectional area by reinforcement and concrete elastic modulus ratio.

In the case of normal cracks as it follows from the diagram in Fig. 1 exclude concrete elements which are within the crack. In this case, the determination of the gravity center and torsion center in a cross-section is easy to perform in resistance to materials known method.

3 Conclusions and Research Prospects

An algorithm for obtaining points on the "Torque moment is relative twisting angle" diagram of a reinforced concrete rod element is given. Its advantage from a similar technique is that shear torsions are determined directly from the concrete shear diagram, and not according to the formulas of the elasticity theory. The above algorithm enables to calculate reinforced concrete rod elements using any concrete shear diagrams obtained both experimentally and theoretically, as well as elements of any cross-section.

In the future, it is planned to develop a computer program and experimental verification of the developed methodology.

References

Azizov T (2010) The use of approximation finite elements in the calculations of structures. News Odessa State Power Acad Archit 39(1):4–9

Azizov TN (2009) Stiffness of reinforced concrete elements in torsion and its influence on the spatial work of bridges NAS of Ukraine. Physico-mech. Institute of Im .. Karpenka, VG, pp 576–590

Cowan GJ (1972) Torsion in ordinary and prestressed concrete: Trans. from English Publishing house of literature on construction, p 104

Harutyunyan NKh (1963) Torsion of elastic bodies. Publishing house of physical and mathematical literature, p 686

Jurkowska N (2018) Considering nonlinear properties of concrete in the design of reinforced concrete structures for torsion. Paper presented at the IOP conference series: materials science and engineering, vol 364, no (1). https://doi.org/10.1088/1757-899x/364/1/012030

Jurkowska N (2019) Presentation of the exact technique for calculation of the torsional constant for the T-section. Paper presented at the IOP conference series: materials science and engineering, vol 473, no (1). https://doi.org/10.1088/1757-899x/473/1/012034

Karpenko NI (1976) The theory of reinforced concrete deformation with cracks. Stroiizdat, p 208

Khmil R, Tytarenko R, Blikharskyy Y, Vegera P, (2018) Development of the procedure for the estimation of reliability of reinforced concrete beams, strengthened by building up the stretched reinforcing bars under load, Eastern-Eur J Enterp Technol 5/7(95)

Kochkarev D, Azizov T, Galinska T (2018) Bending deflection reinforced concrete elements determination. Paper presented at the MATEC Web of conferences, p 230. https://doi.org/10.1051/matecconf/201823002012

Krainskyi P, Blikharskyy Y, Khmil R, Vegera P (2018) Influence of loading level on the bearing capacity of RC columns strengthened by jacketing. Matec web of conferences, p 230

Paramonov DJu (2012) Rigidity and strength of reinforced concrete elements with normal cracking when torsion-bending, p 290

Sribyanik NM (2009) Rotational rigidity of reinforced concrete elements of overlappings with normal cracks: author's abstract. dis … Cand. tech Sciences 05.23.01; Odessa State Academy of Civil Engineering and Architecture, p 23

Vildanova NR (2015) Modulus of shear of concrete in view of plasticity deformations and its influence on torsional stiffness of reinforced concrete elements: dis. … candidate of technical sciences … 05.23.01 « Building constructions, buildings and structures » , p 214

Yaremenko O, Shkola Yu, (2010) Bearing ability and deformability of reinforced concrete rod elements in a complex stressed state. ODABA, p 136

Azizov T (2007) Determination of reinforced concrete elements torsional stiffness with cracks. Roads and bridges. Scientific Paper Collection, vol 7, no (1), pp 3–8

Medieval Material and Technological Experiment in the Construction of King Daniel Romanovich's Residence in Cholm

Lucjan Gazda[1] and Mykola Bevz[2,3](✉)

[1] Department of Geotechnics, Faculty of Civil Engineering and Architecture, Lublin University of Technology, 40 Nadbystrzycka Street, 20-618 Lublin, Poland

[2] Department of Build Conservation, Faculty of Civil Engineering and Architecture, Lublin University of Technology, 40 Nadbystrzycka Street, 20-618 Lublin, Poland

m.bevz@pollub.pl

[3] Department of Architecture and Conservation, Lviv Polytechnic National University, 12 Bandery Street, Lviv 79013, Ukraine

Abstract. Materials obtained during archaeological excavations on Wysoka Gorka in Cholm (today Chełm in Poland), in the area of king Daniel's 13th century residential complex, have been subject to examination. They come in the forms of bricks and glazed tiles, as well as sizeable shapeless and purposeless accumulations indicating post-manufacturing remains. The materials are white, green and multi-colored. They refer to Halytsian alabasters and green glauconite from Cholm which were originally used there. The examination of white materials was performed. They were diagnosed by means of SEM microscope and X-ray diffraction. Furthermore, a micro area chemical analysis was conducted by means of SEM microscope with EDS module. The conducted examination indicated that the materials in question were manufactured using other than ceramic technologies, but similar to the ones used to produce silicate materials nowadays. As raw materials, chalk and biogenic silica obtained from a horsetail were used. The petrification procedure was conducted in hydrothermal conditions. As a result a material structurally similar to marble was obtained. The successful production of the materials translated into possibility of the implementation of the ideological assumptions of the structures having been constructed under the explicit influence of the style of the 13th century Tuscany School.

Keywords: Artificial stone · Chalk · Alchemy · Construction of residence · Cholm (Chełm) · 13th century

1 Introduction

The foundation of capital city of Cholm (Chełm, in polish language) in the thirties of the 13th century by Daniel Romanowich (1201–1264) was most likely realized from scratch, within the period of stormy political and military turbulences in Europe taking

Z. Blikharskyy et al. (Eds.): CEE 2019, LNCE 47, pp. 7–16, 2020.
https://doi.org/10.1007/978-3-030-27011-7_2

place in the first part of 13th century. The occupation of Constantinople by crusaders of the fourth crusade resulted in exerting strong influences of Latin culture on the Halych-Wolodymyria Duchy virtually from all the sides (Voytovych 2014). In the architecture, it was the time of fully matured Romanesque and the breakthrough of Gothic technical and technological solutions. The use of white and green stone in the architectural detail of St John Chrysostom's Church (Halych-Wolhynia Chronicle) within Cholm acropolis was not accidental. The excerpt from Halych-Wolhynia Chronicles describing St Chrysostom's Church: "... He built St. John's Orthodox Church to be outstanding and flawless. Its structure was as following. Arcades 4, there was a vault from every corner, and their foundation on four human heads created by some master. Three windows, embellished by Roman glass (stained glass windows). Approaching the altar, there are two columns, all made from stone, on which there are an arcade and a dome decorated by golden stars in a glaze background. Its inner floor (it relates to the orthodox church) was cast in copper and pure lead gleaming like a mirror. Its two doors, ornamented by Halych white and green hewn stone from Cholm, were sculptured by a master called Avdiy (with) ornaments (in) all colors and gold. At their front (i.e. in the west portal) there was Savior, and at the north part, St John who aroused the feeling of astonishment in viewers" (Buko 2016, p. 224).

The description in chronicle proves that the architectural and structural solution of a building was made at the highest technical an artistic level of its own time. The use of stained glass windows, stone sculpture (heads on columns hewn from whole stone), gold-plated polychrome, Cholm hewn green and Halych white stones in the decoration of door portals, ornaments in all colors and gold, the creation of floor in the unique way from plates cast in copper and lead - demonstrate that it was architectural and artistic action focused on obtaining the ideological result. It may be possible that the location itself in the area where green glauconite meets white chalk was the stimulus for the implementation of this program. The sacral, royal and representative objects of early Byzantium Christianity, the Empire of Charlemagne, or finally, architecturally perfect Roman cathedrals, especially the ones that were inspired by the Pisan School formed in the second part of 12th century, all were arranged in green and white colors. Consequently, in such a context, the special attention shall be paid to the scale and direction of urban and building actions of Daniel Romanowich who founds the series of new towns, builds a new capital city (the action having no analogy in the east European history of the 13th century), builds in a very short time four new temples in Cholm, and after the fire in 1256 rebuilds them. The very significant professional and technical issue described in the chronicle which caused rapid development and high technological level of civil engineering in Cholm – "... Daniel saw that God helps this place, he started calling craftsmans—Germans, Ruthenians, different tribes, Poles went day by day and youths, and all kinds of masters escaped Tartars, saddlers and archers, and thulnics (producers of quivers), blacksmiths of iron and copper, and silver, and there appeared life and they filled courts around the town, fields, villages. He built the St John Orthodox Church to be beautiful and ideal" (Buko 2016, p. 224). This church could be made as a result of construction, technological and artistic experience connections of professionals from Ruthenia, Poland and other parts of Europe. It is most likely that there was the atmosphere of healthy competition and experimentation (Bevz et al. 2015). The chemical and petrographic research on mortars used in the construction of other King Daniel's facility in Cholm – the Cathedral Church of the Blessed Mother of God (Hutzuliak and Shevchenko 2015, pp. 196–206), confirms the experiments with construction materials in the 13th century.

Architectural decoration connecting the green of Cholm glauconite and the white of Halych alabaster brought Cholm to the sophisticated Byzantium patterns and Roman Europe, familiar to Daniel from direct and indirect connections (Dąbrowski 2012). Materials, used to the implementation of this architectural and technical establishment, were yet poorly known (Halych), and in the case of glauconite, not verified in use at all. Alabaster turned out to be not resistant to high temperatures of fire which devoured acropolis facilities in 1256. Similarly, glauconite proved to be not resistant to atmospherical factors, intensified by the action of fire. The problem in the recreation and maintenance of the concept of these ideologies during the reconstruction after the destruction, was considerable dismissal of Cholm from the sources of approved construction stone materials applied to Roman buildings preserved until now. For instance, white and green marbles as well as Tuscan green serpentinites. As far as, there were the resources of green glauconite at the disposal of then investors, a white stone posed the problem. It was most likely that the expensive transportation of alabaster was not repeated from the regions of Halych, which failed in the first uses, yet it could possibly be unavailable due to political reasons (Dąbrowski 2012).

Last archeological studies of foundation relics of the Cathedral Orthodox Church of the Blessed Mother of God in Krylos-Halych also demonstrate the alteration of construction approach in this facility in the 12th–13th centuries. Within the first stage, the cathedra was built from the white alabaster stone, then it was widen and rebuilt using the white limestone. This technological change took place as result of state evaluation and failure of the first material (Lukomskyi and Bevz 2012, p. 26).

The lack of white stone resource in Cholm caused the demand for experimenting with petrification of chalk – soft rock abundantly available locally. The beginning of intensive exploitation of chalk in Cholm came in the 13th century. The use of limestone from baked chalk to mortars gave the knowledge that traditional heating or sintering did not bond loose structure of chalk, but it led to the creation of quicklime, air binder. There was the demand for 'the philosopher's stone' necessary to transmutation. Since the 11th century, white porcelain, which had to inspire the creation of artificial white stones, has been known in Europe through the contacts with the Muslim world. White materials, found during archeological studies (Gołub 2013) were considered so far to be ceramic, however, during more thorough analysis, they turned out to be materials obtained from chalk, and not from white firing clays.

The research study on petrification and mineral materials filling this "alchemic" retort is performed in this article. According to the obtained results, it can be stated that within the frame of conducted experiments, the petrification of chalk with the use of active biogenic silica contained in a horsetail could be carried out, conducting the process in the hydrothermal conditions.

2 The Object of Research

The materials obtained from the eastern profile of excavation no. 23 were subjected to analysis, where the object of distinctly outlined rectangular shape and vague purpose was initially documented (Fig. 1, from the left), which is accompanied by intentional accumulation of chalk (Fig. 1, from the right). This object is located beyond residential arrangements at the east bank. Archeologically, they can be dated for the 2nd or 3rd functioning phase of Daniel's residential and sacral complex, after the destruction of outer walls and raising the ground level, so after the fire from 1256.

Fig. 1. Eastern wall of trench no. 23 within the residential and sacral complex at Wysoka Gorka in Cholm (photo S. Golub). Retort synthesis of limestone from chalk (left) and chalk raw material (right).

The samples from the object, which were operationally called as retort (Fig. 1, from the left), were taken to the research. At the upper part of the object, there is little modified white and grey chalk (Fig. 2a), below light white and grey raw material, harder and more compact than chalk, with distinct replicas of plants stems of segment structure and longitudinal ribs. These replicas have the form of longitudinal negatives as well as longitudinal and transversal cross-sections of stems and leaves of plants morphologically corresponding to a horsetail (Equisetum L.) (Fig. 2b). The organic substance has not been preserved. Inner and bottom parts of retort are filled with hard, structurally homogeneous (microcrystal) light grey and white material, cracked of distinct shell and an irregular sharp-edged burring fracture (Fig. 2c).

This material at the macroscopic view (structure, sound during impact, strength to stroke and hardness) can be treated as well sintered white ceramic. However, the reaction itself with hydrochloric acid clearly indicates that this material is almost purely carbonate. Thus, this material has got the form similar to rocky limestone, obtained by no means in the technological process, it is not a rock but an artificial stone. The sequence of these materials, within the area of the same object (retort) is most likely, the record of local petrification technology of writing chalk, useless as a construction stone, hard stone material, similar to the properties of rocky limestone applied to contemporaneous objects e.g. of Cracow or imported marbles. Without a doubt, they could also replace white Halych stones (alabaster) which were subject to destruction in the fire of the first buildings of complex (Gazda 2016).

a b c

Fig. 2. Sequence of materials and materials from the retort: writing chalk, partially modified chalk with horseshoe replicas (b), microcrystalline carbonate (c).

This material was applied to the fabrication of homogeneously white building and archeological formats found in destructs of Wysoka Gorka (Fig. 3) as well as most likely to the production of white glazed tiles (Figs. 4 and 5) found in Cholm, Stolpie and Bielawin.

b

a

Fig. 3. Building materials obtained from materials synthesized from Cholm chalk. Fragment of a brick or architectural detail (a), glazed tiles (b).

3 Analysis

The materials taken from retort underwent the research: writing chalk, partially modified chalk with replicas of horsetail as well as microcrystal carbonate material (Fig. 2). Phase studies were performed with the use of polarizing optical microscope, electron microscope SEM and X-ray diffraction. The chemical analysis in the micro area with the use of electron microscope SEM with EDS module was also carried out.

Cholm writing chalk is a typical loose biocalcarenite of dominating hole saw bioclasts in its structure (Fig. 4). In mineral terms, it is a pellite calcite (above 95% of CaCO) with an insubstantial admixture of loamy minerals, quartz and iron sulphides. It is a soft rock, subject to liquification after adding water and prone to be molded in the plastic state.

Fig. 4. Biomorphic structure of Cholm chalk writing in the image of scanning microscope

Fig. 5. Replica of the surface of a horsetail in a carbonate matrix. Image from a scanning microscope

The material, in the middle part of retort (Fig. 2b), is in light grey and white, harder and more bound than the chalk as well as it is not susceptible to plasticity after adding water. There are a lot of stem replicas of plants with segmental structure and longitudinal ribs. In the view of scanning microscope, the structure of horsetail surface is clearly legible (Fig. 5).

In the chemical content of this material, a considerable increase of silica share is noticeable (Fig. 6).

In the microscopic view, strong amorphization of chalk structural elements (bioclasts), cement binds creation and embryonal crystallization of calcite are significant components (Fig. 7).

In the microscopic view, the steric capillary silica mineralization (max. 0,1 μm in diameter and 2–3 μm in length) could also be observed, they bind microcrystals of calcite of dimensions to max. 5 μm (Fig. 8). Silica is opal and microcrystalline in its nature. This petrographic material can be defined as silificated limestone having its counterpart in Jurassic massive limestone.

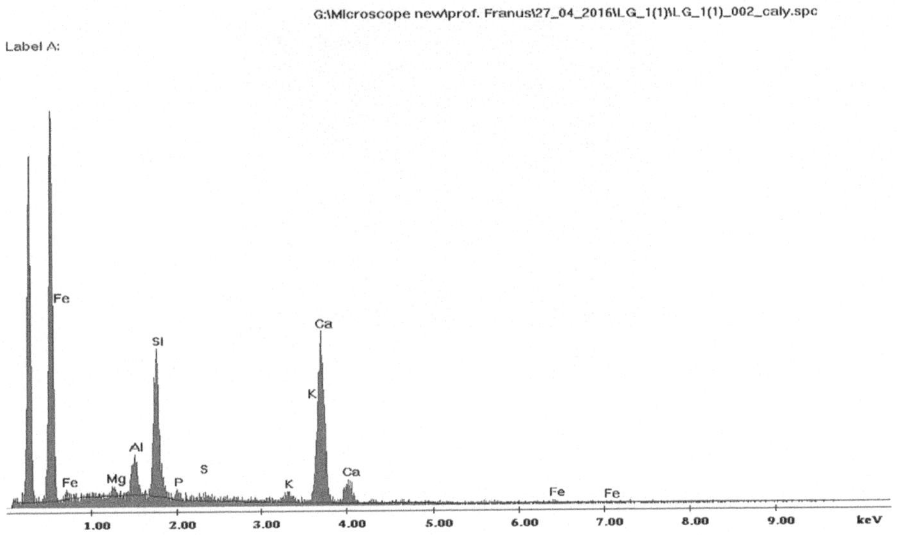

Fig. 6. Results of chemical analysis of EDS SEM plastic with replicas of horsetail stems

The material from the bottom part of retort (Fig. 2c) is macroscopically light yellow and white without distinct structural and textural principles. It is cracked along with the shell surface of unbundling. The material is hard, while stroking it makes soundless noise, In addition, it is fully durable in water.

Fig. 7. Amorphisation of chalk and initial calcite crystallization

Fig. 8. Steric silica mineralization on calcite crystals

In the view of scanning microscope, structurally this material is fully crystallized (Fig. 9). Calcite microcrystals of the dimensions 5–20 µm automorphically adhere tightly to each other.

In the spaces between calcite crystals, there is silica grouting of distinct crystalline structure and cellular texture (Fig. 10).

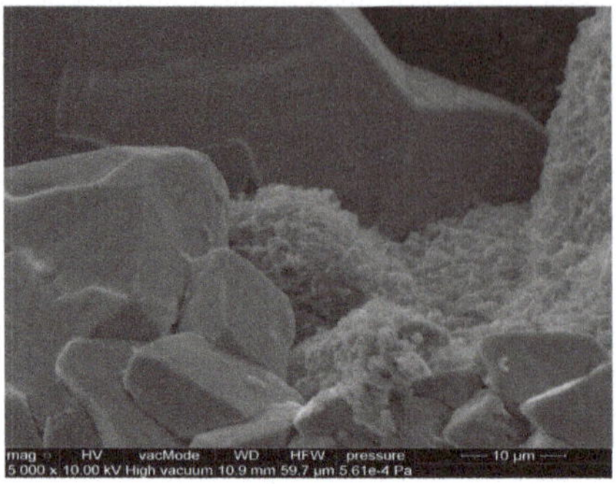

Fig. 9. Crystalline construction of the material obtained by petrified chalk writing.

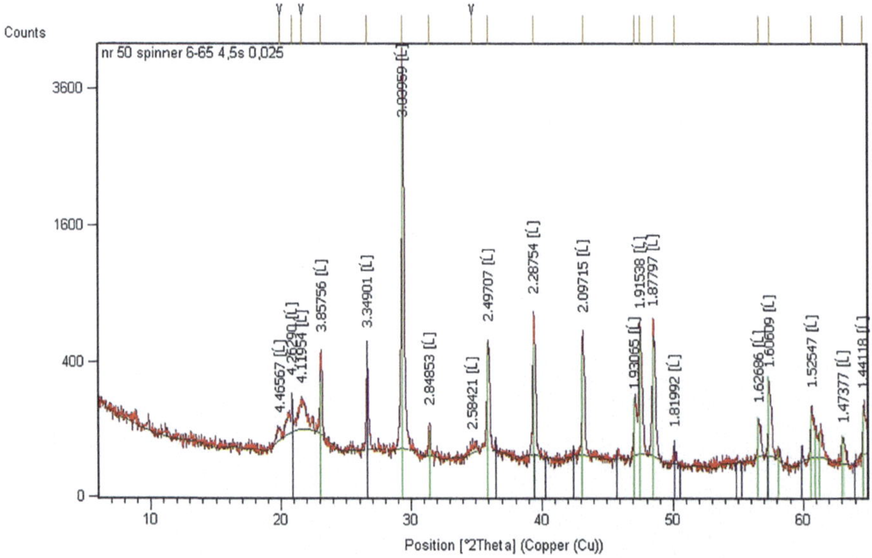

Fig. 10. Diffractogram of hard material.

Calcite has got clear and symmetric diffractive lines, which indicates its well ordered structure. Silica within the angle range of 20–25^{0}2Θ produces diffractive lines characteristic for opal-CT (cristobalite/tridymite) (Jones and Segenit 1971). Semi-quantitative determination of contents of mineral phases on the basis of their diffractive lines enables establishing the calcite content of about 90% and silica 10%. This material can be petrographically (minerally and structurally) defined as silificated fine-blastic marble. Hypothetical reconstruction of main green-white portal of St. John's Orthodox Church is showed in Fig. 11. The Hypothetical reconstruction is performed taking into account the results of archaeological research.

Fig. 11. Hypothetical reconstruction of main portal of St. John's Orthodox Church of the king Daniel's residence in Cholm. Author M. Bevz.

4 Conclusions

Performed phase studies on materials taken from the object which was operationally defined as retort, confirm their carbonate nature. They are the transformation recordings of soft dissolveable writing chalk into the material similar to rocky limestone until obtaining the material with technical values close to marble. It can be inferred from the conducted archeological observations and phase studies that the aim of alchemic activities was to obtain (transmutation) from the soft white chalk, the material which could be a substitute for, unavailable in this region, white building stones (alabaster, limestone, marble).

These experiments were performed also using other raw materials that enabled obtaining similar materials tinged with green or red. Undoubtedly, this process was carried out in the hydrothermal conditions. The forms of construction materials (brick, tiles) were obtained by casting from a slurry or plastic moulding. These fittings gained technically enough strength after complete cooling and evaporation. Full strength and durability were obtained after longer maturation. Among found construction materials, medium hard materials dominate, thus from the phase of biogenic calcite amorphization. The extension of the process, or increase of parameters (e.g. pressure) resulted in obtaining harder materials, however, subject to crack destruction, which could be observed in the material from the bottom part of retort. The process itself was most likely performed in an open tub, thus in the conditions of normal pressure. Nevertheless, it is not excluded that the process was conducted in a closed autoclave (e.g. copper or lead). This can be explained by the continuation of archeological research in this area and more thorough material studies.

The model for green and white buildings created here before the fire in 1256 as well as the attempts to recreate them, was the implementation of co-called Pisan School from the 12th century but in other architectural forms.

References

Bevz M, Buko A, Gazda L, Golub S, Lukomskyi Y (2015) Materialy i technologii w architekturniy biografii mista (na prykladi mista Cholm u Polshchi). Zberezhennia istorychnoji zabudowy tsentra Odesy shliachom vkluchennia do osnovnoho spysku Vsesvitnioji spadshchyny UNESCO. Materialy II mizhnarodnoji naukovo-praktychnoji konferentzii. 24–26 chervnia. Odesa, Astroprint, 136–147

Buko A (2016) Źródła pisane i archeologia: przykład góry katedralnej w Chełmie. Kwartalnik Historyczny. Rocznik CXXIII, 2 PL ISSN 0023-5903

Dąbrowski D (2012) Daniel Romanowich. Król Rusi (ok. 1201–1264). Biografia polityczna. Wydawnictwo Avalon

Gazda L (2016) Wyniki badań surowców i materiałów budowlanych. Od cerkwi katedralnej króla Daniela Romanowicha do bazyliki pw. Narodzenia NMP w Chełmie (Red)

Buko A, Gołub S - Chełm: Muzeum Ziemi Chełmskiej im. Wiktora Ambroziewicza w Chełmie, 213–234

Gołub S (2013) Dokumentacja z badań archeologicznych w latach 2010–2012. Projekt badawczy Ministerstwa Nauki i Szkolnictwa Wyższego. Zespół rezydencjalno – sakralny na Górze Katedralnej w Chełmie. Chełm

Jones JB, Segenit ER (1971) The nature of opal. Nomenklature and constituentphases. J Geol Soc Australia 18

Hutzuliak R, Shevchenko N (2015) Chimiko-petrografichni doslidzhennia budivelnych materialiv z katedry Bohorodytzi u misti Cholmi (1 etap). Korol Danylo Romanowych: kulturna i derzhavotvorcha spadshchyna yoho doby. L'viv: L'vivska Politechnika. Vyd-vo: Rastr-7, S 196–206

Lukomskyi Y, Bevz M (2012) Badania archeologiczne fundamentow dawnej katedry w Krylosie-Haliczu. Problem konserwacji I muzeumfikacji. XXVIII Konferencja Sprawozdawcza "Badania archeologiczne w Polsce Srodkowo-Wschodniej, zachodniej Bialorusi I Ukrainie w roku 2011". – Lublin: 29–30 marca 2012 r. Lublin, UMCS, S 26

Voytovych L (2014) Korol Danylo Romanowych: davni i moderni napriamky falsyfikatzii portretu. Korol Danylo Romanowych: kulturna i derzhavotvorcha spadshchyna yoho doby. L'viv: L'vivska Politechnika. Vyd-vo: Rastr-7. 2014, S 11–32

Crack Resistance of RC Beams on the Shear

Z. Blikharskyy[1], R. Vashkevych[1(✉)], P. Vegera[1], and Y. Blikharskyy[2]

[1] Department of Building Constructions and Bridges, Lviv Polytechnic National University, Karpinskoho Str. 6, Lviv 79013, Ukraine
Rostyslav.V.Vashkevych@lpnu.ua

[2] Department of Highways and Bridges, Lviv Polytechnic National University, Karpinskoho Str. 6, Lviv 79013, Ukraine

Abstract. Most important parameter of serviceability limit state (SLS) for reinforced concrete (RC) beams is crack resistance. These include width of crack opening, cracking distribution, load at which is achieved limit and maximum width of crack opening. This is especially important for shear resistance. In this article, new data of crack resistance of rectangular RC beams are described. Changing parameters were shear span (acquiring values 1, 1.5, 2) and strengthening with FRCM (Fiber Reinforced Concrete Matrix) system in the supporting area. Eight samples for experimental researches were carried out: six once from them without strengthening but with different shear span and two once with composite system. Using the limit width of the crack opening as criteria of exhaustion of the bearing capacity can conclude that remains about 19–26% of load caring capacity before the danger of shear failure of the beam. When the shear span decreases on the 25% the maximum width of the crack opening decreases only on the 7% and on the 40$ when shear span decreased by 50%. External strengthening system reduced maximum crack opening on the 43%: from 0.75 to 0.43 mm and values exceed limit insignificant.

Keywords: RC beam · Shear · Inclined crack · Crack resistance · Strengthening · FRCM

1 Introduction

Reinforced concrete is one of the most common building materials. It has main advantages such as durability, high corrosion resistance of elements. However, it also has significant disadvantages: low tensile strength, bearing elements are large cross-sectional dimensions (in comparison with the steel construction), cracks opening, which lead to corrosion of the internal reinforcement (Blikhars'kyi and Obukh 2018; Blikharskyy et al. 2019). This is especially important for constructions with difficult stress strain state (Bobalo et al. 2018). In view of its prevalence, the study of the main parameters of a stress-strain state is an important issue. This is especially true of external parameters, which arise in reinforced concrete beams on the shear and determine their safety of operation (Khmil et al. 2018; Krainskyi et al. 2018).

In the article (Chen et al. 2018), a novel cracking strut-and-tie model (CSTM) is developed to better predict the shear strength of a deep beam. The proposed CSTM

Z. Blikharskyy et al. (Eds.): CEE 2019, LNCE 47, pp. 17–24, 2020.
https://doi.org/10.1007/978-3-030-27011-7_3

divides the diagonal strut into two parts according to the location of the critical shear crack (CSC).

In the paper (Gherdaoui and Guenfoud 2018) is described the results of the testing of reinforced concrete slabs to the comprehension and control of the phenomenon of transmission of effort and reinforcement, by composite materials, of reinforced concrete slabs subjected to punching. The purpose of this study is the experimental analysis of reinforced concrete slabs that are control and reinforced (repaired) by composite materials [carbon fiber reinforced polymer (CFRP)] subjected to punching after degradation at various levels of load.

Next article (Imjai et al. 2016) proposes a practical semi-empirical method for determining shear crack-induced deformations in Glass Fibre Reinforced Polymer (GFRP) Reinforced Concrete (RC) beams. To calculate the component of deflection due to shear action and crack opening, the proposed model uses a 'single fictitious inclined crack' with a width equal to the sum of the individual effective shear crack widths.

The development and advancement of cement-based composites has shown promise in recent years (Aljazaeri and Myers 2017). Aim of this study is to investigate the effectiveness and the performance of the polyparaphenylene benzobisoxazole fiber-reinforced cementitious mortar (PBO-FRCM) system for shear strengthening and to study the shear performance of the PBO-FRCM system in terms of the availability and absence of internal transverse shear reinforcements.

Deep beams are considered important structural components due to their high using in civil engineering, including in tall buildings, silos (Aghayari and Rahimi 2018). The presence of openings in the deep beam will reduce the shear capacity of the beam, so it is important to use the new methods such as the use of resistive shape memory alloys. Shape memory alloys (SMA) are a series of metals that can permanently restructure deformations and strains and eventually reinstate in their original form.

The study (Guadagnini et al. 2006) investigates the shear behavior of concrete beams reinforced with fiber-reinforced polymer - FRP reinforcement. Different shear span to depth ratios, ranging from 1.1 to 3.3, were analyzed in order to study the variation in the shear behavior of beams characterized by different types of shear failure. No shear reinforcement was provided in the first phase of testing, while in the second phase, just enough glass and carbon shear reinforcement was provided to enable failure due to shear.

One of the main parameters, which determine the operational suitability of reinforced concrete beams, is the width of the opening and spread of cracks in the section (Vegera et al. 2018). Given into account the widespread using of reinforced concrete structures, this is an important task. This is especially true of inclined or diagonal cracks that arise from shear. This is the main indicator in determining the technical condition, as well as the residual bearing capacity of reinforced concrete structures.

2 Aim of the Research

Aim of the research is getting new data about crack resistant of RC beams with different shear span with and without external composite strengthening.

3 Research Methodology

For testing, rectangular, reinforced concrete (RC) beams are chosen. The testing program involves testing four samples. For research, a technique has been developed for testing each support area separately, which described in our article (Vegera et al. 2015). All samples are identical sizes; the deviation is less than 2%.

Tested samples were designed the way that the samples destruction took place by shear, according to DBN B.2.6 – 98. (2011). To accomplish it, these beams were designed with a significant margin of tensile rebar. This is typical for such researches (Alzate et al. 2015; David et al. 1998).

Beams mark follows BO – beam ordinary, BSC – strengthened beam by composite material, the first digit – serial number, the second digit – prototypes number and the third digit – the section number. For example, BO 1.2-2 means that the tested example from the first series of the second beam of the second section.

Tested beams were 2100 mm length, 100 mm width, and 200 mm height of cross section. Transverse reinforcement – A240C Ø 8 mm rebar, with step 100 mm, located in the supporting area and with step 250–300 mm in the area without transverse force (Vegera et al. 2017). Concrete of testing beams is C32/40.

The variable parameter was shear distance (or relative span to effective depth ratio) which acquired the values a/d = 2, 1.5, 1. Last beam was strengthened with FRCM system in the supporting area. Strengthening system is made by stripes width 70 mm, with step 100 mm (Fig. 1).

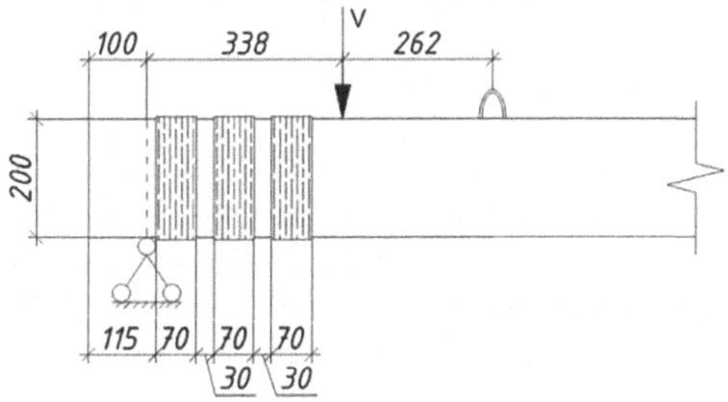

Fig. 1. Scheme of strengthening by FRCM system

4 Results of the Experimental Researches

4.1 The Overview of Carrying Capacity of Testing Samples (ULS)

The bearing capacity of reinforced concrete beams on the shear strength was for BO 2.1 – V_{Ed} = 148.5 kN, BO 2.2 – V_{Ed} = 182.5 kN, for BO 1.3 – V_{Ed} = 258.5 kN, for BSC 2.4 – V_{Ed} = 185.5 kN. The obtained results indicate an increase of the bearing capacity with a decrease of relative span to the effective depth ratio. Shear strength of RC beam strengthened by composite materials is the same as for sample BO 2.2, which had lesser distance to effective depth ratio. The difference between the experiments of twin beams is 5–9%, which indicates the high reliability of the results and the effectiveness of the proposed research methodology.

All samples were destroyed by shear. Unstrengthen samples crushed by cutting compressed zone of the concrete and chipping concrete particles. Inclined crack spread throughout the height of the beam. Destroying was sudden and fast.

The tested samples after destroying are shown in Fig. 2.

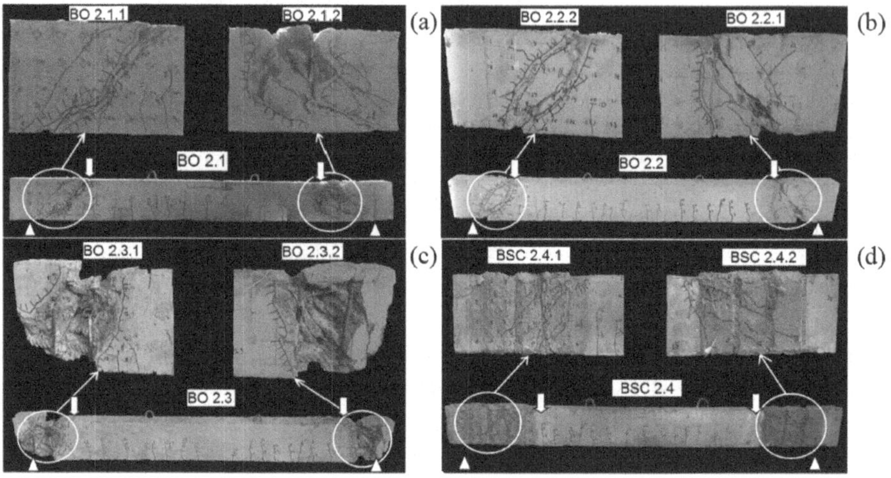

Fig. 2. General view of the RC beam after exhaustion of the bearing capacity: (a) BO 2.1; (b) BO 2.2; (c) BO 2.3; (d) BSC 2.4

Strengthened beam crushed at the time of segregation concrete compressed area with a sharp lengthening strip and impairment the protective layer of system FRCM at the area of spreading inclined cracks. Destroying was more slowly and without chipping concrete compressed zone (Gherdaoui and Guenfoud 2018; Blikharskyy et al. 2018).

4.2 The Overview of Inclined Cracking of Testing Samples (SLS)

At each stage of the load, we fixed the distribution of cracks, measured the width of the crack opening at the location of the transverse reinforcement, and the average width of the opening shear crack. Measurements were made using a microscope with accuracy 0.05 mm.

It was noted the spread of cracks occurring in the following order:

- opening first inclined crack at an angle of 45° to the longitudinal axis of the beam at a load equal to $V_{Ed} = 50 - 60$ kN;
- distribution of the inclined crack along the height of the beam, to the compressed concrete zone with the width of the opening $w_k = 0.05 - 0.15$ mm;
- opening destructive inclined crack from the point of support to the applying load;

- distribution the destructive crack, throughout the height of the section, achieve the limit values of the width of crack opening.

This type of crack distribution was fixed for sample BO 2.1. In the beams BO 2.2 and BO 2.3, only one crack was opening. That was destructive. Distribution of cracks are showed in Fig. 3.

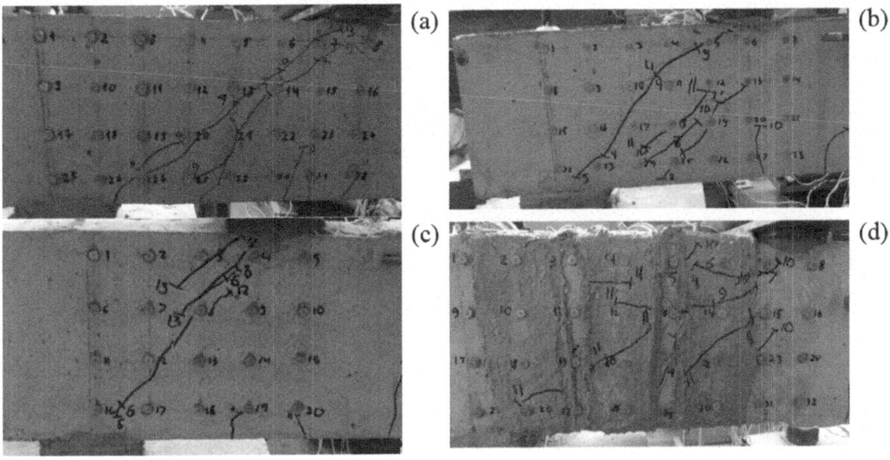

Fig. 3. The appearance and distribution of inclined crack of RC beam: (a) BO 2.1; (b) BO 2.2; (c) BO 2.3; (d) BSC 2.4

The destruction of strengthened sample BSC 2.4 was in accordance with next steps:

- opening of the inclined crack of the limit width ($w_k = 0.4$ mm) on the concrete surface;
- distribution the inclined crack to the compressed zone of concrete and the appearance of a grid of cracks with the width of opening $w_k = 0.05 - 0.2$ mm on the surface of the strengthening system;
- plastic deformation of the transverse reinforcement, destruction of the concrete of the compressed zone, significant deformations of the composite reinforcement, which can be visually observed due to impairment of the protective layer.

5 Scientific Innovation

As a result of tests, the sample BO 2.1 has reached the limit of operational suitability with the value of the transverse force of 120 kN. The serviceability of the samples BO 2.2 increased 1.16 times with a decrease shear span on 25%, and for BO 2.3 – 1.58 times, with lesser shear span on 50%. For sample BSC 2.4, limit value of operational suitability is 150 kN, which is 1.25 times larger than for beam BO 2.1, with the same shear span (Table 1).

Table 1. Bearing capacity of tested beams at the limit width of crack opening

Types of the beam	Types of the beam's cross section	Shear span ratio, a/d	Decreasing of the shear span	Inclined crack width, mm	Experimental shear strength, kN	Average experimental shear strength, kN	Increasing of shear strength
BO 2.1	BO 2.1.1	2	-	0.4	120	120	-
	BO 2.1.2				120		
BO 2.2	BO 2.2.1	1.5	0.75		140	140	1.16
	BO 2.2.2				140		
BO 2.3	BO 2.3.1	1	0.5		180	190	1.58
	BO 2.3.2				200		
BSC 2.4	BSC 2.4.1	2	1		160	150	1.25
	BSC 2.4.2				140		

The boundary value of the crack opening width is fixed at 74–81% of the load carrying capacity of the samples (Table 2).

Table 2. The comparison between ULS and SLS for tested beams

Types of the beam	Types of the beam's cross section	Shear span ratio, a/d	Decreasing of the shear span	Inclined crack width, mm	Average experimental shear strength, ULS	Average experimental shear strength, SLS	Relationship between SLS and ULS
BO 2.1	BO 2.1.1	2	-	0.4	148.5	120	0.81
	BO 2.1.2						
BO 2.2	BO 2.2.1	1.5	0.75		182.5	140	0.77
	BO 2.2.2						
BO 2.3	BO 2.3.1	1	0.5		258.5	190	0.74
	BO 2.3.2						
BSC 2.4	BSC 2.4.1	2	1		185.5	150	0.81
	BSC 2.4.2						

For samples BO 2.1 and BSC 2.4, which have similar shear distance a/d = 2, were the same percent of loading, when cracks opening width was 0.4 mm.

Using the limit width of the crack opening as criteria of exhaustion of the bearing capacity (SLS) we saw that remains about 19–26% of load caring capacity before the danger of shear failure of the beam. The effect of increasing shear strength by SLS is similar to the increasing of the bearing capacity and close by the value.

Maximum measuring value of crack opening is described in the Table 3.

When the shear span decreases on the 25% the width of the crack opening decreases only on the 7% (for sample BO 2.2) and on the 40% when shear span decreased by 50% (for sample BO 2.3). External strengthening system reduced maximum crack opening on the 43%: from 0.75 to 0.43 mm and values exceed limit insignificant. Therefore, FRCM system is a good method of enhancement crack resistant.

Table 3. Maximum value of width of inclined crack opening

Types of the beam	Types of the beam's cross section	Shear span ratio, a/d	Decreasing of the shear span	Shear strength, kN	Maximum values of inclined cracks width, mm	Average values of inclined cracks width, mm	Decreasing of maximum crack opening width
BO 2.1	BO 2.1.1	2	-	140	0.70	0.75	-
	BO 2.1.2			140	0.80		
BO 2.2	BO 2.2.1	1.5	0.75	170	0.85	0.70	0.93
	BO 2.2.2			170	0.65		
BO 2.3	BO 2.3.1	1	0.5	220	0.50	0.45	0.6
	BO 2.3.2			220	0.40		
BSC 2.4	BSC 2.4.1	2	1	160	0.45	0.43	0.57
	BSC 2.4.2			160	0.40		

6 Conclusions

Based on the data above, the following conclusions can be drawn:

1. Using the limit opening of the width crack as criteria of exhaustion of the bearing capacity (SLS) we saw that remains about 19–26% of load bearing capacity before the danger of shear failure of the beam.
2. When shear span was decreased on the 25% (from a/d = 2 to a/d = 1.5) bearing capacity of tested beams at the limit width of crack opening increased on the 16%, and for shear span a/d = 1 (decreasing on the 50%) bearing capacity increased on the 58%.
3. FRCM system increased shear strength in 1.25 times (ULS) but exhaustion of the SLS shear strength was at the 81%, the same as for control samples.
4. External strengthening system reduced maximum crack opening on the 43%: from 0.75 to 0.43 mm and values exceed limit insignificant.

References

Aghayari R, Rahimi F (2018) The novel method for rising up the shear strength and limiting the growth of cracks in deep beams with SMA. In: International conference on new horizons in the engineering science, Istanbul, Turkey, 19 March 2018

Aljazaeri ZR, Myers JJ (2017) Strengthening of reinforced-concrete beams in shear with a fabric-reinforced cementitious matrix. J Compos Constr 21(5):04017041

Alzate A, Arteaga A, De Diego A, Cisneros D, Perera R (2015) Shear strengthening of reinforced concrete members with CFRP sheets [Refuerzo externo a cortante con láminas de CFRP en elementos de hormigón armado]. Materiales de Construccion 63(310):251–265

Blikhars'kyi Y, Obukh Y (2018) Influence of the mechanical and corrosion defects on the strength of thermally hardened reinforcement of 35GS steel. Mater Sci. 54. https://doi.org/10.1007/s11003-018-0183-2

Blikharskyy Z, Selejdak J, Blikharskyy Y, Khmil R (2019) Corrosion of reinforce bars in RC constructions. Syst Saf Hum-Tech Facil Environ 1: 277–283. https://doi.org/10.2478/czoto-2019-0036

Blikharskyy Z, Vegera P, Vashkevych R, Shnal T (2018). Fracture toughness of RC beams on the shear, strengthening by FRCM system. In: MATEC web of conferences, vol 183, p 02009. https://doi.org/10.1051/matecconf/201818302009

Bobalo T, Blikharskyy Y, Vashkevich R, Volynets M (2018) Bearing capacity of RC beams reinforced with high strength rebars and steel plate. In: MATEC web of conferences, vol 230, p 02003. https://doi.org/10.1051/matecconf/201823002003

Chen H, Yi WJ, Hwang HJ (2018) Cracking strut-and-tie model for shear strength evaluation of reinforced concrete deep beams. Eng Struct 163:396–408

David E, Djelal C, Buyle-Bodin F (1998) Repair and strengthening of reinforced concrete beams using composite materials. In: 2nd international PhD Symposium in civil engineering, Budapest, pp 23–34

DBN B.2.6 – 98 (2011) Design of concrete structures. National Standard of Ukraine, 84 p

Gherdaoui M, Guenfoud M (2018) Repairing reinforced concrete slabs by composite materials. J Mater Environ Sci 9(2):701–708

Guadagnini M, Pilakoutas K, Waldron P (2006) Shear resistance of FRP RC beams: experimental study. J Compos Constr 10(6):464–473

Imjai T, Guadagnini M, Garcia R, Pilakoutas K (2016) A practical method for determining shear crack induced deformation in FRP RC beams. Eng Struct 126:353–364

Khmil R, Tytarenko R, Blikharskyy, Y, Vegera P (2018) Development of the procedure for the estimation of reliability of reinforced concrete beams, strengthened by building up the stretched reinforcing bars under load. East-Eur J Enterp Technol 5/7(95):32–42

Krainskyi P, Blikharskyy Y, Khmil R, Vegera P (2018) Influence of loading level on the bearing capacity of RC columns strengthened by jacketing. In: MATEC web of conferences, vol 230, p 02013. https://doi.org/10.1051/matecconf/201823002013

Vegera P, Khmil R, Blikharskyy Z (2017) Shear strength of reinforced concrete beams strengthened by P.B.O. fiber mesh under loading. In: MATEC web of conferences, vol. 116, p 02006. https://doi.org/10.1051/matecconf/201711602006

Vegera P, Vashkevych R, Blikharskyy Z (2018) Fracture toughness of RC beams with different shear span. In: MATEC web of conferences, vol 174, p 02021. https://doi.org/10.1051/matecconf/201817402021

Vegera PI, Khmil RE, Blikharskyy ZZ (2015) Optimization of the methodology of experimental research of inclined sections of reinforced concrete beams. Theory Build Pract 823:38–43. (in Ukranian)

Serviceability of RC Beams Reinforced with High Strength Rebar's and Steel Plate

Taras Bobalo[1], Yaroslav Blikharskyy[2(✉)], Nadiia Kopiika[1], and Mykhailo Volynets[1]

[1] Department of Highways and Bridges, Lviv Politechnic National University, Karpinskogo 6, Lviv 79013, Ukraine

[2] Department of Building Constructions and Bridges, Lviv Politechnic National University, Karpinskogo 6, Lviv 79013, Ukraine

yaroslav.z.blikharskyy@lpnu.ua

Abstract. Nowadays steel plate reinforced concrete beams are widely used in construction around the world. The use of such structures allows high-quality welding, saving materials' saving and avoiding complex joints.

Decking structures contain solid or lattice steel beams, on which the flooring of steel profiled sheets is laid, which simultaneously acts as both formwork and the lower stretched rebar for monolithic concrete. After analysing the special literature and practical issues of steel plate reinforced concrete structures, it could be concluded that such structures have increased bearing capacity and cracking resistance, as well as more economical characteristics comparing with reinforced concrete structures. The high-strength steels' usage in steel-concrete structures makes it possible to achieve even greater savings. Combination of the properties of steel plate reinforced concrete beams reinforced with S275 steel plates with the strength of high-strength rod reinforcement of class A1000 ensures the possibility to achieve additional reduction of metal consumption with simultaneous observance of permissible strain and crack resistance requirements.

Keywords: Deformability · Serviceability · Combined reinforcement · High-strength rebar · Steel plate

1 Introduction

Nowadays concrete structures are one of the most popular materials. However, due to the negative environmental impacts, such structures could be damaged and need to be strengthened (Blikhars'kyi and Obukh 2018; Selejdak et al. 2018; Strieška and Kotes 2018; Kotes et al. 2018); Blikharskyy et al. 2018a; 2018b; Sherrawi et al. 2018; Kos et al. 2017; Zhang et al. 2015). Today most of reinforced concrete structures during the process of their using need to be strengthened (Khmil et al. 2018; Krainskyi et al. 2018a, b). One of the most popular traditional ways to do it is to use reinforced concrete and steel hoops (Bobalo et al. 2018). One of more modern methods is the usage of various composite materials for strengthening both longitudinal (Blikharskyy et al. 2018a, b; Brozda et al. 2017) and cross section (Blikahrskyy et al. 2018a, b; Vegera et al. 2018). Therefore, it is important to manufacture new reinforced concrete

Z. Blikharskyy et al. (Eds.): CEE 2019, LNCE 47, pp. 25–33, 2020.
https://doi.org/10.1007/978-3-030-27011-7_4

structures with the usage of high-strength steel, high-strength concrete, which would provide high strength reserves.

Reduce of metal consumption and improvement of RC structures manufacturing technology is an important issue of modern construction. Today a number of works for cements' modifications in order to reduce the concrete price are conducted (Fadhil et al. 2018; Sobol et al. 2014; Tahey et al. 2013). However, the durability of such structures still has to be ensured. One of the approaches which provides for metal consumption reduce is the use of external steel plates and high-strength steels. The use of such rebar in steel-concrete elements increases the cross-section effective height. In addition it eliminates the necessity of multi-row rods' location along the cross-section height, which leads to steel savings.

Concentrated sheet reinforcement ranking on the external most stressed planes of steel-concrete structures provides better usage of materials' properties. Therefore economical characteristics are increased by 30–40% comparing with reinforced concrete structures.

The previous tests of steel-concrete structures strengthened with the combined reinforcement confirmed high economic efficiency of such approach. It was proved that comparatively with reinforced concrete structures with steel bars, beam elements with sheet reinforcement have increased bearing capacity, stiffness, crack resistance and reduced metal consumption. Usage of steel plates (S275 steel) in combination with high-strength rod (class A1000) without prestress, allows to achieve metal cost reduction ensuring strength, deformability and cracking resistance requirements for structures. Unlike reinforced concrete structures, steel plate reinforced concrete structures allow to optimally allocate the effective plate reinforcement and eliminate the necessity for multi-row rods location. Thus, the technology of manufacturing and laying of concrete mixture is simplified, as well as the metal cost is reduced.

2 Technology of Test Samples' Manufacturing

The process of steel plate reinforced concrete test specimens' manufacturing consisted of two stages:

(1) manufacturing of frames; (2) concreting of steel-concrete specimens.

At the first stage the U-shaped rod anchors were welded with a certain step to the smooth S275 class steel plate.

The ends of rod anchors were directed upward and it served as a transverse armature. The weld seam length corresponded to the width of the plate. Therefore a reliable ad-hesion of external steel plate reinforcement with concrete was obtained. The trans-verse armature height was equal to 220 mm. The next step was to weld rigid butts (Fig. 1).

At the end of the frame manufacturing process, a high-strength reinforcement was in-stalled with strain gauges glued on it in order to measure deformations.

Steel plate reinforced concrete specimens' concreting was carried out in specially made metallic shuttering. Concrete mixture consolidation was made with the use of deep vibrator. Simultaneously with specimens series' concretization, cubic specimens

Fig. 1. Frames of steel plate reinforced concrete test beams with adhesion of steel plate reinforcement to concrete

with 150 mm rib and 100 × 100 mm cross-section 400 mm length prisms were manufactured in special forms.

Manufactured beam specimens, prisms, and cubes were kept for 28 days. In order to avoid concrete shrinkage deformation, the specimens were periodically moistened with water, therefore 100% moisture during hardening was ensured.

During the entire concrete hardening period, moisture and temperature control was ensured in the laboratory room.

3 Research Methodology

For the research purpose realisation six specimens 2600 mm length with 2400 mm span were designed and experimentally investigated. Experimental specimens were made of heavy concrete with the design class C40/50.

Test specimens material characteristics obtained as the result of the laboratory test are presented in (Bobalo et al. 2018).

Beams' testing was performed on the test stand. The load was applied with the use of 1000 kN capacity hydraulic jack. With the use of the distributing traverse to the upper face of the beam load was applied. The load was realized in the form of two concentrated forces applied symmetrically to the middle of the beam at a 1/3 estimated span distance.

The load on the beam during the test was applied gradually on stages, close to 0.1 of the destructive load, ensuring 30 min delay at each level. After loading time, records of all devices were fixed, as well as cracks formation, opening width and development. The magnitude of the load was controlled with the use of earmark manometer on the jack, as well as by the size of the support reactions, measured by two ring dynamometers. Simultaneously these ring dynamometers served as pinned and roller supports.

The beams' deflection was measured using three clock-type indicators with a resolution scale of 0,01 mm (Fig. 2). Two indicators were installed at the 70 cm distance from the support, and the third one- in the middle of the span.

The high-strength rebar strain was measured using glued strain gauges with 20 mm base.

Fig. 2. Placing of strain gauges on the beam

On purpose of moisture and mechanical damage protection of the sensors at the initial stage of concreting, the glued sensors surface was protected by the layer of epoxy resin. In addition by the use of soft plasticine the possibility of gauge scraping from the reinforcing bar due to the concrete strain development was eliminated.

Steel plates' strain was measured using clock type micro-indicators with 200 mm base and electrical strain gauges'chain with 20 mm base. On the lateral surface of concrete in order to determine the main tensile and compression stresses strain gauges with 50 mm base were glued. All strains were measured in a pure bending zone. On purpose of control, clock type micro-indicators with 0,001 mm division value and 200 mm base were installed.

In order to measure strain in compressed zone of concrete on the specimens' upper plane, clock-type indicators with a base of 200 mm base and strain gauges with 50 mm base were used.

4 Results and Discussion

The experimental specimens respond on load before the cracks' formation (elastic stage) could be characterized by elastic strain both in the effective armature and in the compressed and stretched zone concrete.

The results of experimental studies, namely: the concrete strain magnitude in the stretched and compressed zones, the stretched rebar's strain, the deflection before the crack's formation and corresponding loads are given in the Table 1.

From the experimental results it is evident at this stage of loading strain of the high-strength A1000 class rebar in beams with different reinforcement ratios do not differ significantly. However, certain effect on the beams' deflection due to increase of the high-strength rebar percentage could be identified.

The formation of normal cracks in the beam span determines the transition to the second stage of a stress-strain state. On the early stages, cracks tend to form near the most stretched concrete plane. Therefore the determinative factor of its formation is tensile strain of the concrete due to the action of bending moments.

Table 1. Results of beams' experimental studies before the crack's formation

Beams' marking	M, kN*m	$\varepsilon_{s,A1000}*10^{-5}$	$\varepsilon_{s,S275}*10^{-5}$	$\varepsilon_{concrete}*10^{-5}$	Deflection f, mm
B - I - 1	6,9	7	10	9	0,60
B - I - 2	7,15	9	12	10	0,68
B - II - 1	8,00	16	18	25	1,08
B - II - 2	7,46	12	16	31	1,11
B - III - 1	6,42	18	–	16	1,09
B - III - 2	5,6	15	–	11	0,83

The locations of cracks and the distance between it in these beams coincided with the anchors' location which serve as stress concentrators.

The results of experimental studies including the maximum value of compressed concrete strain, stretched rebar's strain, deflections and corresponding loading in the maximum bending moment zone are given in the Table 2.

Table 2. Tests results for beams in the case of loading equal to $0{,}7M_{max}$

Beams' marking	$0{,}7M_{max}$, kN	$\varepsilon_{s,A1000}*10^{-5}$	$\varepsilon_{s,S275}*10^{-5}$	$\varepsilon_{concrete}*10^{-5}$	Deflection $f_{0,7}$, mm
B - I - 1	44,93	105	127	94	4,55
B - I - 2	46,06	117	144	102	5,08
B - II - 1	48,10	259	304	147	9,44
B - II - 2	47,90	251	292	135	9,56
B - III - 1	49,38	439	–	221	20,32
B - III - 2	49,25	445	–	227	21,57

It should be noted that the yield stress of plate was reached faster than those for high-strength rebar, even for beams in which steel plate was acting without adhesion to concrete. It was accompanied with the increase strain and deflection growth. However, the experimental specimens continued to perceive the additional load even after the yield point for the plate, up to the moment, when yield stress of the high-strength rebar was reached.

Therefore, the availability of three limit states for the bended element during the strength calculation should be noted. The first one could be identified when stresses in the steel plate are equal to the yield strength. After this point there is an increase in the strain and deflection growth but the beam continues to perceive an additional load. The second one corresponds to the achievement of the yield point by the high-strength rod reinforcement. On this stage the beam can not perceive the additional load, further loading of the specimens caused considerable yield strain in the rebar, sharp cracks' development and deflections' growth. The third limit state is physical destruction, namely the destruction of the compressed concrete zone.

The beams were designed in such a way that the yield stress of high-strength rod rebar was reached more quickly than the destruction of the compressed concrete zone.

Specimens collapsed according to the plastic law. After the yield stress of the high-strength rod rebar was reached at the next loading stage, cracking of the compressed zone of concrete took place. When the yield stress was reached in the steel plate, the stresses in it did not fall, the sheet continued to carry the load even after the achievement of the yield point. Strengthening of the plate after reaching the yield point could not be achieved, even in the beams where the it has adhesion to concrete.

The comparative calculations of curvature and deflections according to normative sources were performed for conditionally accepted load steps equal to 0.4Mmax and 0.7Mmax.

The experimental beams' deflections are shown in Fig. 3.

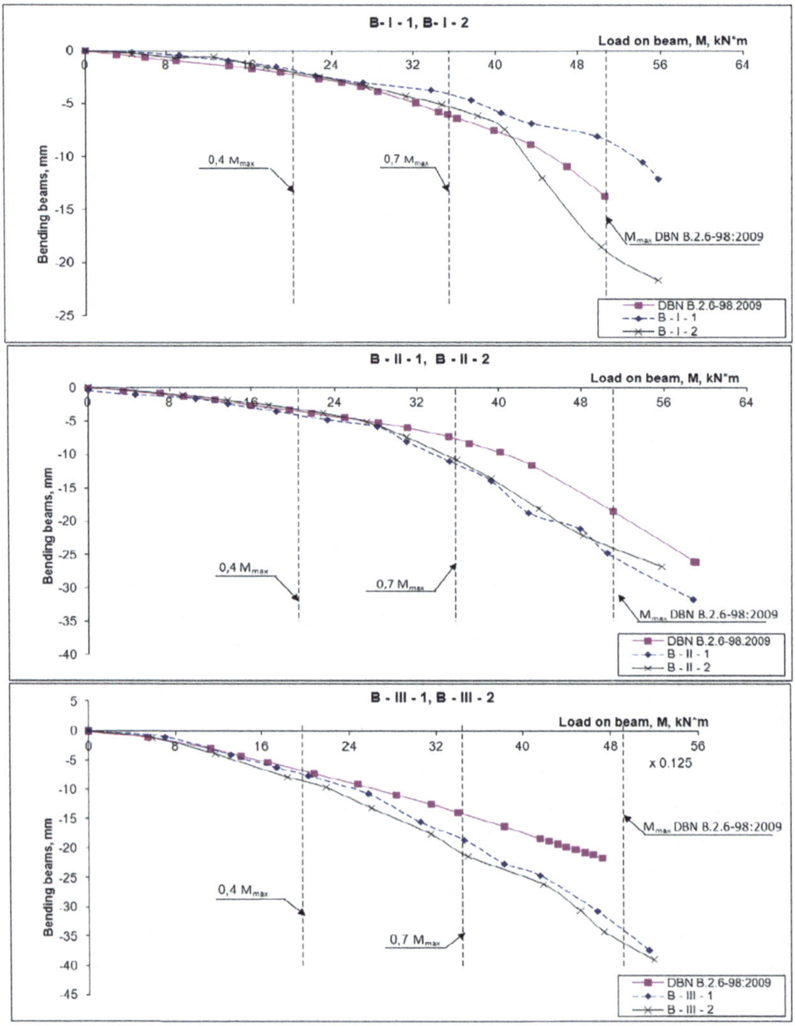

Fig. 3. Experimental and estimated according to DBN V.2.6-98: 2009 deflection graphs for beams.

Numerical values of the experimental and theoretical deflections are given in Table 3.

From the graphs it could be seen that under load less than those which cause cracking, the deflections increase proportionally to loading. Graphs' zones which correspond to this specimens' stage of work are almost linear. The deflections' values for the twin-beams before the cracks' formation are almost the same and retain the linear dependency.

After the formation of cracks in the stretched zone concrete (break of the graph's line), the deflection growth of increased, which could be identified by graph's contortion.

As can be seen from the graphs the beginning of the steel plate yielding is accompanied by a further increase in the strain and deflection growth. The higher the plate reinforcement percentage, the sharper deflection growth could be observed after its yield point. It depends on the difference in load which is perceived by plate and rod rebar.

As the maximum permissible deflection was accepted the value of 1/250 of the beam span equal to 9.6 mm according to DSTU B.V.2.6-156: 2010 [1].

The research results prove that increasing the percentage of high-strength rebar reinforcement correspond to the deflection values' increase. For example, the beams B - I - 1, B - I - 2 (ratio (fyk, A1000As, A1000)/(fyk, plate As, plate) = 31,2%/68,8%; reinforcement percentage amounted 2.77%). The deflections which correspond to 0,7Mmax loading were: 4,55 mm for the beam B - I - 1 and 5,08 mm for the beam B - I - 2, which amounted 47,4% and 52,9% respectively of the maximum permissible (9.6 mm).

Beams B - II - 1, B - II - 2 (fyk, A1000As, A1000)/(fyk, plate As, plate) = 61,9%/ 38,1%; rein-forcement ratio for the beam- 1.97%). Comparatively with the previous series' beams, deflections at load equal to 0,7Mmax increased in 2,1 times, and amounted 9,44 mm - for the beam B - II - 1 and 9,56 mm - for beam B - II - 2. This corresponds to 98, 3% and 99.6% of the maximum allowable deflection.

Beams B - III - 1, B - III - 2 - concrete reinforced with high-strength class A1000 steel bars, with reinforcement percentage equal to 0,89%. Comparatively with the first series' beams, the deflections at 0,7Mmax load increased in 4,5 and 4,2 times

Table 3. Experimental and theoretical values of the tested beams' deflection

Beams' marking	Deflection values, mm					
	$0{,}4M_{max}$			$0{,}7M_{max}$		
	Experimental f_1^{exp}, mm	DBN B.2.6:2009 f_1^{DBN}, mm	$f_1^{DBN} - f_1^{exp} / f_1^{exp}$, %	Experimental f_2^{exp}, mm	DBN B.2.6:2009 f_2^{DBN}, mm	$f_2^{DBN} - f_2^{exp} / f_2^{exp}$, %
B - I - 1	2.21	2.15	−2.7	4,55	5,5	20,9
B - I - 2	2.35		−8.5	5,08		8,3
B - II - 1	4.81	4.5	−6.4	9,44	10,4	10,2
B - II - 2	3.98		13.1	9,56		8,8
B - III - 1	8.25	8.9	7.9	20,32	21,8	7,3
B - III - 2	9.64		−7.7	21,57		1,1

respectively, and amounted 20,32 mm for the beam B - III - 1 and 21,57 mm for the beam B - III - 2. This exceeds the maximum permissible deflection by 211.7% and 224.7% respectively.

5 Conclusions

Using combined reinforcement of steel-concrete beams reinforced with S275 steel plates in combination with a high-strength A1000 rods ensures the transferring of 25–55% of the tensile strength to high-strength rebar. Such approach makes it possible to reduce the percentage of reinforcement by 15–30%. This effect is due to the higher resistance of high-strength rebar, ensuring requirements of durability, deformability and crack resistance. U-shaped anchors are welded in T-bars to steel plate, serve as stress concentrators and cracking sites. The usage of external steel plate reinforcement can effectively strengthen steel-concrete structures (with a higher percentage of rebar) with limited cross-sectional dimensions.

References

Al Sherrawi M, Lyashenko V, Edaan E, Sotnik L (2018) Corrosion as a source of destruction in construction. Int J Civ Eng Technol 9:306–314

Blikhars'kyi Z.Ya, Obukh Y (2018) Influence of the mechanical and corrosion defects on the strength of thermally hardened reinforcement of 35GS steel. Mater Sci 54:273–278. https://doi.org/10.1007/s11003-018-0183-2

Blikharskyy Y, Khmil R, Blikharskyy Z (2018a) Research of RC columns strengthened by carbon FRP under loading. Matec Web of Conf 174:1–8. https://doi.org/10.1051/matecconf/201817404017

Blikharskyy Z, Brózda K, Selejdak J (2018b) Effectivenes of strengthening loaded RC beams with FRCM system. Arch Civ Eng 64:3–13. https://doi.org/10.2478/ace-2018-0025

Blikharskyy Z, Selejdak J, Blikharskyy Y, Khmil Roman (2019) Corrosion of reinforce bars in RC constructions. Syst Saf: Hum - Tech Facility - Environ 1:277–283. https://doi.org/10.2478/czoto-2019-0036

Blikharskyy Z, Vegera P, Vashkevych R, Shnal T (2018c) Fracture toughness of RC beams on the shear, strengthening by FRCM system. Matec Web Conf 183:02009. https://doi.org/10.1051/matecconf/201818302009

Bobalo T, Blikharskyy Y, Vashkevich R, Volynets M (2018) Bearing capacity of RC beams reinforced with high strength rebars and steel plate. Matec Web of Conf 230:02003. https://doi.org/10.1051/matecconf/201823002003

Brózda K, Selejdak J, Koteš P (2017) The analysis of beam reinforced with FRP bars in bending. Procedia Eng 192:64–68. https://doi.org/10.1016/j.proeng.2017.06.011

Fadhil B, Al-Rumaithi A, Al-Sherrawi M (2018) Properties of reactive powder concrete with different types of cement. Int J Civ Eng Technol 9:1313–1321

Khmil R, Tytarenko R, Blikharskyy Y, Vegera P (2018) Development of the procedure for the estimation of reliability of reinforced concrete beams strengthened by building up the stretched reinforcing bars under load. Eastern-Eur J Enterp Technol 5(7):95. https://doi.org/10.15587/1729-4061.2018.142750

Kos Ž, Dmitrović LG, Klimenko E (2017) Developing a Model of a Strain (Deformation) of a Damaged Reinforced Concrete Pillar in Relation to a Linear Load Capacity. Tech J 11 (4):150–154 https://hrcak.srce.hr/190990

Koteš P, Strieška M, Brodnan M, Odrobinak J, Gocál J (2018) Rapid tests of corrosion in corrosion chamber. IOP Conf Ser: Mater Sci Eng 365:052013

Krainskyi P, Blikharskyy Y, Khmil R, Blikharskyy Z (2018a) Experimental study of the strengthening effect of reinforced concrete columns jacketed under service load level. Matec Web of Conf 183:1–5. https://doi.org/10.1051/matecconf/201818302008

Krainskyi P, Blikharskyy Y, Khmil R, Vegera P (2018b) Influence of loading level on the bearing capacity of RC columns strengthened by jacketing. Matec Web of Conf 230:02013. https://doi.org/10.1051/matecconf/201823002013

Selejdak J, Khmil R, Blikharskyy Z (2018) The influence of simultaneous action of the aggressive environment and loading on strength of RC beams. Matec Web of Conf 183:1–6. https://doi.org/10.1051/matecconf/201818302002

Sobol K, Blikharskyy Z, Petrovska N, Terly'a V (2014) Analysis of structure formation peculiarities during hydration of oil-well cement with zeolitic tuff and metakaolin additives. Chem Chem Technol 8(4):461–465. https://doi.org/10.23939/chcht08.04.461

Strieška M, Koteš P (2018) Corrosion map of zinc in Slovakia. Pollack Periodica 13:129–136. https://doi.org/10.1556/606.2018.13.2.13

Tayeh B, Bakar BH, Megat J, Megat A, Voo Y (2013) Evaluation of bond strength between normal concrete substrate and ultra high performance fiber concrete as a repair material. Procedia Eng 54:554–563. https://doi.org/10.1016/j.proeng.2013.03.050

Vegera P, Vashkevych R, Blikharskyy Z (2018) Fracture toughness of RC beams with different shear span. Matec Web of Conf 174:1–8. https://doi.org/10.1051/matecconf/201817402021

Zhang Q, Mol'kov YV, Sobko YM, Blikhars'kyi YZ, Khmil RE (2015) Specific fracture energy of thermally hardened reinforcement. Mater Sci 50(6):824–829. https://doi.org/10.1007/s11003-015-9789-9

The Mechanism of a Penetrative Action for Portland Cement-Based Waterproofing Compositions

Andrii A. Plugin, Oleksii A. Pluhin, Olga S. Borziak(✉), and Olena V. Kaliuzhna

Ukrainian State University of Railway Transport (UkrSURT), Feuerbach sq. 7, Kharkiv 61050, Ukraine
borziak.olga@gmail.com

Abstract. Theoretical and experimental investigations of the studies of the mechanism of a penetrative action of waterproofing compositions have been carried out. It was theoretically justified that the mechanism of penetration of the salts of the complex chemical additive of analyzed mixture into the concrete (mortar) of the structure carried out according to the diffusion mechanism. The diffusion of the salts of the complex chemical additive into the concrete with different porosity was calculated theoretically. For the water-saturated noncapillary pores after two hours diffusion depth attains 9 mm, and after three hours it is already 14 mm; for the loose concrete it occurs at a slower rate and after 1 week its depth attains 3 mm; for the dense concrete after 2 weeks it is within 0.2 mm. It was experimentally confirmed that the salts of the complex chemical additive included into dry Portland cement-based mixtures penetrate the loose concrete with the water resistant less than *W*2 the depth of 5 to 10 mm, and it results in the colmatation of the porous space with the crystal hydrates of *AFm*- and *AFt*-phases at the same depth.

Keywords: Waterproofing composition · Diffusion · Colmatation · Porosity of concrete · Penetration

1 Introduction

Modern materials science uses various ways to improve the waterproofing properties of structures (Kropyvnytska et al. 2019). During the last decades, the dry compositions based on Portland cement and complex chemical additives (CCA) used for the waterproofing coatings of the structures made of concrete and stonework have gained an increasing popularity (Plugin et al. 2018). The coatings have a high adhesion to their surfaces including under the conditions of the adverse (separation) water pressure, Babushkin et al. (2003) suggest that the adhesion and water repellency of the coatings are enhanced due the colmatation of the pores of surface layer of the structure by the hydration products of cement that contains the CCA. However, the structure surface layer compaction degree and the coating cohesion and repellency values depend on a great extent on the rate and depth of the penetration of the colmatating components of mixture. In their turn, the penetration rate and the depth are defined by physical and

Z. Blikharskyy et al. (Eds.): CEE 2019, LNCE 47, pp. 34–41, 2020.
https://doi.org/10.1007/978-3-030-27011-7_5

chemical processes that require more detailed studies. Therefore, the studies of the mechanism of a penetrative action of waterproofing compositions are a topical scientific problem. This scientific paper specifies and continues the research which data was given by Demina et al. (2017) and Plugin et al. (2018).

2 The Analytical Review of Literature Sources

To regulate the performance properties of concrete structures and masonry, additives of different composition and nature of action are used (Krivenko et al. 2018), they are particularly effective when used in the form of complex additives (Smirnova 2019; Kovalchuk et al. 2018). Plugin et al. (2018) analyzed mixtures with the complex chemical additive, which includes nitrates, chlorides, sulfates, carbonates and the calcium and sodium hydroxides. In this case Portland cement hydration products, in particular C_3A and C_4AF are the crystalline hydrates of calcium hydronitro- (Demina et al. 2017), hydrochloride- (Rapin et al. 2002; Plugin and Runova 2018), hydrosulfate- and hydrocarboaluminates (Matschei et al. 2007) of mono (*AFm-*) or triple (*AFt-*) types (Balonis et al. 2011). Demina et al. (2017) and Plugin et al. (2018) have shown that the main colmatating components of hydrating mixture are *AFm-* and *AFt*-phases. The diffusion of substance under the action of concentration gradient can be a mechanism of the spontaneous mass transfer in the absence of pressure and temperature gradients. According to Schukin (as cited in Plugin et al. 2018), solid-phase ions, the colloidal particles of *AFm-* and *AFt*-phases in the liquid medium and the ions out of which these phases are crystallized in the solution can be transferred in the analyzed system. In addition, the diffusion is limited in the porous space due to the tortuosity of capillaries, action of their walls and a decrease in the section area of the solid phase (Plugin et al. 2018). The most probable mechanism of the transfer of the colmatating components of mixture to the structure material is the diffusion of ions dissolved in the mixing water of clinker minerals and salts contained by the CCA.

3 The Basic Test Material

It was assumed that the salts of the CCA of analyzed mixture penetrate into the concrete (mortar) of the structure according to the diffusion mechanism. The diffusion occurs under the action of the difference in the concentrations of salts dissociated in the porous electrolyte of the composition and the porous electrolyte of the structure concrete (Fig. 1a). Since the diffusion of ions proceeds in water solutions, a mandatory condition for the penetration of the salts is the water-saturated state both pores and capillaries of the concrete and structure mortar and that of applied coating. The diffusion of salts proceeds in the conditions of the saturation of porous electrolyte with calcium hydroxide.

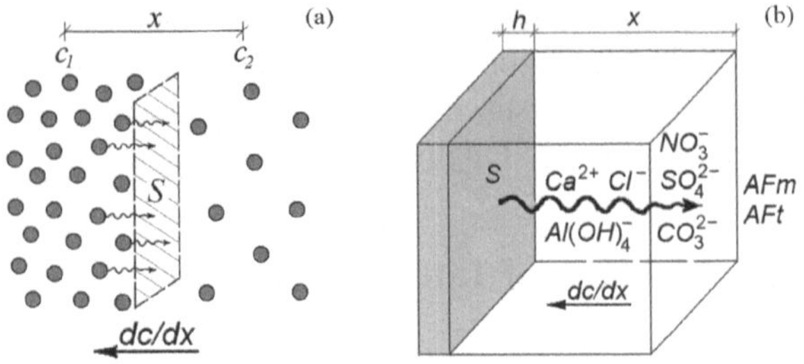

Fig. 1. The diffusion diagram: (a) the ions according to their concentration gradient; (b) the ions of the salts of CCA and aluminate ions that diffuse from the coating with the thickness of h deep into the concrete with the formation of AFm and AFt-phases

The diffusion occurs according to the Fick law, where J is the density of the diffusion flow of substance, mole/(m^2 · s) is the quantity of substance dm, mole that diffuses through the unit of area dS, m^2 for the time unit dt, s:

$$J = -D\frac{dc}{dx}; \qquad J = \frac{dm}{dS} \cdot \frac{1}{dt}, \tag{1}$$

where D is the diffusion coefficient, m^2/s; dc/dx is the concentration gradient (Fig. 1), mole/(m^3 · m).

The diffusion coefficients D of the salts that are contained by the composition of CCA at 25 °C make up m^2/s: $CaCl_2$ – $1.39 \cdot 10^{-9}$; $NaNO_3$ – $1.57 \cdot 10^{-9}$; Na_2SO_4 – $1.23 \cdot 10^{-9}$; on average $1.4 \cdot 10^{-9}$.

However, these coefficients deal only with the diffusion in continuous liquid media. As for capillary-&-porous bodies the diffusion of ions and colloidal particles occurs through the solution that fills their porous space. Alekseev et al. (1990) state that in this case Fick equations can be written as

$$\frac{dm}{dS} \cdot \frac{1}{dt} = -D * \frac{dc}{dx}, \tag{2}$$

where $D*$ is an effective coefficient of the diffusion through the capillary and porous body. Using this equation, we can define the diffusion depth x when the section area S (Fig. 1b), the effective diffusion coefficient $D*$, the diffusion duration t, and the concentration gradient $dc/dx = \Delta c/x = (c_1 - c_2)/x$ (Fig. 1) are known values:

$$x = -\frac{StD * \Delta c}{m}. \tag{3}$$

In the case, when the diffusion occurs through large water-saturated noncapillary pores, the effective diffusion coefficient value may attain the value of diffusion coefficient for the continuous medium. For the salts that are part of CCA we have $D* \sim D = 1.4 \cdot 10^{-9}$ m/s. For capillary-&-porous bodies the effective diffusion coefficient $D*$ is defined by the value of diffusion coefficient D in the continuous medium and by the parameters that characterize their porous structure (Alekseev et al. 1990): $D^* = a_1 a_2 a_3 D$ where a_1, a_2, a_3 are the parameters that take into account the waviness of capillaries (a_1); the effect of capillary walls (a_2); a decrease in the diffusion flow (section area) due to the solid phase of the system (a_3).

To analyze the influence of the porous space of concrete on the diffusion of ions through it we will represent the composite structure model as an ordinary cubic packing of structure-forming elements (the particles of hydration products, the fillers, etc., Fig. 2a). If the porous space is rather opened and branched, the parameter a_1 that takes into account the waviness of capillaries may approach a maximum value of 1. In the case of extreme maximum waviness of the capillaries (Fig. 2a) the ions have to bypass all the structure-forming elements (Fig. 2a). Hence, due to the bypass a minimum value of the parameter a_1 will be defined by the extension of diffusion pass $a_1 = L_{min}/L_{max}$. It can be seen from the model in Fig. 2a that if we express L_{min} and L_{max} in terms of the average size of structure-forming elements d, we will obtain $a_1 = d/\pi d = 0.32$.

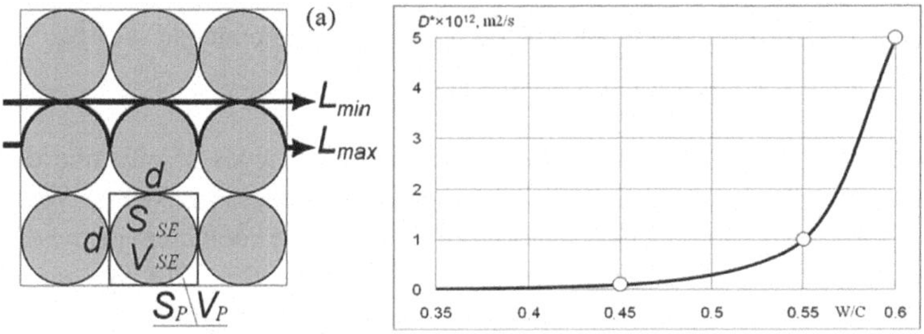

Fig. 2. The diffusion of ions through the concrete: (a) is the diagram of the model of concrete structure used for the diffusion analysis; (b) is the dependence of the coefficient of diffusion of chlorine ions D^* into the dense concrete on the water-to-cement ratio W/C

Hence, the actual values of the parameter a_1 can vary in the range of 0.32 to 1. If we assume that the capillaries in the repaired structure concrete are rather large and the effect of their walls can be neglected, the parameter a_2 will be equal to 1. The parameter a_3 that characterizes a decrease in the diffusion flow (the section area) due to the solid phase will be equal to the value of the surface porosity of composite P_S. The surface porosity can be derived from the volumetric porosity P_V that is easily defined by standard tests. Using the model in Fig. 2, in particular the unit cube with the side d, the section area of $S = d^2$ and the volume of $V = d^3$ that includes one structure-forming element with the diameter d, the section area $S_{SE} = \pi d^2/4$ and the volume $V_{SE} = \pi d^3/6$ we can define the area and volume of the pores in it, accordingly:

$$S_p = S - S_{SE} = d^2 - \frac{\pi d^2}{4} = d^2\left(1 - \frac{\pi}{4}\right); \quad V_P = V - V_{SE} = d^3 - \frac{\pi d^3}{6} = d^3\left(1 - \frac{\pi}{6}\right) \tag{4}$$

Let's define the surface porosity P_S and the volumetric porosity P_V:

$$P_S = \frac{S_P}{S} = \frac{d^2(1-\pi/4)}{d^2} = 1 - \frac{\pi}{4}; \quad P_V = \frac{V_P}{V} = \frac{d^3(1-\pi/6)}{d^3} = 1 - \frac{\pi}{6}. \tag{5}$$

We will get the relation between the volumetric porosity and the surface porosity:

$$P_S = P_V \frac{1-\pi/4}{1-\pi/6} = 0.45 P_V \quad \text{or } a_3 = 0.45 P_V. \tag{6}$$

The volumetric porosity of concrete will be taken proceeding from the value of its density, in particular the average density of $\rho_a = 2100$ kg/m^3 (for the old concrete that yielded to the leaching), and the true density of $\rho = 2650$ kg/m^3 (for the fillers dominating in volume): $P_V = 1 - (2100/2650) = 0.21$.

If we assume that all the pores in the concrete that yielded to the leaching are available for the diffusion, the parameter a_3 and the product of the parameters $a_1 a_2 a_3$ will be equal accordingly: $a_3 = 0.45 \cdot 0.21 = 0.095$; $a_1 a_2 a_3 = 0.32 \cdot 1 \cdot 0.095 = 3 \cdot 10^{-2}$.

The effective diffusion coefficient of the additive in the concrete will be:

$$D^* = 3 \cdot 10^{-2} \cdot 1.4 \cdot 10^{-9} = 4.2 \cdot 10^{-11}\ \text{m}^2/\text{s}.$$

Plugin et al. (2018) have defined experimentally the values of effective diffusion coefficient for the dense concretes of water resistance grades, i.e. W4 to W8. Using these values we constructed the dependences of the effective coefficient of the diffusion of chlorine ions in the concrete on the water-to-cement ratio (Fig. 2b). We can see, the values of D^* for dense concretes are lower by one order of magnitude.

Let's give consideration to the diffusion of the salt of CCA from the coating of 1 m^3 into the concrete. Let's assume that the salts are quickly dissolved in the mixing water, diffuse in-depth of the concrete together with the Ca^{2+} and $Al(OH)^{4-}$ ions of the aluminate phases of cement, bind with them forming crystalline hydrates already in the concrete; therefore in the porous concrete electrolyte their concentration will remain to be equal to zero. The volume of the coating with the area of 1 m^2 and 3 mm thick (0.003 m) is equal to $V = 0.003$ m^3. For the coating density of 2300 kg/m^3 its mass will be equal to $2300 \cdot 0.003 = 6.9$ kg. When the content of CCA in the mix is equal to 1 mass% the content of each salt can be taken as 0.2% (0.002). A maximum amount of salt that can diffuse will be equal to $m = 6.9 \cdot 0.002 = 1.38 \cdot 10^{-2}$ kg. Since the dissolution of salts is rather high $CaCl_2$ – 740 kg/m^3; $NaNO_3$ – 876 kg/m^3; Na_2SO_4 – 192 kg/m^3, let's define their initial concentration in terms of the consumption of CCA. A maximum concentration of each salt in the porous space of coating will be:

$$c = \frac{m}{V P_V} = \frac{1.38 \cdot 10^{-2}}{0.003 \cdot 0.18} = 25.6\ \text{kg/m}^3 \tag{7}$$

where the coating porosity was defined using the values of its average density $\rho_a = 2300$ kg/m^3 and the true density of the mix component dominating in volume i.e. the quartz sand $\rho = 2650$ kg/m^3: $P_V = 1 - (2300/2650) = 0.18$.

Since the salt concentration in the coating electrolyte will decrease due to the diffusion from a maximum value to a zero and in the concrete electrolyte it will be equal to zero during the diffusion, the concentration for the entire diffusion period Δc can be assumed to be equal to $0.5c = \Delta c = 0.5 \cdot 25.6 = 12.8$ kg/m^3.

By substituting the values of $S = 1$ m^3, $\Delta c = 12.8$ kg/m^3, $m = 1.38 \cdot 10^{-2}$ kg into Eq. (3) we will get

$$x = -tD * \cdot \frac{S\Delta c}{m} = -tD * \frac{1 \cdot 12.8}{1.38 \cdot 10^{-2}} = -928D * t. \tag{8}$$

Let's study the dependence Eq. (8) for the penetration of the salts of CCA into the large noncapillary pore ($D^* = 1.4 \cdot 10^{-9}$ m/s), loose concrete (prepared with a high W/C ratio or the concrete whose water resistance decreased due to the leaching and weathering attaining the value less than that of the minimum grade W2, $D^* = 4.2 \cdot 10^{-11}$ m^2/s), and the dense concrete ($D^* = 1 \cdot 10^{-12}$ m^2/s). Since the diffusion is expected to occur prior to the salt binding to form *AFm*- and *AFt*-phases that are already formed at the beginning of the hardening of Portland cement with the CCA we will study this dependence during the period of two days (Fig. 3).

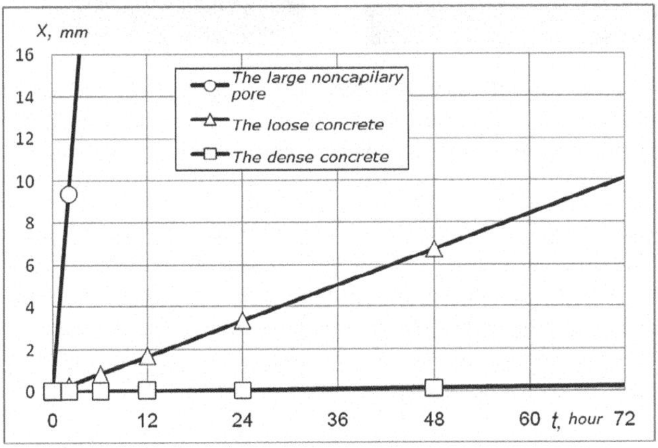

Fig. 3. The diffusion kinetics (dependence of the depth x on time t) of the salts of CCA in the concrete: for the large noncapilary pore ($D^* = 1.4 \cdot 10^{-9}$ m/s); loose concrete ($D^* = 4.2 \cdot 10^{-11}$ m^2/s), and the dense concrete ($D^* = 1 \cdot 10^{-12}$ m^2/s)

Figure 3 shows that the diffusion of salts through the water-saturated noncapillary pores occurs quickly and after two hours its depth attains 9 mm, and after three hours it is already 14 mm. In the loose concrete the salt diffusion occurs at a slower rate and after 1 week its depth attains 3 mm and after 2 weeks it is equal to 7 mm. In the case of the dense concrete the salt diffusion occurs at a very slow rate and after 2 weeks it is within 0.2 mm. The obtained kinetics of the diffusion of the salts of CCA and as a consequence the depth of their penetration into the concrete and the opportunity of the formation of *AFm*- and *AFt*-phases at that depth was confirmed by the following experiment:

The dry mixture was mixed using no pure water; instead of it we used for this purpose the water solution of luminophore – naphthalene benzimidazole. The mixture was applied as a layer of 3 mm thick on the face of previously water-saturated specimens in the form of the bars with the size of 160 × 40 × 40 mm made of the cement-sand solution with ratio of 1:3, W/C = 0.75, at an age of 28 weeks. After the 3 weeks the specimens were split perpendicular to the coating. The obtained section was inspected in white and ultraviolet rays and the pictures were taken (Fig. 4).

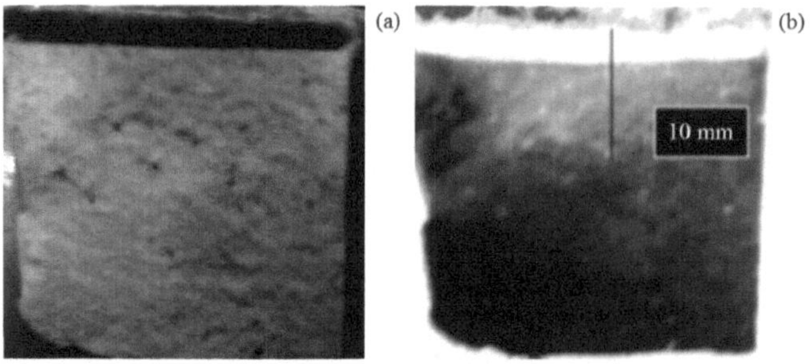

Fig. 4. The lateral fracture of the specimen made of the cement-sand mortar and coated with the mixture of a penetrative action that was mixed using the luminaphore solution: (a) is the image in the white light; (b) is the image in ultraviolet rays (the contrast is enhanced)

In the ultraviolet rays, the concrete with penetrated luminaphore and as a matter of fact the salts was glowing. Figure 4 shows the images of the section of the specimen of coated concrete. We can see that in the middle of the specimen where the influence of the external surface of the specimen is minimum luminaphore and CCA salts penetrated to the depth of approximately 7 mm that complies with the kinetics of their penetration into the loose concrete (Fig. 3).

4 Conclusion

The theoretical and experimental investigations showed that the salts of the complex chemical additive included into dry Portland cement-based mixtures intended for waterproofing coatings penetrate the loose concrete with the water resistance less than W2 attaining the depth of 5 to10 mm. It results in the colmatation of the porous space with the crystal hydrates of *AFm*- and *AFt*-phases at the same depth. Dense water-permeable concretes show no penetration.

Acknowledgements. The activity presented in the paper is part of the research grant «Theoretical and experimental bases of determining, predicting and ensuring the carrying capacity and durability of transport facilities under aggressive conditions».

References

Alekseev SN, Ivanov FM, Modryi S, Shissl P (1990) The durability of reinforced concrete in aggressive environments. Stroyizdat, Moskow (in Russian)

Babushkin VI, Kostuk TA, Kondraschenko EV (2003) Methodological approach to the question of determining the penetration depth of penetrating compositions. Naukovyj visnyk budivnycztva 22:43–46 (in Russian)

Balonis MM, Medala M, Glasser FP (2011) Influence of calcium nitrate and nitrite on the constitution of the AFm and AFt cement hydrates - experiments and thermodynamic modeling. Adv Cem Res 23(3):129–143

Demina OI, Plugin AA, Dedenyova EB, Bondarenko DO, Kostuk TA, Bondarenko AI (2017) Interaction of Portland cement hydration products with complex chemical additives containing fiberglass in moisture-proof cement compositions. Funct Mater 24(3):415–419

Kovalchuk O, Grabovchak V, Govdun Y (2018) Alkali activated cements mix design for concretes application in high corrosive conditions. Matec Web Conf 230:03007

Krivenko P, Petropavlovskyi O, Kovalchuk O, Lapovska S, Pasko A (2018) Design of the composition of alkali activated portland cement using mineral additives of technogenic origin. East Eur J Enterp Technol 4/6(94):6–15

Kropyvnytska T, Semeniv R, Kotiv R, Kaminskyy A, Hots V (2019) Studying the effect of nanoliquids on the operational properties of brick building structures. East Eur J Enterp Technol 5/6(95):27–32

Matschei T, Lothenbach B, Glasser FP (2007) Thermodynamic properties of Portland cement hydrates in the system CaO-Al_2O_3-SiO_2-$CaSO_4$-$CaCO_3$-H_2O. Cem Concr Res 37:1379–1410

Plugin AA, Kostyuk TO, Proshhyn OYu, Bondarenko DO, Pluhin OA, Borziak OS, Arutyunov VA (2018) Waterproofing cement composites. Kolegium, Kharkiv (in Ukrainian)

Plugin AA, Runova RF (2018) Bonding calcium chloride and calcium nitrate into stable hydration portland cement products: stability conditions of calcium hydrochloraluminates and calcium hydronitroaluminates. Int J Eng Res Afr 36:69–73

Rapin JP, Elkaim E, Francois M, Renaudin G (2002) Structual transition of Friedel's salt $3CaO \cdot Al_2O_3 \cdot CaCl_2 \cdot 10H_2O$ studied by synchrotron powder diffraction. Cem Concr Res 32:513–519

Smirnova O (2019) Compatibility of shungisite microfillers with polycarboxylate admixtures in cement compositions. ARPN J Eng Appl Sci 3(14):600–610

Preservation and Maintenance of the Odesa's Historical Building Heritage – Nowadays Problem

Olena Chernieva[1(✉)] and Gennadiy Plahotny[2]

[1] Department of Reinforced Concrete Structures and Transport Facilities, The Odesa State Academy of Civil Engineering and Architecture, Didrihsona 4, Odesa 65029, Ukraine
spring.cherneva@gmail.com

[2] Department of Architectural Structures, The Odesa State Academy of Civil Engineering and Architecture, Didrihsona 4, Odesa 65029, Ukraine

Abstract. The article is devoted to the problem of Odesa historical buildings' preservation. According to the Law of Ukraine about Protection of the Cultural Heritage (2000), we can call the Vorontsov Lighthouse, National Academic Theatre of Opera & Ballet and the Factory of sparkling wines the objects of historical and cultural heritage of Odesa. The article presents the results of the preservation monitoring (based on the reports) of these buildings. Reports were performed by employees of the Odessa State Academy of Civil Engineering and Architecture during 1990–2019. A huge number of negative factors had an impact on historical buildings. One of the most important structures of the building is its underground part. The variability of the soil is one of the main causes of structural damage. Foundations' settlement monitoring and control of the structures durability with non-destructive and micro-invasive testing methods are being conducted. The article describes the historical facts of building maintenance, which explain the reasons for the appearance of deformations.

Keywords: Historical building heritage · Differential settlement · Monuments · Substructure

1 Introduction

Maintenance and preservation of architectural heritage require a balance between the structural safety needs and the respect for their architectural and cultural value. Analysis of recent researches shows us that this problem is being considered all over the world. The Italian researches use Structural health monitoring (SHM) (De Stefano et al. 2016) & Building Information Modelling (BIM) (Biagini et al. 2016). Portugal researchers use the diagnostic procedure, which includes and details in situ and laboratory testing (Masciotta et al. 2016). The three most important design-related failure causes are weather impact, impact from occupants, and loads and moisture from the wet areas (Chong and Low 2006). We cannot disprove this; just add the impact of emergencies that affect the preservation of the historical heritage of the city.

Z. Blikharskyy et al. (Eds.): CEE 2019, LNCE 47, pp. 42–49, 2020.
https://doi.org/10.1007/978-3-030-27011-7_6

Aim is to show the preservation history of the famous Odesa buildings and marine structures and their nowadays condition.

Methods. The research of rare archival data, visual and instrumental surveys of the buildings' preservation, non-destructive and micro-invasive testing methods, engineering monitoring.

Scientific novelty and Practical significance. The results of the research will help to choose the optimal method for the restoration and reconstruction of Odesa historical buildings in the future.

2 The Vorontsov Lighthouse

The Vorontsov Lighthouse is a main symbol of marine Odesa. It presents the city from the sea side. Nowadays it is the red-and-white, 27.2 m tall lighthouse in the Black Seaport of Odessa, Ukraine. It was named after Prince Mikhail Semyonovich Vorontsov, one of the governors-general of the Odessa's region. If you go back to the history of the city, you can call the Vorontsov lighthouse the Odesa's historical monument.

Unlike numbers of European ports, the port of Odessa has rather difficult conditions for escorting ships to their berths and bringing them to the open sea. The presence of four entrance gates to the port, each of which has its own navigation situation (for example, Vorontsov's entrance requires a fourfold change of course), leads to significant waste of time and effort, and most importantly, objectively create a significant danger for navigation.

Because of this, in 1860, under the leadership of engineer N. Gora, the old Quarantine Jetty was extended by 1030 m. The end of this jetty was additionally extended by 617 m, thus forming the Roadstead Mole (Jetty), which protects the roadstead from dangerous winds (Fig. 1a and b). Moles were interconnected by stone breakwater in 1882(Gleb-Koshanskii 2007).

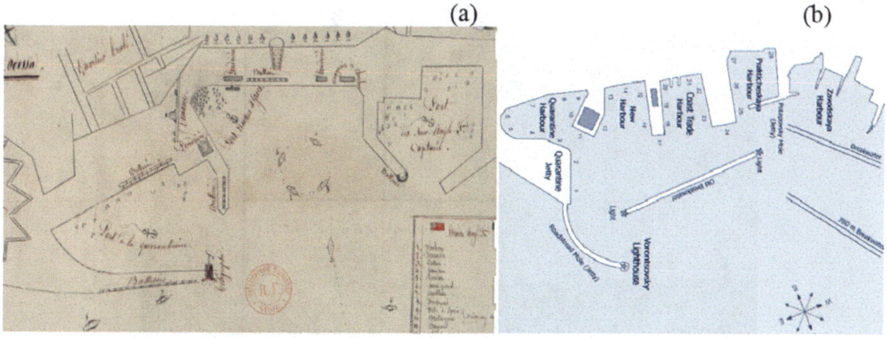

Fig. 1. Port of Odesa before 1860 (a) and after (b)

In 1845 a wooden lighthouse was installed in the head of the Roadstead Mole, which in 1863 was replaced by a cast-iron lighthouse. In 1888 the height of the lighthouse was increased, and its trunk was made of reinforced concrete and solid brick. Such a lighthouse was blown up in 1941, depriving the fascists of the point of artillery fire adjustment.

The modern lighthouse, as a symbol of Odessa, was built in 1954 according to the "ChernomorNIIproekt" project by the Port building organization. The lighthouse with a height of 29 m is installed on a solid stone foundation.

History knows many examples of collisions and even the death of ships near the port. The danger of navigation especially increased during peak hours, at night, and also when visibility deteriorates when the lights of Odessa lighthouses are extinguished by fog. On the night of 04/01/1994, the Israeli vessel "Pan Antwerpen", leaving the port water area, with its bow struck the junction between the lighthouse and breakwater, which led to the appearance of vertical local cracks up to 2 mm thick in the trunk. The trunk of the lighthouse was tilted to the north-west at an angle of $5^{o}20'$. The visual and instrumental examination of the lighthouse's technical condition was carried out by a group of specialists from the Research and Production Center "Reconstruction, Restoration and Ecology of the Environment", with the participation of specialists from the Odessa State Academy of Civil Engineering and Architecture. The cracks were covered by strips (crack tell-tales). Four months later, the survey results showed that no further deformations of the lighthouse bearing structures were observed, which made it possible to develop a project of lighthouse's strengthening using a strapping reinforced concrete belt and anchoring it with the body of the stone breakwater. The cracks in the trunk were injected with expanding cement solution. In 2017 ice forces on bottom-founded structures were identified from response measurements according to (Nord et al. 2016). As a result of further observations the entire construction of the Vorontsov lighthouse in Odessa is in a satisfactory technical condition (Fig. 2).

Fig. 2. Nowadays condition of the Vorontsov Lighthouse

3 The Odesa National Academic Theatre of Opera and Ballet

The geological conditions of the Northern Black Sea region are represented by a thick upper layer of loess soils, underlain by a layer of dense red-brown clay and loam and located below the Pontian sediments of shell limestone (Pronin 2009). The specificity of loess soils is their ability to additionally deform when soaked, being in a tense state due to external load and its own weight. The main residential buildings of Odessa, which were built in the 19th century, were built on strip foundations on a loess base. Some of these buildings had basements up to 3 m high, which were rarely used as utility rooms. Buildings of that period had a one or two-story underground part. For example, under a part of the Opera and Ballet Theater there are two-level basements.

The history of the Odesa Opera and Ballet Theater is closely connected with the history of the city. In 1804, a Petersburg architect of French origin, Jean-François Thomas de Thomon, designed the theater in the form of a rectangular ancient temple, which was erected in 1809. On the night of January 1 to 2, 1873, the building burned down, a fire arose from a gas lamp that illuminated the clock on the facade. The world competition for a new theater project did not reveal a winner. The development of the project was entrusted to the Viennese architects Ferdinand Fellner and Herman Helmer (Yeksarova and Yeksarov 2018; Dobrolyubskii et al. 2016). During the construction of the theater, the local architects A. Bernardazzi, K. Dmytrenko and others refined the project. The construction of the new theater began on September 16, 1884. The opening of the theater took place on October 1, 1887. The building has a frameless structural scheme with bearing exterior and interior walls that form the main hall, the scenic part of 500 m^2 and a three-tiered area for visitors.

The main facade of the building, directed to Deribasovskaya Street, has a semi-circular shape, in which there are an entrance and two staircase nodes. From the outside of the facade there is a two-tiered central arch with three sculptural groups and masks - symbols of the theater. The basement and ground floor have a strict surface of the walls without any special decorations. Above them there is a sophisticated second floor with balconies, columns and pilasters of the Ionic order. Sculpture busts are installed in the round niches. The central place on the attic of the arch is reserved for Melpomene, one of 9 Greek muses, the patroness of tragedies. She is racing in a chariot with four panthers. The sculptures located on high pedestals on both sides of the main entrance were created based on the ancient Greek drama: tragedy and comedy. The exterior design of the theater is made in the Baroque style. The interior design of the theater and its premises is made in the late Rococo style. The special effect of the hall decoration is provided by the use of precious facing materials, gilding, stucco decorations, sculptures, lamps. A large bronze chandelier with crystal pendants forms a composite center. Under the rectangular part of the building there are two-level basements, in which there are storage and utility areas.

The substructure's upper layer is a loess collapsible loam, below which there is a layer of red-brown clay and a Pontian stage of limestone. The strip foundations of the building, located at different heights, are made of rubble stone, concrete and bricks. There is no basement under the semicircular part of the theatre. Under the rest of the building there are basements, which are covered with vaulted structures. According to archival sources, in the process of laying the foundations there was a heavy downpour that moistened the subsiding loams. During the construction process, the foundations and the underground part were repeatedly flooded with stormwater. This made an influence to the maintenance of the building. As a result, in 1903, the western part of the building was 18 cm lower than the eastern part. Then the broadening of the foundations of the western part was performed. The inclined metal dome was jacked out to its original state. Vertical and inclined cracks were found in the outer walls, which were later caulked with cement-sand mortar. In March 1905, during the performance, a fire broke out on the stage, as a result the hall and adjoining corridors suffered. In the near future the theater was restored. The next deformations appeared during the war and occupation. The bomb and shells' explosions have caused additional differential settlement of foundation sand led to increasing of cracks and the emergence of new ones. It is noteworthy that on April 10, 1944, the flag of the liberators was hoisted on the central balcony of the front part of the theater. As soon as the theater was cleared of mine (leaving, the fascists wanted to blow it up) a performance was staged for the soldiers-liberators. However, the theater building was in poor technical condition. Therefore, in 1955–1956, single-solution of silica was performed for consolidation of soils. This method of amplification was not sufficiently reliable due to the complexity of the quality control of work. The silicified soil blocks could not create a solid subfoundation. In 1967, a major restoration was carried out in the theater. The restorers washed with acids the old gilding. In eighty cases out of 100 the gilding turned green. This meant that not pure gold had been used, but a surrogate. As a result, restorers had to additionally use 10 kg of pure gold.

Considering the deformed state of the theater, public authorities of Ukraine decided to entrust a group of employees of the Scientific Center of the Odessa State Academy of Civil Engineering and Architecture to investigate the technical condition of the theater. From 1991 to 1997, a visual-instrumental examination of the state of the load-bearing structures was carried out and the dynamics of the further development of deformations were monitored. 150 wall crack tell-tales were installed on the building, which made it possible to determine the nature of the walls deformation's development. During the restoration campaign, non-destructive and micro-invasive analyses (Tortora et al. 2016) were performed in order to determine the detachments of sculptures surfaces. Further development of cracks was observed in the front semicircular part of the building, where there are no basements. Cracks conditionally divided it into seven separate blocks. The width of the crack opening was within 1–3 mm. Internal walls and vaulted ceilings of the underground part were also deformed.

Experts have proposed three options for strengthening the underground part of the theater:

I - installation of injection pile foundations on the base slab. It was proposed to arrange the slab under the existing foundations.

II - installation of a pile ring from the jack piles around the building with special devices for adjusting the settlement and moving of the building.

III - installation of a three-row pile pre-stressed casing, which is pressed tightly against the external walls at the level of the overlap of the upper basement.

Monolithic reinforced concrete belt was recommended to installing the attic of the building. Summarizing all the proposals, during the restoration in 1999–2000, the foundations and soils of the theater were strengthened using paired injection pile foundations with reinforced concrete beams - lintels. 1900 piles and 950 support beams were loaded. The condition of the building structures is constantly monitored now.

4 The Odesa Factory of Sparkling Wines

The main building of the Odesa Factory of sparkling wines on French Boulevard is a two-story building with a cellars in two levels, each 5 m high. Cellars' ceilings are represented by longitudinal brick vaults. The foundations under the external and internal longitudinal bearing walls are strip and brick with a bottom grade - 10.0 m. The subfoundation is dense red-brown clay and loam. The level of groundwater, mainly of technogenic character, is at the level of −9.3 m. Thus, the base of the foundations is located 0.7 m below the level of groundwater. The design pressure on the foot of the outer foundations was 0.68 MPa, and on the base of the middle longitudinal ones - 0.72 MPa. In the fifties of the twentieth century, reinforced concrete amphoras were installed in the lower cellar, measuring 3.0×4.0 m for storing raw materials. Maintenance of amphora tanks was difficult because of their partial flooding with industrial waters, which led to their deformations.

To make a decision on its rational use, experts from the Odessa State Academy of Civil Engineering and Architecture were invited. They offered to install 42 metal containers of 25 m^3 each for the storage of wine raw materials. The tanks were installed on metal beams, which were supported on longitudinal bearing walls. The pressure from the weight of the tanks on the foundation base increased by 0.02 MPa. According to the calculation, when wine was fully loaded with containers, the total pressure on the base of the foundations additionally increased by 0.07 MPa, thus the pressure on the bottom of the extreme foundations was 0.77 MPa, medium - 0.81 MPa. To create a uniformly distributed operating load to the foundations, the test loading of tanks with water was carried out in ten steps, which were maintained for seven days until the foundations' differential settlement stabilized. The total additional load on the main part of the building amounted to 15000 kN. To monitor the uniformity of the building's relative settlement during the sequential loading and unloading of tanks, a system of water levels and wall marks was installed at the corners of the lower cellar for geodetic leveling of the settlement. Thirty hydraulic locks of vertical movement were attached to the outer and inner bearing walls. When each load step was applied, the difference in relative settlements did not exceed 2 mm. When tanks were fully loaded, the total settlement of the main part of the building was 7 mm. Tank unloading was performed in reverse order of loading. Due to the absence of settlement joints, vertical cracks up to 5 mm thick were formed in the places of junction of the main part of the building to its extensions. Their disclosure stopped at the full stabilization of the building's settlement. Subsequently, since 1998, the operational loading and unloading of containers with wine in the upper cellar has been carried out according to a scheme similar to the testing of the building by trial loading and unloading. Further cracking is not observed now, they are completely caulked with elastic insulation and plastered. The rise of the groundwater level in the lower cellar is controlled automatically using a water pump located in a pit up to 1.5 m deep.

The formation of new cracks that may pose a danger to the state of the building is not found at the moment. There are separate cracks above and below window areas up to 3 mm wide (Fig. 3). Works were done according to DBN V.3.2-1-2004, 2005.

Fig. 3. Separate cracks above and below window areas

5 Conclusions

For more than 28 years, employees of the Odessa State Academy of Civil Engineering and Architecture have been monitoring the historical buildings of Odessa. It is necessary to continue monitoring the condition of buildings to prevent the formation of irreversible deformations. Traditional methods are often time-consuming and barely efficient, because they are related to the exchange of two-dimensional and paper-based support information. Therefore, we need to use the European experience.

Acknowledgements. The activities presented in the article are part of the scientific-restoration report about technical condition of the Odesa's objects of architectural and historical heritage. Report results are not printed material and are not freely available. Further monitoring of the state of buildings and structures was carried out personally by the authors of the article and was not previously published.

References

Biagini C, Capone P, Donato V, Facchini N (2016) Towards the BIM implementation for historical building restoration sites. Autom Constr 71(1):74–86. https://doi.org/10.1016/j.autcon.2016.03.003

DBN V.3.2-1-2004 (2005) State Building Norms of Ukraine. Restoration, Conservation and Repair Works of Cultural Heritage Buildings, p 120

Defectography of the Odessa State Opera and Ballet Theater (1990–1995). Album of drawings. Research and Production Center "Reconstruction, Restoration and Ecology of the Environment"

De Stefano A, Matta E, Clemente P (2016) Structural health monitoring of historical heritage in Italy: some relevant experiences. J Civ Struct Health Monit 2016:83–106. https://doi.org/10.1007/s13349-016-0154-y

Dobrolyubskii A, Ieksarova N, Yeksarov V (2016) Smart port city of Odessa: brilliant embodiment of Vitruvius' theories of beauty. Int J Archit Technol Sustain 2016(1):1–10

Gleb-Koshanskii N (2007) Port and Odesa. New print, p 344

Law of Ukraine. On Protection of the Cultural Heritage (Bulletin of the Verkhovna Rada of Ukraine (BBP), 2000, No 39, 333)

Masciotta MG, Roque J, Ramos L, Lourenco P (2016) A multidisciplinary approach to assess the health state of heritage structures: the case study of the Church of Monastery of Jeronimos in Lisbon. Constr Build Mater 2016(116):169–187. https://doi.org/10.1016/j.conbuildmat.2016.04.146

Nord T, Oiseth O, Lourens E (2016) Ice force identification on the Nordstromsgrund lighthouse. Comput Struct 169:24–39. https://doi.org/10.1016/j.compstruc.2016.02.016

Pronin KK (2009) Natural caves of the Black Sea-Azov and Moldavian-Podolsk karst regions. Simferopol-Odesa "SONAT", p 130

Tortora M, Sfarra S, Chiarini M, Daniele V, Taglieri G, Cerichelli G (2016) Non-destructive and micro-invasive testing techniques for characterizing materials, structures and restoration problems in mural paintings. Appl Surf Sci 2016(387):971–985. https://doi.org/10.1016/j.apsusc.2016.07.023

Chong W-K, Low S-P (2006) Latent building defects: causes and design strategies to prevent them. J Perform Constructed Facil 20(3):213–221. https://doi.org/10.1061/0887-3828.20.3.213

Yeksarova N, Yeksarov V (2018) Genesis of the formation of the cultural and social core of Odessa. Vitruvio-Int J Archit Technol Sustain 3(2):67–77. https://doi.org/10.4995/11018

Coefficient of Flow Rate of Inlet Cylindrical Nozzles with Lateral Orthogonal Inflow

V. V. Cherniuk[1,2(✉)], V. V. Ivaniv[1], I. V. Bihun[1], and Ja. M. Wojtowicz[1]

[1] Department of Hydraulics and Sanitary Engineering, Lviv Polytechnic National University, 12, Bandera Street, Lviv 79013, Ukraine
v.cherniuk@ukr.net

[2] Catholics University of Lublin named after John Paul II, Lublin, Poland

Abstract. Cylindrical nozzles each of which has a lateral orthogonal inlet of fluid into it we recommend to apply in pressure distributive pipelines. By means of rotating these nozzles about their longitudinal axes, we can change the values of the angles β between the direction of the flow of fluid inside the distribution pipeline and the direction of the inflowing into the nozzle stream. The coefficient μ of flow rate through the nozzle with lateral orthogonal inlet is a function of the angle β. By means of changing the angles β, we regulate the non-uniformity of fluid dispensation from the pipeline along the path. It is experimentally established that the values of the coefficient μ for such nozzles depend on the angle β, on the squared ratio $(d/D)^2$ of the cross-section area of the nozzle d to the cross-section area of the distributive pipeline D, as well as on Reynolds' criterion Re which is calculated for the water stream in the nozzle. As operating fluid, we used water. The angles β were assigned the following values: 0, 45, 90, 135, and 180°. In the experiments, we used the distributive pipelines whose diameters D were equal to 11.28; 16.13; 20.18; 26.01 mm. The diameters d of the nozzles were 4.83; 6.01; 8.02; 8.99 mm. ratios $(d/D)^2$ of cross-section areas were 0.0887; 0.119; 0.158; 0.183; 0.251. The greatest values of the coefficient μ of flow rate of the nozzles were obtained for $\beta = 0^{\circ}$; the least for $\beta = 90^{\circ}$. The greatest range of regulation of flow capacity (throughput) of nozzles has been obtained for $(d/D)^2 = 0.251$; the values of μ variated by 23%.

Keywords: Cylindrical nozzles · Coefficient of flow rate of nozzle

1 Introduction

Pressure distributive pipelines (DPs) are widely used in irrigation, in water supply; in power generation; in drainage; in ventilation; in agricultural aviation (plants sprinkling); in mechanical engineering (distributive fuel lines of multi cylinder engines of internal combustion), etc.

Manufacturing processes, in overwhelming majority of cases, call for ensuring of uniformity of fluid dispensation along the DP.

The method of reduction of non-uniformity of fluid dispensation along DP whose application does not call for change in geometrical parameters of DP in the course of

Z. Blikharskyy et al. (Eds.): CEE 2019, LNCE 47, pp. 50–57, 2020.
https://doi.org/10.1007/978-3-030-27011-7_7

operation is interesting. This method is based on the ability of hydro-solvable high-molecular polymers whose molecules are of chain-like structure to reduce hydraulic frictional resistance in pipes in turbulent flow and on their ability to regulate local losses of head in pipes; this was shown by Povkh and Chernyuk (1986, 1989).

Chen and Sparrow (2009a, 2009b) numerically investigated the decrease in non-uniformity of fluid dispensation from an equipped with rectangular holes DP along the stream. DPs with longitudinal continuous rectangular slot were also investigated.

Hassan et al. (2014) have found that the non-uniformity of water dispensation along the path from a conic-shaped DP is less than that in the case of cylindrical DP.

Lee et al. (2012) experimentally investigated the influence of the wall thickness t of a DP of rectangular cross-section upon the non-uniformity of the water dispensation from it along the path. With the increase in the wall thickness t of DP and with the decrease in the value of $\Sigma\omega_{or}$, the non-uniformity of water dispensation from DP along the path decreased.

Wang et al. (2011), Zeng et al. (2012), Kim and Byun (2013) experimentally investigated the influence of geometric parameters of jet branch from DP in a heat exchanger for non-uniformity of dispensation of working fluid from a DP along the path.

Abubakar (1977) theoretically shown that the value of the coefficient μ depends on the angle β of water outflow through an orifice.

Zemlyanaya and Gulyakin (2017, 2018) have come to the conclusion that the least non-uniformity of fluid dispensation from pressure DPs is achieved in the confuser DP with continuous slot in its wall.

The aforesaid works concern outflow of different fluids from pressure DPs through orifices and slots. However, neither of them ensures regulation of fluid dispensation from DP along the path in the course of operation, since their geometric parameters are constant.

Cherniuk and Ivaniv (2017) have invented a way of regulation of fluid dispensation by means of changing the angle β. For this purpose, nozzles are rotated about their longitudinal axes. Here, β is the angle between the direction of the water stream and that of the outflowing through the nozzle jet. In literature, data concerning values of the coefficient μ of flow rate of cylindric nozzles with lateral inflow are lacking.

Aim of the Work: to experimentally establish the value of the coefficient μ of flow rate of cylindrical nozzles with lateral orthogonal inflow of stream depending on the angle β between the direction of water stream in a distributive pipeline and the direction of the outflowing jet.

2 Procedure of the Experiment

The working head in the investigated nozzles was varied within the range of 0 through 17 m. The values of head which were less than 2.2 m were measured by a hydro piezometer whose scale division value was 1.0 mm. The values of head which were greater than 2.2 m were measured by a spring pressure gauge whose class of accuracy was 0.6 and the scale division value 0.02 kg/cm^2 (Fig. 1*a*) The working fluid was water. The flow rate of water which was flowing through the investigated nozzles was determined by means of volumetric method.

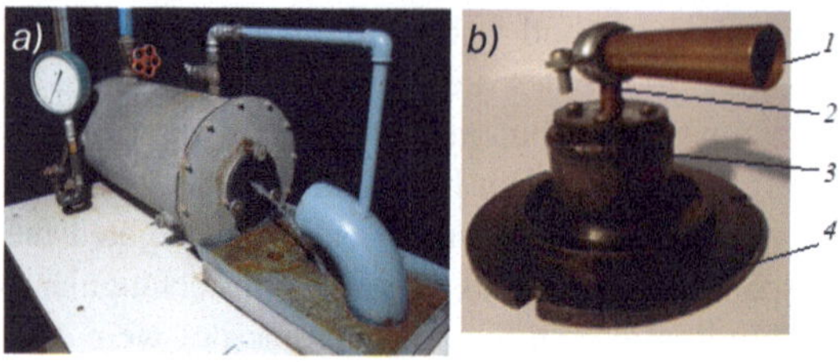

Fig. 1. Experimental setup: (a) pressure tank with nozzle in its butt wall; (b) assembly of fixation of investigated nozzle: 1 – terminal segment of DP; 2 – investigated nozzle; 3 – hollow cylinder; 4 – inner side of butt lid of pressure tank.

For ensuring such conditions of water inflow into the cylindrical nozzles with lateral inflow of stream as the conditions in a distributive pipeline (DP), the inlet nozzle was installed in the blind pipe, in this case the latter is presented by a terminal segment of DP (Fig. 1*b*). The nozzles were installed in the inner side of the butt wall of the cylinder-shaped pressure tank. The outlet terminal of a nozzle was tapped outside the tank.

Cylindrical nozzles with lateral inflow of stream (Fig. 2*a*) are fixed in a pressure distributive pipeline providing the possibility of their rotation by a fixed angle about their axes (Fig. 2*b*). In the course of the experiments, the angle β between the direction of water stream in PD and the direction of the outflow jet were assigned the values within the range of 0° … 180° (see the Table 1 and Fig. 3).

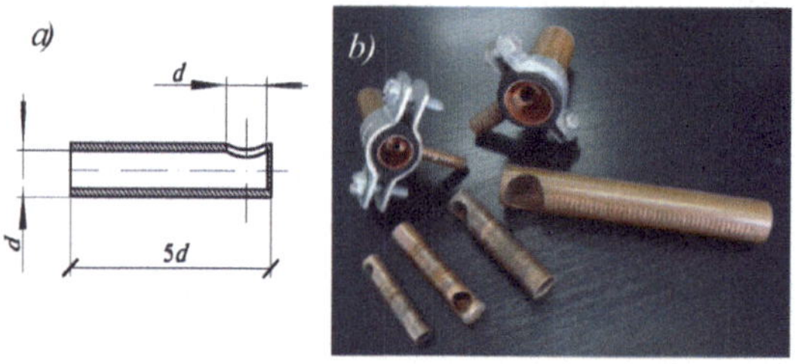

Fig. 2. Cylindrical nozzles with orthogonal lateral inflow: a – schematic diagram; b – general view (in left upper corner two segment of DP with installed in them nozzles are shown)

Table 1. Geometric characteristics of the investigated nozzles

Diameters, *mm*		Squared ratio of cross-section areas of nozzle and PD, $(d/D)^2$
Distributive pipelines, D	Nozzles, d	
20.18	6.01	0.0887
26.01	8.99	0.119
20.18	8.02	0.158
11.28	4.83	0.183
16.13	8.08	0.251

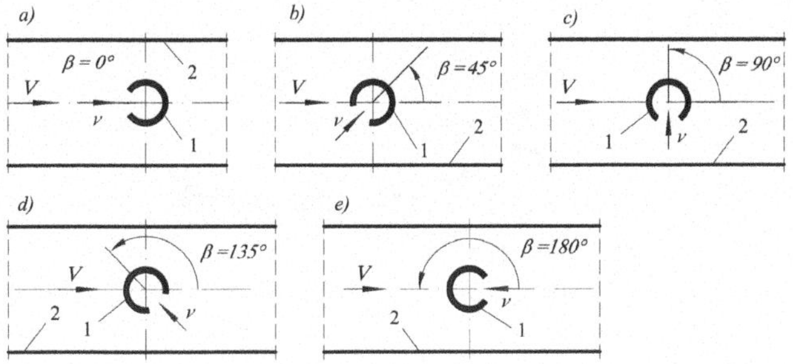

Fig. 3. Parts of DP with nozzles installed at different angles β of stream inflow: $\beta = 0^{\circ}$ - (*a*), 45° - (*b*); 90° - (c); 135° - (d); 180° - (e); 1 – nozzle (cross-section); 2 – wall of DP; V – velocities of water stream; v – velocities of stream inflow into nozzles

3 Mathematical Procession of Results of Experimental Data

The volumetric flow rate of water was calculated according to the formula $Q = W/t$, where W is the volume of water which inflow into the measuring tank during the time t The coefficient μ of flow rate was calculated according to the formula $Q = \mu\omega\sqrt{2gH}$, where Q is the flow rate of water through the nozzle, m^3/s; ω is the live cross-section area of the jet in the inlet hole and in the body of the nozzle, m^2; g is the gravity acceleration, g = 9.81 m/s^2; H is the head of water at the inflow into the nozzle, m.

The values of Reynolds' criterion for the stream in a nozzle were determined by the formula $\mathrm{Re} = vd/v$, where v is the speed of water in the nozzle; m/s; d is the diameter of the inlet hole and of the body of the nozzle, m; v is the kinematic viscosity of water, which was determined according to the formula $\psi = (\mu_{i^{\circ}} - \mu_{0^{\circ}}/\mu_{0^{\circ}}) \cdot 100\%$ were, $\mu_{0^{\circ}}$ and $\mu_{i^{\circ}}$ are coefficients of flow rate of the nozzle for the angles $\beta = 0°$ and $\beta = i°$, respectively; where $i° = 45°$; $90°$; $135°$; $180°$.

4 Results of Experimental Investigations

The mathematical procession of experimental data has been carried out for all the five values of the squared ratio $(d/D)^2$. For each of the squared ratios $(d/D)^2$, for five different values of the angle β, we obtained graphical dependences of the coefficient μ of flow rate of outlet cylindrical nozzles with lateral inflow of stream line into the nozzle on the values of Reynolds' criterion Re_d. In Fig. 4, the dependences $\mu = f(\mathrm{Re}_d, \beta)$ for $(d/D)^2$ = 0.0887 and 0.251 are presented. From these graphs, it can be seen that the values of the coefficient μ essentially depend on the angle β (Fig. 5).

The dependences $\mu = f[(d/D)^2, \beta]$ are of similar character within the investigated range of Reynolds' criterion, the later was varied from 10^3 through $100 \cdot 10^3$. From Fig. 6, it can be seen that for all the values of β the greatest value of the coefficient μ of

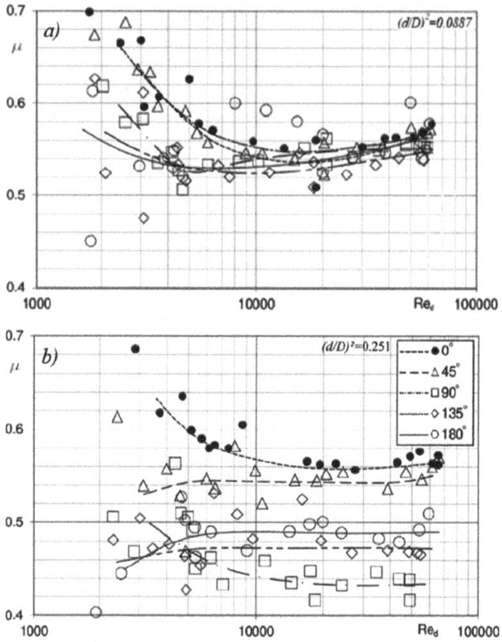

Fig. 4. Dependence $\mu = f(\text{Re}, \beta)$ for nozzles whose $(d/D)^2$ are 0.0887 (a); 0.251 (b), and for different values of β.

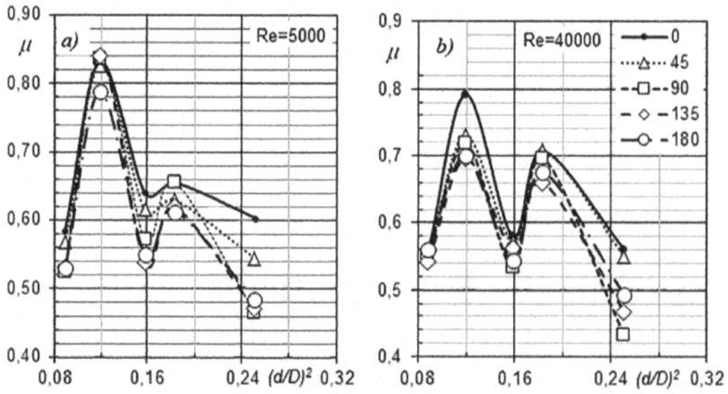

Fig. 5. Dependence of coefficient μ of flow rate of nozzles on $(d/D)^2$ for Re_d = 5000 (a); 40000 (b) and for different values of β

water flow rate through the nozzles was obtained for $(d/D)^2$ = 0.119 and 0.183. With the increase in the squared ratio $(d/D)^2$, a tendency of increase in the difference between the greatest and the least values of the coefficient μ is observed. For each of the squared ratio $(d/D)^2$, the greatest value of the coefficient μ was obtained for $\beta = 0°$, and the least one at $\beta = 90°$. The cause of the least value of the coefficient μ for $\beta = 90°$ is the greatest deformation of the stream at its inflow into the nozzle (Fig. 7).

The graphical dependences $\mu = f[\beta, (d/D)^2]$, which are presented in Fig. 7, indicate that the greatest range of variation of the μ is obtained for $\beta = 90°$ and for $(d/D)^2 = 0.251$. The comparison of relative variation of the coefficient μ for $Re_d = 10^4$ for $\beta = 0°$ and $\beta = 90°$ and for different values of $(d/D)^2$ of nozzles of DP is presented in Table 2.

Table 2. Relative variation of coefficient μ for $Re_d = 10^4$ and for $\beta = 0°$ and $\beta = 90°$ for different values of $(d/D)^2$.

Ratio of cross-section areas of nozzle and of PD, $(d/D)^2$	Coefficients μ of flow rate of nozzle for angles β		Relative change ψ of coefficient μ, %
	0°	90°	
0.0887	0.552	0.540	2.17
0.119	0.811	0.738	9.00
0.158	0.605	0.529	12.6
0.183	0.663	0.678	−2.26
0.251	0.570	0.444	22.1

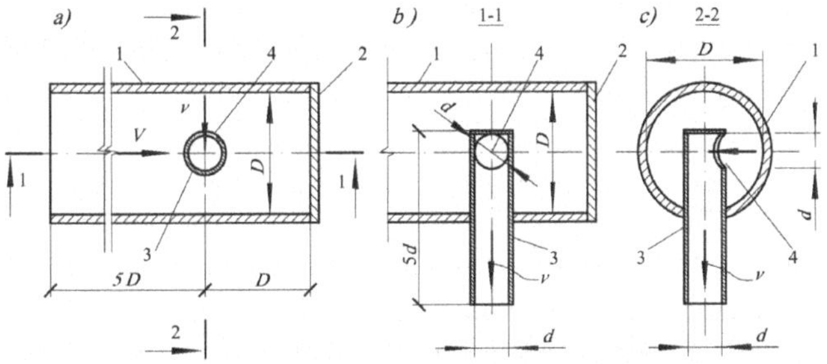

Fig. 6. Schematic diagram of outlet cylindrical nozzle with orthogonal lateral inflow for $\beta = 90°$ positioned at the end of DP: a – longitudinal section; b – section 1-1; c – section 2-2; 1 – wall of DP; 2 butt blind of DP; 3 – nozzle; 4 – inlet hole of nozzle; 5 – blind near inlet butt of nozzle

From results of calculations, which are presented in Table 2, it is found that the maximal value of ψ of relative variation of the coefficient μ of flow rate amounts to 22.1%. Such an effect has been obtained for a single nozzle. In DP, there are installed from several to some tens of nozzles. It should be expected that in operation of series of nozzles in DP the effect of regulation of fluid dispensation will be enhanced. Thus, by means of the variation of values of angles β we can achieve essential regulation of non-uniformity of fluid dispensation along the path.

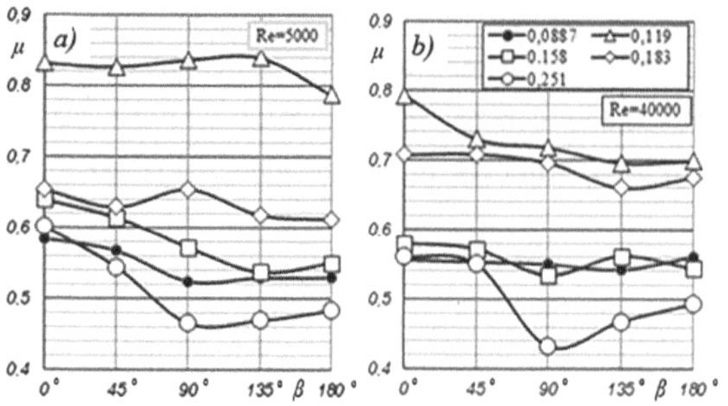

Fig. 7. Dependence of coefficient μ of flow rate in nozzles on β for Re_d = 5000 (a); 40000 (b): and for different values of $(d/D)^2$

5 Conclusions

The dependence of values of the coefficient μ of flow rate in cylindric nozzles with lateral orthogonal inflow on the angle β between the direction of water stream in a distributive pipeline and the direction of the outflowing jet has been experimentally established. The nozzles were investigated for five values of the squared ratio $(d/D)^2$ of the cross-section areas of the nozzle and of the pipeline; each value of $(d/D)^2$ was investigated for five different values of the angle β. The squared ratios $(d/D)^2$ were within the range of 0.0887 through 0.251. The angles β were assigned the values 0°; 45°; 90°; 135°, and 180°. Essential dependence of the values of the coefficient μ on the angle β has been established. For each the investigated value of $(d/D)^2$, the highest value of the coefficient μ was obtained for at $\beta = 0°$, and the least at $\beta = 90°$. The cause of the least value of the coefficient at μ $\beta = 0°$ is the greatest deformation of stream at its inflowing into the nozzle in this case. The maximal value of the relative change of the coefficient μ amounts to 22.1%. It is obtained for $(d/D)^2 = 0.251$. It should be expected that in simultaneous operation of series of nozzles in a distributive pipeline the effect of regulation of fluid dispensation from the distributive pipeline will be enhanced. Then, by means of variation of the angles β at nozzles along the distributive pipeline we can essentially change the law of fluid dispensation along the path.

References

Chen AW, Sparrow EM (2009a) Turbulence modelling for flow in a distribution manifold. Int J Heat Mass Transf 5–6:1573–1581

Chen AW, Sparrow EM (2009b) Effect of exit-port geometry on the performance of a flow distribution manifold. Appl Therm Eng 13:2689–2692

Cherniuk VV, Ivaniv VV (2017) Patent for invention № 115840 of Ukraine, IPC G05D 7/00, F17D 1/02, F17D 1/08. State Patent Office of Ukraine, Kyiv (in Ukraine)

Hassan JM, Mohamed TA, Mohammed WS, Alawee WH (2014) Modeling the uniformity of manifold with various configurations. J Fluids Vol Article ID 325259, 8 p. http://dx.doi.org/10.1155/2014/325259

Kim N, Byun H (2013) Effect of inlet configuration on upward branching of two-phase refrigerant in a parallel flow heat exchanger. Int J Refrig 36(3):1062–1077

Povkh IL, Chernyuk VV (1986) Experimental investigation of the influence of polyacrylamide admixtures on the resistance of diffusers. Eng-Phys J 51(3):357–361 (in Russian)

Povkh IL, Chernyuk VV (1989) Resistance of confusors in turbulent flow of water with additives of poliakrilamide. Eng-Phys J 57(5):709–712 (in Russian)

Abubakar SS (1977) Factors affecting orifice discharge in a multi-outlet irrigation pipe. A Master's thesis. Manhattan, Kansas, pp 1–84

Lee S, Moon N, Lee J (2012) A study on the exit flow characteristics by the orifice configuration of multi-perforated tubes. J Mech Sci Technol 26(9):2751–2758

Zemlyanaya NV, Gulyakin AV (2017) Analysis of causes of non-uniform flow distribution in manifold systems with variable flow rate along length. In: International conference on construction, architecture and technosphere safety. Held at the Far Eastern Federal University, Vladivostok, Far Eastern Federal University, 22–23 September 2017. IOP conference series: materials science and engineering, vol 262, pp 1–7

Zemlianaia NV, Gulyakin AV (2018) Parametrization of the water distribution process in perforated pipelines. Bull Sch Eng Far East Fed Univ 2(35):77–85. www.dvfu.ru/en/vestnikis (in Russian)

Wang CC, Yang KS, Tsai JS, Chen IY (2011) Characteristics of flow distribution in compact parallel flow heat exchangers, part II: modified inlet header. Appl Therm Eng 31(16):3235–3242

Zeng D, Pan M, Tang Y (2012) Qualitative investigation on effects of manifold shape on methanol steam reforming for hydrogen production. Renew Energy 39(1):313–322

The Utilization of Cherry Wood Sawdust for Heavy Metals Removal from Wastewaters

Stefan Demcak(✉), Zdenka Kovacova, and Magdalena Balintova

Faculty of Civil Engineering, Institute of Environmental Engineering, Technical University of Kosice, Vysokoskolska 4, 042 00 Kosice, Slovakia
stefan.demcak@tuke.sk

Abstract. The presence of inorganic pollutants such as heavy metal ions in industrial effluents is a worldwide issue for the environment. These pollutants are not only hazardous in exceeding concentrations but due to the property of accumulation in living organisms it is urgent to look for the plausible solutions for their reducing and treatment from wastewaters.

In this study, cherry wooden sawdust were used for removal Cu(II), Zn(II) and Fe(II) ions from model solutions with using the static and kinetic adsorption experiments. Infrared spectrometry of cherry wooden sawdust confirmed the presence of the functional groups which correspond with hemicelluloses, cellulose and lignin. At static adsorption was achieved approximately of 70% efficiency for all treated model solutions that is comparable with the efficiency of the adsorption processes reached after 5 min at kinetic experiments. The highest efficiency of Fe(II) removal (89%) was observed after 120 min of intensively shaking at frequency of 100 rpm. The mechanism of ion exchange on the beginning adsorption process by the changes of pH values was indicated.

Keywords: Wastewater treatment · Heavy metals · Adsorption · Wooden sawdust

1 Introduction

Water pollution by the heavy metals is a result of the industrial activities as mining, mineral processing and metallurgical operations. The heavy metals present in the wastewater are persistent, non-biodegradable in nature, accumulate in living organisms and thus they are entering the food chain, where are causing serious health diseases and disorders (Demcak et al. 2017a; Singovszka et al. 2016). From these reasons, treatment of heavy metals from the industrial wastewaters is necessarily before their discharging into the aquatic environment. The most commonly used methods for heavy metals removal from contaminated water includes chemical precipitation, ion exchange, membrane filtration, reverse osmosis, and electro dialysis (Fu and Wang 2011; Ngah and Hanafiah 2008). These methods are effective, on the other hand are costly, produce large volumes of toxic sludge and require high energy input. They are associated with the generation of toxic sludge, disposal of which renders it expensive and non-ecofriendly in nature (Tripathi and Ranjan 2015). Based on these facts, the safe and economical treatment of heavy metals from the wastewater in the last decades was

Z. Blikharskyy et al. (Eds.): CEE 2019, LNCE 47, pp. 58–65, 2020.
https://doi.org/10.1007/978-3-030-27011-7_8

investigated. Use of adsorption has emerged out to be better alternative to conventional methods of water treatment (Rahmani et al. 2009; Shah et al. 2009). It was observed that natural low-costs adsorbents, due to their wide abundance in nature can be used as cheap alternative to industrially produced sorbents (Bailey et al. 1999).

The adsorption by nonconventional materials is a highly effective method because it is a simple and cost effective method for recovering and eliminating heavy metal ions from wastewaters (Dehghani et al. 2016; Siti et al. 2013). The natural adsorbent should have high selectivity to facilitate quick separations, favorable transport and kinetic characteristics, stability (thermal, mechanical, chemical), regeneration capacity and low solubility in the liquid in contact to ensure the competitive for the commercial (Tripathi and Ranjan 2015). The mostly used low-cost natural adsorbents with the above parameters are agricultural waste (rice husk, plant bark, waste tea, and walnut shel), industrial by-products (fly ash, blast furnace sludge, lignin, bark, and sawdust), natural materials (zeolites, clay, peat moss and chitin) or modified biopolymers (Tripathi and Ranjan 2015; Crini 2006; Bailey et al. 1999). In addition, the main benefit of adsorption over the conventional methods is an absence of generation of sludge and its subsequent storage.

The wooden by-products or wastes as wooden sawdust are promising adsorbent materials due to their cheap costs production, processing and possibility of their regeneration (Ahmaruzzaman 2011). The wooden sawdust are perspective for removing metal ions, some types of acid and basic dyes, and other unwanted compounds from contaminated waters but the adsorption efficiency closely depends also on the composition of the wastewater (Balintova et al. 2016b; Keränen et al. 2016). The wooden sawdust are forming of complex compounds of metal cations with sorbent functional groups that are the main metal binding sites (Crini 2006; Gardea-Torresdey et al. 1990). The use of non-conventional adsorbents as wooden sawdust may contribute to the sustainability of the environment and offer a lot of promising benefits for commercial purpose in the future (Ghaedi and Mosallanejad 2013).

The present study deals with the sorption properties of cherry wooden sawdust for copper, zinc and iron removal from model solutions. Cherry sawdust was analysed by infrared spectrometry for characterization of functional groups, which can be responsible for the heavy metals binding. Efficiency of heavy metals removal was analysed by colorimetric method and changes of pH values were also measured.

2 Materials and Methods

As the adsorbent material was used the locally available wooden sawdust of cherry trees that was dried and sieved. For the adsorption experiments the fractions with a particle size under 2.0 mm were used. The FTIR measurements of the wooden sawdust were performed on a Bruker Alpha Platinum-ATR spectrometer (BRUKER OPTICS, Ettingen, Germany). A total of 24 scans were performed on each sample in the range of 4,000 to 400 cm^{-1}.

The single component model solution with initial concentrations of Cu(II), Zn(II) and Fe(II) $10 mg \cdot L^{-1}$ respectively was prepared by dissolving of appropriate sulphate salts in deionised water. The initial and residuals concentrations of the Cu(II), Zn(II) and Fe(II) were determined using the colorimetric method with a Colorimeter DR890, (HACH LANGE, Germany) and the appropriate reagent and pH values of model solutions by pH meter inoLab pH 730 (WTW, Germany) were also measured.

The batch adsorption experiments were carried out in static and kinetic conditions. In both adsorption experiments, 1 g of cherry wooden sawdust was mixed with 100 mL of model solutions. In static mode, the sorbent-sorbate interaction time was 1 day. In the kinetic mode, to determine the contact time required for equilibrium adsorption the samples were analysed in different time intervals (5, 10, 15, 30, 45, 60 and 120 min). In both modes, at the end of the adsorption experiments, the cherry wooden sawdust was removed by filtration through a laboratory filter paper. The residual concentrations of heavy metals ions in solutions were determined by colorimetric method and pH changes were also tracked. The efficiency of ion removal η was calculated using the following equation (Eq. 1):

$$\eta = \frac{(c_0 - c_e)}{c_0} \cdot 100\%, \tag{1}$$

where η is efficiency of ion removal (%), c_0 is the initial concentration of appropriate ions ($mg.L^{-1}$) and c_e in an equilibrium concentration of ions ($mg \cdot L^{-1}$).

3 Results and Discussion

3.1 The FTIR Cherry Sawdust Characterisation

The heavy metal removal by the wooden materials from waters is influenced by various factors. Important role has the surface structures hydroxyl, carboxyl, carbonyl, amine, and amino functional groups that are present in organic materials that are capable to bind metal ions (Ricordel et al. 2001). The structure of sawdust is primarily formed by cellulose, hemicellulose, and lignin (Kidalova et al. 2015). The functional groups of cherry wooden sawdust were determined using FTIR spectroscopy that the IR spectrum is shown in Fig. 1. The FTIR spectroscopy reveal several major intense bands that can be divided into the following three significant areas of wavenumbers: 3,650 to 3,000 cm^{-1} (-OH and -NH functional groups), 3,000 to 2,800 cm^{-1} (symmetric and asymmetric vibrations of CH_3 functional groups), and 1,750 to 800 cm^{-1} (C = O, -OH, -COOH, $-NH_x$, and haloalkanes functional groups) (Demcak et al. 2017b; Balintova et al. 2016a; Zhang et al. 2015; Schwanninger et al. 2004).

3.2 Static Adsorption Study

The results of the static absorption experiments for model solutions with cation concentrations of 10 $mg.L^{-1}$ are shown in Table 1. The cherry wooden sawdust used in the absorption experiments was able to remove copper, zinc and iron ions from the solution. The best efficiency (89.4%) of ion removal was observed for Fe(II) adsorption. Results correspond to research of Shukla et al. (2002) and observed that wooden sawdust as a promising adsorbent for heavy metal removal from wastewater.

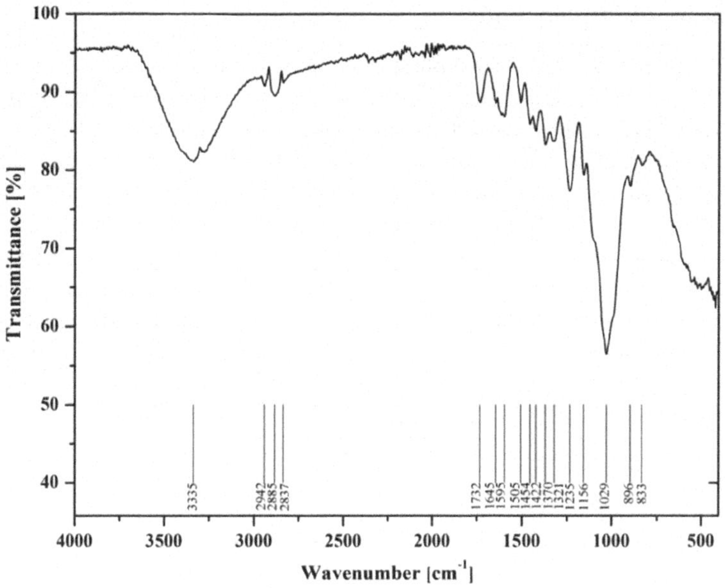

Fig. 1. Infrared spectrum of cherry wooden sawdust

They observed that selected wooden sawdust are potentially more economical than the common heavy metal removal processes.

Table 1. Results of static sorption experiments with cherry wooden sawdust

Ion	Input values		Adsorption experiments		
	c_0 [mg · L^{-1}]	pH	c_e [mg · L^{-1}]	η [%]	pH
Cu(II)	10.0	6.3	3.47	65.3	5.2
Zn(II)	10.0	6.2	2.57	74.3	5.3
Fe(II)	10.0	5.9	1.06	89.4	5.0

The pH monitoring is an important parameter in the characterisation of adsorption processes. Due to the different input pH values of the model solutions, the adsorption took place in different pH ranges. In all causes decrease of pH values was observed. Decrease of pH values is due to the higher concentration of H^+ ions present in the reaction mixture which compete with the dissolved metal ions for the adsorption sites at wooden sawdust (Yu et al. 2003).

3.3 Kinetic Adsorption Study

Based on results from static adsorption condition, the cherry sawdust was used for detailed study under dynamic adsorption mode. The efficiency of Cu(II), Zn(II), and Fe (II) removal and pH changes over the experimental time are shown in Figs. 2, 3 and 4. The curves indicate the rapid progress of ion removal in 5 min of sorbent-sorbate interaction accompanied by a decrease pH values, where approximately 70% (Cu(II) and Zn(II)) and 84% (Fe(II)) ions were removed from the solution. The residual time of experiment can be considered as a relative settled with slower growth of ions removal efficiency. Based on these results can be supposed that the ions removal might be interpreted as a two-stage process consisting of ion-exchange (at the beginning interaction) and adsorption (in rest of interaction). After 45 min was obtained the maximum efficiency of Zn(II) removal from model solution 78.7%. In 120 min were reached the highest efficiency of Cu(II) (84.9%) and Fe(II) (88.6%) removal.

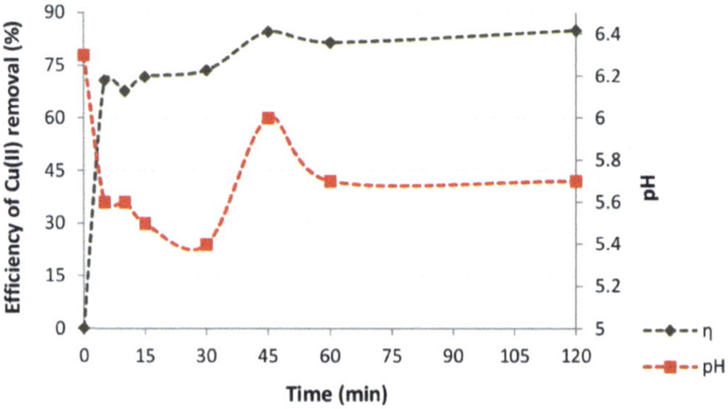

Fig. 2. Comparison of Cu(II) sorption efficiencies and changes of pH values over the experimental time

Changes of pH values in solutions were also observed. A significant change of pH value was revealed in case of Cu(II) removal. Holub et al. (2013) observed that an intensive pH change is caused by the high initial concentration of heavy metals in solution that are involved in the intensive ion exchange with the adsorbent. The change of pH was recorded after 45 min of the adsorption where pH values increase from 5.4 to 6.0. It could be caused by the mechanism of ion exchange between Cu(II) and chemical elements in the cherry sawdust. After the completion of the ion exchange, the pH began to decrease gradually.

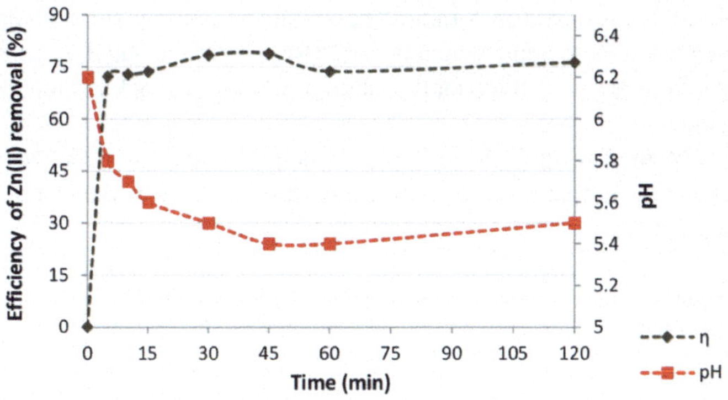

Fig. 3. Comparison of Zn(II) sorption efficiencies and changes of pH values over the experimental time

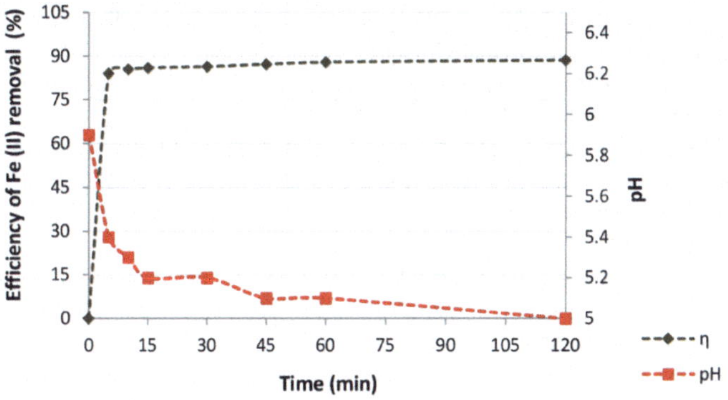

Fig. 4. Comparison of Fe(II) sorption efficiencies and changes of pH values over the experimental time

4 Conclusions

Sawdust appears to be a promising adsorbent for removal of Cu(II), Zn(II) and Fe(II) from model solution. Sawdust is economical sorbent because of its low cost, easy availability, renewability and high affinity for heavy metals. Thus, this process is more economical than current process technology.

The Fourier transform infrared spectroscopy (FT-IR) spectra of cherry sawdust can proved the functional groups, like; -OH, C = O and $-NH_x$ which are responsible for metal binding and suggested the main mechanisms involved in the removal of heavy metals ions might be the ionic exchange and complexation.

The adsorption of Cu(II), Zn(II) and Fe(II) is dependent on its contact time and pH of the water. The static adsorption study proved that the efficiency of Cu(II), Zn(II) and Fe(II) removal was height than 65% during experiment. The best efficiency (89%) was observed for iron removal from model solution. The kinetic adsorption study, was determined the rapid progress of ion removal in 5 min of beginning of experiment accompanied by a decrease pH values, where approximately 70% (for Cu(II) and Zn(II)) and 84% (for Fe (II)) ions were removed from the aquatic solution. Changes of pH are due to adsorption and ion exchange.

From the obtained results, the use of wooden sawdust has benefits to the environment because these materials are value sorbents, and are suitable for wastewater purification.

Acknowledgements. This work has been supported by the Slovak Grant Agency for Science (Grant No. 1/0419/19).

References

Ahmaruzzaman M (2011) Industrial wastes as low-cost potential adsorbents for the treatment of wastewater laden with heavy Metals. Adv Colloid Interface Sci 166:36–59

Bailey SE, Olin TJ, Bricka RM, Adrian DD (1999) A review of potentially low-cost sorbents for heavy metals. Water Res 33(11):2469–2479

Balintova M, Demcak S, Pagacova B (2016a) A study of sorption heavy metals by natural organic sorbents. Int J Energy Environ 10:189–194

Balintova M, Demcak S, Singovszka E, Pagacova B (2016b) Study of inorganic pollution sorption from acidic solutions by natural sorbents. Chem Eng Trans 53:55–60

Crini G (2006) Non-conventional low-cost adsorbents for dye removal: a review. Bioresour Technol 97:1061–1085

Dehghani MH, Sanaei D, Ali I, Bhatnagar A (2016) Removal of chromium (VI) from aqueous solution using treated waste newspaper as a low-cost adsorbent: kinetic modeling and isotherm studies. J Mol Liq 215:671–679

Demcak S, Balintova M, Hurakova M, Frontasyeva MV, Zinicovscaia I, Yushin N (2017a) Utilization of poplar wood sawdust for heavy metals removal from model solutions. Nova Biotechnologica et Chimica 16:26–31

Demcak S, Balintova M, Demcakova M (2017b) Study of heavy metals removal from model solutions by wooden materials. In: IOP conference series: earth and environmental science, vol 92, pp 1–6

Fu F, Wang Q (2011) Removal of heavy metal ions from wastewaters: a review. J Environ Manag 92:407–418

Gardea-Torresdey JL, Becker-Hapak MK, Hosea JM, Darnall DW (1990) Effect of chemical modification of algal carboxyl groups on metal ion binding. Environ Sci Technol 24:1372–1378

Ghaedi M, Mosallanejad N (2013) Removal of heavy metal ions from polluted waters by using of low cost adsorbents. J Chem Health Risks 3:7–22

Holub M, Balintova M, Pavlikova P, Palascakova L (2013) Study of sorption properties of zeolite in acidic conditions in dependence on particle size. Chem Eng Trans 32:559–564

Keränen A, Leiviskä T, Zinicovscaia I, Frontasyeva MV, Hormi O, Tanskanen J (2016) Quaternized pine sawdust in the treatment of mining wastewater. Environ Technol 37:1390–1397

Kidalova L, Stevulova N, Terpakova E (2015) Influence of water absorption on the selected properties of hemp hurds composites. Pollack Periodica 10:123–132

Ngah WW, Hanafiah MAKM (2008) Removal of heavy metal ions from wastewater by chemically modified plant wastes as adsorbents: a review. Bioresour Technol 99:3935–3948

Rahmani K, Mahvi AH, Vaezi F, Mesdaghinia AR, Nabizadeh NR, Nazmara S (2009) Bioremoval of lead by use of waste activated sludge. Int J Environ Res 3:471–476
Ricordel S, Taha S, Cisse I, Dorange G (2001) Heavy metals removal by adsorption onto peanut husks carbon: characterization, kinetic study and modeling. Sep Purif Technol 24:389–401
Schwanninger M, Rodrigues JC, Pereira H, Hinterstoisser B (2004) Effects of short-time vibratory ball milling on the shape of FT-IR spectra of wood and cellulose. Vib Spectrosc 36:23–40
Shah BA, Shah AV, Singh RR (2009) Sorption isotherms and kinetics of chromium uptake from wastewater using natural sorbent material. Int J Environ Sci Technol 6:77–90
Shukla A, Zhang YH, Dubey P, Margrave JL, Shukla SS (2002) The role of sawdust in the removal of unwanted materials from water. J Hazard Mater 95:137–152
Singovszka E, Balintova M, Holub M (2016) Heavy metal contamination and its indexing approach for sediment in Smolnik creek (Slovakia). Clean Technol Environ Policy 18:305–313
Siti N, Mohd H, Md LK, Shamsul I (2013) Adsorption process of heavy metals by low-cost adsorbent: a review. World Appl Sci J 28:1518–1530
Tripathi A, Ranjan MR (2015) Heavy metal removal from wastewater using low cost adsorbents. J Bioremediation Biodegrad 6:1–5
Yu LJ, Shukla SS, Dorris KL, Shukla A, Margrave JL (2003) Adsorption of chromium from aqueous solutions by maple sawdust. J Hazard Mater 100:53–63
Zhang P, Dong SJ, Ma HH, Zhang BX, Wang YF, Hu XM (2015) Fractionation of corn stover into cellulose, hemicellulose and lignin using a series of ionic liquids. Ind. Crop. Prod. 76:688–696

Experimental Investigation and First Application of Lightweight Abutment Backfill Made of Used Tyre Bales

A. Duda(✉) and T. Siwowski

Faculty of Civil and Environmental Engineering and Architecture, Rzeszow University of Technology, Rzeszow, Poland
aduda@prz.edu.pl

Abstract. This article describes the tests of pressure distribution on a bridge abutment, generated by the backfill made of recycled car tyres compressed in the form of bales with interfaces of medium sands and a buffer layer with tyre shreds or gravel. The backfill pressure distribution was measured continuously over a period of three months using pressure sensors. Static and dynamic load tests of abutment were conducted. The results of pressure field measurements were compared with the analytical calculations from medium sand backfilling according to the EC7. Based on the field measurements, the reduction of pressure on the bridge abutment was found if the tyre bales had been used, in comparison with the conventional backfilling of sand or gravel. The article also describes the first Polish application of abutment backfill from the tyre bales in a road bridge.

Keywords: Used tyre · Tyre bale · Bridge abutment · Backfill · Recycling

1 Introduction

Tyre bales from compressed used car tyres are a new recycled building material which is a product of tyre recycling process. Tyre compressing and packaging is a solution resulting from the necessity to limit the sizes of landfill sites and to reduce the risk of spontaneous combustion of tyres there. The compressed tyres are bound with plastic bands or galvanized steel wires of high strength. The tyre bale looks like a cuboid (Fig. 1) which height corresponds to the maximum height of compressed tyres, width results from the size of press, and length depends on the number of compressed tyres (Simm et al. 2004; Winter et al. 2006; Zornberg et al. 2005). The process of tyre pressing and packaging has been presented in detail in the BSI specification (PAS 108 2007), describing the production method of tyre bales used in construction.

Within the comprehensive R&D project on the innovative recycling materials, enhancing the sustainability of roads and bridges, the Department of Roads and Bridges at the Rzeszow University of Technology, Poland, has conducted laboratory and field tests on the innovative backfilling material made of used car tyre bales and various aggregates.

Z. Blikharskyy et al. (Eds.): CEE 2019, LNCE 47, pp. 66–73, 2020.
https://doi.org/10.1007/978-3-030-27011-7_9

Fig. 1. Tyre bale (left) and galvanized steel wires spanning tyre bales (right)

The R&D project comprised, among others, the research on shear strength of geocomposites made of tyre bales and filling materials such as a rubber aggregate, expanded clay aggregate and natural aggregate and on pressure distribution of geocomposites made of tyre bales and various backfill materials on abutment backwall, under static and dynamic load (Duda and Siwowski 2018; Duda and Siwowski 2019).

This paper describes the pressure distribution investigation on bridge abutment generated by the backfill made of used car tyres. The pressure field measurement had been carried out over a period of three months using pressure sensors. The results of pressure measurement were compared with the analytical calculations taking into account a conventional backfilling according to the EC7. Based on the field experiment, the reduction of pressure on the bridge abutment was found if the tyre bales had been used, in comparison with the conventional backfilling of sand or gravel. The paper also shortly presents the first Polish application of abutment backfill made of tyre bales in a road bridge.

2 Materials Parameters

Used tyre bales were the basic filling material of abutment backfill. Table 1 presents the basic properties of tyre bales, which are mostly more favourable than the analogous properties of natural aggregates, showed also in this table for comparison. One of the crucial drawbacks of tyre bales is their high compressibility, which results in the small values of deformation modulus. This problem can be minimized by means of a suitable design and adequate construction solutions.

Table 1. Comparison of tyre bales properties acc. to the (PAS 108 2007) and tyre shreds (Geosyntec Consultants 2008) with natural aggregates.

Parameter	Tyre bales	Tyre shreds	Minerals (sand, gravel)
Unit weight (kN/m^3)	5 ÷ 6	4 ÷ 6	18 ÷ 20
Friction angle (°)	35 ÷ 36	19 ÷ 38	35 ÷ 40
Elastic modulus (MPa)	0,8 ÷ 1,3	0,8 ÷ 1,3	40 ÷ 100
Permeability coefficient (cm/s)	2 ÷ 4	2 ÷ 10	0,1 ÷ 1

To reveal the potential backfill pressure reduction of novel sustainable solution, the experimental results were compared to the analytically determined pressure distribution of conventional backfill made of medium sand. Table 2 presents the basic material parameters assumed in the analytical model of pressure according to the EC7 (PN-EN 1997). The unit weight of pavement structure of 4.32 kPa was also included in the pressure analysis.

Table 2. Material parameters of medium sand assumed in the analytical model of pressure.

Friction angle (°)	Unit weight (kN/m^3)	Coefficient of active/at-rest pressure (-)	Pressure distribution	Angle of shear resistance between ground and wall	Adhesion/cohesion (kPa)
35	18,5	0,271/0,426	linear triangular	ignore	0

3 Large-Scale Stand for Field Investigations

The field investigations of pressure on the retaining wall or bridge abutment with the backfill made of tyre shreds were performed in the past (Humphrey et al. 1998; Tweedie et al. 1998). The field investigations of bridge abutment with the backfill made of tyre bales were carried out on a special test stand simulating the actual abutment and approach road section. The test stand included a special chamber with the dimensions of 6.1 × 7.75 m in plan on the outer contour and 5.0 × 6.5 m on the internal contour. The chamber was made of steel sheet piles of GU16N type with the length of 8 ÷ 10 m, topped with the reinforced concrete cap (Fig. 2).

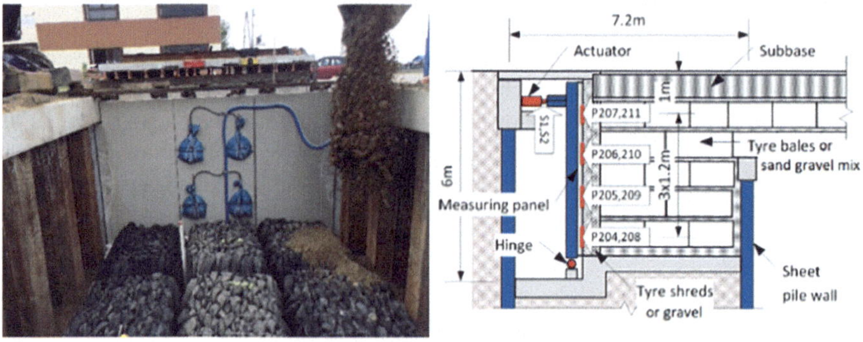

Fig. 2. Test chamber and arrangement of pressure sensors

The chamber consisted of two parts: test part and measurement part, separated by a movable wall. In the measurement part, with the dimensions of 1.5 × 5.0 m, the instruments (force sensors, displacement sensors) and pressure actuators were installed.

The test part of chamber with the dimensions of 5.0 × 5.0 m was used for various backfilling placement. The chamber depth was 6.0 m and 5.38 m in the measurement and test part respectively. The movable wall was made of 0.15 m thick reinforced concrete slab with the height of 5.0 m, and two steel sections HEB 340, spaced at 1.35 m and a composite with the slab. The steel sections were attached to the bottom plate of chamber with hinges created by a bolt connection. The wall was divided into three panels with the height of 5.75 m and the width of 1.66 m. On the middle part the pressure sensors were installed.

The approach road with the total length of 22 m had the 3 m wide traffic lane and was divided into three sections depending of their subgrades:

- section 1: 10 m long with the built-in two layers of tyre bales,
- section 2: 8 m long with the built-in one layer of tyre bales,
- section 3: 4 m long reference section without bales.

On the layers of tyre bales, a 0.7 m thick transmission layer of the natural aggregate was built, followed by a cement-sand base course with the thickness of 0.1 m and a pavement with concrete cubs with the thickness of 0.08 m (Duda and Siwowski 2019).

4 Test Program and Set up

In the paper only a part of the full investigation program is presented (Duda and Siwowski 2018). This research simulated a typical case of bridge abutment or retaining structure backfilling, in which the active or at-rest pressure was generated (the moving wall was blocked). It comprised the field pressure measurement of abutment backfill made of tyre bales and two 0.4 m thick buffer layers. Two buffer layers were applied in the tests: the first one made of rubber shreds and the second made of gravel. The rubber shreds made of used tyres were applied in the sake of further sustainability of novel abutment backfilling.

During the tests, the following measurements were made (Fig. 2):

- pressure on the abutment backwall (blocked movable wall in this case) (8 sensors),
- actuator force (2 sensors),
- temperature in the backfill (10 sensors),
- horizontal wall deflection (2 sensors),
- deformation of the pavement structure on approach section (geodetic measurement).

The measurement data was collected automatically every three minutes using the OMNIA CAB2 acquisition system.

The analytical calculations of active and at-rest pressure assuming the conventional backfill material were performed according to the EC7 (PN-EN 1997).

5 Results and Discussion

The field measurement of pressure conducted during seven research days is presented in Figs. 3, 4, 5 and 6. The pressure of backfill made of tyre bales and two buffer layers (EXP-A and EXP-B) on four backwall depths (sensors P-204/208, P-205/209, P-206/210 and P-207/211) is showed and compared with analytical results obtained for the conventional backfill.

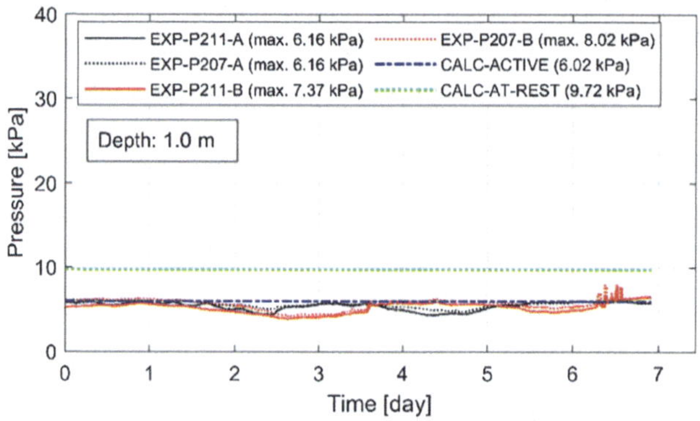

Fig. 3. Measured pressure distribution on the depth of 1.0 m for two buffer layers (EXP-A and EXP-B) compared with analytical results (CALC-ACTIVE and CALC-AT-REST).

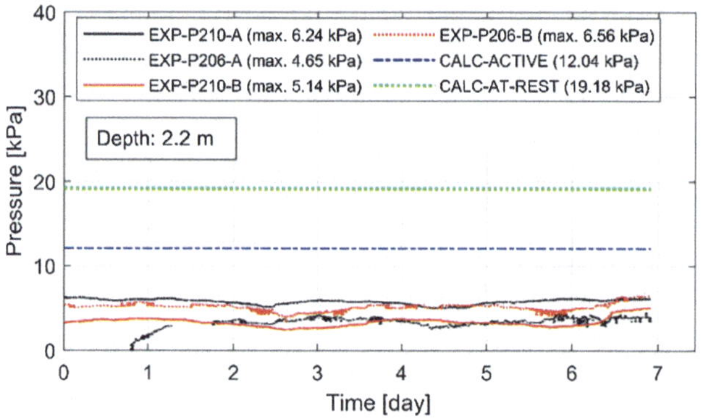

Fig. 4. Measured pressure distribution on the depth of 2.2 m for two buffer layers (EXP-A and EXP-B) compared with analytical results (CALC-ACTIVE and CALC-AT-REST).

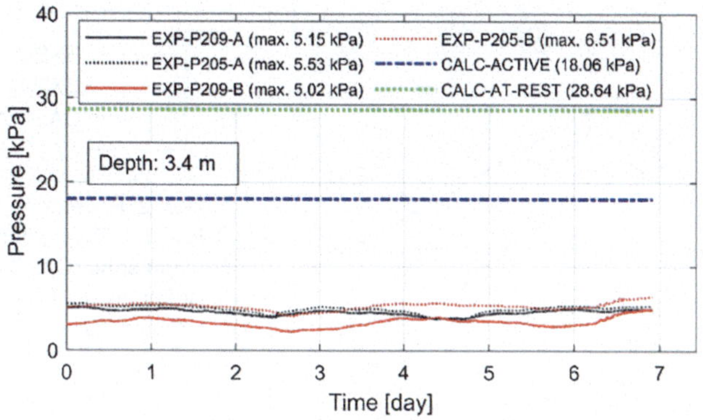

Fig. 5. Measured pressure distribution on the depth of 3.4 m for two buffer layers (EXP-A and EXP-B) compared with analytical results (CALC-ACTIVE and CALC-AT-REST).

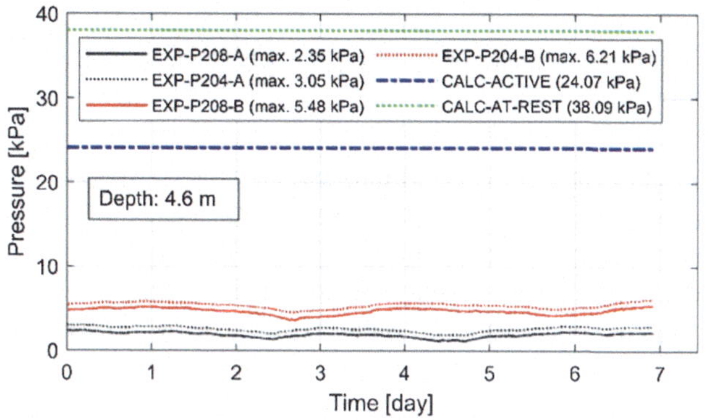

Fig. 6. Measured pressure distribution on the depth of 4.6 m for two buffer layers (EXP-A and EXP-B) compared with analytical results (CALC-ACTIVE and CALC-AT-REST).

Figure 7 shows a summary of the results obtained experimentally and analytically on the full depth of abutment backwall.

The comparison of experimental measurements and analytical results showed in Figs. 3, 4, 5, 6 and 7 reveals a significant pressure reduction for the backfill with tyre bales as compared to the conventional backfill with medium sand. Depending on the depth level, quantitative differences of backfill pressures between the sustainable and conventional backfill are as follows:

- about 2 to 3 times difference at the depth of 2.2 m below the ground level,
- about 3 to 4 times difference at the depth of 3.4 m below the ground level,
- about 4 to 5 times difference at the depth of 4.6 m below the ground level.

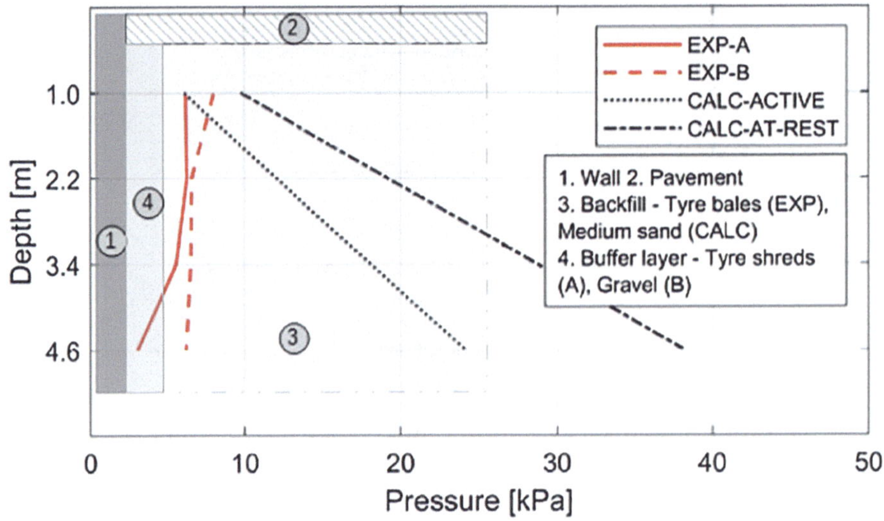

Fig. 7. Summary of pressure values obtained experimentally and analytically on the full depth of abutment backwall.

No particular difference of the backfill pressure with two different buffer layers was observed in the test. However, the rubber shred buffer layer further extends the sustainability of novel abutment backfilling made of recycled materials, therefore its application in the final solution should be considered.

Finally, basing on the obtained results of this part of R&D project it can be concluded that:

- pressure on the bridge abutment generated by the backfill made of used tyre bales with interfaces of medium sands has the character of a silo pressure,
- retaining structure backfill made of used tyre bales can be treated in the calculations as a stable block, which does not generate a pressure.

6 First Polish Application of Abutment Backfill from Tyre Bales

Within the aforementioned R&D project the first Polish application of abutment backfill from tyre bales has taken place. Currently, a composite steel-concrete bridge is being built over the San River in Sielnica with the carrying capacity of 40 tonnes and the total length of 157.0 m (Fig. 8). Tyre bales have been used in backfilling of both abutments to relieve the foundations and reduce pressure, due to the complex state of stresses induced by the span anchorage in the abutments. In each abutment, thirty tyre bales were used in the backfill. The pressure monitoring is planned after the bridge will be completed.

Fig. 8. Single layer of tyre bales before backfilling (left) and bridge construction (right).

Acknowledgements. The tests have been performed within the R&D project called "*ReUse - Innovative Recycling Materials, Enhancing the Sustainability of Bridge Facilities*" (Innotech No. K3/ IN3/ 38/228116/ NCBiR/ 15) and co-financed with the European Regional Development Fund. It has been implemented by the Polish industry-science consortium: Promost Consulting Sp. z o.o. Sp. k (leader), Remost Debica sp. z o. o., Geotech Rzeszow sp. z o. o. and Rzeszow University of Technology.

References

Duda A, Siwowski T (2018) Research on the recycled tyre bales backfill pressure on bridge abutment. ACTA Scientiarum Polonorum seria Architektura 17(1):123–134. https://doi.org/10.22630/ASPA.2018.17.2.21

Duda A, Siwowski T (2019) Experimental and numerical distribution of pressure on the abutment with backfill from tyre bales. MATEC Web Conf 262:05005. https://doi.org/10.1051/matecconf/201926205005

Geosyntec Consultants. Guidance Manual for Engineering Uses of Scrap Tires. Geosyntec Project No.: ME0012-11 (2008)

Humphrey, DN, Whetten, N, Weaver, J, Recker, K, Cosgrove, TA (1998): Tire shreds as lightweight fill for embankments and retaining walls. ASCE Geotechnical Special Publication No. 79, ASCE, Reston, Virginia

PAS 108:2007 - Specification for the production of tyre bales for use in construction. British Standards Institution (2007)

PN-EN 1997-1:2008 Eurocode 7: Geotechnical design - Part 1: General rules

Simm, JD, Wallis, MJ, Collins, K (2004): Sustainable re-use of tyres in port, coastal, and river engineering: guidance for planning, implementation, and maintenance. HRW Report SR 669, Wallingford: HR Wallingford

Tweedie JJ, Humphrey DN, Sandford TC (1998) Full scale field trials of tire shreds as lightweight retaining wall backfill, at-rest condition. Transp Res Board 1619:64–71

Winter, MG, Watts, GRA, Johnson, PE (2006): Tyre bales in construction. TRL PPR 080. Crowthorne: TRL Limited

Zornberg, JG, Christopher, BR, Oosterbaan, MD (2005): Tire bales in highway applications: feasibility and properties evaluation. Colorado department of transportation, Report No. CDOT-DTD-R-2005-2, Denver, Colorado

Fourth Generation of District Heating and Centralized Heating Supply Systems of Ukraine

N. M. Fialko(✉), N. P. Tymchenko, and Ju. V. Sherenkovskiy

Institute of Engineering Thermophysics of the NAS of Ukraine,
2A, Kapnist Street, Kyiv 03057, Ukraine
fialko.n.m@gmail.com

Abstract. This article investigates possibilities and expediency of transformation of existing Ukra'inian District Heating Systems (DH-UA) into the modern district heating of the 4th Generation (4G-DH). Three research problems were tasked to be solved. The first concerns studying the current state of existing DH-UA. The second one was connected with the analysis of the requirements to the last generation of district heating systems that begin to spread in the EU. The third problem was searching of ways to transform of DH-UA to 4G-DH. The survey of the current state of Ukrainian heating networks shows that existing systems DH UA correspond to 2G-DH–3G-DH levels. These systems from one hand are on the critically low technical level and on the other hand have the great potentials of improvements of energetic, economic and ecological characteristics of consumption of primary energy sources in Ukraine. The analysis of the opportunities, prospects and pace of convergence of existing DH-UA to the requirements of the last generation of European DH systems was carried out. Implementation of 4G-DH as an energy-efficient, low carbon, competitive heating system is a key issue for reforming the housing and utilities sector in Ukraine.

Keywords: District Heating Systems · District heating of the 4th generation · Housing and utilities sector · Energy, economic and environmental efficiency

1 Introduction

One of the main tasks of the housing and utilities sector of Ukraine (HUSU) is to supply the population with heat energy. Heating services –space heating and domestic hot water supply (DHW) - relate to life supporting concern the majority of the population, require the consumption of large volumes of expensive energy resources (high exergy primary energy resources, thermal and electric energy).

Until recently, in order to meet the energy needs of the housing, domestic and socio-budgetary sectors of the economy, the most commonly used traditional fossil carbon-contained types of primary energy resources - coal, petroleum products, and natural gas. But over time, their energetic usage as a whole has become the main (up to 80%) source of greenhouse gases and other harmful emissions. Anthropogenic pollution of the atmosphere with Carbon dioxide is believed to be the main cause of modern

Z. Blikharskyy et al. (Eds.): CEE 2019, LNCE 47, pp. 74–86, 2020.
https://doi.org/10.1007/978-3-030-27011-7_10

global warming. Its temperature indicators include, for example, variations of the curve of the average annual anomalous global temperature Δt_{anom} as a universal indicator of the average temperature of the land and the ocean. Corresponding data are provided by National Aeronautics and Space Administration, Goddard Institute for Space Studies "GISS Surface Temperature Analysis (v3)" (2017). The basic temperature to build the curve of Fig. 1 was calculated as the average value of a 30-year period (1951–1980). The function is smoothed by using LOWESS (LOcally WEighted Scatterplot Smoothing) method of non-parametric regression analysis, which is based on the k-nearest neighbouring point model. To analyse the function, a part of data corresponding to a ten-year data window with effective approximately of near five-year smoothing was used. Triangles on the time axis of Fig. 1 - mark large volcanic eruptions, as one-time powerful sources of Carbon dioxide emissions into the Earth's atmosphere.

Anomalous curve Δt_{anom} values with acceleration are approaching the critical value of Δt^{cr}_{anom}, beyond which an irreversible degradation of the Earth's biosphere is expected. The critical excess of the average annual global anomalous temperature is estimated by the value of $\Delta t^{cr}_{anom} \geq 3$ K. The International Energy Agency (IEA) considers the 2DS ("2 Degree Scenario") as the main scenario of global energy development. According to 2DS, by 2050, there must be actions implemented that restrict growth of temperature Δt_{anom} no more than on 2 K. Data were provided in Evaluation report 5 (AR5) of Intergovernmental Panel on Climate Change (IPCC) "Climate change 2014. Climate change mitigation" (2018). The Paris Agreement of 2015, ratified by almost 200 countries, including Ukraine, aims to stabilize the global rise of average temperature at the $\Delta t_{anom} = \text{const} = 1.5$ K mark from the average data for the last decades of the previous preindustrial period. Greater restraint of the Δt_{anom} threshold to 1.5 K will significantly reduce the risks and consequences of the warming of the climate. For 2014, about a thousand diverse geoclimatic scenarios have already been developed according to Evaluation report 5 (AR5) of Intergovernmental Panel on Climate Change (IPCC) "Climate change 2014. Climate change mitigation" (2018). Most of their authors converge on the need for an urgent response to the climate challenge.

Based on the empirically recorded Δt_{anom} growth rate over the past 40 years, you can estimate the time remaining before the realization of this climatic threat. If we do not take extraordinary outrunning measures to decarbonize energetics, then in 2035, humanity will be faced the problem of the consequences of catastrophic *aridization* (desertification) of the territories at middle latitudes and melting Earth's glaciers on the polar caps of the planet.

Thus, the priority of a non-carbon economy is the necessary requirement for sustainable world development. As a result, the need to develop a strategy of the national energy transition (ET). Among other things, the ET includes a plan to replace traditional carbon-contained fossil primary energy sources with renewable low-exergy energy sources - wind, solar, biomass, geothermal, hydrogen, wave, etc. The deadlines and volumes of efforts of this plan are determined by the nature and pace growth of the curve in the Fig. 1. It is natural that the problem of energy efficiency, economical efficiency and ecological efficiency of the housing and utilities sector of Ukraine, as the

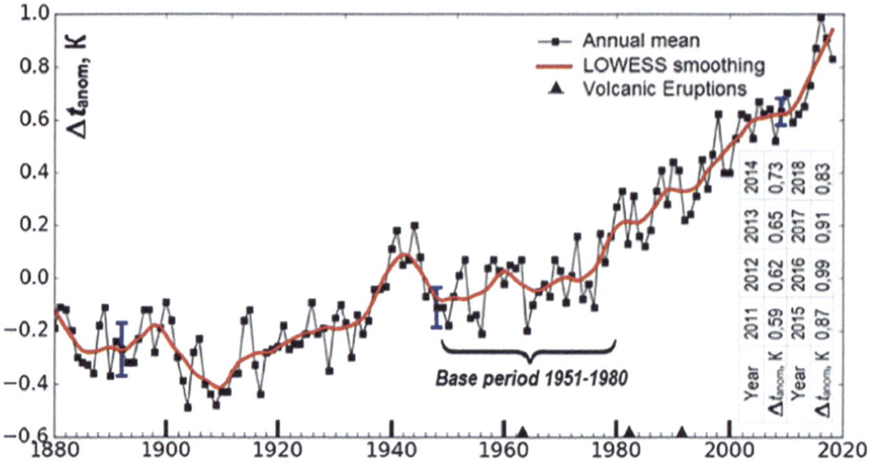

Fig. 1. The deviations curve from the average annual global temperature of the global temperature Δt_{anom}. The bottom to the right is the data for the last 8 (2011–2018) years. Triangles on the axis of time the large eruptions of volcanoes mark. Data were provided by "GISS Surface Temperature Analysis (v3). National Aeronautics and Space Administration. Goddard Institute for Space Studies.", (2017)

economy's branch which consumes the largest share of "toxic" primary energy sources and with very low energy efficiency, is one of the priority areas of the research.

2 The Purpose of the Work

The purpose of the work is to study the prospects of adaptation and modernization of existing district heating systems of Ukraine (DH-UA) to the level of requirements of modern energy-efficient smart systems of non-carbon heat supply in the residential sector. To achieve the goal, it is necessary to consider two research problems. The first one concerns the current state of existing DH-UA, its energy efficiency and the place in the balance of primary energy consumption of the country. The second one is devoted to the analysis of the opportunities, prospects and pace of transformation of existing DH-UA to the requirements of district heating systems of the last generation of district heating systems that begin to spread in the EU.

3 Research Methodology

Contained in the paper methods include, first, a comparative and typological analysis of the DH-UA weaknesses. Second, survey of some examples of district heating systems which were developed most recently. Proposed methodology is intended to estimated potential opportunities only of general approach to develop and implement 4GDH concept in Ukraine.

4 Results of the Study

4.1 Current State of DH-UA

After almost complete privatization of the greatest part of the housing stock of Ukraine, the development of housing and communal services, especially the heat supply sector, has gone for a long time into a phase of certain socio-economic stagnation. The situation in the DH-UA is characterized by the low energy, economic and ecological efficiency at all stages of the energy technological process of obtaining primary energy source: the generation of energy, its transportation (with large electricity consumption for coolant pumping through heat networks), distribution and final consumption. In general, the situation of the DH-UA can be defined as the term "degradation" (Geletukha et al. 2019). Following are some main interconnected signs, causes and effects of such situation.

1. High unit costs of primary energy sources per unit of heated space at the level of end consumers. The actual values of heat energy consumption by the housing stock are 3–4 times higher than the normative or actually determined domestic indicators. At the same time, the majority of consumers live in multi-flat residential buildings.
2. In turn, the indicators specified in point 1 are very far from the standards of a passive house (NZEB format), which, after 2020, should be guided by new construction in the EU countries and countries associated with the EU, including Ukraine.
3. Low heat-loss indicators of building envelopes (enclosures) and imperfect traditional indoor heating systems, especially in the part of equipment with low-temperature automatic devices of heating and their control systems.
4. Low availability of technical and software means of home automation and management of heating, ventilation and air conditioning to Industry 4.0 platforms, IoT in the primary (household level) of the multilevel network of end users of energy according to automation standards "EN 15232 Energy Performance of Buildings - Impact of Building Automation Control and Building Management" (2017)
5. Outdated architecture, structure and composition of DH-UA, which is oriented towards the use of traditional fossil carbon-contained primary energy sources (correspondingly, a negligible share of RES in the energy balance of the country), with a relatively small share of cogeneration in the energy supply balance. In the technological warehouse, DH-UA equipment and energy storage systems are also absent.
6. Practical inability of final-users of energy to become active participants in a liberalized energy market to realize their right to freely (on a competitive basis) choose of an energy service provider.
7. Inability to realize on-line balancing of response and demand in distributed generation and energy consumption processes within the framework of the concept of mass active final consumer of energy of multilevel smart power networks.
8. The impossibility of implementing electricity demand management, when consumers are oriented towards the tracing of the nondispatchable naturally intermittent and stochastic nature of the energy production schedule of the most available types of RES-generation (that is, demand adjustment, rather than electricity supply adjustment).

9. Unsatisfactory technical level: more than a third ($\sim$ 38%) of the heating networks, most of which are in the emergency ($\sim$ 15%) or transient state, and the losses in energy transport on the average reach 19% versus $\sim$ 6% technically predetermined.
10. Physical and moral outdated equipment of enterprises of the thermal communal energy (TCE): 20% of boilers exceeded their time of life, and cogeneration units are single.
11. The equipment of DH-UA is under loaded (installed capacity of heat generating units is 71 GW; actual load, for example, is estimated at 26 GW in 2015), and is characterized by low indicators of reliability, energy, economic and environmental efficiency. The situation is aggravated by the fact that the volumes of energy consumption are continuously decreasing (and will decrease with the thermal modernization of old buildings and the development of distributed renewable generation, the implementation of energy efficient technologies in housing and communal services) and the negative tendencies of further deterioration of the technical state of power units and heating networks.
12. Insufficient attention to the use of potential of energy recovery technique (secondary energy resources) for the production of heat energy as commodity for DH-UA.

Table 1 provides indicators and data on the DH of 15 countries in which district heating has gained the significant development in recent years. The last row contains information about the DH-UA. Almost all of the parameters (population coverage for DH services, total capacity of DH, total heat output) Ukraine is in the top 5 list. But in today's conditions the key indicator is the share of RES in the national energy balance of the consumed primary energy sources. This indicator is characterizing the competitiveness of the DH in the ecological sense. The share of RES in Ukraine (0.89% in 2017) is several dozen less than in Iceland (85 times), Norway (69), Denmark (52), France (44), Switzerland (35).

It is significant that energy final energy consumption statistics of the leading economies of the world (PRC, USA, EU) are divided into three main sections: the housing and utilities sector (HUSU); the industrial sector (IS) and the transport sector (TS). Further, HUSU includes the cost of energy in the social and budgetary sector and providing services.

In the EU (Table 2) in 2017 the most energy consumption had the HUSU (38.6/41.7%, data according to the old methodology for calculating energy balance of final energy consumption, in the denominator - according the new one) in the numerator. Transport sector had 33.7/30.8% with the lion's share of auto vehicles - 27.3/28.9% and IS had 25.2/24.6%. A similar structure of final energy consumption is observed in Ukraine: HUSU - 42.5/46.1%; TS - 17.6/18.8% (share of auto vehicles - 13.3/14.4%); IS - 35,8/30,7%. Agriculture, forestry, fishing and all other energy consumption sectors in the EU is accounted <3%, in Ukraine <5% of final energy consumption.

Table 1. Values of some indicators of district heating. According to Report "Decentralization, heating in Ukraine: potential and ways of introduction" (2017), and Statistics Overview (2016), and Bova (2017), and Sigal (2018)

Names indicators		Population coverage for DH services	Total capacity of DH (GWt)	Increasing in the DH pipes length, 2009–13	Total heat output (EJ)	Share of RES (except for CHPP)
Ranking of countries[a]	1	Iceland (92%)	China (463)	Italy (58%)	China (3,2)	Iceland (76%)
	2	Latvia (65%)	Poland (56,5)	Norway (53%)	Germany (0,26)	Norway (61%)
	3	Denmark (63%)	Germany (49,7)	Switzerland (52%)	Poland (0,25)	Denmark (46%)
	4	Estonia (62%)	South Korea (30)	China (43%)	Sweden (0,18)	France (39%)
	5	Lithuania (57%)	Finland/Czech Republic (23)	Sweden/ Austria (21%)	South Korea (0,17)	Switzerland (31%)
Ukraine[a]		(77%)-2009[b] 76,3%-2017 (−0,7%)	71 (installed)[c]	−25%	0,21 (2015)	0,89%

Notes: [a]Assessments, 2017. [b]Autonomous heating: large cities 21, 8%, small cities 59.4%. [c]Maximum load 26 GW (2015).

Table 2. Structure of balances (%) of energy consumption in the EU and Ukraine (according to Eurostat)

Energy supply sector	EU, 2017	Ukraine, 2016
1. The housing and utilities sector	38,6/41,7	42,5/46,1
2. Transport sector	33,7/30,8	17,6/18,8
With auto vehicles	27,3/28,9	13,3/14,4
3. Industrial sector	25,2/24,6	35,8/30,7
4. Others sectors	2,56/2,84	4,1/4,43
Total	100/100	100/100

Data from Table 2 show that the share of primary energy sources consumed in the domestic sector of Ukraine is noticeably prevailing for all other (including industrial and transport) sectors of the economy. At the same time, Ukraine's energy consumption is extremely inefficient, besides, with the obvious (in the light of the Paris Agreement of 2015) environmental threat and the background of poor-quality services with their high costs and tariffs, especially with regard to primary energy services. Since the main costs of primary energy sources (both environmental pollution) are related with functioning of the DH, the focus problem must be about prospects of the DH UA. There are two polar answers to it. The first is rapid total reconstruction, replacement of networks, physically and morally obsolete heat generating equipment. But for financial reasons (very great capital investments) and the lack of investors, the first answer is unrealistic.

The second answer implies a complete rejection of the existing DH UA, replacing them with individual, autonomous or other types of heating. Such a policy was partly "understandable" in the context of a temporary cross-subsidization of population spending on heat supply. A well-known example is the two cities of Mukacheve and Uzhhorod, where individual gas boilers with individual discharge of combustion products were installed in each apartment of multiflat buildings. Also, new construction often envisaged the installation of a modular boiler house instead of connecting the building to an existing DH network.

But in Ukraine, energy price policy in HUSU in recent years, beginning from 2015, has been radically changed. Natural gas tariffs for all categories of consumers have lost their social aspect, started to equalize and grow in the direction of economically predetermined values. As a result, the cost of individual heating services has grown multiple times, and the motivation for the district heating has, on the contrary, begun to revive. The last wave of abandonment of DH was observed only in small and medium-sized cities with a very disrupted (more precisely, emergency) municipal economy. An example are cities of Marganets, Pokrov, Kam'yans'ke, with spontaneous attempts of citizens to transition to direct electric heating. Therefore, the second polar answer ("Complete Denial of DH") to the question "What to do with obsolete DH-UA?" is also wrong, and the practice of its implementation is vicious. Experts from the Ministry of Regional Development of Ukraine, academic and educational institutions have expressed the opinion that from the many points of view - ecology, efficiency and energy security - there is no alternative to centralized heat supply (Timchenko and Fialko 2017; Report "Decentralization, heating in Ukraine: potential and ways of introduction" 2017; Sigal 2018; Malinovsky et al. 2009; Geletukha et al. 2016).

The above-mentioned answers are also unsatisfactory due to the fact that in both cases the use of high exergy nuclear and fossil carbon-contained kinds of fuel remains. But their use is planned to be cut by 2050 so that annual greenhouse gas emissions to reduce by 80% compared to 1990 levels. The reduction will be encouraged, for example, by environmental taxes, as is the case, for example, in Denmark and Germany.

In general, with the introduction of economically predetermined tariffs and modern energy-efficient technologies, the benefits of a centralized combined power generation and its network distribution in conditions of liberalization of the energy market become again attractive to the population. It is obvious that the modernization of the existing DH-UA, its greening should be carried out in such a way as to preserve its useful features, to escape of the above-mentioned shortcomings and to provide new features inherent to the technologies of the last, fourth generation of district heating that developed in EU during the last 10 years.

4.2 Analysis of the Opportunities, Prospects and Pace of Transformation of Existing DH-UA to the Requirements of the Last Generation of DHs

Functionally, DH consists of heat sources, for example - a boiler; working fluid (heat carrier, working substance) for charging from the source of thermal energy and its transfer to the sets of end users; the corresponding pipeline systems (networks) for the transportation and distribution of the energy carrier between the sets of final consumers of energy; equipment for adjusting the heat-supply parameters of the heat carrier according to operating modes throughout the path of its passage from the generation stage to the heating devices installed at the end-user. Today, the most part of the DH in the world corresponds to the 2G-DH and 3G-DH levels (Fig. 2). The rationale, structure and composition of the 4G-DH, their evolution is described in detail in the works of Henrik Lund, his co-authors (Lund et al. 2014, 2017, 2018a, b, c) and numerous followers. Modern DHs (4GDHs) provide their functioning in the "environment" of energy efficient buildings, the integration of centralized heating and electricity supply into the smart energy system (SES), the dominance of RES in the balance of consumption of primary energy sources. A small number of energy-efficient homes in Ukraine with highly effective thermal insulation, advanced home automation (including domestic heating plant, or local heat distribution and metering station) and equipped with modular house boilers meet the requirements of even 3G-DH +.

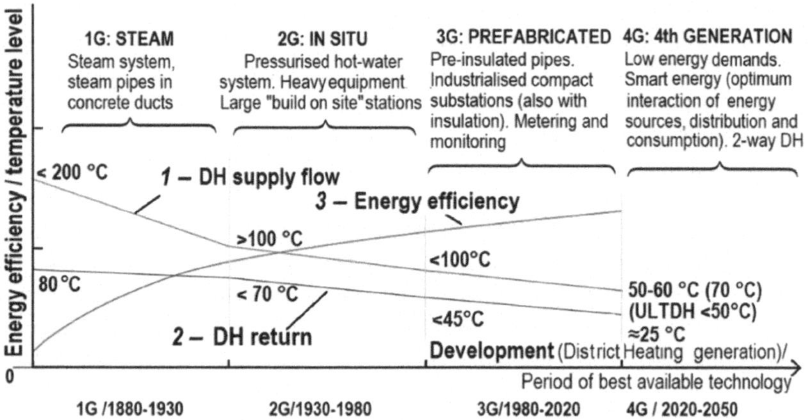

Fig. 2. Illustration of concepts of four generations of DH: typical changes in the coolant temperatures in the supply and return pipelines of DH (curves 1, 2) and energy efficiency of heat supply processes (curve 3) at different stages of the development of DH. (Thorsen et al. 2018; Lund et al. 2011).

In systems of 4G-DH, instead of fossil carbon-contained primary energy sources (e.g. coal, natural gas), are used predominantly RES, usually biomass, low-energy solar, wind, and geothermal resources. Figure 2 shows the classification and scheme of the evolution of DH in the course of their technological transition from the first generation of 1G-DH to the modern 4G-DH. Table 3 identifies the features, indicators, components and elements of each type of DH. Each of the listed in Table 3 properties and conditions for 4G-DH, although it improves energy efficiency or environmental friendliness, is only necessary but not sufficient. Only a complex of all components provides the quality of the fourth-generation system.

The full-scale transformation of the existing DH-UA to the 4G-DH requirements is inexpedient due to the fact that the project capacities of the DH-UA (power of energy units, capacity of heat transfer of distribution networks, etc.) were calculated for the old main housing stock. Most of the houses of this housing stock is characterized by increased heat losses, which in turn were caused by low thermal resistance of building envelopes (enclosures), large infiltration losses, lack of devices for local heat distribution and metering both house and apartment levels, lack of means of billing for actually consumed energy and, consequently, payment of heating services by gross indicators (area, room size, number of persons, etc.). In addition, some of them are morally and physically obsolete. An example of this commonly used 2G-DH equipment is the lift-heating unit elevator assembly. These widely spread devices, despite simplicity, reliability, have significant disadvantages:

- absence of the possibility of controlling the temperature of the coolant at the outlet;
- large dimensions of the mixing chamber;
- the need for a pressure difference between the lines of supply and return of coolant 0,08–0,2 MPa.

Here we can see detailed and expended description of some key indicators and 4G-DH elements, which are almost mandatory and which should be involved in upgrading the existing DH-UA to 4G-DH (U). This list has a methodological value and has an importance as the starting point for developing a real project or roadmap for the modernization of existing DH-UA.

The most commonly used coolant for 4G-DH (UA) would be water, but with its lower operating parameters: temperatures - from 30 to 70 °C, pressure $\leq$ 0.2 MPa. Due to lower temperatures, it will be possible to use pre-insulated, flexible, double, may be triple polymer pipes with small diameter pipelines. Forced circulation will be carried out by less powerful pumping systems not only in the individual heat point at home, but also at the apartment level. This will allow for the implementation of a quantitative individual automatic mode for controlling temperature heating regimes. Plate heat exchangers, meters of working and emergency parameters of modes, smart sensors, automatic shut-off and regulating equipment, information and switching equipment and technologies, including wireless mesh technologies, integrated in WLAN, WAN, will be widely used for providing of services both heating and DHW. It is believed that the degree of electrification of 4G-DH will be high, especially if the integration of electric and thermal networks into a multilevel smart energy system (SmEM). Electricity produced at the PV-stations, wind farms in the 4G-DH (UA) will be widely used for electric heating. Low-temperature (radiators < 50 °C) and ultra-low temperature ("warm" floor, walls, ceiling) heating devices will replace cast-iron radiators. To get hot water in low-temperature systems (up to 30 C) heat pumps will be used. They are embedded in the return pipeline and provided with a buffer tank. Network water is heated to an acceptable temperature through a heat exchanger. To prevent the threat of infection with the legionella, periodic special preventive thermal regimes of water paths are provided.

Table 3. Features of the development of district heating systems over the past 140 years

Main features	1G: steam	2G: in situ	3G: prefabricated	4G: 4th generation
Timeline	1880–1930	1930–1980	1980–2020	Nearest Future
General description	Steam system, steam pipes in concrete ducts	Pressurised hot-water system. Large "build on the site" stations	Pre-insulated pipes Industrialised compact substations (also with insulation)	Low energy demands Smart energy (optimum interaction of energy sources, distribution and consumption) 2-way DH
Energy carrier	Pressed steam	Pressed hot water	Pressed hot water	Hot water, air
Control and management		quality	quality; quality+ quantitate, metering and monitoring	+DSM, RSM, Apartment microlevel, Smart Energy Systems, Meshnet
Pressure in networks	0,17 MPa	<1.2 MPa	≤0,1 MPa	
DH flow	<200 °C	>100 °C	<100 °C	50–60 °C (70 °C) (ULTDH <50 °C)
DH return	<80 °C	<70 °C	<45 °C	≈25 °C
Heating devices	High and middle temperature		Low temperature	Ultra Low temperature
Energy efficiency	Low	Middle	High	Very high
Gas Capacity	Very low	Low	High	Low
CHP		Coal, Gas[1]	Coal, Oil, Gas[2]	Gas[3], Biomass,
Fuel, fuel mix	Coal	Coal, oil incineration	Mix: Gas, Waste, Oil, Coal, RES	Mix of RES
Share of FCC Fuels	100% (without NPP)	100% (with NPP)	75%[3]	≈0% (without NPP\)
RES (structure)	0	0	25%[4]	≈100%
RES (kinds)	-	Industry surplus Waste incineration Geothermal	+Large scale PV & solar Large wind farms (incl. off-shore ones) Mix (gas+ biomass +waste) THP	+Wind surplus CHP waste CHP future energy sources Biomass conversion Wave 2-way District Heating Electricity

(*continued*)

Table 3. (*continued*)

Main features	1G: steam	2G: in situ	3G: prefabricated	4G: 4th generation
Heat Storage	Steam	Hot water	Hot water, TES	Hot water Seasonal heat storage
Other Energy Storage			Batteries (e.g. Li-Ion)	Batteries (e.g. Li-Ion);
Centralised DC	No	No	No	Cold storage District & cooling grids
DH and scale of DH	Local, HH	Local, DH1	Local, DH, City blocks	Local, DH, municipalities

4G-DH technologies will be implemented not only in new buildings, but also during the modernization of housing stock. Its basis now forms houses of old building. The zone-time-profiled microclimate of rooms will be implemented with minimization of return water temperature or according to more complex optimization tasks (for e.g. implementation of demand or response side management). Energy, unlike previous generations of DH, will be produced mainly from RES. Coal steam boilers, occasionally CHP, that were characteristic for 1G-DH; coal, oil-fired CHP, natural gas heat and power plant, typical for most countries of former Soviet Union (2G-DH), will remain in the past.

Qualitative, qualitative-quantitative, quantitative control systems inherent to 2G-DH, 3G-DH, will be replaced by an automatic system with a target multi-level function of minimizing the primary energy sources consumption per of unit area. This will take into account the consumption of primary energy sources throughout the technological chain of energy generation, its transport, distribution and final consumption.

4G-DH systems will also be integrated with centralized or local cooling systems (for example, absorption chillers) and seasonal thermal energy storage.

During 4G-DH (UA) implementation, there will be a transition period when the old technological base will begin to update by 4G-DH elements. In particular, generating systems using traditional fuels will be gradually supplemented by distributed generation, cogeneration techniques. Synchronically, fuel for 2G-DH, 3G-DH boilers which were originally designed for traditional fossil primary energy sources, will begin continuously "enriched" with a mix of biomass, waste, secondary energy resources.

At the transition stage, 3G - DH + $\rightarrow$ 4G - DH (UA), thermal energy will be partially obtained using energy efficient electric heating technologies. Electrification of DH is one of the promising directions of the modern energy strategy of the EU. Such attention to the high-exergy clean energy is typical for countries with a surplus of electricity production. The share of energy produced on the nuclear power plants (NPP) in the electricity balance of Ukraine is more 50%. At the same time, the development of technologies and means of energy accumulation, including seasonal storage, is a prerequisite for the electrification of heat supply services.

Another aspect of the 4G-DH UA electrification is the implementation of heat pump installations. At the transition stage, 3G-DH + will be widely used heat pumps, both in the form of powerful centralized stations and small-scale networks. Over time, the prevalence and total capacity of heat pumps will reach the degree when their integration into the single network system will bring a great energy effect. This technology was named as the fifth-generation of DH technology (Buffa et al. 2019). 5GDHC will be able to implement bi-directional (2-way, Fig. 2, either from the network or into the network) and dual mode (heating/cooling) within the Smart Energy (power + heating) Systems paradigm.

5 Conclusions

1. Implementation of 4G-DH as an energy-efficient, low carbon, competitive heating system is a key issue for reforming the housing and utilities sector in Ukraine.
2. The introduction of 4G-DH in Ukraine should be carried out by gradually adding its elements to existing heating supply systems through measures such as increasing the share of renewable energy sources, developing technologies and means of energy storage, integration of heat and power networks, multi - level monitoring of heat supply, and the like.
3. The urgent task is to develop a road map for the implementation of 4G-DH in Ukraine.

References

Bova Y (2017) Denial of centralized heating is like walking on a razor's edge. https://energo.delo.ua/energo-government-policy/koli-mi-vidmovilis-vid-cenratlizovanogo-opalennja-vidchuvav-scho-335487/

Buffa S, Cozzini M, D'Antoni M, Baratieri M, Fedrizzi R (2019) 5th generation district heating and cooling systems: a review of existing cases in Europe. Renew Sustain Energy Rev 104:504–522

Central heating: where and why it needs to be saved (2017) https://energo.delo.ua/energo-for-companies/centralizovannoe-otoplenie-gde-i-pochemu-ego-nuzhno-sohranit-334850/

Climate Change (2014) Climate change mitigation. Evaluation report 5 (AR5). Intergovernmental Panel on Climate Change (IPCC). https://www.ipcc.ch/site/assets/uploads/2018/03/WGIIIAR5_SPM_TS_Vol._ru-1.pdf

Decentralization, heating in Ukraine: potential and ways of introduction (2017) Report in the framework of the project "Secretariat and Expert hub on energy efficiency. https://zhytlo.in.ua/ua/napryamok/zhkp/decentralzovane_opalennya_v_ukran-_potencal_ta_shlyahi_vprovadzhennya.html

EN 15232 energy performance of buildings - impact of building automation control and building management (2017) http://www.cres.gr/greenbuilding/PDF/prend/set4/WI_22_TC-approval_version_prEN_15232_Integrated_Building_Automation_Systems.pdf

Geletukha GG, Kramar V, Oleinik EM, Antonenko VA (2019) Analysis of the possibilities of preservation and development of centralized heat supply in Ukraine. Thermophys Therm Power Eng 1(41):53–58

Geletukha GG, Zheleznaya TA, Bashnyi AI (2016) Analysis of models of functioning of the centralized heat supply sector of the countries of the European Union. Part 1. Ind Heat Eng 4 (38):63–70

GISS Surface Temperature Analysis (v3) (2017) National aeronautics and space administration. Goddard Institute for Space Studies. https://data.giss.nasa.gov/gistemp/graphs/

Malinovsky AA, Turkovsky VG, Muzichak AZ (2009) Centralized heat supply has a perspective in Ukraine. Probl Gen Energy 19:50–56

Lund H, Duic N, Østergaard P, Mathiesen B (2018a) Smart energy and district heating: special issue dedicated to the 2016 conference on smart energy systems and 4th generation district heating. Energy 160:1220–1223

Lund H, Duic N, Østergaard P, Mathiesenc B (2018b) Future district heating systems and technologies: On the role of smart energy systems and 4th generation district heating. Energy 165(A):614–619

Lund H, Hvelplund FK, Mathiesen BV, Østergaard PA, Christensen P, Connolly D, Schaltz E, Pillay JR, Nielsen MP, Felby C, Bentsen NS (2011) Coherent energy and environmental system analysis. Department of Development and Planning, Aalborg University, p 90

Lund H, Werner S, Wiltshire R, Thorsen J, Hvelplund F, Mathiesen B (2014) 4th Generation District Heating (4GDH): integrating smart thermal grids into future sustainable energy systems. Energy 68:1–11

Lund H, Østergaard P, Connolly D, Mathiesen N (2017) Smart energy and smart energy systems. Energy 137(C):556–565

Lund H, Østergaarda P, Changa M, Werner S, Svendsenc S, Sorknæsa P, Thorsend J, Hvelplunda F, Ole B, Mortensene G, VadMathiesenf B, Bojeseng C, Duich N, Zhangi X, Mölleraj B (2018c) The status of 4th generation district heating: research and results. Energy 164:147–159

Sigal AI (2018) Ways of stagnation of district heating systems in Ukraine. In: Problems of ecology and operation of energy facilities. Institute of Industrial Ecology, Kyiv, pp 9–16

Statistics Overview (2016) http://www.euroheat.org/wp-content/uploads/2016/03/2015-Country-by-country-Statistics-Overview.pdf

Thorsen JE, Lund H, Mathiesen BV (2018) Progression of district heating – 1st to 4th generation. https://vbn.aau.dk/ws/portalfiles/portal/280710833/1_4GDH_progression_revised_May2018.pdf

Timchenko NP, Fialko NM (2017) Multi-storey buildings as active consumers of intellectual electric grids. Work Inst Electrodyn Natl Acad Sci Ukr 48:15–22

Fialko NM, Timchenko NP, Khalatov AA, Sherenkovsky JuV (2016) Intelligent energy systems for heat supply of buildings. Bull Natl Univ “Lviv Polytechnic” “Theory and practice of construction” 844:203–209

Influence of Aggregate on Properties of Semi-dry Concrete in Hardened State

Robert Figmig(✉)

Department of Material Engineering, Institute of Environmental Engineering, Faculty of Civil Engineering, Technical Universtiy of Kosice, Vysokoskolska 4, 04200 Kosice, Slovakia
robert.figmig@tuke.sk

Abstract. The article deals with the influence of various aggregate compositions used to semi-dry concrete on some properties of hardened concrete. The concrete mix compositions were aimed at using in precast production of manholes and pipes for sewerage systems. Five fractions of aggregate (0/1, 0/2, 0/4, 2/4 and 4/8, respectively) were used, while grains of fraction 0/1 had rounded shape and fractions 0/2 and 0/4 were pre-crushed. Fraction 2/4 and 4/8 included half of rounded as well as half of crushed grains. Concrete mix compositions also varied within the content of cement paste according to experiences from production plant. The concrete surface (from −2 to 1), density of hardened concrete in saturated and dried state (from 2265 to 2416 and from 2142 to 2336 kg.m^{-3}), splitting strength (from 4.9 to 6.2 MPa) and water absorption (from 3.4 to 5.8 wt%) were observed.

Keywords: Semi-dry concrete · Mix composition · Aggregate · Durability

1 Introduction

The key role of the modern concrete technology is designing durable and sustainable concrete meeting the technical standard requirements with minimum costs (Swamy 2008; Silfwerbrand 2012; Alexander et al. 2017). Concrete that is intended for precast production of manholes and pipes for sewerage systems can be made either in S0 consistency (slump 0–10 mm, semi-dry) following with vibrating and pressing or as SCC. This type of concrete has to meet water absorption and tensile strength requirements as well as esthetic aspect. It is necessary to provide accurate fresh concrete mix composition to achieve these properties. That includes selection the aggregate with suitable grading and grains shape that should ensure the minimalization space for cement paste to decrease mainly the water absorption (Figmig and Kovac 2019; Wassermann et al. 2009; Lindquist et al. 2015). The most suitable for this purpose should be well grading but also gap-grading aggregate with rounded grains because of ideal arrangement of grains in composite structure, what causes minimalization of voids between them (Malewar et al. 2017; Ince and Çetin 2018). On the other hand, there is a lack of fineness in this type of aggregates, what leads to using additives or cementitious materials to supplement them. Other type of aggregate originates from crushing process, where grains with angular shape and considerable fineness are

Z. Blikharskyy et al. (Eds.): CEE 2019, LNCE 47, pp. 87–94, 2020.
https://doi.org/10.1007/978-3-030-27011-7_11

created. Using this type of aggregate to concrete needs much water or cement paste to fill the space (voids) between its grains caused by their angularity, what can be connected with increasing the porosity of concrete inner structure (Bentz et al. 2017; Mehta and Monteiro 2006; Alexander and Mindess 2005).

This experiment was focused on varying the different fraction and type of local aggregate composition as well as varying the amount of cement paste, while final concrete surface, water absorption and splitting strength were investigated.

2 Materials and Methods

2.1 Input Materials

For the experiment, these materials were used:

- Cement – CEM I 42,5 R
- Aggregate - 0/2, 0/4, 4/8 – pre-crushed
- Aggregate - 0/1, 2/4 – washed siliceous sand and coarser fraction
- Tap water
- Lignosulfonate plasticizer for semi-dry concrete – liquid (1,20 g/cm^3, 40% of solid content)

The key properties of various aggregates for this experiment as density in dry state, bulk density, voidance, water absorption and grains shape are listed in Table 1 and gradings of each fraction are listed in Table 2 as well as in the Fig. 1.

Table 1. Some properties of used aggregates

Fraction	Density dry [kg.m^{-3}]	Bulk density [kg.m^{-3}]	Voidance [%]	Water absorption [wt%]	Grains shape
0/1	2630	1780	32.3	0.3	rounded
0/2	2610	1815	30.5	1.1	angular
0/4	2610	1935	25.9	1.5	angular
2/4	2650	1650	37.7	0.4	ang./round.
4/8	2610	1615	38.1	0.8	ang./round.

Table 2. Grading of used aggregates (passing through the sieve in %)

Sieve (mm)	0.063	0.125	0.25	0.5	1	1.4	2	2.8	4	5.6	8	11.2	16
0/1	0.2	1.7	12.2	47.5	98.1	99.8	100	–	–	–	–	–	–
0/2	2.3	4.9	15.5	39.4	68.2	–	91.1	97.8	100	–	–	–	–
0/4	5.6	8.6	13.8	23.9	40.3	–	64.6	–	95.1	99.8	100	–	–
2/4	0.0	0.0	0.0	0.5	0.6	–	5.2	–	85.2	100	–	–	–
4/8	0.7	–	0.8	–	–	–	1.2	–	5.1	35.6	92.3	100	100

Fig. 1. Grains shape of used aggregates [author]

2.2 Mix Compositions

In this study, overall 8 various mix compositions were tested (M00 + M00P – M06). Batch M00 was mix composition using at the concrete plant, M00P was the same batch with adding the plasticizer and re-mixed. Batches M01-M06 was laboratory prepared with various proportions (relative) of cement, water, plasticizer and various aggregate fraction. Compositions were based on experiences with aim to decrease cement content, while achieving the same or better technical requirements, what should lead to costs save as well as to make the concrete more environmental. Fresh concrete compositions for each batch are listed in Table 3. The water in the table means total content of water, amount of plasticizer is related to amount of cement.

Table 3. Fresh concrete mix compositions

Batch	Cement	Water	Plasticizer	0/1	0/2	0/4	2/4	4/8
M00	1	0.34	0.0%	0.0	0.0	3.6	0.0	0.0
M00P	1	0.31	2.1%	0.0	0.0	3.6	0.0	0.0
M01	1	0.42	0.0%	0.0	0.0	3.7	0.0	1.0
M02	1	0.42	0.0%	0.0	0.0	4.0	0.0	0.6
M03	1	0.40	0.0%	0.0	2.0	0.0	2.0	0.6
M04	1	0.35	0.5%	1.0	0.0	0.0	3.5	0.4
M05	1	0.39	0.5%	0.0	2.0	0.0	2.5	0.4
M06	1	0.44	0.5%	0.0	0.6	4.2	0.0	0.6

2.3 Methods

Fresh concrete for each mix composition was homogenized and put into the cube forms (a = 100 mm – fit to wall thickness of manhole or pipe) while vibrating on the table and pressing with handram until the complete compaction. The top of sample was leveled by flat bar. For finding out the fresh concrete density (ρ_{rmc}), weight of each sample was measured. Based on the ρ_{rmc}, absolute compositions of fresh concretes were calculated.

The samples were cured under the water till the 28th day, then the weights were again measured and concrete density in saturated state (ρ_{sat}) for each batch was calculated. The samples that underwent water absorption test were put into the oven and dried at the temperature of 105 ± 5 °C until their mass was not changed. Based on results, concrete density in dried state (ρ_{dry}) as well as water absorption in weight percent (WA) was calculated.

Before performing split tensile strength (f_{ct}) test, the samples surfaces (S) were assessed, while the top layer was considered as key. The number rating was chosen as follows: from 1 for flat surface to −2 for too porous surface.

3 Results

The results of fresh concrete density (ρ_{rmc}), concrete density in saturated state (ρ_{sat}), concrete density in dried state (ρ_{sat}), absolute final composition of fresh concretes, surface rating (S), water absorption (WA) and split tensile strength (f_{ct}) are given in Table 4.

Table 4. Results of concrete properties

Batch	ρ_{rmc}	ρ_{sat}	ρ_{dry}	C	W	Pl	0/1	0/2	0/4	2/4	4/8	S	WA	f_{ct}
	[kg.m^{-3}]			[kg]								[-]	[wt%]	[MPa]
M00	2366	2371	2244	479	163	0.0	0	0	1724	0	0	1	5.7	6.1
M00P	2342	2347	2219	477	148	10.0	0	0	1717	0	0	1	5.8	5.9
M01	2395	2399	2295	391	164	0.0	0	0	1448	0	391	1	4.6	5.9
M02	2321	2331	2220	386	162	0.0	0	0	1542	0	231	−1/0	5.0	5.8
M03	2345	2347	2246	391	156	0.0	0	782	0	782	235	0	4.5	6.0
M04	2411	2416	2336	386	135	1.9	386	0	0	1350	154	1	3.4	6.2
M05	2312	2318	2224	368	143	1.8	0	735	0	919	147	0	4.2	5.8
M06	2238	2265	2142	327	144	1.6	0	196	1374	0	196	−2	5.7	4.9

The densities ρ_{rmc}, ρ_{sat} and ρ_{sat} varied from 2238, 2265 and 2142 kg.m^{-3} for M06 to 2411, 2416 and 2336 kg.m^{-3} for M04. It could be consequence of lack of binder (in form of cement paste) and lack of rounded grains in M06 causing too porous inner structure. On the other hand, enough cement paste in combination with aggregate with almost entire rounded grains in M04 could cause well dense concrete microstructure.

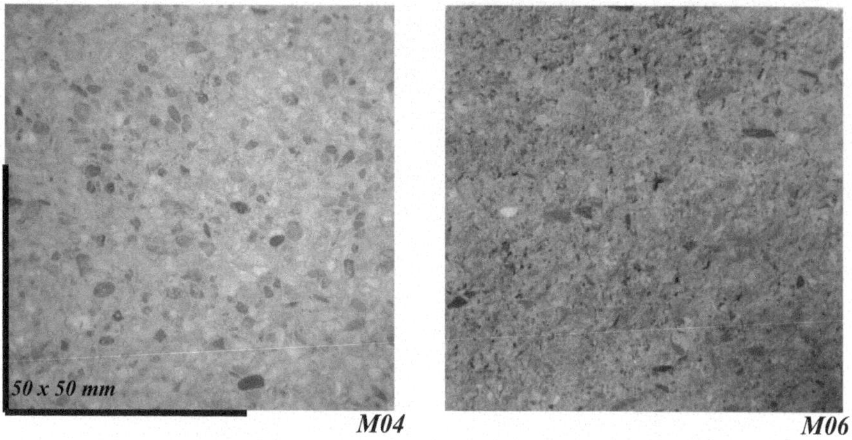

Fig. 2. Inner structure of samples M04 and M06 [author]

The comparison of composite structure both of mix compositions M04 and M06 is in the Fig. 2.

Surface rates were as follows: 1 for M00, M00P, M01 and M04, 0 for M03 and M05, 0/−1 for M02 and −2 for M06. The visual aspect of surface of semi-dry concrete may be connected with amount of fineness as well as with amount of cement paste and ease of its movement among aggregate grains. Surfaces of samples of each batch are shown in the Fig. 3.

The results of water absorption test varied from 3.4 wt% for M04 to 5.8 wt% for M00P. The higher values of WA are observed within batches, where higher cement content or crushed aggregate is used. The significance can be seen with M00 (also M00P) and M06 that achieved the same value. If the split tensile strength would be sufficient as achieved with M06, it is possible to save 30% amount of cement while the WA stays the same or similar.

The values of split tensile strength (f_{ct}) varied from 4.9 MPa for M06 to 5.8–6.2 MPa for another mix compositions. The key role of tensile strength could play the structural porosity/density of composite, what could be influenced by overall composition of concrete composition.

The graph rendered in the Fig. 4 shows the dependence of water absorption and tensile strength on the density of concrete as cement composite. There are obvious two trends, that with increasing of density, water absorption declines while tensile strength inclines, what is in accordance with composite theory.

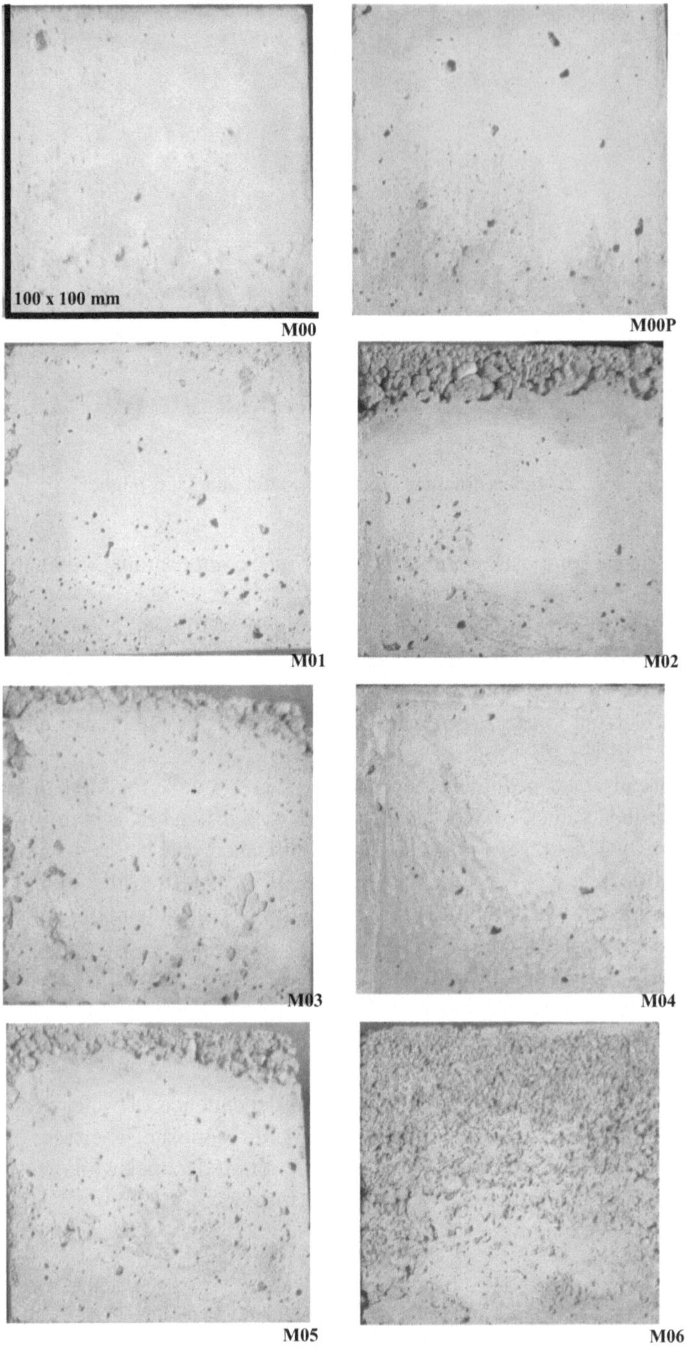

Fig. 3. Surface of hardened concrete samples [author]

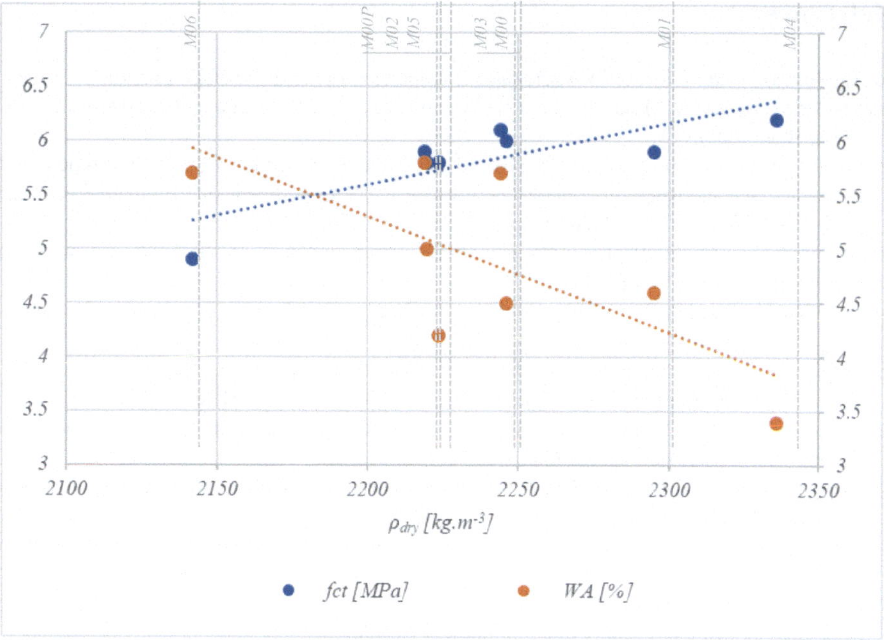

Fig. 4. Dependence of WA and f_{ct} on density

4 Conclusions

This experimental study was focused on investigation of semi-dry concrete and its selected properties in hardened state, while composition of fresh concrete from local materials of concrete plant was changed. Based on result can be deduced as follows:

- it seems to be necessary, especially within semi-dry concrete, to select the most suitable aggregate from the fraction granularity and grains shape point of view to minimize content of voids in composite,
- then the overall composition of semi-dry concrete should be designed as the maximum amount of aggregate (filler) and minimum amount of cement paste (binder), while workability and surface structure should be taken into consideration,
- minimizing content of cement paste, either by decreasing the amount of cement or water could lead to improve (decrease) the water absorption (also to costs save and sustainable production), what is easier when aggregates with rounded grains are used.

Acknowledgements. The activity presented in the paper is part of the research grant VEGA 1/0648/17 and VEGA 1/0524/18.

References

Alexander M, Mindess S (2005) Aggregates in concrete. Taylor & Francis, London

Alexander M, Bentur A, Mindess S (2017) Durability of concrete: design and construction. CRC Press, Boca Raton

Bentz DP, Arnold J, Boisclair MJ, Jones SZ, Rothfeld P, Stutzman PE (2017) Influence of aggregate characteristics on concrete performance. US Department of Commerce, National Institute of Standards and Technology

Figmig R, Kovac M (2019) The influence of aggregate gradation on the fresh and hardened concrete properties. Young Sci 11:1–8

Ince R, Çetin SY (2018) Effect of grading type of aggregate on fracture parameters of concrete. Mag Concr Res 1–9

Lindquist W, Darwin D, Browning JA, McLeodd HAK, Yuane J, Reynoldsf D (2015) Implementation of concrete aggregate optimization. Constr Build Mater 74:49–56

Malewar Y, Saleem S, Titiksh A (2017) Gap grading of aggregates & its effect on the inherent properties of concrete. J Ceram Concr Sci 2(1):1–9

Mehta PK, Monteiro PJ (2006) Concrete: microstructure, properties and materials, 3rd edn. McGraw-Hill, New York

Silfwerbrand J (2012) Sustainable concrete is more than just durable concrete. Struct Concr 13 (1):1–2

Swamy RN (2008) Sustainable concrete for the 21st century concept of strength through durability. Jpn Soc Civ Eng Concr Committee Newsl, 13

Wassermann R, Katz A, Bentur A (2009) Minimum cement content requirements: a must or a myth? Mater Struct 42:973–982

Evaluation of Stress-Deformed Condition Level of Glued Materials for the Without Anchor Steel-Concrete Joint

Serhiy Zolotov[1], Pavlo Firsov[2(✉)], and Hamze Muhamad[1]

[1] Department of Building Structures, O.M Beketov National University of Urban Economy, Marshala Bazhanova Street 17, Kharkiv 61002, Ukraine
zolotovsergej963@gmail.com, amirkobani@gmail.com
[2] Department of Architecture and Engineering, Luhansk National Agrarian University, Alchevskih Street 44, Kharkiv 61002, Ukraine
pavelfirsov1991@gmail.com

Abstract. The article is devoted to the deformation definition of the proposed acrylic modified glued compositions for the development of reliable, high-strength steel-concrete joints, using the Lode-Nadai coefficient, which determines the stress condition form of the medium. The numerical calculation, according to the principles of materials energy rationalization using the Vasil'kov-Shmukler criterion, is carried out, which testifies that the proposed acrylic compositions for the without anchor joint have a substantial structural strength margin (for the stress – at 6 times, for the energy of deformation – at 36 times). Diagrams of adhesive material deformation (σ-ε diagrams) under the influence of corresponding exertional efforts (tension, compression, shear) were constructed. According to the deformation diagrams, linear deformations and the boundary value of the deformation energy density were calculated. The obtained results of experimental research data can be used for attachment joints projecting and industrial technological lines development, as well as for further improvement of polymeric adhesive materials, widely applied in industrial and civil engineering.

Keywords: Stress-deformed condition · Acrylic modified compositions

1 Introduction

The most common types of technological equipment and production lines fastenings are welded, bolted and anchor joints. An alternative type of joint arranging is the without anchor way of fastening (Zolotov and Firsov 2015). Fastening by this method is carried out by gluing technological equipment fastening units or supporting parts directly to the concrete structure surface (Morkovskaya 2011). In order to arrange this type of glued joint, industrial production adhesive mixtures, which have the necessary adhesion level to concrete and steel, as well as their own high physical and mechanical characteristics (Chamochín et al. 2010), are widely used. The main material in the without anchor glued fastenings are epoxy and acrylic compounds, because their chemical properties are corresponding to all the necessary requirements for adhesives

Z. Blikharskyy et al. (Eds.): CEE 2019, LNCE 47, pp. 95–102, 2020.
https://doi.org/10.1007/978-3-030-27011-7_12

(del Real-Romero et al. 2006) to create reliable, high-strength steel-concrete joints (Firsov 2017; Zolotov 2016).

At the same time, the current development direction of the construction industry leads to a direct search of constructive systems solutions, which develop in two parallel directions:

- creation of elements (structures, fastenings) with the maximum ability to carry load and/or stiffness with the given cost of materials;
- creation of elements (constructions, fastenings) with minimal materials expense with the given resource.

In both cases, the problem is reduced to the formulation and solution of nonlinear problems of rationalization (Vasil'kov 2008). The concept of "rational structure" succeeded to concrete and transform with the help of Genrikh Vasil'kov and Valeriy Shmuckler scientific research (Shmukler 2005). According to the mentioned scientists, the ideology of the direct design tasks formulation involve the reflection of two circumstances:

- introduction of a single goal function as a criterion;
- introduction of isoperimetric condition restriction, associated with the control of the system stress-deformed condition.

In this case, a significant simplification of the tasks formulation, for rationalizing the structure parameters, can be achieved (Shmukler et al. 2008). It is assumed that the same record of the purpose function reflects the possibility of an approximate equivalent replacement of the traditional criteria (cost, weight, volume etc.) by one, related at the same time to the stress-deformed condition character and features (Shmukler 2017). Also, it is important to note that not only the interacting bodies system, but also a separate deformed body can have the potential energy of deformation.

2 Purpose

The main purpose of the research is in the complex study of the stress-deformed condition of acrylic modified adhesive compositions, which is held on the basis of modern principles of materials energy rationalization.

3 Methods

Experimental lab methods of modified acrylic adhesives samples testing, methods of theoretical and structural mechanics for assessing the stress-deformed condition of proposed compositions for glued steel-concrete joints designing and theoretical analysis of obtain results are used in present research.

4 Results

Consequently, each of the potential energy deformation density parts of is determined by the method of calculation proposed by Valeriy Shmuckler "Eqs. (1), (2), (3), (4) and (5)":

$$e_{nc} = \frac{9}{2} K \times \varepsilon_0^2 + \int_0^{\varepsilon_{nc}} \sigma_c d\varepsilon_c; \tag{1}$$

$$e_{nt} = \frac{9}{2} K \times \varepsilon_0^2 + \int_0^{\varepsilon_{nt}} \sigma_t d\varepsilon_t; \tag{2}$$

$$e_{nsh} = \int_0^{\varepsilon_{nsh}} \tau d\gamma; \tag{3}$$

$$K = \frac{E_0}{3 \times (1 - 2 \times v)}; \tag{4}$$

$$\varepsilon_0 = \frac{1}{3} I_1(T_\varepsilon). \tag{5}$$

where K is the volumetric deformation module, $I_1(T_\varepsilon)$ is the first invariant of the deformation tensor and E_0 is the initial deformation module of the first material kind.

In this case, the process of the material deformation is represented as a polynomial of the seventh (odd) degree "Eq. (6)":

$$\sigma = \alpha_1 \varepsilon + \alpha_3 \varepsilon^3 + \alpha_5 \varepsilon^5 + \alpha_7 \varepsilon^7. \tag{6}$$

The Lode-Nadai strain and deformation parameters are directly defined to the characteristics of the stress-deformed condition "Eq. (7) main normal stresses, Eq. (8) main linear deformations":

$$\chi_\sigma = \frac{2\sigma_2 - \sigma_1 - \sigma_3}{\sigma_1 - \sigma_3}; \sigma_1 \geq \sigma_2 \geq \sigma \tag{7}$$

$$\chi_\varepsilon = \frac{2\varepsilon_2 - \varepsilon_1 - \varepsilon_3}{\varepsilon_1 - \varepsilon_3}, \varepsilon_1 \geq \varepsilon_2 \geq \varepsilon_3 \tag{8}$$

A distinctive feature of the Lode-Nadai coefficient application in mathematical models of solids mechanical properties is the constant transformation in the framework of algebraic inequality $\sigma_1 \geq \sigma_2 \geq \sigma_3$ (Valovoi et al. 2013). The same applies to the major deformations irregularities. If the one of the stresses magnitude violates the algebraic inequality, it is necessary to renumber the existing stresses and change the numbering of stress axes. In this case, the volumetric stressful condition turns out to be a variable kind of load and goes into the classifications of the complex loading.

The value of the Lode-Nadai parameter determines the kind of volumetric stressed condition (Krul et al. 2019). According to the Lode-Nadai parameter magnitude the deformation form for the realized deformation is determined.

χ_ε is the Lode-Nadai parameter for the deformed condition, which is in the range $(+1 \geq \chi_\varepsilon \geq -1)$ with:

- in the case of a "generalized" tension $\chi_\varepsilon = -1$, $e_u = e_{nt}$;
- in the case of a "generalized" compression $\chi_\varepsilon = +1$, $e_u = e_{nc}$;
- in the case of a "generalized" shear $\chi_\varepsilon = 0$, $e_u = e_{nsh}$.

After determining α1, α3, α5 and α7 "Eq. (6)" gets the following form "Eq. (9)":

$$e_n = 0{,}5\chi_\varepsilon^2[(\chi_\varepsilon + 1)e_{nc} - (\chi_\varepsilon - 1)e_{nt}] + (1 - \chi_\varepsilon^2)e_{nsh}, \tag{9}$$

where e_n is the normalized density of potential deformation energy and e_{nc}, e_{nt}, e_{nsh} are the normalized values of the potential energy deformation density in the case of tension, compression and shear.

Equation (9) is a criterion that determines the operational state of a rationalized element in a local zone, point and finite element.

On the basis of materials energy rationalization modern principles, laboratory tests of the proposed adhesive samples were carried out, for further without anchor fastening structural elements designing, under the influence of corresponding external loads.

Samples for tension tests in the form of scapula with a total length of 150 mm, a working part width of 10 mm and a working part length of 60 mm were mechanically sharpened (Fig. 1a). Compression test specimens of 50 mm width and 50 mm height (length) were also mechanically sharpened from the workpieces used for tension tests (Fig. 1b). For shear tests, the tubular specimens with a total length of 250 mm, 120 mm working length, internal diameter of the working part of 30 mm, and an outer diameter of the working part of 34 mm also were sharpened mechanically. Mechanical shear tests were carried out on the test machine "Instron-1122" (Fig. 1c).

The method of determining the shear value consists in the torsion of test samples. Then, the conditional (sheared) fluidity limit and destructive shear effort are used as indicators. Shear deformation is defined as the product of the relative angle of torsion on the radius of the sample. Relative shear deformation is defined as the ratio of shear deformation to the basis of the measurement (Malkin et al. 1978) according to the "Eq. (10)":

$$\varepsilon = \frac{r\varphi}{l} \times 100\%, \tag{10}$$

where r is the radius of the sample, l is the sample base and φ is the torsion angle on the selected base.

Destructive shear stress is determined according to the "Eq. (11)":

$$\sigma = \frac{M}{2\pi r_{sh}^2 \delta}, \tag{11}$$

Fig. 1. Adhesive material block samples lab tests for constructing σ-ε diagrams under the influence of tension (a), compression (b) and shear (c)

where M is the torque at the measuring point, δ is the thickness of the sample working part wall and r_{sh} is the average radius of the sample in the working part, which is determined according to the "Eq. (12)":

$$r_{sh} = \frac{(r_0 + r_i)}{2}, \tag{12}$$

where r_i is internal radius of the sample working part and r_0 is the outer radius of the sample working part.

After conducting appropriate laboratory tests σ-ε diagrams, under the influence of tension (Fig. 2a), compression (Fig. 2b) and shear (Fig. 2c), were constructed.

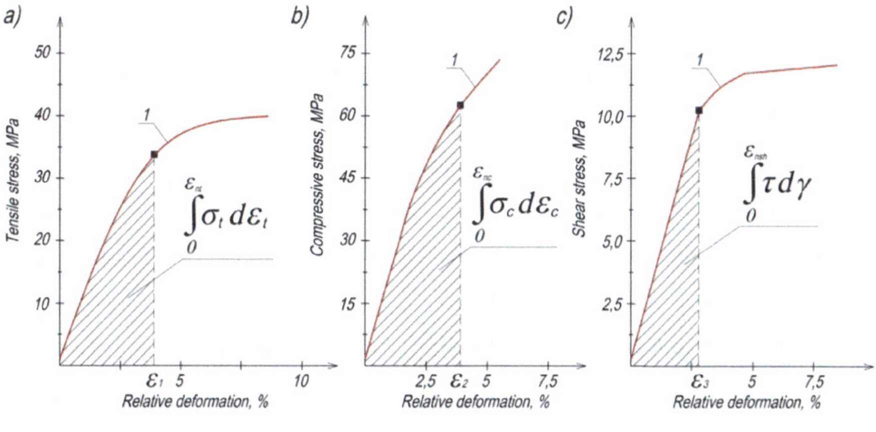

Fig. 2. Deformation diagrams of the proposed acrylic compositions under the influence of tension (a), compression (b) and shear (c): *1* – adhesive material strain curve

Linear deformations were calculated according to the deformation diagrams of the adhesive material (Fig. 2). In this case, e_n is the limiting value of the deformation energy density, which corresponds to the boundary value of the adhesive material deformation plasticity beginning. Consequently, according to (Fig. 2) $\varepsilon_1 = 0.042$; $\varepsilon_2 = 0.040$; $\varepsilon_3 = 0$ - as there is no deformation in the direction of the third main axis. The condition $\varepsilon_1 \geq \varepsilon_2 \geq \varepsilon_3$ is executed, the renumber of the stresses is not required (Fig. 3).

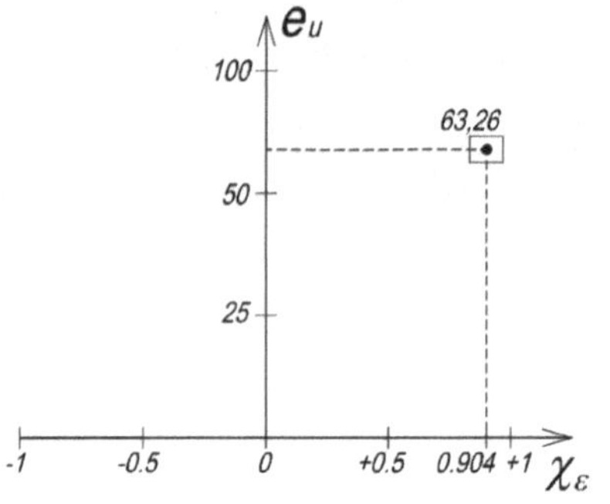

Fig. 3. Energy parameter of the proposed acrylic modified composition at the point of the deformation plasticity beginning

Now it is necessary to numerically calculate each of the potential deformation energy density parts by processing the experimental data according to the "Eqs. (1), (2), (3), (4) and (5)":

$$e_{nc} = \frac{9}{2}K \times \varepsilon_0^2 + \int_0^{\varepsilon_{nc}} \sigma_c d\varepsilon_c = 71{,}42; \qquad e_{nsh} = \int_0^{\varepsilon_{nsh}} \tau d\gamma = 16{,}50;$$

$$e_{nt} = \frac{9}{2}K \times \varepsilon_0^2 + \int_0^{\varepsilon_{nt}} \sigma_t d\varepsilon_t = 122{,}02; \quad K = \frac{E_0}{3 \times (1 - 2 \times \nu)} = 7{,}1\,\text{GPa};$$

$$\varepsilon_0 = \frac{1}{3}I_1(T_\varepsilon) = \frac{1}{3}(\varepsilon_1 + \varepsilon_2) = \frac{1}{3}(0{,}042 + 0{,}040) = 0{,}027.$$

Calculation of the Lode-Nadai parameter value for the proposed adhesive material is done through the linear deformations, according to the "Eq. (8)":

$$\chi_\varepsilon = \frac{2\varepsilon_2 - \varepsilon_1 - \varepsilon_3}{\varepsilon_1 - \varepsilon_3} = \frac{2 \times 0{,}040 - 0{,}042 - 0}{0{,}042 - 0} = 0{,}904.$$

Consequently, the normalized density of the deformation potential energy e_n will be equal according to the "Eq. (9)":

$$e_n = 0{,}5\chi_\varepsilon^2[(\chi_\varepsilon + 1)e_{nc} - (\chi_\varepsilon - 1)e_{nt}] + (1 - \chi_\varepsilon^2)e_{nsh} = 0{,}5 \times 0{,}904^2 \times [(0{,}904 + 1) \times 71{,}42 - (0{,}904 - 1) \times 122{,}02] + (1 - 0{,}904^2) \times 16{,}50 = 63{,}26\,\text{GPa}.$$

The required plasticity is a function of the technological parameters and properties of the material. If it is "located" under the magnitude of plasticity e_O, then the material can withstand a technological operation without destruction. Otherwise, it is necessary to either increase the material plasticity (for example, by heating), or change the required plasticity. This is achieved, for example, by applying hydrostatic pressure.

It is envisaged that material deformation occurs without destruction if the accumulated deformation satisfies to the inequality "Eq. (13)":

$$e_0 < e_n(\chi_\varepsilon). \tag{13}$$

It is assumed that in an elastic zone with a linear connection between stress and deformation "Eq. (14)":

$$e_0 = \frac{1}{2} \times \sigma \times \varepsilon = \frac{\sigma^2}{2E_0}. \tag{14}$$

In our case, at a point with maximum stress according to the "Eq. (13)":

$$e_0 = \frac{3{,}67^2}{2 \times 4{,}26 \times 10^3} = 0{,}86 \times 10^3\,\text{MPa}$$

$0{,}86 \times 10^3$ MPa < $63{,}26 \times 10^3$ MPa - the condition is fulfilled.

5 Scientific Novelty

For the first time, a numerical evaluation of the proposed acrylic adhesive compositions stress-deformed condition was conducted, taking into account the potential deformation energy density, which indicates that the tested material has a substantial structural strength and can be evidently used in the designing of anchor and/or without anchor glued joints.

6 Practical Significance

Practical significance for the engineering industry is in the application of improved acrylic adhesive compositions for the designing of anchor and without anchor reliable steel-concrete joints of industrial equipment and technological communications.

7 Conclusions

In the present study performed calculation, which is based on the modern principles of materials energy rationalization (taking into account the deformation potential energy density), suggests that the proposed acrylic composition for without anchor designing has an essential reserve of structural strength:

- for the stresses - at 6 times;
- for the deformation energy - at 36 times (because the magnitude of plasticity e_0 under elastic deformation is proportional to the stress square σ).

References

Zolotov MS, Firsov PM (2015) The influence of timber knot constructive factors on the strength of glued joint without anchors. Acad J Resource Sav Mater Constr Build Struct of National University of Water and Environmental Engineering 30:126–132

Morkovskaya N (2011) Tehnologija sozdanija bezankernyh kreplenij na akrilovyh klejah. Sci Tech Collect Communal Econ Cities Eng Sci Arch Ser of O.M. Beketov National University of Urban Economy 101:47–50

Chamochín R, Cano M, Abenojar J, Pantoja M, Ballesteros Y, del Real-Romero JC (2010) The effect of surface treatment on the behavior of toughened acrylic adhesive/GRP (epoxy) composite joints. J Adhes Sci Technol 24(11–12):1903–1916

del Real-Romero JC, Cano M, Abenojar J, Martínez MA (2006) Adhesive bonding of aluminium with structural acrylic adhesives: durability in wet environments. J Adhes Sci Technol 20 (16):1801–1818

Firsov PM (2017) Steel-glued without anchor joint short-term strength experimental research on acrylic modified compositions. Collect Sci Work of Ukrainian State University of Railway Transport 167:4–17

Zolotov SM (2016) Akrilovye klei dlja soedinenija stroitel'nyh konstrukcij. O.M. Beketov National University of Urban Economy, Kharkiv

Vasil'kov GV (2008) Jevoljucionnye zadachi stroitel'noj mehaniki. Sinergeticheskaja paradigma. LKI, Moscow

Shmukler VS (2005) Evolutionist approach in rationalization of building structures. In: Third international structural engineering and construction conference, ISEC-03, Shunan, Japan

Shmukler VS, Klimov Y, Burak NP (2008) Framework systems of facilitative type. Zolotyie stranitsy, Kharkiv

Shmukler VS (2017) New energy principles of structures rationalization. Collect Sci Work of Ukrainian State University of Railway Transport 167:54–69

Valovoi OI, Yeromenko OIu, Valovoi MO (2013) Kryterii mitsnosti betonu dlia zahalnoho vypadku napruzhenoho stanu. Acad J Ind Mach Build Civ Eng of Poltava National Technical Yuri Kondratyuk University 4(1):55–63

Krul Y, Kaplin OIu, Delyavskyy M (2019) Rationalization of the parameters of composite reinforced concrete superstructures under conditions of multi-criterion. In: Scientific session on applied mechanics X, AIP conference proceedings vol 2077, pp 020031-1–020031-6

Malkin G, Askadsksij AA, Kovriga VV (1978) Metody izmerenija mehanicheskih svojstv polimerov. Himiya, Moscow

Geometrical Simulation of Optimized Vacuum-Condensation Spraying Technology for Titanium Nitride on Structural Steel

O. Gumen(✉), I. Bilyk, and M. Kruzhkova

National Technical University of Ukraine "Igor Sikorsky Kyiv Polytechnic Institute", pr. Peremogy 37, Kiev 03056, Ukraine
fmf_ikg@ukr.net

Abstract. The work is aimed at studying and solving the problems of vacuum-condensation spraying for titanium nitride. The nature of the work is applicable because the proposed advanced technology can be applied to the production of any parts with coating. The applied technology will reduce a cost of manufacturing wear-resistant parts and therefore tends to save energy. The promise of the advanced vacuum-condensation spray technology is due to a decrease in the number of defects on the surface of the spray product. The novelty of the work is to improve the coating properties by reducing the defects of the structure.

Keywords: Vacuum-condensation spray · Droplet phase · Titanium nitride · Structural steel · Linear parameters · Geometrical simulation

1 Introduction

In the conditions of the current pace of development of science and technology, the requirements to the properties of details of mechanisms and materials for their manufacture are continuously increasing. However, most of the materials are subject to corrosion; wear during operation, or low strength. The priority direction in the field of increasing wear resistance is the creation of coated materials. The use of coatings makes it possible to produce parts of machines and mechanisms of the appropriate designation from ordinary structural steels on the working surfaces of which spray a special wear-resistant material. Gas-thermal and vacuum-condensation coating methods have got especially wide opportunities.

Spraying is the process of coating some materials by others, during which the atomic materials of the material, which is being sprayed, settle on the surface of the parts, forming a coating. There are different methods of coatings, but in this work for the sake of the experiment, the reaction method of the vacuum-condensation method of atomization of titanium nitride to structural steel was used. Some compounds cannot be sprayed onto the parts by direct evaporation or ion spraying due to partial or complete decomposition. These compounds, including elements whose partial vapor pressure is very different (for example, nitrides, carbides). Therefore, in the form of coatings they are obtained by the reaction between the atoms of the steam flow of metals and atoms of specially introduced into the chamber of chemically active gases. The slight heating

Z. Blikharskyy et al. (Eds.): CEE 2019, LNCE 47, pp. 103–110, 2020.
https://doi.org/10.1007/978-3-030-27011-7_13

and slight deformation of the base during the spraying process, as well as the possibility of coating only on the specified parts of the part, determine the high technological efficiency of the application of the vacuum-condensation spray method to increase wear resistance.

However, products with spray coatings have a superficial defect - a droplet phase, which manifestation essentially depends on the technological parameters of the spraying process and leads to additional machining of the finished product, and therefore unnecessary energy and resource consumption. This prompted a solution to the problem of eliminating the defect.

2 The Goal

The purpose of the work is to study the dependence of the droplet phase on the surface of the impregnated products, namely the linear parameters of the formed voids, on the strength of the arc spray current as the main technological parameter for optimizing the process and improving the quality of the applied coating.

3 Methodology

The research work is based on theoretical (research of scientific sources on the study of the droplet phase, the development of an experimental model, analysis of the results) and experimental (production of samples at different vortex arc current, surface microstructure studies and droplet phase manifestations in samples using light and electron microscopes) research methods.

4 Vacuum-Condensation Spray of Titanium Nitride on Steel

The coating of the samples by the vacuum-condensation method of reaction spraying was carried out in accordance with the basic requirements of the spray technology in two stages: preparation of the surface of the samples and their spraying. Consider each stage of spraying in more detail.

The surface of the sample should be with a minimum height of micronutrients and completely nonfat. Fat pollution will reduce the effectiveness of the process, which will adversely affect the surface of the coating. Therefore, before the spraying process, the surface of the sample was ground and degreased with alcohol.

To prevent the influence of air molecules on the chemical reaction between the vapor flow of titanium and nitrogen atoms, a vacuum was created in the chamber.

In the vacuum chamber of the "Bulat VU-1" installation under the general pressure in a chamber of 12 Pa, samples of titanium nitride were sprayed. To reduce the defects created and optimize the process, the current strength of the arc was changed in the following order: 80 A, 100 A, 120 A, 140 A. We have received 4 samples with different droplet manifestations.

To determine the dependence of the manifestation of the droplet phase and its parameters on the current strength of the arc discharge, the study of the surface of the samples was carried out on different microscopes, which made it possible to detect defects in the form of emptiness, to investigate their nature, to make linear measurements of tumors, to investigate the structure of the surface.

5 Results

Surface Research on Neophot Light Microscope
Experiments on the determination of the linear parameters of the formed voids and the nature of their formation were carried out on a light microscope "Neofot" with an increase of 800 times (Appendix B). The study made it possible to compare the surface of the sprayed specimens. The results of experiments for each sample are presented in Table 1.

Table 1.

№	I, (A)	Diameter of the droplet phase, (mkm)
1	80 A	54,592
2	100 A	53,889
3	120 A	38,447
4	140 A	67,264

For the sample number 1, which was sprayed with the lowest values of the current strength of the arc – 80 A, quite large voids are characteristic, that is, the droplet phase is well shown (Fig. 1).

For the sample number 2, which was sprayed by the current strength of the arc 100 A, there is a characteristic manifestation of the droplet phase to a lesser extent, but defects are also monitored (Fig. 2).

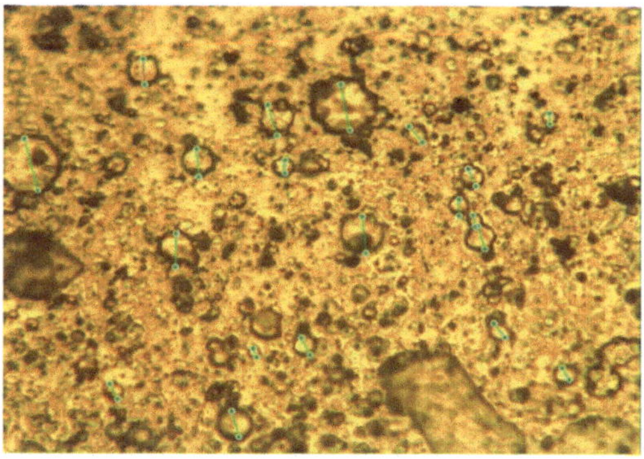

Fig. 1. Surface structure of the sample №1

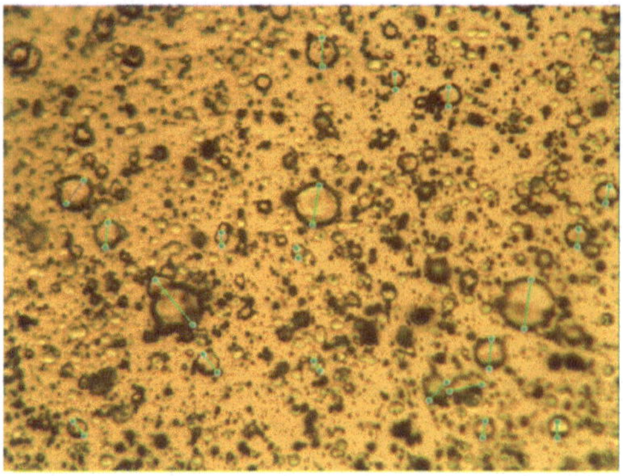

Fig. 2. Surface structure of the sample №2

For the sample number 3, which was sprayed with indicators of current strength of the arc 120 A, there is a characteristic of the smallest manifestation of the droplet phase. Defects have the smallest size compared with other samples and are more evenly distributed (Fig. 3).

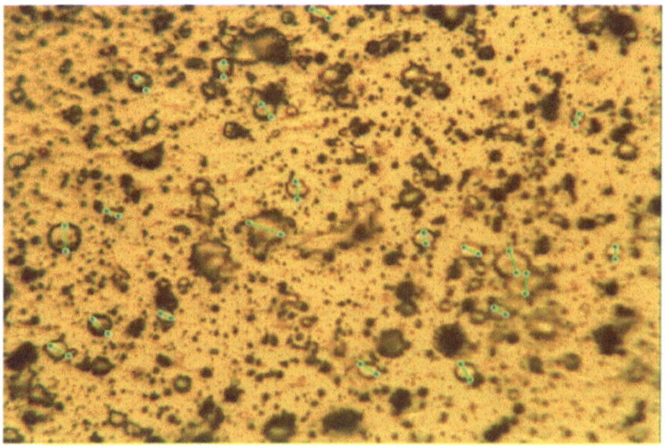

Fig. 3. Surface structure of the sample №3

For the sample number 4, which was sprayed with indicators of current strength of the arc 140 A, is characterized by the largest manifestation of the droplet phase. Defects have the largest dimensions in comparison with other samples, are uneven, metal foundations on the surface are observed. This is due to the high temperature on the surface of the cathode, caused by the current strength of the arc, which focuses on the cathode. In this case, the evaporation of metal atoms passes quickly. This causes a disturbance in the equilibrium between the number of atoms of the vapor flow of the metal (titanium) and the atoms of the chemically active gas (nitrogen). Some titanium atoms, due to the high rate of precipitation on the surface, do not react with nitrogen, so a droplet phase is observed on the surface in the form casting of titanium (Fig. 4).

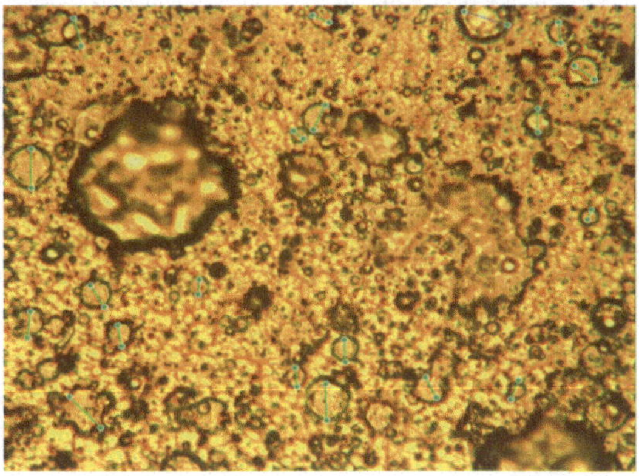

Fig. 4. Surface structure of the sample №4

A study on an optical microscope showed that the size of the formed defects depends on the current strength of the arc spraying. This is presented on the chart (Fig. 5).

On the graph there is a decrease in the size of the formed droplet phase with an increase in the force of the arc of the atomization arc to 120 A. This is due to an increase in the concentration of pulses on the surface of the cathode, which destroys the droplet phase in its formation. A sharp increase sizes of the formed voids spraying with a current of 140 A is due to the high temperature on the surface of the cathode.

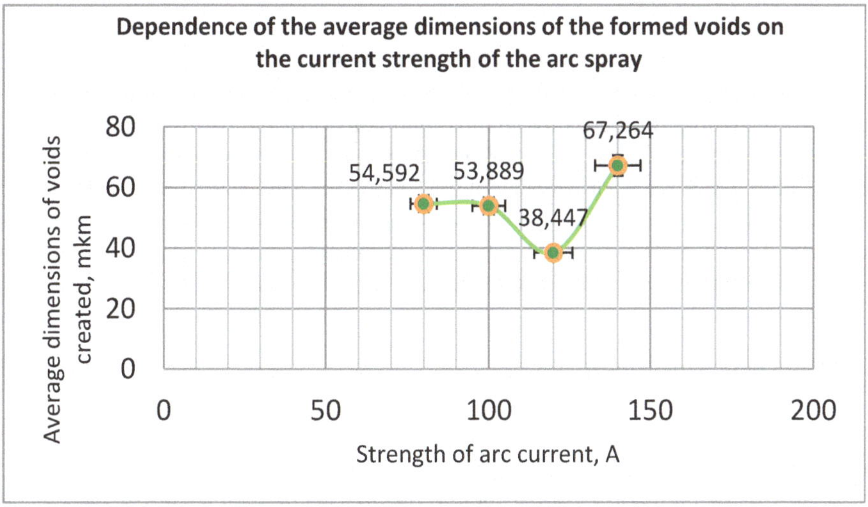

Fig. 5. Graph of the dependence of the size of droplet formation at different values of current strength: 80 A, 100 A, 120 A, 140 A

According to the results of the study, the number of defects per unit area of surface for each sample was calculated on a light microscope and a chart was constructed (Fig. 6). The decrease in the number of formed voids is observed with increasing the arc force current up to 120 A and a sharp increase in the number of defects when spraying for an arc current of 140 A.

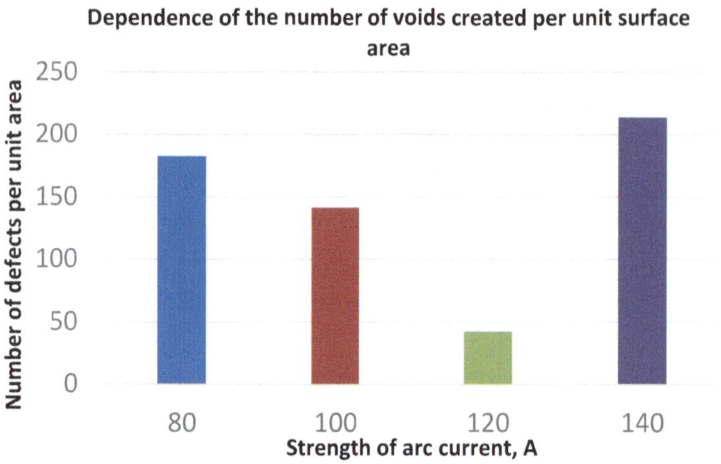

Fig. 6. Dependence of the number of created voids per unit surface area

6 Scientific Novelty

Today, the droplet phase is a poorly-studied phenomenon due to the impossibility of directly influencing its manifestation. Therefore, the scientific novelty of the work is to improve the coating properties by reducing the defects of the structure.

In order to reduce the roughness of the TiN coating, it is recommended to polish the surface, after which the wear intensity of the coating is significantly reduced. However, this process requires additional energy and resource costs, which increases the cost of the finished product. Therefore it is expedient optimize the process of reaction titanium nitride sputtering to reduce the surface defect - droplet phase.

7 Practical Significance

The applied technology will reduce the cost of manufacturing wear-resistant parts, and therefore tends to save materials and save energy. The promise of the advanced technology of vacuum-condensation spray is due to a decrease in the number of defects on the surface of the spray product.

The nature of the work is applicable, because the proposed technology of manufacturing materials can be used in the production of any parts of the mechanism, which with the passage of time are damaged. This will make the parts more economical and will extend the life of the works.

The main disadvantage of this method of spraying is the formation of a droplet phase on the surface of the product. In order to solve the problem of reducing the defects on the surface of the impregnated products, the force of the current of the arc of the spray was changed.

8 Conclusions

The use of the coating method of some other materials in the manufacture of parts is promising, since it allows you to create parts with increased wear resistance properties, using fewer resources. The technology of vacuum-condensation spraying allows the application of both pure materials and mixtures: this is the main advantage of the method of spraying compared with electrolytic coating. The spraying process takes place fairly quickly, evenly and is energy-saving compared with other coating methods.

During the work spray of titanium nitride onto steel was carried out by the reactive method of vacuum-condensation spraying for various indicators of arc current - 80 A, 100 A, 120 A and 140 A, a manifestation was revealed the droplet phase on the surface of the spray surface, the microstructure of the spray surface was investigated, the linear measurements of the formed defects were made, their number per unit surface area was calculated and their character was determined.

It was found out that the manifestation of the droplet phase, the size of voids, their number per unit area depends on the strength of the current arc spraying. The smallest sizes of the droplet phase were formed when spraying for an arc current of 120 A, which can be used as optimal.

The promise of the advanced technology of vacuum-condensation spray is due to a decrease in the number of defects on the surface of the spray product.

The proposed technology for the manufacture of materials can be used in manufacturing for the manufacture of sprayed surfaces, which are damaged with the passage of time. This will make the parts more economical and will extend the life of the works.

References

Krechmar E (1966) Dusting of metals, ceramics and plastics. Mashinostroenie, 234 p

Linnik V, Pekshev P (1985) Modern gas thermal coating technology. Mashinostroenie, 85 p

Lakhtin YuM, Moiteev VA (1974) Metal studies. Mashinostroenie, 116 p

Ulyanov VA, Magazinnikov VP (1985) Recommendations for the restoration of crankshafts of automobile engines by flame spraying. GOSNIITI, 32 p

Paton BE (1979) The technology of electric welding of metals and alloys by melting. Mashinostroenie, 203 p

Vinokurov VA (1979) Welding in mechanical engineering. Mashinostroenie, 327 p

Dubovoy AN, Khachaturov EB, Loy SA (1986) Technology of applying gas-thermal powder coatings. NKI, Nikolaev, p 43

Lisnikov VN, Ukrainian VS, Bogatyrev GF (1985) Plasma spraying of coatings in the manufacture of products in electronic technology. SSU, Saratov, p 132

Runkov MN (1966) Reference master surfacing site. Mashinostroenie, 200 p

Denisova NE, Shorin VA, Gontar IN (2006) Tribotechnical materials science and tribotechnology. Penza PSU, 248 p
Poltaev NK (1980) Labor protection in engineering. Higher School, 340 p
Dubovoy AN (1980) Textbook for students. NCI, Nikolaev, p 56
Runge Y (ed) (1984) Welding technology. Metallurgy, 549 p
Sytnik KM, Brion AV (1987) Biosphere, ecology, nature conservation. Naukova Dumka, 522 p
Kudinov VV, Ivanov VM (1981) Plasma spraying of refractory coatings. Mashinostroenie, 192 p

Hot Asphalt Concrete with Application of Formaldehyde Modified Bitumen

Volodymyr Gunka[1], Iurii Sidun[2(✉)], Serhiy Solodkyy[2], and Nataliya Vytrykush[3]

[1] Institute of Chemistry and Chemical Technology, Lviv Polytecnic National University, St. Bandery 12, Lviv 79013, Ukraine
[2] Institute of Building and Environmental Engineering, Lviv Polytecnic National University, St. Bandery 12, Lviv 79013, Ukraine
siduniurii@gmail.com
[3] Viacheslav Chornovil Institute of Sustainadle Development, Lviv Polytecnic National University, St. Bandery 12, Lviv 79013, Ukraine

Abstract. The chemical modification of oxidized bitumen BND 70/100 by formaldehyde, or rather its aqueous solution, formalin, was carried out for use in hot asphalt pavement technologies. The chemical modification conditions of bituminous binder by formaldehyde are given. There is shown considerably increases of the heat capacity (softening point) and adhesion properties (adhesion to glass) in the chemical modification process of oxidized bitumen by formaldehyde. The adhesion determining methods of binder with granite crushed stone and glass plates as variants of adhesive ability estimation of bitumen are given. It has been established that the modification by formalin in the amount of 10% by weight dramatically reduces the bitumen viscosity, so it showed necessary to carefully monitor the source bitumen and formalin ratio content during the modification. A hot dense fine-grained asphalt mix was selected. The influence of modified bitumen on physical and mechanical properties of asphalt concrete, such as water saturation, border compressive strength at 20 and 50 °C and the long-term water resistance coefficient after 15 days.

Keywords: Bitumen · Chemical modification · Formaldehyde · Formalin · Hot asphalt concrete

1 Introduction

At present, the problem of restoration of old and construction of new road coverings has sharply become complicated due to the lack and high cost of quality bitumen and mineral fillers. The need for these materials in road construction organizations is satisfied by less than a half. As the bitumen content in road bitumen-mineral compositions is only 5–9% by weight of the mineral components, the problem of saving both components is solved by extending the service life of the coating. Therefore, the creation durable bitumen and materials on their basis resistant to cracking, aging and atmospheric influences is an actual and timely task. The bulk of bitumen for road use is obtained by petroleum residuals oxidation with air oxygen in various types oxidizing

Z. Blikharskyy et al. (Eds.): CEE 2019, LNCE 47, pp. 111–118, 2020.
https://doi.org/10.1007/978-3-030-27011-7_14

reactors (Pyshyev et al. 2015). This energy consuming and long (10 to 24 h) process associated with the release of toxic components and does not always allow receiving of bitumen that meet the requirements of standards. But even the use of oxidized bitumen that meet the standards does not always provide the required quality of asphalt concrete coatings. The main reason for this circumstance is the insufficient deformability most oxidized bitumen, weak adhesion to mineral materials (especially acidic), low resistance to aging processes. The above facts became the preconditions for work in the direction of creating such bitumen, which would be thermostable, deformative and resistant to aging and had sufficient adhesion with acid rocks. Now the most popular bitumen modifiers in the world are polymers. The most popular are styren-butadiene-styrene type of thermoelastoplastic. Perspective is the use of reactoplastics, which contain either dual (unsaturated) bonds or chemically active groups. Their molecules can react with each other or with bitumen molecules under certain conditions (heating, irradiation or addition of hardener substances). Reactoplastics include epoxy, phenol-formaldehyde, carbamide, organosilicon and other resins (Nykypanchuk et al. 2013; Pyshyev et al. 2015; Zhu et al. 2014; Demchuk et al. 2018; Çubuk et al. 2014; Saha and Suman 2017; Pyshyev et al. 2019). The main disadvantage that restrains increasing pace in output of polymer modified bitumen is their high cost. Adding 2.5–3.5% of thermoelastoplast or 1–1.5% of terpolymers increases bitumen cost in 1.5–2.5 times (Pyshyev et al. 2017). Therefore, it is important to search for inexpensive substances or technologies, the implementation of which would improve the working characteristics of petroleum bitumen. It is known that aromatic hydrocarbons, especially highly condensed, containing heteroatoms, interact with formaldehyde (polycondensation reaction), thus obtaining a large gamma of aromatic hydrocarbon formaldehyde resin, including heavy petroleum residues (tar, distilled and oxidized bitumen) - formalites (Moshchinskaya 1969; Higashihara and Okoshi 2014). But in the literature there is no data on the modification of bitumen binders by formaldehyde, therefore, these studies have been conducted to determine the effect of formaldehyde (formalin) on the quality indicators of bitumen and hot asphalt concrete for use in road construction. This method of obtaining commercial road bitumen from oil residues is more expedient to use at oil refineries, what will allow replacing energy-intensive oxidation and vacuum distillation processes. The using of cheap (US $ 380–400 per ton) and available formalin as a modifier of petroleum residues directly at the refinery will allow to obtain for cheaper and better quality abrasive materials for road pavement. For comparison, the cost of non-modified road bitumen is US $ 600–650 per tonne, and polymer additives (for example, Kraton D 1192) are used for their modification, costing from US $ 4.000–5.000 per ton.

2 Experimental

2.1 Initial Materials

To produce asphalt concrete mixes there were used chips grade 5–10, 10–20 mm and chips screenings 0–5 mm (granite quarry JSC « Polonskiy Gorniy Combinat » .) and mineral limestone powder (fraction < 0.071 mm). To obtain modified bitumen there was used crude-oil road viscous oxidized bitumen grade 70/100 from JSC Mozyr Refinery, Belarus, Gomel Region, town of Mozyr (BND 70/100). The operation characteristics of taken bitumen are presented in Table 1.

Table 1. Characteristics of the bitumen

Property	Values
Penetration at 25 °C (0.1 mm)	80
Softening point (°C)	47
Ductility at 25 °C (cm)	100
Adhesion to glass (%)	62
Adhesion to gravel (mark)	3

A 37% formaldehyde solution (formalin) was used for chemical modification.

2.2 Experimental Procedure

The bitumen modification was carried out with stirring for 2 h at 80 C in a three-input reactor. At first the bitumen was dissolved in benzene, the solvent content was 30% by weight. The solvent was used to reduce the reaction mixture viscosity. After the bitumen-benzene mixture heating to 80 °C, the required amount of formalin and concentrated hydrochloric acid (catalyst) was added, and then recorded the beginning of the process, which was carried out for 2 h. At the end of the modification process, solvent, water and unreacted components was distilled, and then modified bitumen dried in a cabinet for 2 h at 190 °C. The output of modified bitumen was determined by the weighing results of raw materials and modified bitumen.

2.3 Analysis of Raw Material and Products

Softening temperature was determined according to (EN 1427:2015), penetration at 25 C – (EN 1426:2015), ductility at 25 °C – (EN 13587:2016).

Adhesion to glass surface was determined by the following method: testing is done on prepared glass plates and bitumen. The surface of rectangular glass plates (made of sheet glass) is wiped by solvent, washed by water with household soap and thereafter – by distilled water. Prepared in this way plates are boiled in distilled water during 30 min, while thereafter they are fixed in vertical position at room temperature or in drying cabinet at (105 ± 2) °C during (35 ± 5) min. The glass surface is wiped by filter paper before bitumen is applied upon it. After cooling a thin rectangular contour (70 by 24 mm) is marked by pencil or ball-pen upon one side of the glass plate, within the boundaries of which the glass is covered by bitumen film. The bitumen is applied on the opposite to the contour side of the glass plate. Before testing the bitumen is dewatered by heating to 105 °C, with gradual mixing by glass rod. Thickness of bitumen film on the glass surface shall be (0.2 ± 0.01) mm. For each glass plate (or for each rectangular contour applied upon the glass plate) a portion of bitumen (0.35 ± 0.01 g) is weighed on a scale. Distribution of bitumen over the area of glass plate, marked by rectangular contour, is done by means of heating under the filament lamp or other heating device at (125 ± 5) °C during 20 min. In course of bitumen distribution they prevent its overflow outside the boundaries of rectangular contour marked on a glass or stone plate. The glass plates with bitumen films applied upon them are conditioned during 30 min in horizontal position in heating cabinet at the temperature (80 ± 3)°C higher than the bitumen softening temperature. After that the plates are air-cooled during 30 min at room temperature, and the photos are made. On the bottom of a water bath they put a ceramic support (as high as not less than 40 mm) for the plates. Distilled water is filled into the bath and its temperature is raised till (85 ± 0.5) °C. The end of thermometer shall be immersed into the water till the level of the surface of support. The prepared glass plates with bitumen are thoroughly placed horizontally upon the support. The thickness of the water layer upon the bitumen surface shall be 40 mm. The distance between the glass plates shall be 10 mm. The glass plates with bitumen are conditioned in water at (85 ± 0.5) °C during 25 min. Bitumen separated from the surface of glass or stone plates is removed by filter paper. After conditioning of plates with bitumen – water is gradually filled-up into the water bath in the quantity, providing for decreasing the water temperature below the bitumen softening temperature by 10 °C. The plates are removed from water and brought upon a leaf of cardboard, where it shall stay during 15 min, while thereafter the photo is shot for the second time. After that, on the basis of photos made before and after the testing (and using the applied programs) the area of retained bitumen film on the glass (in percent) is determined.

The adhesion with chips' surface was determined by the following method: from the medium sample of chips there was chosen six chips-grains sized not less than 10 mm and dried in drying cabinet at 105–110 °C. Each chips-grain was thread-tied or thin-wire-tied (diameter of not more than 0.5 mm) and heated in drying cabinet. Heating temperature for the chips shall be: when using viscous bitumen 130–250 °C, when using liquid bitumen – 80–100 °C. After 1 h the heated chips-grains were immersed (one at a time) for 15 s into bitumen binder heated till 160 °C, while after that they were removed and hanged on the clamp stand – to let the excessive bitumen flow down. The testing was carried out not sooner than 1 h after chips-grains treatment by bitumen. To provide that a beaker was filled for 2/3 of the volume by distilled water, put on electric stove, sand bath or asbestos grid above the burner flame and the water was brought to boiling (not allowing the stormy boiling). Each of the chips-grain, hanged on the clamp stand, was put down (one at a time) into the middle of the beaker – so as not to let it touch the bottom and walls of the beaker, and was kept in the boiling water: when using viscous bitumen – for 30 min, when using liquid bitumen – for 3 min. After the completion of the time-period indicated – the bitumen, which separated from the chips surface (during water-boiling) and rose to the surface, was removed by filter paper. The chips-grain was taken out of the beaker and immersed into the glass with cold distilled water for 1–3 min – for cooling and strengthening of the bitumen film, which remained on the chips surface. The cooled chips were taken out of water and put on a filter paper. The surface of the chips-grains was examined and estimated on the quality of bitumen adhesion to the chips – as per the degree of retaining the binder film according to Table 2. The compression tensile strength at 20 °C and 50 °C of asphalt concrete was determined on mechanical presses with press-plate movement speed of (3.0 ± 0.1) mm/min. Before the testing the samples are thermostat-conditioned in a vessel with water during (60 ± 5) min at the temperature: (50 ± 1) °C, (20 ± 1) °C. The samples for testing compression tensile strength at 50 °C are placed (before the thermostat-conditioning) into the tight polyethylene bags – to prevent their contact with water. The average density of asphalt concrete was determined by hydrostatic weighing. Water-saturation was determined by quantity of water, which is absorbed by a sample at pre-set mode of saturation in vacuum unit. The factor of durable water-resistance was determined due to decrease of compression tensile strength at (20 ± 1) °C, while the samples of asphalt concrete after the water influence during 15 days were compared with the samples air-conditioned at (20 ± 1) °C.

Table 2. Estimation of quality of bitumen binder adhesion to the chips surface.

Characteristics of bitumen film on the chips surface	Estimation of adhesion quality
Binder film completely remains on the surface, while its thickness can be somewhere decreased	Excellent (mark five)
Binder film completely remains on the surface, but partly separated from the sharp angles and ribs	Good (mark four)
Binder film completely remains on the surface for more than 50%	Satisfactory (mark three)
Binder film completely remains on the surface for less than 50%. On the uncovered surface there are observed separate bitumen drops	Bad (mark two)

3 Results and Discussion

The bitumen chemical modification process is based on the copolycondensation reactions of bitumen aromatic fragments and formaldehyde (Moshchinskaya 1969; Higashihara and Okoshi 2014). The BND 70/100 modification process was carried out under the conditions given in Table 3. These parameters were selected to establish the petroleum bitumen by formaldehyde modifying feasibility. The influence of these conditions on the modification process will be considered more in further studies. Benzene was used as a solvent to reduce the viscosity of the reaction mixture. Concentrated hydrochloric acid was used as a catalyst for the copolycondensation process of bitumen aromatic fragments and formaldehyde. The basic physical and mechanical parameters (Table 4) of the initial oxidized bitumen BND 70/100 brand and formalin modified bitumen: FMBO1 (5% by weight of formalin per bitumen) and FMBO2 (10% by weight of formalin per bitumen) were compared, to confirm the positive effect of bitumen modifying by formaldehyde.

The results presented in Table 4 shows that during the bitumen modification by formalin, the bitumen softening temperature significantly increases from 47 to 62 and 70 °C, while the plastic performance of modified bitumen decreases (penetration decreases from 80 to 45 and 23 mm^{-1}). FMBO1 according to EN 14023:2010

Table 3. Conditions for the chemical modification of bitumen by formaldehyde

Parameter	Value
Temperature, °C	80
Modification time, hour	2
Reaction mixture composition, % wt. on the bitumen	
oxidized bitumen	100
benzene (bitumen/benzene = 70/30)	44.4
formalin	5/10
including formaldehyde	1.9/3.7
concentrated HCl	1.3/2.5

Table 4. Physical and mechanical parameters of bitumen

Property	Values		
	BND 70/100	FMBO1	FMBO2
Penetration at 25°C (0.1 mm)	80	45	23
Softening point (°C)	47	62	70
Ductility at 25°C (cm)	100	58	28
Adhesion to glass (%)	62	96	98
Adhesion to gravel (mark)	3	5	5

corresponds to the polymer modified bitumen of PMB 25/55–60, and FMBO2 - PMB 10/40–65 brands. During the modification, the road bitumen adhesive properties are also significantly improved, adhesion to the glass surface increases from 62 to 96–98%, and adhesion to rubble from 3 to 5 points. The content of formalin during modification greatly affects the resulting binder hardness (penetration), because with a large amount of formalin, the modified bitumen viscosity significantly increased (substantially decreasing penetration). We recommend the use bitumen for modification by formalin in a significant amount (> 5% by weight to bitumen) and road construction with penetration > 100 mm^{-1}. Chemical modification by formalin greatly improves the conditioned bitumen heat-resistant properties; therefore, using such a modification, it is possible to obtain bitumen binder products from substandard raw materials, for example: tar, black oil, asphalts of deasphalting or other by-products of oil refining. Therefore, our further research will continue in this direction. The next stage of the study was to determine the effect of modified bitumen on the properties of hot asphalt concrete (Table 5). Hot asphalt concrete was prepared only on FMBO1, since the FMBO2 binder is too viscous, as evidenced its penetration at 25 °C. By means of grading curves of dense-graded continuous mixes there was designed the chip-grain carcass of asphalt concrete mix. The asphalt concrete mix was designed based on the following characteristics: hot fine-grained dense-graded asphalt concrete with residual porosity from 2% to 5%, with quantity of chip-grains sized more than 5 mm – 45–55% and the maximum grain-size up to 20 mm. The binder content for asphalt concrete was 6.5% by weight.

Table 5. Properties of designed composition of hot asphalt concrete with application of BND 70/100 and FMBO1

No. Name of index	Actual indexes of asphalt concrete samples on bitumens:	
	BND 70/100	FMBO1
1. Average density, g/cm^3	2.36	2.35
2. Water-saturation, % by volume	1.0	0.9
3. Compression tensile strength, MPa, at:		
20 °C	3.5	7.4
50 °C	1.5	3.1
4. Durable water-resistance factor after 15 days	0.95	0.96

It is seen (Table 5) that FMBO1 binding in comparison with the initial oxidized bitumen (BND 70/100) increases the compression tensile strength of asphalt concrete samples more than twice (from 3.5 to 7.4 MPa at 20 °C and from 1.5 to 3.1 MPa at 50 °C), while not particularly affecting the material water permeability.

4 Conclusions

The chemical modification of oxidized road petroleum bitumen BND 70/100 was carried out and respectively FMBO1 and FMBO2 were obtained at 80 °C for 2 h with formalin content of 5 and 10% by weight on bitumen and benzene addition as solvent. As a result of chemical modification heat resistant (bitumen softening temperature increases from 47 to 62–70 °C) and adhesive properties of bitumen (adhesion to the glass surface increases from 62 to 96–98%, and adhesion to rubble from 3 to 5 points) was significantly improved. The grain skeleton of the asphalt concrete mixture has been selected by granulometric curves of dense intermittent mixes, as well as asphalt concrete mixtures and cylindrical asphalt concrete samples with application of formaldehyde modified bitumen (FMBO1). Asphalt concrete with FMBO1 application has more than twice higher value of strength index compared to the unmodified at compression and temperature 20 °C (increases from 3.5 to 7.4 MPa) and 50 °C (increases from 1.5 to 3, 1 MPa). In addition, it has been found that modified bitumen does not significantly reduce the saturation by water from 1.0 to 0.9% by volume and increases the asphalt concrete coefficient of long-term water resistance after 15 days (from 0.95 to 0.96).

References

Pyshyev S, Grytsenko Y, Solodkyy S, Sidun I, Vollis O (2015) Using bitumen emulsions based on oxidated, distillation and modified oxidated bitumens for slurry seal production. Chem Chem Technol 9(3):359–366

Nykypanchuk M, Hrynchuk Y, Pyshyev S, Gunka V, Grytsenko Y, Bratychak M (2016) Polymer modified bitumen: review. Chem Chem Technol 10(4):631–636

Pyshyev S, Gunka V, Grytsenko Y, Shved M, Kochubei V (2017) Oil and gas processing products to obtain polymers modified bitumen. Int J Pavement Res Technol 10(4):289–296. https://doi.org/10.1016/j.ijprt.2017.05.001

Zhu J, Birgisson B, Kringos N (2014) Polymer modification of bitumen: Adv Challenges Eur Polym J 54:18–38. https://doi.org/10.1016/j.eurpolymj.2014.02.005

Nykypanchuk M, Hrynchuk Y, Olchovyk M (2013) Effect of modified bitumen on physico-mechanical properties of asphalt concrete. Chem Chem Technol 7(4):467–470. https://doi.org/10.23939/chcht07.04.467

Demchuk Y, Sidun I, Gunka V, Pyshyev S, Solodkyy S (2018) Effect of phenol-cresol-formaldehyde resin on adhesive and physico-mechanical properties of road bitumen. Chem Chem Technol 12(4):456–461. https://doi.org/10.23939/chcht12.04.456

Çubuk, M, Gürü, M, Çubuk, MK, Arslan, D (2014): Rheological properties and performance evaluation of phenol formaldehyde modified bitumen. J Mater Civ Eng, 26 (6). https://doi.org/10.1155/2018/7913527

Saha, SK, Suman, SK. (2017). Characterization of bakelite-modified bitumen. Innovative Infrastructure Solutions, 2(3). https://doi.org/10.1007/s41062-017-0052-0

Pyshyev S, Demchuk Y, Gunka V, Sidun I, Shved M, Bilushchak H, Obshta A (2019) Development of mathematical model and identification of optimal conditions to obtain phenol-cresol-formaldehyde resin. Chem Chem Technol 13(2):212–217. https://doi.org/10.23939/chcht13.02.212

Moshchinskaya, NK (1969): Polimernye matermly na osnove aromatlcheskikh uglevodorodov i formal'degida (Polymeric Materials Based on Aromatic Hydrocarbons and Formaldehyde), Kyiv, p. 266

Higashihara, G, Okoshi, A (2014): Aromatic hydrocarbon formaldehyde resin, modified aromatic hydrocarbon formaldehyde resin, and epoxy resin, and method for producing said resins, European Patent EP3012275A1

EN 1427-2015. European Standard. Bitumen and bituminous binders. Determination of the softening point. Ring and Ball method

EN 1426-2015. European Standard. Bitumen and bituminous binders. Methods of tests for petroleum and its products. Determination of needle penetration

EN 13587:2016. Bitumen and bituminous binders. Determination of the tensile properties of bituminous binders by the tensile test method)

EN 14023:2010. Bitumen and bituminous binders. Specification framework for polymer modified bitumens

Studying Equalization of the Radial Fan's Discharge Flow

B. Hulai(✉), O. Dovbush, B. Piznak, and M. Kasynets

Department of Heat and Gas Supply and Ventilation,
Lviv Polytechnic National University, St. Bandery 12, Lviv 79013, Ukraine
08bogdan1986@ukr.net

Abstract. It has been established that the qualitative parameters of the non-uniformity of the radial fan's discharge flow are energy parameters such as coefficients of kinetic energy and the amount of motion. Compared with the ratio of maximum velocity to the average one, in the case of backflows, the specified indicators reflect the flow characteristic throughout its section. It has been proposed to determine the pressure loss in the diffusers located behind the radial fan, that is, with the non-uniform distribution of the velocities at the inlet, and with the uniform distribution of the velocities at the outlet, using the change in the kinetic energy coefficient.

Keywords: Ventilation system · Air flow · Velocity field

1 Introduction

Reducing operational and capital costs, ensuring the reliability of ventilation systems performance and their characteristics is possible with the use of upgraded designs of the ventilation system elements, especially their aerodynamic improvement. However, it is difficult and costly to investigate in a laboratory environment new structural solutions of the elements of the ventilation equipment, which could improve the efficiency of ventilation systems by equalizing the discharge flow of the radial fan. The available recommendations for connecting radial fans to ventilation systems through equalizing areas or structurally advanced diffusers are controversial and not always ensure their effectiveness (Hulai 2010a, b), especially considering the non-uniformity in velocity and direction of the flow.

Based on the analysis of literary sources, it has been established that the main direction for improvement of the aerodynamic characteristics of ventilation systems and their elements, and hence their effective operation, may be equalization of the discharge flow (Herasymenko et al. 2003; Hulai 2010a, b).

2 Analysis of Literary Sources

The relevance of studying the velocity field and the corresponding influence on the performance of the ventilation equipment and its energy efficiency has been established. In particular, in the studies of the uniformity of the velocity field (Nesterenko

Z. Blikharskyy et al. (Eds.): CEE 2019, LNCE 47, pp. 119–126, 2020.
https://doi.org/10.1007/978-3-030-27011-7_15

and Leshchenko 2015) the laws of distribution of the field of air flow velocities have been obtained, depending on the design parameters and operating modes of the feeding device - air separator. However, the energy indicators of the flow and their influence on the coefficient of local resistance and pressure losses in general were not determined in the paper.

Numerous studies on the distribution of pressure and velocity fields (Liu et al. 2017) have been focused on providing a uniform pressure field without taking into account the centrifugal force. In particular, in the paper (Xua et al. 2018), which took into account the influence of centrifugal forces on the non-uniformity velocity distribution, the main focus of the research was on determining the influence of the velocity of rotation on the velocity of air, without considering the energy indices or the effect of the flow on system elements. Other current studies of airflows and velocity fields do not give estimates for non-uniformity, but are based on numerical simulation (Wang et al. 2018; Duong and Chung 2019).

3 Objectives of Research

Scientific substantiation and analysis of the uniformity of distribution of the radial fan flow on the basis of the analysis of energy indicators and their influence on the coefficient of local resistance and pressure losses in general, for improving and developing new structural solutions of ventilation systems elements. On the basis of the obtained results, it is possible to improve the investigated elements without conducting a considerable amount of experimental research. This is relevant for ensuring energy efficiency and increasing cost-effectiveness of the ventilation system performance in general.

4 Determination of Energy Indicators of the Discharge Flow of the Radial Fan in the Elements of Ventilation System

The estimation of indicators of non-uniformity in the distribution of velocities in the section is determined by the ratio of the maximum flow rate to its average value in the section: $\overline{v}_{\text{макс}} = v_{\text{макс}}/v_{\text{к}},$, however, this ratio does not take into account the value of non-uniformity in the entire plane of the section (Zhukovskyi and Labai 2003).

A more precise estimation of the indicator of non-uniformity in the distribution of velocities can be estimated by the magnitude of the overall flow characteristic, for example, by the ratio of the true kinetic energy value, or the amount of flow motion that passes through the given section to their mean values in this section.

It is generally known that for a uniform and non-uniform flow (Fig. 1) at constant pressure in a section given sudden expansion, the equation of the amount of motion takes the form:

$$(p_1 - p_2)F_2 = \int_{F_2} \rho v^2 dF - \int_{F_1} \rho v^2 dF \quad (1)$$

where v - is a variable velocity in the section, m/s; ρ - gas density, kg/m^3; p_1, p_2 - total pressures in the first and the second sections, respectively, Pa; F_1, F_2 - areas in the first and the second sections, respectively, m^2. When designated:

$$\beta_1^{\text{к.р}} = \frac{\int_{F_1} v^2 dF}{v_1^2 F_1} = \frac{1}{F_1} \int_{F_1} (\frac{v}{v_1})^2 dF \text{ and } \beta_2^{\text{к.р}} = \frac{\int_{F_{12}} v^2 dF}{v_2^2 F_2} = \frac{1}{F_2} \int_{F_2} (\frac{v}{v_2})^2 dF$$

where $\beta_1^{\text{к.р}}, \beta_2^{\text{к.р}}$- coefficients of the amount of gas motion.

v_1, v_2 - average velocity in the corresponding section, m/s.

"Equation (1)" after transformation takes a form:

$$p_2 - p_1 = \beta_1^{\text{к.р}} \rho v_1^2 \frac{F_1}{F_2} - \beta_2^{\text{к.р}} \rho v_2^2 \quad (2)$$

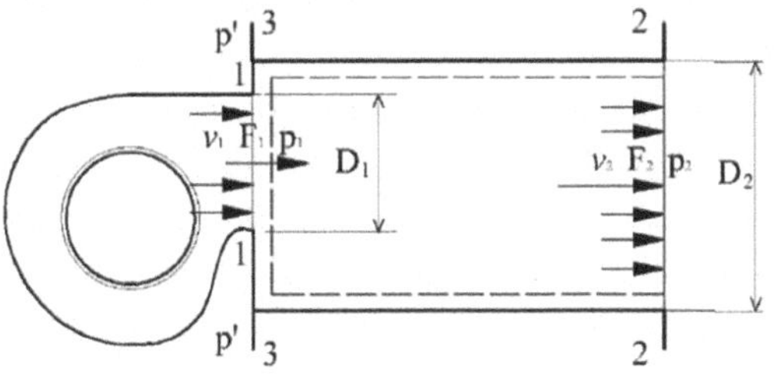

Fig. 1. Sudden flow expansion

In the general case, for a flow with non-uniform distribution of velocities, as well as in case of its twist or deviation, the equation of one-second energy for a section where there is a sudden expansion, as well as with uniform distribution of pressures across the section, has the form:

$$p_1 \int_{F_1} v\, dF - p_2 \int_{F_2} v\, dF + \frac{\rho}{2} \int_{F_1} c^2 v\, dF - \frac{\rho}{2} \int_{F_2} c^2 v\, dF - \Delta E = 0 \quad (3)$$

Let us introduce the designation: $\alpha_1^{\text{к.е}} = \frac{1}{F_1} \int_{F_1} c^2 \frac{v}{v_1^3} dF$ та $\alpha_2^{\text{к.е}} = \frac{1}{F_2} \int_{F_2} c^2 \frac{v}{v_2^3} dF$
where $\alpha_1^{\text{к.е}}, \alpha_2^{\text{к.е}}$- coefficients of kinetic energy (Coriolis coefficient).

If the flow is not twisted, and the resulting velocity coincides with the axial one, then $c = v$, in this case:

$$\alpha_1^{\text{к.е}} = \frac{\int_{F_1} v^3\,dF}{v_1^3 F_1} = \frac{1}{F_1}\int_{F_1}\left(\frac{v}{v_1}\right)^3 dF \text{ та } \alpha_2^{\text{к.е}} = \frac{\int_{F_2} v^3\,dF}{v_2^3 F_2} = \frac{1}{F_2}\int_{F_2}\left(\frac{v}{v_2}\right)^3 dF$$

Given the continuity equation we obtain:

$$p_2 - p_1 = \frac{\rho}{2}\alpha_1^{\text{к.е}} v_1^2 - \frac{\rho}{2}\alpha_2^{\text{к.е}} v_2^2 - \frac{\Delta E}{Q} \tag{4}$$

Equating "Eqs. (2) and (4)", and also by designating $\frac{F_2}{F_1} = n$, and expressing $v_2 = v_1 \frac{F_1}{F_2}$, we receive an expression

$$\Delta p_m = \frac{\rho v_1^2}{2}\left(\frac{2\beta_2^{\text{к.р}} - \alpha_2^{\text{к.е}}}{n^2} + \alpha_1^{\text{к.е}} - \frac{2\beta_1^{\text{к.р}}}{n}\right) \tag{5}$$

If we introduce the designation $\bar{v} = v/v_k$ of deviation of velocity from the average value in the section, and from the determination of the average velocity $\frac{1}{F_k}\int_{F_k}\bar{v}\,dF = 0$, then one can, for example, express the kinetic energy coefficient as follows:

$$\alpha_{\text{к}}^{\text{к.е}} = 1 + \frac{3}{F_k}\int\limits_{F_k}\bar{v}^2 dF + \frac{1}{F_k}\int\limits_{F_k}\Delta\bar{v}^3\,dF \tag{6}$$

If we neglect $\frac{1}{F_k}\int_{F_k}\Delta\bar{v}^3 dF$ due to the sign variability of the function, and with a relatively small non-uniformity of the flow compared with the first integral expression, it will be negligible, then:

$$\alpha_{\text{к}}^{\text{к.е}} = 1 + \frac{1}{F_k}\int\limits_{F_k}\bar{v}^2\,dF = 1 + \delta_v \tag{7}$$

Similarly, one can express the coefficient of the amount of motion M_k.

$$\beta_{\text{к}}^{\text{к.р}} = 1 + \frac{1}{F_k}\int\limits_{F_k}\bar{v}^2\,dF = 1 + \delta_v' \tag{8}$$

It is obvious that based on the deviation of these coefficients from the unit we can judge about the non-uniformity of the flow. If the coefficient of kinetic energy is determined as:

$$\alpha_{\text{к}}^{\text{к.е}} = 3 + \frac{3}{F_k}\int\limits_{F_k}\bar{v}^2\,dF - 2 = 3\left(1 + \frac{1}{F_k}\int\limits_{F_k}\bar{v}^2\,dF\right) - 2 \tag{9}$$

That, taking into account "Eq. (8)", you can also obtain a dependence:

$$\alpha_{\text{к}}^{\text{к.е}} \approx 3\beta_{\text{к}}^{\text{к.р}} - 2 \tag{10}$$

However, this dependence does not fully characterize the flow with significant non-uniformity, therefore, in cases of the discharge flow of the radial fan, the following ratio should be used:

$$\alpha_{\text{к}}^{\text{к.е}} = 1 + \frac{3}{F_k}\int\limits_{F_k} \bar{v}^2\, dF + \frac{1}{F_k}\int\limits_{F_k} \bar{v}^3\, dF = 3\beta_{\text{к}}^{\text{к.р}} - 2 + \frac{1}{F_k}\int\limits_{F_k} \bar{v}^3 dF \tag{11}$$

5 Determination of Pressure Loss with a Sharp Expansion Given Non-uniform Flow

It is known that local air pressure losses in a turbulent flow regime are proportional to the kinetic energy of a flow and by taking them to a smaller cross-section of an airline. Equating the right-hand sides of the equations, and reducing them by dynamic pressure, the coefficient of local resistance ξ is determined, which characterizes the fraction of local energy losses from the kinetic energy of the flow N and the amount of motion M:

$$\xi = \frac{2\beta_2^{\text{к.р}} - \alpha_2^{\text{к.е}}}{n^2} + \alpha_1^{\text{к.е}} - \frac{2\beta_1^{\text{к.р}}}{n} \tag{12}$$

where $n = \frac{F_2}{F_1}$ - exponent of expansion.

In general cases, (Idelchik 1975; Sokolov 1999; Kharev 1999), with uniform distribution of velocities in all sections of the considered site ($\alpha_1^{\text{к.е}} \approx \beta_1^{\text{к.р}} \approx 1$ and $\alpha_2^{\text{к.е}} \approx \beta_2^{\text{к.р}} \approx 1$), the coefficient of local resistance will be determined by "Eq. (11)".

$$\xi = \left(1 - \frac{1}{n}\right)^2 \tag{13}$$

However, in this case, with unsteady turbulent motion in the section of the discharge nozzle of the radial fan, I-I coefficients of kinetic energy $\alpha_1^{\text{к.е}}$ and the amount of motion M_1 acquire certain values, depending on the non-uniformity of distribution of full pressure and velocities in the area of this section. In the section II-II, which is at a sufficient distance to equalize the total pressure and the surface of velocities distribution $N_2 \approx M_2 \approx 1$, then the formula will look like:

$$\xi = \frac{1}{n^2} + \alpha_1^{\text{к.е}} - \frac{2\beta_1^{\text{к.р}}}{n} \tag{14}$$

If we formulate that $\beta^{\text{к.р}} = (\alpha^{\text{к.е}} + 2)/3$ then the previous expression will look like

$$\xi = \frac{1}{n^2} + \alpha_1^{\text{K.e}} - \frac{2\alpha_1^{\text{K.e}} + 4}{3n} \tag{15}$$

Hence

$$\alpha_1^{\text{K.e}} = \left(3\xi n^2 - 4n + 3\right) / \left(3n^2 - 2n\right) \tag{16}$$

Thus, knowing or having determined the coefficient of local resistance experimentally for a sharp expansion of the radial fan located after the discharge nozzle of the radial fan, it is possible to determine the coefficients of kinetic energy $\alpha_1^{\text{K.e}}$, respectively, the coefficient of motion amount $\beta_1^{\text{K.p}}$ of the flow, for their further application in the calculation of local resistance of the ventilation elements in the same location.

6 Determination of Pressure Loss in Diffusers Given the Non-uniform Flow

Works and experimental studies (Kharev 1999) showed that the velocity field at the outlet of the radial fan is non-uniform in all modes of its operation. This non-uniformity is manifested in determining the longitudinal and transverse velocities in the plane of the outlet nozzle of the fan, in addition, the fields of transverse and longitudinal velocities are different. Therefore, in a diffuser located behind the radial fan, the loss of pressure is always greater than in diffusers located isolated, which have been studied given more uniform field of velocities.

It is known that the pressure loss in the diffusers is determined according to the formula:

$$\Delta p_{\text{d}} = \xi_d \rho \frac{{v_1}^2}{2} \tag{17}$$

However, the given formula does not take into account the non-uniformity of the input flow into the diffuser (Fig. 2).

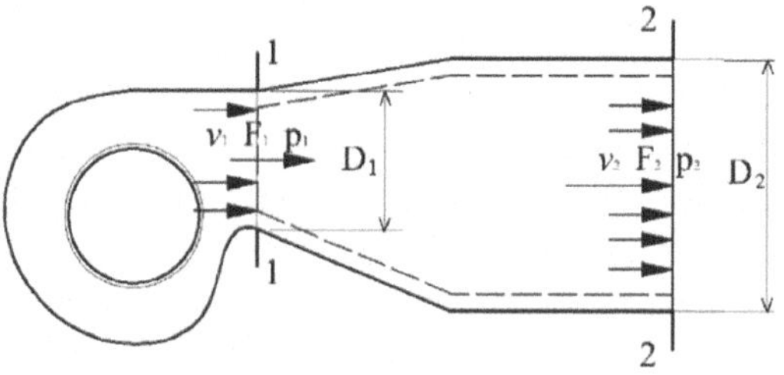

Fig. 2. Diffuser after radial fan

It is proposed to determine the pressure loss in the diffusers located behind the radial fan, that is, with the non-uniform distribution of the input velocities, and in the non-uniform distribution of the velocities at the outlet, using a change in the coefficient of kinetic energy. That is, taking into account "Eqs. (3), (10), (13) and (15)", and comparing them we obtain

$$\xi'_{\text{d}} = \xi'_{exp} + \xi'_{\text{Tp}} = (\varphi_{exp} + \frac{\lambda}{8\sin\frac{\alpha_d}{2}})(\frac{4-\alpha_2^{\text{к.е}}}{3\overline{F}_d^2} + \alpha_1^{\text{к.е}} - \frac{2\alpha_1^{\text{к.е}}+4}{3\overline{F}_{\text{d}}} \tag{18}$$

where φ_{exp} – coefficient of expansion of the diffuser.

$$\varphi_{exp} \approx k_D \cdot k_{form}(tg\frac{\alpha_{Д}}{2})^{1,25} \tag{19}$$

where k_D, k_{form} - respectively, the coefficients of the diffuser angle of expansion, and its shape, which are determined analytically in accordance with (Zhukovskyi and Labai 2003).

7 Results of Analytical Research

In determining the loss of pressure in the diffusers given a non-uniform flow at the inlet and uniform one at the outlet, the following dependence is proposed:

$$\xi'_d = (\varphi_{exp} + \frac{\lambda}{8\sin\frac{\alpha_d}{2}})(\frac{1}{\overline{F}_d^2} + \alpha_1^{\text{к.е}} - \frac{2\alpha_1^{\text{к.е}}+4}{3\overline{F}_d} \tag{20}$$

However, with the use of flow equalization means in the diffuser, the pressure loss should decrease, so "Eq. (17)", taking into account "Eqs. (18) and (19)" can be recorded as follows:

$$\Delta p_{\text{d}} = \xi'_d \rho \frac{v_1^2}{2} k_{\text{B}} = (\varphi_{exp} + \frac{\lambda}{8\sin\frac{\alpha_d}{2}})(\frac{4-\alpha_2^{\text{к.е}}}{3\overline{F}_{\text{d}}^2} + \alpha_1^{\text{к.е}} - \frac{2\alpha_1^{\text{к.е}}+4}{3\overline{F}_d})\,\rho\frac{v_1^2}{2}k_{\text{B}}) \tag{21}$$

where $k_{\text{в}}$ – a new flow equalization coefficient, which is determined experimentally, and takes into account the reduction of pressure losses by means of equalization; $\xi'_{\text{д}}$ - coefficient of local resistance of the diffuser with non-uniform flow at the inlet, which can be determined by "Eqs. (18) and (20)".

The scientific novelty is the proposed dependence of determination of the coefficient of local resistance and the loss of pressure in the diffusers given the non-uniform flow at the inlet and the uniform one at the outlet.

8 Practical Value

On the basis of the obtained results, it is possible to improve the studied elements without conducting a considerable amount of experimental research. This is relevant for the aerodynamic improvement of ventilation elements, and therefore ensuring energy efficiency and increasing cost-effectiveness of the ventilation systems performance in general.

9 Conclusions

It has been established that the qualitative parameters of the flow non-uniformity are the following energy parameters: coefficients of kinetic energy and the amount of motion, which, in comparison with the ratio of maximum velocity to the average one, in case of backflows, reflect the characteristic of the flow throughout its section better. The dependence for determining the coefficient of local resistance and pressure loss in the diffusers given the non-uniform flow at the inlet and the uniform one at the outlet has been proposed.

References

Hulai B (2010a) Influence of structural means of equalizing the discharge flow of the radial fan on the ventilation system. Engineering support and construction mechanization. Bull Donbas Natl Acad Civ Eng Arch 13:83–89

Heramsimenko V, Riazantsev N, Animov Yu (2003) Influence of the flow non-uniformity at the inlet on the centrifugal compressor performance. Aerosp Eng Technol 41(6):18–22

Hulai B (2010b) Improving the efficiency of the radial fan performance in the ventilation system. Sci Bull Natl For Univ Ukr 20(4):93–96

Nesterenko O, Leshchenko S (2015) Investigating the non-uniformity of air flow in the aspirating channel at multilevel introduction of grain. Bulletin named after Petro Vasylenko 11:115–123

Liu H-H, Cheng C-H, Hsueh K-L, Hong C-W (2017) Modeling and design of air-side manifolds and measurement on an industrial 5-kW hydrogen fuel cell stack. Int J Hydrog Energy 42 (30):19216–19226

Xua C-Y, Wub L-L, Chen T (2018) Development and modeling for process control purposes. J ZheJiang Univ (Eng Sci) 56:7886–7894

Wang H, Nie W, Cheng W, Liu Q, Jin H (2018) Effects of air volume ratio parameters on air curtain dust suppression in a rock tunnel's fully-mechanized working face. Adv Powder Technol 29(2):230–244

Duong XQ, Chung JD (2019) Module integration in an adsorption cooling system. Appl Therm Eng 155(5):508–514

Zhukovskyi S, Labai V (2003) Aerodynamics of ventilation: manual. Publ. by Lviv Polytechnic National University, Lviv, p 372

Idelchik I (1975) Reference book on hydraulic resistances. Mechanical Engineering. Moscow, 554 p

Sokolov V (1999) Aerodynamics of gas flows in the channels of complex ventilation systems. EUNU, Luhansk, p 200

Kharev A (1999) Local resistance of mine ventilation networks. Ugletekhizdat, Moscow

Analysis of the Utilization of Plastics in Water and Wastewater Systems of Poland's Podkarpackie Region, and Methods for Their Identification

G. Kalda[1], I. Piegdoń[1(✉)], and W. Gaweł[2]

[1] Department of Water Supply and Sewage Systems, Faculty of Civil and Environmental Engineering and Architecture, Rzeszow University of Technology, Al. Powstańców Warszawy 12, 35-959 Rzeszów, Poland
piegi@prz.edu.pl
[2] GUMAT, Sędziszów Małopolski, Poland

Abstract. The article shows analysis of the utilization of plastics and method for their identification in facilities in Subcarpathian region. Paper presents types of plastics which have been used in the most facilities. Single-layered pipes from homogeneous plastics as polyethylene or polypropylene were applied in the past. Now pipes made of polypropylene are used mainly in external sewage system or domestic waste disposal system. In installation of heating system polybutene for example wall heating are also widely used. In relation to other plastics used in this country for installation polybutene is not the most important material. In Subcarpathian region exist landfills which do not carry out waste compaction. This landfills are conventional, compacted and where waste are compressing by bulldozers. Identification of plastic have been mostly executed in specialist laboratory located in industrial plants. However there is necessity of carry out specific analyses in terms of its quantity and quality because they are still rare.

Keywords: Utilization · Identification · Plastics · Analysis

1 Introduction

Plastics are a group of materials whose basic components are high molecular-weight organic compounds called polymers. These are in practice combined with auxiliary compounds, for example to raise the aesthetic value; to modify thermal, mechanical and dielectric features; and also to improve or impart characteristics of a more specialized nature. While the auxiliary components referred to may be plasticizers, stabilizers, dyes, fillers, pigments, flame retardants and so on, their incorporation does nothing to change polymer structure, whether it takes place during the preparation process, or as semi-finished products take shape. Subsequently, plastic wastes are typical assigned to the municipal or industrial categories, in line with the place of generation. If single polymers are then to be extracted from the municipal component of waste, a procedure allowing for their identification is required.

Z. Blikharskyy et al. (Eds.): CEE 2019, LNCE 47, pp. 127–134, 2020.
https://doi.org/10.1007/978-3-030-27011-7_16

Moreover, this procedure needs to be feasible economically. This paper therefore details methods of utilization applying to plastics produced and used throughout the region of Poland under analysis, including in domestic installations, as connected with the common process of polymer production from various branches of industry (Frącz 2011).

2 Types of Plastic Present in Domestic Installations and Water Network in the Podkarpackie Region

In Poland's Podkarpackie voivodeship, the water-supply network and sewerage include pipes made of various materials whose structure changes constantly. As of 2016, Podkarpackie had a water-distribution network 12,492.7 km in length, as well as sewer pipes extending to 8786.4 km. However, by 2018 the respective figures had risen to 13,460.9 and 13,780 km. Plastics represent the primary materials along these networks, often thermoplastic plastics in pipes, but also with fittings of polyethylene and polyvinyl chloride in wide use (Tomboulian et al. 2004; Kowalska et al. 2011; Mortezania and Othman 2012).

Ever-greater lengths of polyethylene piping are to be noted in water-supply systems in particular, and use of this material is increasing more than that of others. Thus, replacement or renovation processes typically see old pipes made of cast iron, asbestos-cement and steel exchanged for polyethylene (PE) pipes. Old water pipes made of cast iron and asbestos cement fail the most frequently. As a result of this failure process, the safety and reliability of water supply to customers is reduced (Rak 2003; Zimoch and Łobos 2012; Boryczko and Tchórzewska-Cieślak 2013; Szpak and Tchórzewska-Cieślak 2016; Tchórzewska Cieślak and Piegdoń 2016; Piegdoń and Tchórzewska-Cieślak 2018; Pietrucha-Urbanik and Studziński 2019).

Polyvinyl chloride (PVC) is only used less willingly today, though it does find application, both as new networks are constructed and old ones replaced. This leaves the most typical situation as one in which networks comprise single-layer pipes made of homogeneous materials, such as PE, or polypropylene (PP). In Podkarpackie, the latter are mainly used in external and internal pipes carrying wastewater. Fewer companies are involved in making the PP pipes used in the Podkarpackie region, as opposed to those made of other plastic materials. The most-favoured polyethylene again gains use domestically in the region, while also being present more widely in systems of sewers and drains.

An example of PVC and cast-iron water pipes making way for those made of PE is provided by the town of Ropczyce. However, elements of the water-supply networks in Jarosław, Rzeszów, Krosno, Stalowa Wola, Sanok, Leżajsk, Łańcut and Tarnobrzeg all include this plastic. PE is present in sewer pipes in such localities as Sanok, Leżajsk, and Rzeszów. In Mielec, polyethylene installations account for 10.4% of all installations. Polyethylene of both medium and high densities is to be found (MDPE and HDPE). The latter is selected for its resistance to corrosion, as well as thermal resistance in the face of constant operation at low temperatures, and related mechanical resistance. Polyethylene is also characterized by a greater flexibility useful in damping vibrations, and therefore ensuring quietness of operation of installations (Stachurek 2012). PVC is mainly present in the sewer pipes of Krosno, Sanok, Brzozów, Łańcut, and Rzeszów. It is the material in pipes produced by the largest number of companies. These are also present in (for example cold-water) domestic installations in Jarosław and Łańcut, as well as widely in sewerage and drainage installations, and in tele-technical and power cables. PVC pipes can also be used in hot-water installations, given their ability to withstand temperatures as high as 100 °C.

However, below-zero temperatures cause this material to crumble and become unusable (Stachurek 2012), hence the near-homogeneous focus in Podkarpackie on water-distribution networks that use PVC and PE. Furthermore, trends connecting material with year of construction are not as visible as might be expected. Plastic does not occur as frequently as other materials in the region's oldest water-supply networks, but otherwise the main materials are PVC, or else PE in the group of towns with 10–20,000 inhabitants. These data are presented in Fig. 1, which also presents relationships between pipe sizes and lengths (Bergel et al. 2013).

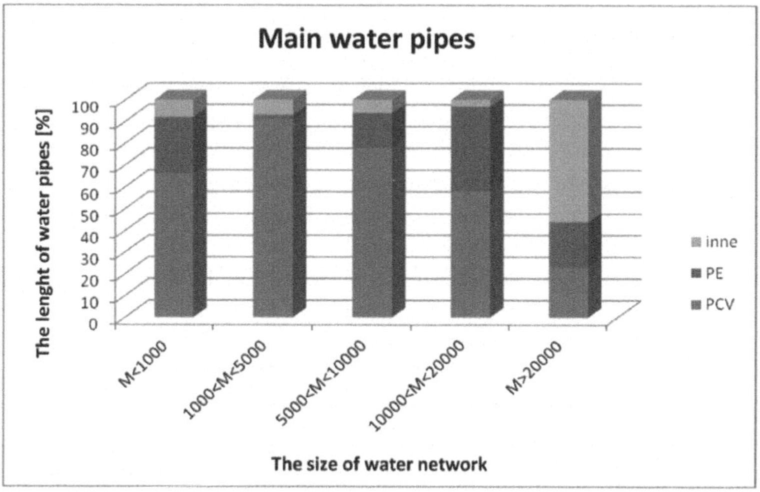

Fig. 1. Dependence of the size of water networks on the length of the pipes (on the basis on (Bergel et al., 2013))

Analysis of this diagram shows how PVC and PE have been used to construct most facilities of the first four categories, though the largest water networks show somewhat more limited use of these materials. Where localities have 5–10,000 inhabitants, more than 94% of the length of water-supply network is of plastic (Bergel et al. 2013).

PP-B materials offer flexibility in the range 1100–1300 MPa, and high impact strength, but a new formula also available involves PP-HM polymer. These materials have high stiffness and impact resistance even at low temperatures, due to the crystalline structure of PP, and the elastic phase, which is well dispersed. These materials also show greater resistance to creep and stress corrosion. Polypropylene PP-R does not interact with chemical organic or inorganic compounds.

Another feature of PP-R is the accumulation of static electricity on the surface of the material, which results in the non-flammable transmission of flammable or explosive liquids. In spite of these features, polypropylene is only used to a limited extent in Podkarpackie's sewer systems.

3 Analysis of Methods of Utilization of Plastics Present in Podkarpacie

Utilization is the process whereby recyclable materials in different kinds of waste are made use of (Jayasekara et al. 2005; Dietrich 2013; Bujak 2015; Awasthi et al. 2017). But plastic recovery is both a complex and very expensive matter. In any case, waste from plastic pipes is still not available in the Podkarpackie region, given the long operational life of the material. Furthermore, there is a tendency to leave used pipes in the ground, given the excessively high costs of disposal.

3.1 Storage of Plastic Waste

There are currently four main ways in which plastic waste is dealt with. The first of these is to store waste at landfills adapted specially for the purpose. There are landfills where waste are not compacted, traditional landfills, compacted landfills and landfills on which bulldozers are in the process of compacting the waste. At present, Podkarpackie has 18 landfills that take plastic waste, including those at Dynów, Leżajsk, Strzyżów, Ostrów, Sokołów Małopolski, Krosno, Bukowisko, Zagórz, Przemyśl, Mielec and Radymno. As of 2014, the capacity of Podkarpackie landfills was 14,000 t/year. However, in 2016 this increased to 31,000 t/a. The number of waste-sorting plants in the region as of 2018 was in turn 17, with locations in Rzeszów, Ustrzyki Dolne, Strzyżów, Paszczyna, Pysznica, Giedlarowa, Krosno, Wolica, Dukla, Sanok, Żarnowiec, Leżajsk, Błażowa, Kozodrza, Jodłowa, Stalowa Wola and Krzeszów (Kalda and Wilk 2014).

Waste pressing on landfills is used when wastes are of the same kind, e.g. PE or PP. Sorted waste is pressed and prepared for storage or transport, while plastic waste not sorted or mixed is baled and stored. Waste crumbling is practiced to reduce volume and facilitate further processing. Shredded plastic waste can likewise be stored or directed for further processing. The processes are achieved at mills equipped with rotary shafts mounted with knives (Oleszkiewicz 1999; Kijeński et al. 2011).

3.2 Incineration of Polymer Waste

Incineration represents another way of utilizing the waste under study. Prior to 2018, Podkarpackie had no municipal waste incineration plant, given a lack of engagement in environmental protection and consequent inclination to simply dump waste at landfills. This method was anyway the cheapest, and required little in the way of investment. However, from 2018, Podkarpackie has had 5 incinerators of industrial waste in operation, at Rzeszów, Pustków, Bratkowice, Jedlicze and Tarnobrzeg. The process of incineration achieves a ca. 60% reduction in the mass of waste decreases around 60%, while in terms of volume only 10% remains. The process takes about 2 h at a temperature of 1000 °C, and the resulting heat can be used directly or used to produce electricity. Polyvinyl chloride has a low calorific value, as it is 56.7% chlorine. Furthermore, the combustion process generates large quantities of dioxins, which can be very dangerous. Toxic waste gases are in any case evolved, and these pose a further major threat to human health and the environment. Difficulties with both organization and public acceptance therefore ensure that storage is the most popular form of plastic waste management (Oleszkiewicz 1999; Kijeński et al. 2011).

3.3 Biodegradation of Installation Materials

Given that the greater part of all plastic waste generated is still at landfills, and occupying large amounts of space there, costs associated with storage are not insignificant. Plastic material ages automatically, with a major impact on biodegradation. However, polymers are durable, and may last even 1000 years. Degradation can be thermal and mechanical, as well as proceeding with the aid of irradiation, oxidation, biological or chemical degradation (Zheng et al. 2005). Unfortunately, the plastics used in water installations are (necessarily) hard to biodegrade (Shimao 2001; Sivan 2011). As of 2016, the mass of waste undergoing biodegradation amounted to 39,421.69 tonnes. By 2018, this had decreased to 20,466.05 tonnes. The amount of biodegradable waste in 2016 was 214,545.203t, as opposed to 174,508.284t in 2018. The decline may reflect increased use of non-biodegradable wastes, such as plastics. The calculations of the Podkarpackie Office for Spatial Planning in Rzeszów take no account of plastics, given the very prolonged nature of the biodegradation process. To put it another way, Podkarpackie does not see biodegradation of installation materials as a plausible method of disposal.

4 Methods of Identifying the Plastics Used in Installations in the Podkarpackie Region

One way of pre-identifying the process that has produced a polymer entails organoleptic testing (Dietrich 2013). This begins with determination of the external appearance, light transmittance, color and surface type of the material, and can be followed by auditory testing involving the dropping of the material onto a rigid base from a certain height, or by hitting the material with a different material. The quality and quantity of sound produced both need to be taken account of. The sounds may be deaf, metallic or damped, and is lower in tone the more elastic the material is. Much also depends on the shape of the object. The sense of touch may also prove useful in assessing the type of plastic, given that PE is soft and it can be scratched very easily, while PP exhibits far greater hardness and stiffness (Frącz 2011).

Another method entails the identification of plastics by reference to their behavior in an open flame. A sample of the plastic should be placed above a colorless flame for about 10 s, and it is not only the behavior of the plastics that should be observed, but also the smell of vapors and fumes emitted. As this method is applied, attention should also be paid to the type of flame and its color, the behavior of the material following removal from the flame and the flammability the material displays.

Another method involves the identification of plastics by reference to their solubility, notwithstanding the fact that limited solubility is a characteristic feature of polymers. PVC will nevertheless dissolve – in chloroform, cyclohexanol and carbon tetrachloride – among other things. In turn, PE can be dissolved by benzene and pyridine, as well as chloroform and carbon tetrachloride. As polymer solubility is dependent on molecular weight, it can be changed: if the crystallinity and molecular weight are higher, the solubility of the polymer is lower.

In the Podkarpackie region, such processes of identification generally require cooperation with specialized laboratories existing close to plants and enterprises. However, it has so far been only a rare necessity for detailed quantitative and qualitative analyses of plastics to be carried out. No specific information on processes of plastic identification in companies in Podkarpackie is available, and nor are methods deployed known in the public domain. The methods referred to above are in common use in various branches of industry across Poland, but there are no precise data on this where Podkarpackie is concerned (Kalda and Kula 2015).

5 Summary

The plastics deployed in water-supply and wastewater installations prove to be very good materials for the purpose. When it comes to the ultimate utilization of waste plastic from such sources, the situation in Poland's Podkarpackie region is primarily one in which polymer materials are recycled and processed into new products. However, it remains unknown what methods are deployed in Podkarpackie to achieve the necessary objective of determining the identities of given kinds of plastic. In general, new developments continue appear as regards both the use of plastics in installations, and the technologies used in both processing and production. More and more emphasis is now put on reducing amounts of wastes going to landfills, and this denotes increased involvement in recycling and recovery. And, while the results in this regard are still not very good, a steady improvement from year to year is to be noted. This goes hand in hand with growing environmental awareness among the people of Podkarpackie, with a consequent steady popularization of the idea of products being made from processed waste materials.

References

Awasthi AK, Shivashankar M, Majumder S (2017) Plastic solid waste utilization technologies: a review. In: 14th international conference on science, engineering and technology, IOP conference series: materials science and engineering. India, Vellore. https://doi.org/10.1088/1757-899x/263/2/022024

Bergel T, Kaczor G, Bugajski P (2013) Technical condition of water supply networks in small waterworks in the Małopolska and Podkarpackie voivodships. Infrastruct. Ecol. Countryside 3(4):291–304 (in Polish)

Boryczko K, Tchórzewska-Cieślak B (2013) Analysis and assessment of the risk of lack of water supply using the EPANET program. In: Environmental engineering IV. Taylor & Francis Group, London, pp 63–68

Bujak JW (2015) Thermal utilization (treatment) of plastic waste. Energy 90(2):1468–1477. https://doi.org/10.1016/j.energy.2015.06.106

Dietrich B (2013) Simple methods for identification of plastics. Carl Hanser Verlag, Munich

Frącz W (2011) Processing of polymer materials. Oficyna Wydawnicza Politechniki Rzeszowskiej, Rzeszów (in Polish)

Jayasekara R, Harding I, Bowater I, Lonergan G (2005) Biodegradability of a selected range of polymers and polymer blends and standard methods for assessment of biodegradation. J Polym Environ 13(3):231–251. https://doi.org/10.1007/s10924-005-4758-2

Kalda G, Kula K (2015) Problems of management of municipal and industrial waste in the Podkarpackie voivodship. J Civ Eng Environ Architect XXXII(62(3)):179–190 (in Polish)

Kalda G, Wilk M (2014) Analysis of industrial waste management in Podkarpacie. J Civ Eng Environ Architect XXXI(61(4)):109–123

Kijeński JK, Błędzki A, Jeziórska R (2011) Recovery and recycling of polymeric materials. PWN Scientific Publisher, Warsaw (in Polish)

Kowalska B, Kowalski D, Kwietniewski M, Musz A (2011) The impact of polyethylene pipe material on the quality of water in the water supply network. Current issues in the treatment and distribution of water, vol 2. Gliwice

Mortezania S, Othman F (2012) Cost analysis of pipes for application in sewage systems. Mater Des 33:356–361. https://doi.org/10.1016/j.matdes.2011.01.062

Oleszkiewicz J (1999) Exploitation of a landfill. Kraków, LEM PROJEKT s.c.(in Polish)

Piegdoń I, Tchórzewska-Cieślak B (2018) Seasonality of water supply network failure in the aspect of water supply safety. In: EKO-DOK 2018. Polanica Zdrój, E3S Web of Conferences, vol 44, pp 1–8. https://doi.org/10.1051/e3sconf/20184400140

Pietrucha-Urbanik K, Studziński A (2019) Qualitative analysis of the failure risk of water pipes in terms of water supply safety. Eng Fail Anal 95:371–378. https://doi.org/10.1016/j.engfailanal.2018.09.008

Rak J (2003) A study of the qualitative methods for risk assessment in water supply systems. Environ Prot Eng 29(3–4):123–134

Shimao M (2001) Biodegradation of plastics. Curr Opin Biotechnol 12(3):242–247. https://doi.org/10.1016/s0958-1669(00)00206-8

Sivan A (2011) New perspectives in plastic biodegradation. Curr Opin Biotechnol 22(3):422–426. https://doi.org/10.1016/j.copbio.2011.01.013

Stachurek I (2012) Problems with biodegradation of plastics in the environment. In: Scientific papers of the college of labor protection management in Katowice, vol 1, no (8), pp 74–108 (in Polish)

Szpak D, Tchórzewska-Cieślak B (2016) Water producers risk analysis connected with collective water supply system functioning. In: Dependability engineering and complex systems. Advances in intelligent systems and computing 470. Proceedings of the eleventh international conference on dependability and complex systems DepCoS-RELCOMEX. Springer, pp 479–489

Tchórzewska Cieślak B, Piegdoń I (2016) The method of identyfication the failure risk on water supply networks. J KONBiN 1(37):73–94

Tomboulian P, Schweitzer L, Mullin K, Wilson J (2004) Materials used in drinking water distribution systems: contribution to taste-and-odor. Water Sci Technol 49(9):219–226. https://doi.org/10.2166/wst.2004.0575
Zheng Y, Yanful EK, Bassi AS (2005) A review of plastic waste biodegradation. Crit Rev Biotechnol 25(4):243–250. https://doi.org/doi.org/10.1080/07388550500346359
Zimoch I, Łobos E (2012) Comprehensive interpretation of safety of wide water supply systems. Environ Prot Eng 38(3):107–117. https://doi.org/10.5277/epe120310

Engineering Method of Calculation of Beam Structures Inclined Sections Based on the Fatigue Fracture Model

V. Karpiuk, Yu. Somina(✉), and O. Maistrenko

Department of Reinforced Concrete Constructions and Transport Structures,
Odessa State Academy of Civil Engineering and Architecture,
Didrihsona Street 4, Odessa 65029, Ukraine
syomina3091@ukr.net

Abstract. The *aim of this paper* is the creation of engineering method of inclined sections of beam structures calculation under the action of cyclic loads, that takes into account the concrete and reinforcement fatigue strength. *Methods of research* are key expressions determining the endurance limits of concrete, longitudinal reinforcement, and anchoring longitudinal reinforcement, which based on the real physical and calculated models of beam structures inclined sections work. *The results of work* links with the creation of expressions determining the boundary forces that can be carried out by structural element under the action of cyclic loading. *Scientific novelty and practical significance* is that presented engineering method takes into account the change of stress-strain state of span reinforced concrete structures at their repeated load, as well as, a change of strength properties of the concrete, reinforcement and their clutch at the moment *t*, and can be used on all range of strength characteristics of materials change- from low-cycle repeated, to repeated cycle load.

Keywords: Reinforced concrete · Beam · Inclined section · Cyclic load · Fatigue strength

1 Introduction

The experience of designing, construction and exploitation of span reinforced concrete structures shows that, practically, all of them work at complex stress-strain state. At the same time, the researchers pay much more attention to the calculation of strength, deformability and crack-resistance of elements in normal sections, than to the calculation of their areas near supports, including the inclined sections. Besides, a lot of span reinforced concrete elements are subjected to the action of low-cycle loads. They lead to such specific features of concrete work as nonlinearity of deformation, micro cracking, accumulation of residual deformations, low-cycle fatigue (fatigue damage), decompression of concrete etc. The results of the numerous researches of Karpiuk et al. (2018a, 2018b), Aslani and Jowkarmeimandi (2012), Trapko and Trapko (2012), Gopinath et al. (2015), Naghibdehi et al. (2015), Gomon (2009), Kuhniuk (2001), Babich (1997), Karpiuk et al. (2018a, 2018b) show that the destruction of structures

Z. Blikharskyy et al. (Eds.): CEE 2019, LNCE 47, pp. 135–144, 2020.
https://doi.org/10.1007/978-3-030-27011-7_17

under the action of low-cycle loads occurs at the smaller stresses, than the destruction of structures under the action of short duration static loads. Therefore, the *aim of author's researches* is the creation of engineering method of calculation of beam structures inclined sections under the action of cyclic loads, that takes into account the concrete and reinforcement fatigue strength.

2 Methods

It is known that the change of the material's strength under cyclic loading in half-logarithmic coordinates $f_{cd} = lgN$ is described by linear dependence. The endurance line is characterized by inclined and horizontal areas. Therefore, the characteristic points of the endurance line are the beginning point and the points of its inflection.

For concrete, the beginning of the endurance line is a point on the axis of stress (Fig. 1) at $N = 1$, which corresponds to its dynamic strength at a one-time load at a speed equal to the speed of cyclic load applied. Herewith, the larger their frequencies, the greater the strength at a one-time dynamic load.

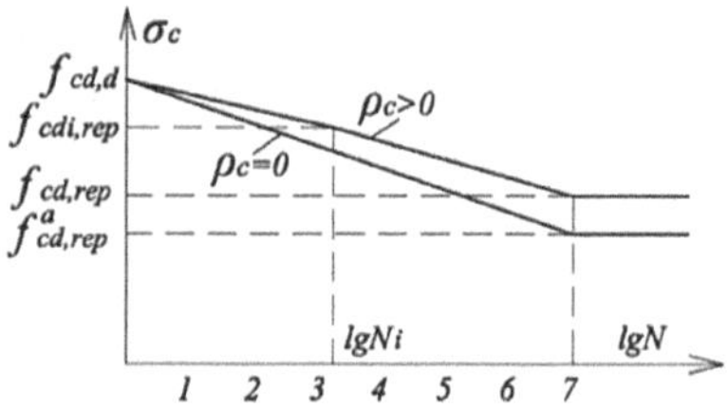

Fig. 1. Calculated endurance concrete lines

In practical calculations, we accept:

$$f_{cd,d} = k_d \cdot f_{cd}, \tag{1}$$

where k_d, f_{cd} – coefficient of dynamic strengthening of concrete and its strength at static loading.

According to the proposition of Kirillov (1978), the coefficient of dynamic strengthening of the concrete is determined by the formula:

$$k_d = 1 + 0,6 \cdot \frac{0,27 + 0,8[th(0,15lg\upsilon) - 0,358]}{1 - 0,358tg(lg\upsilon)}, \tag{2}$$

and taking into account the plastic resource at the previous stages of loading:

$$k_d = 1 + 0,085 lg\upsilon \left\langle lg \left\{ \begin{array}{l} c_\infty(t,\tau) - \dfrac{c_\infty(t,\tau)\sigma_{c1}^{\max}(1-\rho_{c1})\left[1-(1-a)^N\right]10^5}{f_{cd}} + \\ + \dfrac{\sum_{i=2}^{k} \Delta\sigma_c c_\infty(t,\tau)(1-\rho_{b1})\left[1-(1-a)^N\right]10^5}{f_{cd}} \end{array} \right\} \right\rangle, \quad (3)$$

For concrete, the absolute endurance limit at $\rho_c = 0$ in the work of Kirillov (1978) varies within (0,47…0,55)f_{cd}, and coincides with the lower limit of micro cracking. Therefore, in practical calculations, the absolute limit of endurance $f_{cd,rep}^a$ at $\rho_c = 0$ equals 0,5f_{cd}, and the relative endurance limit in this case is:

$$k_{c,rep}^a = f_{cd,rep}^a / f_{cd} = 0,5. \quad (4)$$

The relative values of concrete endurance at tensile, shearing and torsion are taken the same. Since the lower limit of micro cracking f_{crc}^0 depends only on the level of the action stress and the type of stress state, then for any ρ_c, the endurance limit will be the same, and its value will have an effect on the structure endurance, that is, on the cycles number at which the endurance lines will come to an end. As known, the overlap of these lines begins at $N \geq 10^7$. For $\rho_c = 0$, it can be assumed $N = 10^7$, which creates some reserve. For larger values of ρ_c this point moves to the right along the lgN axis, the greater, the larger ρ_c:

$$k_{c,rep} = \frac{f_{cd,rep}}{f_{cd}} = \frac{k_{c,rep}^a}{1 - \rho_c\left(1 - \frac{k_{c,rep}^a}{k_d}\right)}, \quad (5)$$

and endurance strength at $N < 10^7$

$$f_{cdi,rep} = f_{cd,d} - \frac{lgN_i}{7}\left(f_{cd,d} - f_{cd,rep}\right) \quad (6)$$

or in relative values taking into account Eqs. (1) and (6)

$$k_{ci,rep} = 1,3 - \frac{lgN_i}{7}\left(1,3 - \frac{0,5}{1 - 0,616\rho_c}\right). \quad (7)$$

Taking k_d = 1,3 and considering Eq. (4) we get:

$$k_{ci,rep} = 1,3 - \frac{lgN_i}{7}\left(1,3 - \frac{0,5}{1 - 0,616\rho_c}\right). \quad (8)$$

For the reinforcement, Kirillov (1978) recommends to accept the following expression as the beginning of the endurance line:

$$f_{yd,d} = \eta \cdot \sigma_u, \tag{9}$$

and the overlap of the endurance line (Fig. 2) in the point with coordinates

$$f_{yd,rep} = \sigma_u \frac{k_0 \cdot k_c \cdot k_r}{1 - \rho_s \left(1 - \frac{k_0 \cdot k_c \cdot k_r}{\eta}\right)}, lgN = 6,3, \tag{10}$$

where $k_0 = \frac{f_{yd0,rep}}{\sigma_u}$ – relative reinforcement endurance limit at $\rho_s = 0$;

k_c – a coefficient that takes into account the presence of a weld joint or other stress concentrator;
k_r – a coefficient that takes into account the reinforcement diameter;
σ_u – temporary reinforcement resistance to rupture;
$\eta = 1,8$ – empirical coefficient.

The element's fatigue strength for the number of cycles $N < 2 \cdot 10^6$ is:

$$f_{ydi,rep} = f_{yd,d} - \frac{lgN_i}{lg(2 \cdot 10^6)} \left(f_{yd,d} - f_{yd,rep}\right), \tag{11}$$

and taking into account Eqs. (9) and (10) in relative values, it has the form:

$$k_{ydi,rep} = \eta - \frac{\lg N_i}{6,3} \left[\eta - \frac{k_0 \cdot k_c \cdot k_r}{1 - \rho_s \left(1 - \frac{k_0 \cdot k_c \cdot k_r}{\eta}\right)}\right]. \tag{12}$$

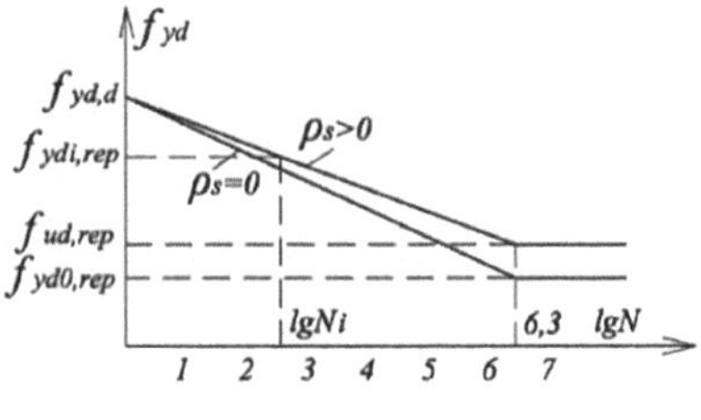

Fig. 2. Calculated endurance reinforcement lines

In practical calculations the stress change, which occurs as a result of development of vibro-creep deformation of compressed concrete, in conditions of complex stress state, takes into account the functions of stress accumulations in concrete H_{σ_c}, longitudinal H_{σ_s} and transverse H_{σ_w} reinforcement. These are functions of concrete vibro-creep deformations which are calculated on the theory of vibro-creep by Bondarenko (1966).

Actual stresses in concrete, longitudinal and transverse reinforcement, at the moment of time *t*, on the base of physical models of Karpiuk et al. (2017), in calculations, are presented in the form:

$$\sigma_c^{\max}(t) = \sigma_c^{\max}(t_0)\cdot H_{\sigma_c};\ \sigma_s^{\max}(t) = \sigma_s^{\max}(t_0)\cdot H_{\sigma_s};\\ \sigma_{sw}^{\max}(t) = \sigma_{sw}^{\max}(t_0)\cdot H_{\sigma_{sw}}, \tag{13}$$

where $\sigma_c^{\max}(t_0)$; $\sigma_s^{\max}(t_0)$; $\sigma_{sw}^{\max}(t_0)$ – initial stresses, corresponding, in concrete, longitudinal and transverse reinforcement.

At average shear spans taking into account Eq. (1) fatigue conditions are presented as:

$$\sigma_{1c}^{\max}(t_0) \le f_{cdc,rep}(t)/H_{\sigma_c},\ \sigma_{sw,\alpha}^{\max}(t_0) \le f_{ydsw,rep}(t)/H_{\sigma_{sw}},\quad \sigma_{s6}^{\max}(t_0) \le f_{ydsq,rep}(t)/H_{\sigma_s},\\ \sigma_s^{\max}(t_0) \le f_{ydan,rep}(t)/H_{\sigma_s}. \tag{14}$$

Fatigue strength of compressed concrete zone over critical inclined crack in reinforced concrete elements is determined respectively:

$$f_{cd,rep} = \frac{f_{ydi,rep}\cdot A_{sw}\cdot\cos\beta\cdot\left(\dfrac{h\cdot ctg\varphi}{l_{\sup}\sin^2\beta} - ctg\varphi\cdot ctg\beta - \cos^2\varphi\right)}{H_{\sigma_{sw\beta}}\cdot b\cdot s\cdot\left\langle 1 - L_\varepsilon\cdot\left(\dfrac{G_c}{\sin^2\varphi} + \dfrac{6E_sI_s\cdot n\cdot\cos(\varphi-\beta)\sin\beta}{b\left(d_s\cdot\sqrt[4]{\dfrac{E_s}{E_c}}\cdot\left(1{,}4+1{,}25\sqrt[4]{\dfrac{a_s}{d_s}}\right)\right)^3\sin\varphi}\right)\cdot\left(\dfrac{1}{E_c}+H_\varepsilon\right)\right\rangle} \tag{15}$$

where $f_{cdi,rep}$; $f_{ydi,rep}$ – fatigue strengths of concrete and reinforcement at tension;
H_ε – function of deformations accumulations in concrete at repeated loads;
E_s, E_c, G_c – modulus of elasticity of reinforcement and concrete, and concrete shear modulus.

Longitudinal reinforcement fatigue strength in the place where it intersects with inclined crack in conditions of flat stress state is determined as:

$$f_{ydq,rep}(t) = \frac{\sigma_u\left\langle k_{ds} - \left\{k_{ds} - k_o k_c k_r/\left[1-\rho_s\left(k_o k_c k_r/k_{ds}\right)\right]\right\}\cdot\lg N_i/6.3\right\rangle}{\sqrt{1+3\left(\tau_s^{\max}/\sigma_{s6}^{\max}\right)^2}}. \tag{16}$$

Longitudinal reinforcement anchoring fatigue strength has the form:

$$f_{yd,an}(t,\tau) = f_{cdi,rep} B_o (d + 2c_r) L / f_{cd} \cdot d^2, \quad (17)$$

where B_o – reference clutch parameter.

Fatigue strengths of concrete and reinforcement in free conditions at axle load at compression by the equation of Kirillov (1978):

$$f_{cdi,rep} = f_{cd} \left\langle k_{dc} - \left(k_{dc} - k^a_{c,rep} \Big/ \left\{1 - \rho_c \left[1 - \left(k^a_{c,rep} \big/ k_{dc}\right)\right]\right\}\right) \cdot \lg N_i \big/ 7 \right\rangle, \quad (18)$$

for concrete at tension:

$$f_{cdti,rep} = f_{cdt} \left\langle k_{dc} - \left(k_{dc} - k^a_{c,rep} \Big/ \left\{1 - \rho_c \left[1 - \left(k^a_{c,rep} \big/ k_{dc}\right)\right]\right\}\right) \cdot \lg N_i \big/ 7 \right\rangle, \quad (19)$$

for reinforcement:

$$f_{ydi,rep} = \sigma_u \langle k_{ds} - \{k_{ds} - k_o k_c k_r / [1 - \rho_s (k_o k_{CT} k_r / k_{ds})]\} \cdot \lg N_i / 6,3 \rangle, \quad (20)$$

where k_{dc} and k_{ds} – factors of dynamic strengthening of concrete and reinforcement;
ρ_c and ρ_s – factors of stress cycle asymmetry in concrete and reinforcement;
$k^a_{c,rep} = f^a_{c,rep} / f_{cd}$ – concrete absolute fatigue strength;
$k_o = f_{yd0,rep} / \sigma_u$ – reinforcement relative fatigue strength at $\rho_s = 0$;
k_{CT} – factor that takes into account the presence of welding or another stress concentrator.

3 Results

Based on conditions of concrete endurance over danger inclined crack, longitudinal and transverse reinforcement we got boundary forces that can be carried out by structural element with ***medium shear span***:

$$P_{lim,2h_0}^{min} \leq \begin{cases} \dfrac{k_{w\beta,rep}\cdot\sigma_u\cdot A_{sw}\cdot cos\beta\cdot\left(\dfrac{h\cdot ctg\varphi}{l_{sup}\cdot sin^2\beta}-ctg\varphi\cdot ctg\beta-cos^2\varphi\right)l_{sup}\cdot sin^2\beta}{H_{w\beta}\cdot s\left\langle 1-L_\varepsilon\left\{\dfrac{G_c}{sin^2\varphi}+\dfrac{6E_s\cdot I_s\cdot n\cdot cos(\varphi-\beta)sin\beta}{b\left[d_s\cdot\sqrt[4]{\dfrac{E_s}{E_c}}\cdot\left(1{,}1+1{,}254\sqrt[4]{\dfrac{a_s}{d_s}}\right)\right]^3 sin\varphi}\right\}\cdot\left(\dfrac{1}{E_c}+H_\varepsilon\right)\right\rangle} \\ or \\ \dfrac{k_{wi,rep}\cdot\sigma_u\cdot\omega_{sw}^1\cdot b\cdot h_0}{H_{w\alpha}\cdot E_{sw}\cdot ctg\alpha\left\{\dfrac{1}{G_c}+\dfrac{2\left(c_0-x_1ctg\alpha-0{,}5l_{sup}\right)}{E_c\cdot h_0\left[(1+\lambda)-0{,}33\xi_1\left(1+\lambda+\lambda^2\right)\right]}\cdot\dfrac{h_0-x}{x}tg\varphi\right\}}, \\ or\quad \dfrac{k_{si,rep}\cdot\sigma_u\cdot A_s}{H_{\sigma_s}\cdot j_1}, \\ or \\ \dfrac{k_{ci,rep}\cdot B_0\left(d+2c_r\right)L\cdot A_s}{H_{\sigma_s}\cdot d^2\cdot j_1}. \end{cases} \tag{21}$$

$$\text{where } j_1=\frac{\kappa_1+\dfrac{\left(c_0-a_1-0{,}5Xctg\alpha+\lambda\cdot z_1 sin2\varphi\right)}{a_1}\left(\kappa_1+\kappa_2\right)+\left(1+\kappa_3\right)ctg\alpha}{\left(1+\lambda\right)\left(\kappa_1+\kappa_2\right)\dfrac{z_1}{a_1}+\left(1+\kappa_3\right)\cdot ctg\alpha}; \tag{22}$$

$$j_2=\frac{\kappa_2+\left(1+\kappa_3\right)\dfrac{\left(c_0-0{,}5Xctg\alpha\right)}{z_1}\cdot ctg\alpha}{\kappa_1+\kappa_2+\left(1+\kappa_3\right)\dfrac{a_1}{z_1}\cdot ctg\alpha}; \tag{23}$$

For determination of boundary forces that can be carried out by structural element under multi-cycle loading it should be used $k_{ci,rep}=0,5$; and $k_{si,rep}$ – taking into account Eq. (10).

Relative fatigue strength of concrete at limited number of cycles $(N<10^7)\,k_{ci,rep}=0,5$ is determined by Eq. (7), and relative fatigue strength of reinforcement $k_{si,rep}$ – by Eq. (12).

Comparison of calculation and actual values of samples-beams inclined sections bearing capacity in the work of Karpiuk et al. (2018a, 2018b) showed that, despite the complete coincidence of physic models and actual destruction pictures, the destructive uploading shear force V_{u3} is 1,91…7,92 times higher than the predicted shear effort of concrete endurance over danger inclined crack and transverse reinforcement $P_{lim,c,sw}$, determined by the recommendations of Mirsayapov (2009) and Kirillov (1978). The reason of such divergence is a higher destructive influence of multi-cycle load, compared to low-cycle repeated shear load, undervaluation of actual influence of different design factors in indicated works.

For this problem solution on common stage of research boundary force $P_{lim,2h_0}^{min}$ it is recommended to multiply by the factor k_{cycl}. The resistance of support zones of span reinforced concrete structures, in particular, calculation value of shear force V_{Rd}, which can be carried by inclined section of such structure is expedient to determine by taking into account its endurance at low-cycle repeated load action by expression:

$$V_{Rd} = k_{cycl} \begin{Bmatrix} P_{lim,h_0}^{min} \\ P_{lim,2h_0}^{min} \\ P_{lim,3h_0}^{min} \end{Bmatrix}, \quad (24)$$

where k_{cycl} – ratio of destructive shear forces to minimum calculation boundary forces of endurance of support zones of samples-beams.

Mathematical model of factor $\hat{k}_{cycl}$, that characterizes the ratio of destructive (uploading) shear force V_{u3} to calculation value of boundary force, at a condition of concrete endurance over danger inclined crack and transverse reinforcement in shear span $P_{lim,c,sw}$ at low-cycle repeated load of testing samples-beams, has the form:

$$\begin{aligned} \hat{Y}(k_{cycl}) = {} & 3,84 + 0,44X_1 - 0,69X_2 - 1,68X_3 - 0,12X_4 + 0,11X_1^2 + 0,06X_2^2 \\ & + 0,31X_3^2 - 0,08X_4^2 - 0,34X_1X_2 + 0,25X_2X_3, \upsilon = 5,1\%. \end{aligned} \quad (25)$$

The highest influence on the value of factor k_{cycl} (Fig. 3) has the quantity of transverse reinforcement, then – concrete class, then – value of relative shear span and, finally, – the level of low-cycle repeated load.

For convenience of the practical application of factor k_{cycl}, it is expediently to change a form of Eq. (25) by changing coding values of testing factors with natural ones:

$$\begin{aligned} k_{cycl} = {} & 3,84 + 0,44\left(\frac{a/h_0 - 2}{1}\right) - 0,69\left(\frac{C - 35}{15}\right) - 1,68\left(\frac{\rho_{sw} - 0,0029}{0,0014}\right) \\ & - 0,12\left(\frac{\eta - 0,65}{0,15}\right) + 0,11\left(\frac{a/h_0 - 2}{1}\right)^2 + 0,06\left(\frac{C - 35}{15}\right)^2 + 0,31\left(\frac{\rho_{sw} - 0,0029}{0,0014}\right)^2 \\ & - 0,08\left(\frac{\eta - 0,65}{0,15}\right)^2 - 0,34\left(\frac{a/h_0 - 2}{1}\right)\left(\frac{C - 35}{15}\right) + 0,25\left(\frac{C - 35}{15}\right)\left(\frac{\rho_{sw} - 0,0029}{0,0014}\right). \end{aligned} \quad (26)$$

Equation (26) is fair, not only inside the change of testing factors, but its extrapolation is possible on the value up to 25% from the values of their intervals.

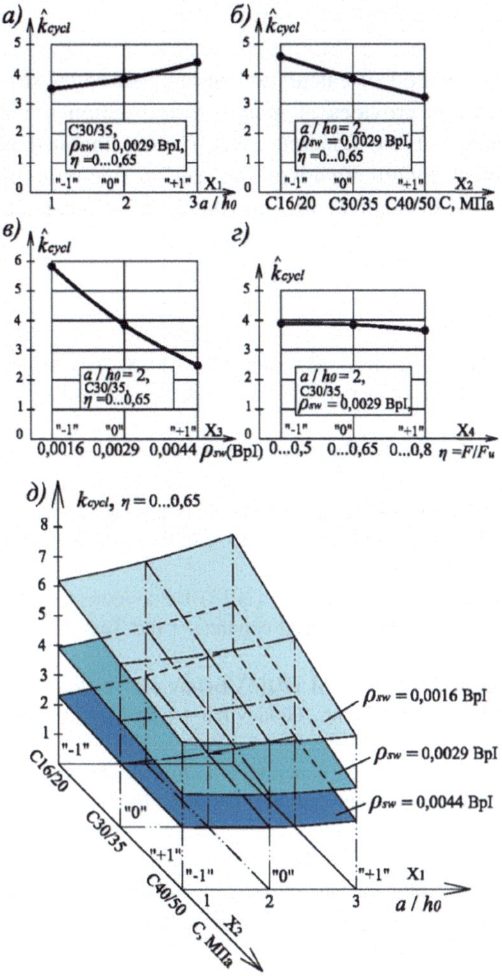

Fig. 3. Factor k_{cycl} dependency on relative shear span (a), concrete class (b), quantity of transverse reinforcement (c), level of low-cycle load of constant sign (d) and joint influence of testing factors on it

4 Scientific Novelty and Practical Significance

Presented engineering method takes into account the change of stress-strain state of span reinforced concrete structures under repeated load, as well as, a change of strength properties of the concrete, reinforcement and their clutch at the moment t, and can be used on all range of material's strength characteristics change from low-cycle load to multi-cycle repeated loading.

5 Conclusions

Proposed engineering method of calculation of support zones strength of beam reinforced concrete structures provides all possible destruction schemes, and allows to predict fatigue strength of the concrete over danger inclined crack, and presents transverse reinforcement, longitudinal working reinforcement on rupture or slippage on this crack because of its inadequate amount (break) or its inadequate anchorage.

References

Karpiuk VM, Somina YuA, Kostiuk AI, Maistrenko OF (2018a) Features of the stress-strain state and calculation of reinforced concrete structures under cyclic loading of high levels. OSABA, Odesa (in Ukrainian)

Aslani F, Jowkarmeimandi R (2012) Stress-strain model for concrete under cyclic loading. Mag Concr Res 64(8):673–685. https://doi.org/10.1680/macr.11.00120

Trapko W, Trapko T (2012) Load-bearing capacity of compressed concrete elements subjected to repeated load strengthened with CFRP materials. J Civ Eng Manag 18(4):590. https://doi.org/10.3846/13923730.2012.701664

Gopinath S, Murthy R, Relyer N, Prabha M (2015) Behaviour of reinforced concrete beams strengthened with basalt textile reinforced concrete. J Ind Text 44(6):924. https://doi.org/10.1177/1528083714521068

Naghibdehi M, Naghipour M, Rabiee M (2015) Behaviour of functionally graded reinforced concrete beams under cyclic loading. Gradevinar 67(5):427–439. https://doi.org/10.14256/JCE.1124.2014

Gomon PS (2009) Work of reinforced concrete T-section beams under the action of repeated loading, new materials, equipment and technologies in industry, Mogilyov, Ukraine (in Russian)

Kuhniuk OM (2001) Influence of low-cycle loads on mechanical characteristics of concrete and on the work of bending reinforced concrete elements (abstract of the dissertation for obtaining the scientific degree of PhD, specialty 05.23.01) "Building constructions, buildings and structures", Rivne (in Ukrainian)

Babich EM (1997) Investigation of the work of beams with mixed reinforcement under the action of low-cycle loads of high levels, actual problems of water economy, Rivne, Ukraine, 21–23 October 1997 (in Ukrainian)

Karpiuk VM, Kostiuk AI, Somina YuA (2018b) Models of bay reinforced concrete elements resistance at action of cycle permanent sign high level forces. Ind Mach Build Civ Eng 1 (50):112–123. https://doi.org/10.26906/znp.2018.50.1066

Kirillov AP (1978) The endurance of hydrotechnical reinforced concrete. Energy, Moscow (in Russian)

Bondarenko VM (1966) About vibro-creep deformations of concrete. Struct Strength Deform Concr 344–351 (in Russian)

Karpiuk V, Kostiuk A, Maistrenko O, Somina Yu (2017) Influence of intermittent cyclic loading on reinforced concrete resistance model. Electron J Fac Civ Eng Osijek Croat 15:59–74. https://doi.org/10.13167/2017.15.6

Mirsayapov IT (2009) Endurance of reinforced concrete structures under the action of shear forces (abstract of the dissertation for obtaining the scientific degree of ScD, specialty 05.23.01 "Building constructions, buildings and structures". Kazan (in Russian)

Assessment of the Working Environment in Terms of Visual Perception

D. Katunsky[1(✉)] and E. Dolnikova[2]

[1] Department of Architecture and Building Structures, Faculty of Civil Engineering, Technical University of Kosice, Vysokoskolska 4, 042 00 Kosice, Slovakia
dusan.katunsky@tuke.sk

[2] Department of Building Physics, Faculty of Civil Engineering, Technical University of Kosice, Vysokoskolska 4, 042 00 Kosice, Slovakia
erika.dolnikova@tuke.sk

Abstract. Natural daylight is a vital element in our life and it is very important at work. Providing natural daylight into the working environment is one of fundamental importance for the comfort, efficiency and safety for the people in that environment. This paper deals with visual comfort evaluation in administrative building, in office spaces. The human eye is able to perceive light at the moment it occurs. It cannot accumulate or move it in the human body for future use. Light perception is a complex process that is described by photometric quantities. It is important to work with variables in the theory and practice of daylight, enabling the creation of a healthy and safe environment for people to work and relax, and for the energy efficient management of buildings. One case study is presented in this contribution. It was measured in real situation an also calculated by the Radiance simulation program and Velux Daylight Visualizer.

Keywords: Daylight · Visual comfort · Glare · Luminance · Radiance

1 Introduction

Light is the only elementary part of life that has become an almost irreplaceable part of life. Because man is more connected with the interior, the task is to ensure that natural daylight in the interior is as high as possible. Lighting in working places is currently at a level that in many cases does not meet the requirements of legislation and standards. Optimal environmental conditions depend on the purpose for which the environment is intended. Since the top lighting is largely involved in the overall lighting climate, the purpose of the paper is to evaluate the visual comfort in a selected administrative room with two types of glazing and positions of observed places. In the study of the measured values in the room were used as boundary conditions in the simulation program RADIANCE and program Velux. The program Radiance evaluated the visual comfort for two types of illumination by the Guth VCP method. The results show that the use of diffuse glazing instead of simple wire glazing makes the brightness and glare levels smaller, but the satisfaction of people with visual comfort is greater with normal glazing. Furthermore, the model of the room can be used to calculate light conditions

Z. Blikharskyy et al. (Eds.): CEE 2019, LNCE 47, pp. 145–152, 2020.
https://doi.org/10.1007/978-3-030-27011-7_18

for other types of illumination as well as for comparison with other types of glazing. An alternative to diffusing glazing could also be verified for side windows.

2 Visual Requirements in Working Places

The current European technical standard for interior workroom lighting STN EN 12464-1 supports complex solutions. The priority is to create optimal conditions for visual performance, in a specific working environment, so that visual comfort is maximized. Daylight is certainly not a special light source for vision, and the link between improved performance cannot be reliably shown (Boyce 2004). Lighting design for office buildings has focused largely on providing sufficient light for visual performance, minimal glare, good colour rendering and energy conservation (Figueiro and Rea 2014). In the work environment where the visual task is performed, the brightness of the surfaces located there is important. It is determined on the basis of reflectivity and surface illumination. Too high brightness contrasts your eyesight. High reflectivity floors have an adverse effect on brightness distribution (Katunsky et al. 2018). The current visual comfort system is mostly focused on office buildings. However, there is limited knowledge about daylight quality and visual well-being in industrial work environments (Tregenza and Wilson 2011) Various factors play an important role in assessing the appropriate working conditions (see Fig. 1) (Linhart and Scartezzini 2011). It depends on the quality of the lighting. It is essential that there is no glare at working place.

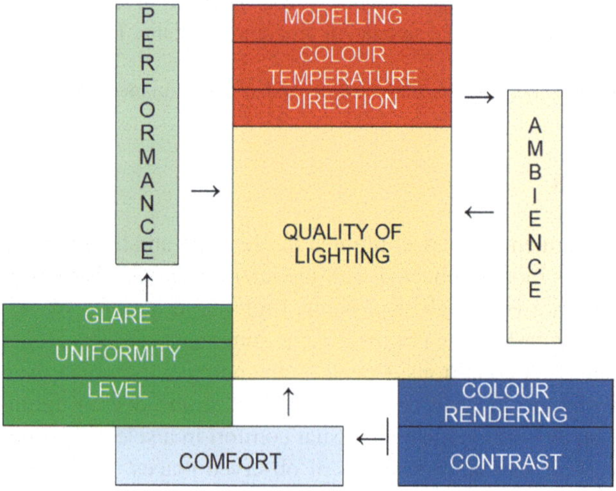

Fig. 1. Factors affecting proper lighting

3 Experiment in Situ, Measurements and Calculations

The Measurement of daylighting was carried out in an office in Kosice, Slovakia. The interior of office was modelled using Radiance with dimensions 3.5 m (width) × 5.5 m (depth) × 2.7 m (height), with the reference workplace at 750 mm. The height of parapet is 900 mm. Light loss coefficient due to window construction was considered $\tau = 0.63$. The room can be used for medium-precision work with different kinds of work and therefore the room is included in III - IV class of visual activities. The office is oriented in the south direction. Vertical height is assumed to be the third floor of an office building. The materials of the walls, ceiling and floor of this model take into account visual comfort with the following values for reflectance: walls 0.7, carpet 0.2, and ceiling 0.7. Light loss coefficient due to window construction was desired $\tau = 0.63$. The basic system consists of a single view window made of double plastic glazing and with dimensions 1200 mm × 1800 mm. The height of parapet is 900 mm. Measurement of daylighting was carried out according to Slovak standard STN 36 0020. The instruments were used two Data Logger "ALMEMO 2690-10 A + Illuminance Sensor ALMEMO FLA 623VL". The instrument was used to measure the Luminance meter LS-110 with an accuracy of 2%. Measurements at office were recorded in April. On selected days, the value of the external horizontal illumination at a uniformly cloudy sky ranged from 10,500 lx to 25,000 lx.

Table 1. The maximum permissible brightness (luminance) of the subject observed and the brightness of the illuminating opening positioned 60° from the normal viewing direction

Class of visual activities	The ratio of the luminance of the observed object to the luminance of the illuminating opening (windows)
I, II, III	1:40
IV	1:100
V, VI, VII	Not intended

To reduce or avoid glare, the brightness of a window must be less than 4000 cd/m^2 and 60 cd/m^2. It can be in the visual field of the observer for side-lit and top-lit illuminations, respectively. The observed ratio of luminance detail in the sky brightness should be less than 1:200th (Wirz and Fournier 2010). The resulting luminance values can be evaluated e.g. according to the criteria of maximum allowed luminance in the field of view of the observer between the observed object and the background, distant dark and light objects, the value of the glare factor and the ratio of the brightness of the observed object to the sky (see Table 1). It should be noted that the brightness of the lighting aperture itself should not exceed the standard value (Matterson 2013). An outstanding feature of the program is its ability to carry out visual comfort analyses. Different glare indexes can be calculated for a given rendered picture, including: DGP, DGI, UGR, VCP and other. In the contribution, the Guth VCP method is computed. Guth VCP calculations result in an index that describes the percentage of the people occupying a certain position. An acceptable value is usually 70% value (Wienold 2004).

4 Results and Discussion

This article presents the results of measurements for two days for points 1–3. The points are 1.75 m apart and 1 m from the wall in the longitudinal direction. In the second direction they are in the middle of the room. Measurements were made at table position “A”. In the table position “B” only the simulation was realized (see Fig. 2).

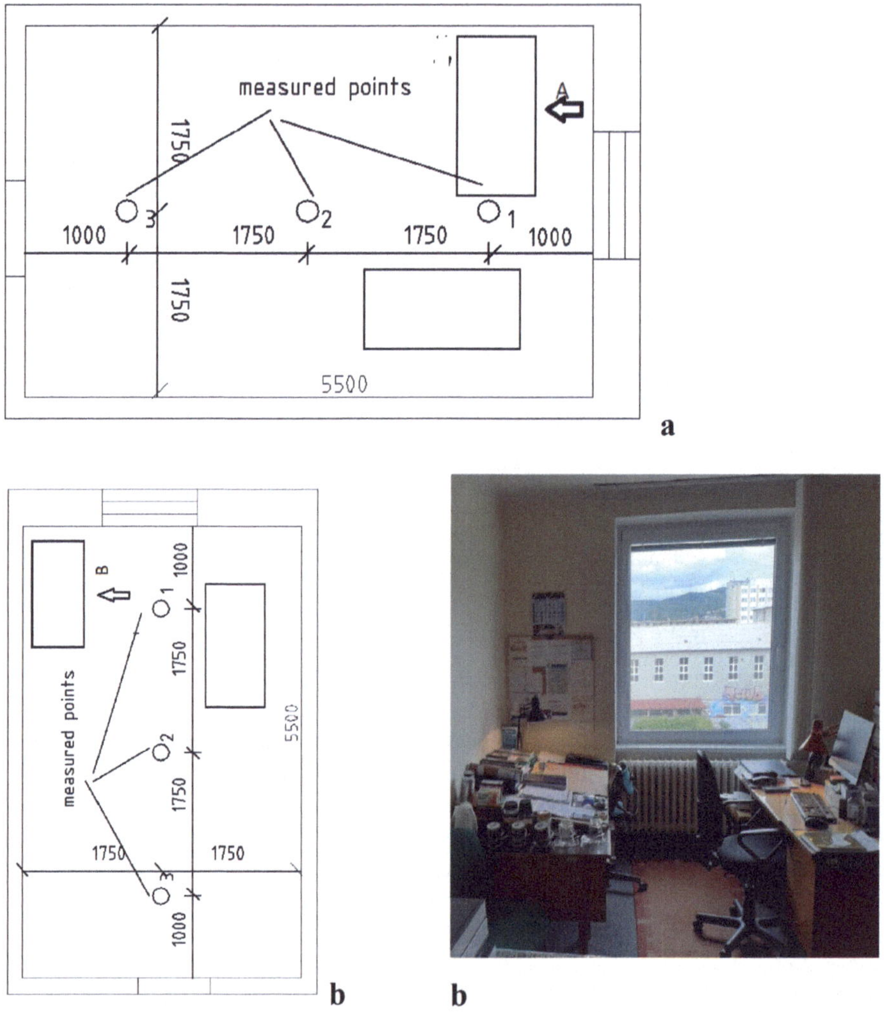

Fig. 2. Measured points in office room; **(a)** table position at A; **(b)** table position at B

For illuminance analysis the Perez sky model was used in software, using weather data as source for sky condition information. Daylighting simulations were performed with Velux Daylight Visualizer 3 and RADIANCE simulation program. Calculation of luminance is based on the model created by simulation and daylighting of the area. The brightness values are then used in the selected glare calculation procedure. The Fig. 3 shows Iso-valor curves using by simulation programs at position A. The Fig. 4 shows Iso-valor curves using by simulation programs at position B. The simulation program Radiance also allows qualification of luminance distribution by providing three dimensional displays that are made according to the eye's sensitivity (non-linear-function).

The Fig. 5 shows Luminance values (cd/m^2). Potential glare sources are identified (they are shown by circles in the Fig. 6). If we compare the brightness results in both programs, there are differences in some places and in some places the values are closer (E.g. window brightness). This is because Velux Daylight Visualizer is a much simpler program and has easier defined boundary conditions. But it can serve as a guideline for visual comfort assessment in working places.

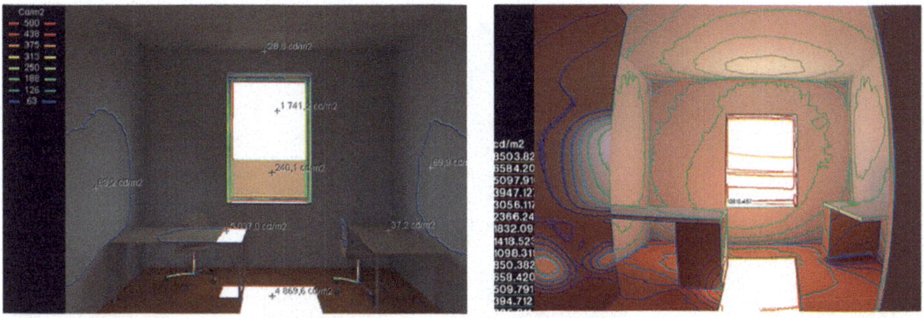

a b

Fig. 3. Results of luminance level (cd/m^2), south view, table position at **A** – Iso-valor curves; (**a**) – calculated by Velux Daylight Visualiser; (**b**) – calculated by Radiance

a b

Fig. 4. Results of luminance level (cd/m^2), south view, table position at **B** – Iso-valor curves; (**a**) – calculated by Velux Daylight Visualiser; (**b**) – calculated by Radiance

It can be concluded from Figs. 3, 4, 5 and 6 that the brightness ratios of the individual jobs under consideration are not within the maximum allowed brightness values. At the same time, there is glaring (see Fig. 6). According to standard the cut-off

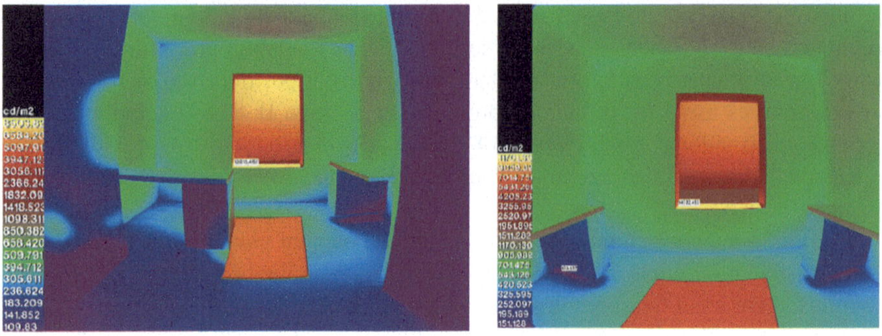

a b

Fig. 5. Luminance values (cd/m^2); (**a**) – position A; (**b**) – position B

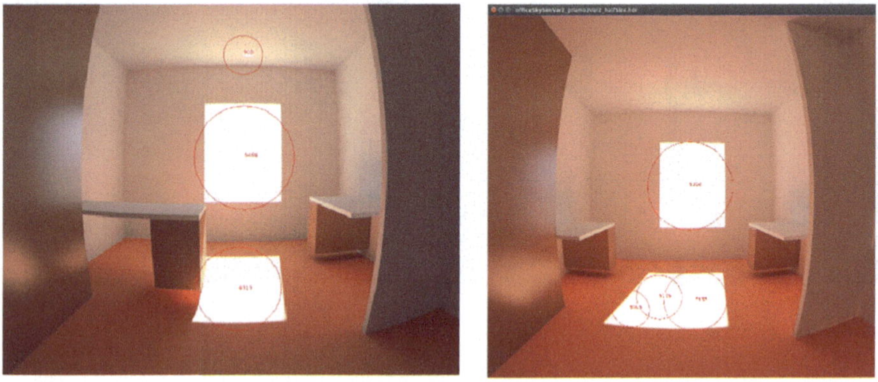

a b

Fig. 6. Potencial glare sources (they are shown by circles); (**a**) – position A; (**b**) – position B

value of the UGR is for offices 19. In the dazzling eyes can also occur from windows in sunny situations or from high brightness in bright clouds. Depending on the glare index of DGP at position A, the satisfaction of people in terms of glare would be 30%, at table position B 23% (see Table 1). Table 1 shows that values of glare index (DGI and UGR) are exceeded, that is, visual comfort is impaired.

Radiance also calculated the Guth VCP Glare Index. For position A, on average, would be satisfied from 5% to 100% according to the direction of view. In position B, this satisfaction would be from 80% to 100% according to the direction of view (Table 2).

Table 2. Calculated glare indices for both A and B

Glare index	Position A	Position B
DGP	0,30	0,23
DGI	23,25	−14,73
UGR	27,33	−10,74

5 Conclusions

Visual well-being and productivity are closely related. It is not surprising, since the better the visual comfort, the higher the productivity of the workers. In addition to visual comfort, it affects the product: size, contrast, brightness of the source and the time it takes. The greater the impacts are the more productivity.

In the calculation, two different table positions were considered. The visual comfort parameters are evaluated by modelling the office space in the software Velux Daylight Visualizer and Radiance. The first software allows calculating the luminance value, illuminance value, Daylight Factor. The second one gives the data about DGP, DGI, UGR, VCP and other. The value of DGP which intolerable glare is higher than 45, for DGI is higher 31 and for UGR is higher than 28. The higher value of UGR, the greater harm to human eyes.

The calculated UGR value exceeds the threshold by up to 70% at point A (limit for office is 19). In some case disturbing glare can occur. Window located in the south orientation leads to critical glare problems introducing the need of sun shading devices. The solution would also be to change the position of the table so as to reduce the level of glare as much as possible.

Acknowledgements. This paper was elaborated with the financial support of the research project VEGA 1/0674/18 of the Scientific Grant Agency, the Ministry of Education, Science, Research, and Sport of the Slovak Republic and the Slovak Academy of Sciences.

References

Bellia L, Bisegna F, Spada G (2011) Lighting indoor environment: Visual and noon-visual light sources with different spectral power distribution. Build Environ 46:1984–1992

Bellia L, Fragliasso F, Stefanizzi E (2017) Daylit offices: a comparison between measured parameters assessing light quality and users' opinions. Build Environ 113:92–106

Boyce PR (2004) Reviews of Technical reports on daylight and productivity. Lighting Research Center, Rensselaer Polytechnic Institute

Carli MD, Giuli VD, Zecchin R (2008) Review on visual comfort in office buildings and influences of daylight in productivity. Indoor Air-2008. Copenhagen, Denmark

Figueiro MG, Rea MS (2014) Office lighting and personal light exposures in two seasons: impact on sleep and mood. Light Res Technol 48:352–364

Integral Lighting (2015) ČSN 360020; czech office of standards, metrology and testing. Czech Republic, Prague

Katunsky D, Dolnikova E, Dolnik B (2018) Daytime lighting assessment in textile factory using connected windows in Slovakia case study. Sustainability 10(3):1–20. https://doi.org/10.3390/su10030655

Linhart F, Scartezzini JL (2011) Evening office lighting – visual comfort vs. energy, efficiency vs. performance. Build Environ 46:981–989

Matterson MLG, Ferra JO, Salom J, Portilla JH (2013) Dynamic daylight simulation and visual comfort survey in mediterranean climate, case study in office building. In: Proceedings of BS 2013

Tagliabue LCH, Buzzeti M, Arosio B (2012) Energy saving through the sun: analysis of visual comfort and energy consumption in office space. Energy Procedia 30:693–703

Tregenza P, Wilson M (2011) Daylighting architecture and lighting design. Routledge, New York

Wirz JA, Fournier C (2010) Light health and wellbeing. Implications from chronobiology for architectural design. World Health Des 3(01):44–49

Wienold J, Reetz CH, Kuhn TE, Christoffersen J (2004) Evalglare: a new radiance – based tool to evaluate glare in office spaces. In: Conference: 3rd international radiance workshop, Fribourg (CH)

EN 12464-1 (2012) Light and lighting-lighting of work places-Part 1: Indoor work places; Slovak Republic office of standards, metrology and testing. Slovakia, Bratislava

STN 730580 (2000) Daylighting in buildings, – 1 basic requirement, 1986 – 2; daylighting of residential buildings, 2000, Slovak Republic office of standards, metrology and testing. Slovakia, Bratislava

Operation Life-Cycle Model of the "Building - Base" System

O. Kichaeva(✉)

O. M. Beketov National University of Urban Economy in Kharkiv, Kharkiv 61002, Ukraine
o_kichaeva@ukr.net

Abstract. The purpose of the work is to create an operation life-cycle model of the "building -base" system. In this model, the numerical reliability parameter is taken as the criterion of the system technical state. This serves as a quantitative integral estimation of the system technical state and risk. The reliability calculation method on the proposed life cycle model is proposed. The process of a system degradation during a life cycle is described by a nonlinear algebraic function. The result is the determination of the reliability lower values in the "building-base" system, taking into account the degraded system state and the safety characteristic β. The scientific novelty of the proposed methodology is that the model takes into account the construction geotechnical conditions, the system maintenance conditions, the system design lifetime. The practical significance of the proposed model is seen in the possibility to objectively determine the system operational state, depending on the calculation reliability.

Keywords: The system limit state · System life cycle · Criterion of technical state · System reliability · System "building - base"

1 Introduction

In the normative document of the Joint Committee on Structural Safety and the book of Melchers R.E., the concept of "durability" is formulated as the probability of achieving the limit state in time. Existing durability models can be divided into the following groups: deterministic, stochastic (random) models of the life cycle cost (Bobalo et al. 2018, Bogdanoff and Kozin 1989; Faber 1988; Frangopol 1999; Frangopol et al. 2004).

Therefore, it is of unquestionable interest to create new life cycle models of the entire "building - base" system, which can take into account the most influential factors that determine the most important aspects of this system work (Khmil et al. 2018; Frangopol and Okasha 2008; Lantoukh-Liachtchenko 2006; Lantoukh-Liachtchenko 2008; Maystrenko 2008).

2 Discussion

The operation life-cycle model of the "building - base" system is offered.

Z. Blikharskyy et al. (Eds.): CEE 2019, LNCE 47, pp. 153–160, 2020.
https://doi.org/10.1007/978-3-030-27011-7_19

Theoretically, the model is based on the following hypotheses:

1. The criterion for the system technical state is a numerical reliability parameter, which serves as a quantitative integral estimation value of the system technical state.
2. The base reliability during the *operation life cycle* is taken as a constant, determined at the design stage.
3. The system degradation process during the operation life cycle is represented by four discrete states with continuous time.
4. The system degradation process during a life cycle is described by a nonlinear algebraic function.

Thus, the only parameter of the model is the system initial reliability P_d value, which is obtained at the design stage. The model is deterministic, time t is continuous throughout the operation life cycle. The life cycle model is reduced to a form parallel to the normative document DSTU-N B V.1.2-18:2016 (Maystrenko 2008. Lantoukh-Liachtchenko 2008) – the discrete form with a division of the life cycle into four discrete operating conditions.

Note that the model ignores the random nature of the time in the life cycle, which means that with the given initial data, the lifetime which is predicted is the only fixed value.

The life-cycle model can be presented as follows:

$$P(t) = P_{\mathrm{d}} \cdot f(t) \quad (1)$$

where P_d is the "building - base" system calculated reliability at the design stage – the composition of two reliability parameters: buildings and structures; *f*(*t*) is indicator function of degradation; *t* is the time, years.

The degradation function can be represented in such a form:

$$f(t) = \left(1 - s \cdot K1 \cdot K2 \cdot P_c \cdot P_d^{-1} \cdot t^2\right). \quad (2)$$

Except that there is a connection between the building and the construction in the offered model, which is determined at the earliest stage (the design), we will introduce the operating coefficient *K1*, which we will take depending on the geotechnical category of operation complexity (according to Eurocode 7) in three variants:

- operation in simple geotechnical conditions $K = 1{,}0$;
- operation in geotechnical conditions of medium complexity $K = 1{,}2$;
- operation in complicated geotechnical conditions $K = 1{,}4$.

K2 is the operation coefficient, which takes the following values depending on the maintenance conditions of the "building - base" system:

- the technical operation organization in accordance with the requirements of the current norms $K = 1{,}0$;
- technical operation with some violation of norms $K = 1{,}05$;
- epy organization of technical operation is absent $K = 1{,}1$;

s – dimensional model coefficient;
$P_c = P_d - P_{lim}$ – reliability, difference between design and limit values in operation.
We will put (2) into the life-cycle model, we will get:

$$P(t) = P_d\left(1 - s \cdot K1 \cdot K2 \cdot P_c \cdot P_d^{-1} \cdot t^2\right) \quad (3)$$

The dimensional coefficient of the model s is expressed through the initial constant - the design building lifetime T_d:

$$s = \frac{1}{T_d^2}, \quad (4)$$

where T_d is the design building lifetime, years.

The value of the established operation time T_{ef} according to DBN B.1.2-14-2009 is the initial building resource. The operational discrete generalized classification states of the "building-base" system is presented in Table 1.

Tables 2, 3, and 4 show the lower reliability values of the "building - base" system, calculated by model (3) in the environment of MathCAD with different combinations of coefficients K1 and K2 = 1.0. Here, K1 takes values 1.0, 1.2 and 1.4. The lower reliability values with different combinations of coefficients K1 and K2 = 1.05, and K1 and K2 = 1.1 were similarly calculated (Figs. 1, 2, 3, 4, 5 and 6).

The limit states listed in the table are determined according to DBN B.1.2-14:2009; categories of technical state – according to DSTU-N B V.1.2-18:2016.

Table 1. Operation discrete states classification of the "building - base" system

Technical state category	Name	Generalized characteristics of the system state
1	Normal	System elements meet all requirements of the project and current operation standards
2	Satisfactory	Elements of the system partially do not meet the project requirements, but the requirements of neither the first nor the second limit states groups are not violated
3	Not suitable for normal operation	The elements of the system partially do not meet the project requirements, but the requirements of the first limit states group are not violated. The requirements of the second limit states group are violated. The system "building-base" is operated in limited mode and requires special control over the state of its elements
4	Emergency	Elements of the system do not meet the first limit states group requirements and the impossibility of meeting the requirements is figured out, which indicates the need to stop the operation of the "building – base" system

Table 2. The lower reliability values of the system "building - base" (K1 = 1,0; K2 = 1,0)

Technical state category	System reliability, P_1	Security Feature, β_i	Deterioration, %
	0,99986	3,63	0
1	0,99674	2,72	1–10
2	0,98739	2,24	11–40
3	0,97181	1,91	41–60
4	0,95000	1,64	>60

Table 3. The lower reliability values of the "building - base" system (K1 = 1,2; K2 = 1,0)

Technical state category	System reliability, P_1	Security Feature, β_i	Deterioration, %
	0,99986	3,63	0
1	0,99612	2,66	1–10
2	0,98490	2,17	11–40
3	0,96620	1,83	41–60
4	0,94003	1,56	>60

Table 4. The lower reliability values of the "building - base" system (K1 = 1,4; K2 = 1,0)

Technical state category	System reliability, P_1	Security Feature, β_i	Deterioration, %
	0,99986	3,63	0
1	0,99550	2,61	1–10
2	0,98241	2,11	11–40
3	0,96060	1,76	41–60
4	0,93006	1,48	>60

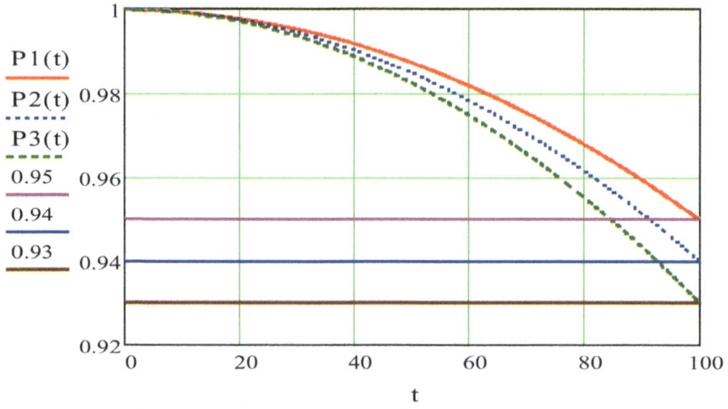

Fig. 1. Life-cycle diagrams of the "building-base" system with K1 = 1,0; 1,2; 1,4, K2 = 1,0

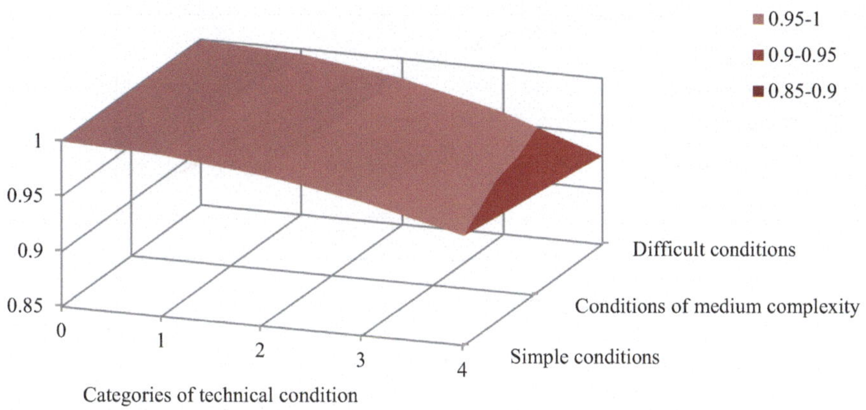

Fig. 2. The surface, which reflects the dependencies between system reliability, operation conditions and (Tables 2, 3 and 4)

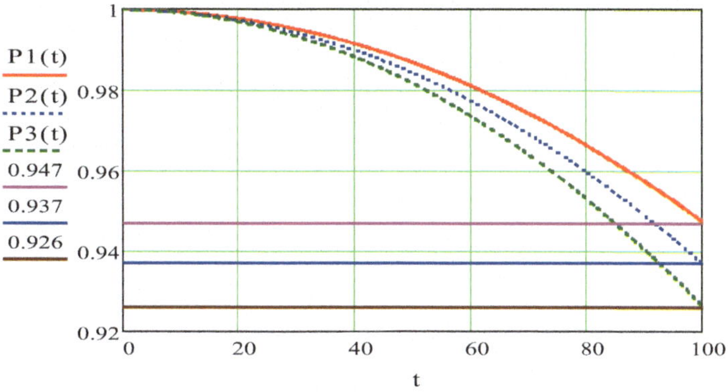

Fig. 3. Life-cycle diagrams of the "building-base" system with K1 = 1,0; 1,2; 1,4, K2 = 1,05

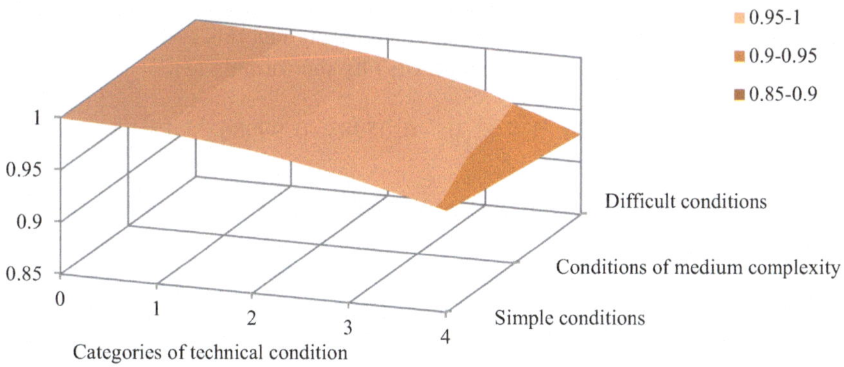

Fig. 4. The surface, which reflects the dependencies between system reliability, operation conditions and technical state categories

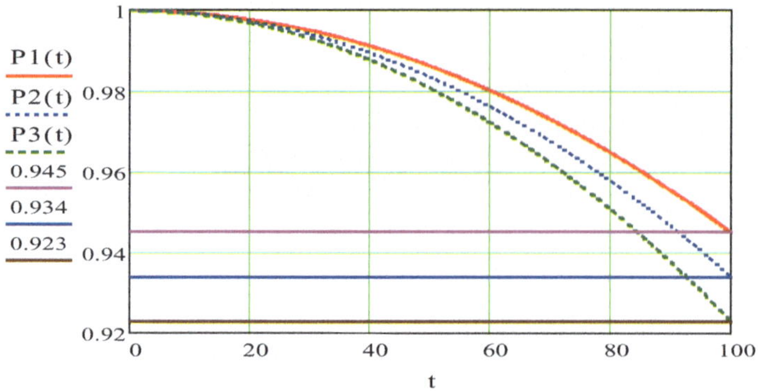

Fig. 5. Life-cycle diagrams of the "building-base" system with K1 = 1,0; 1,2; 1,4, K2 = 1,1

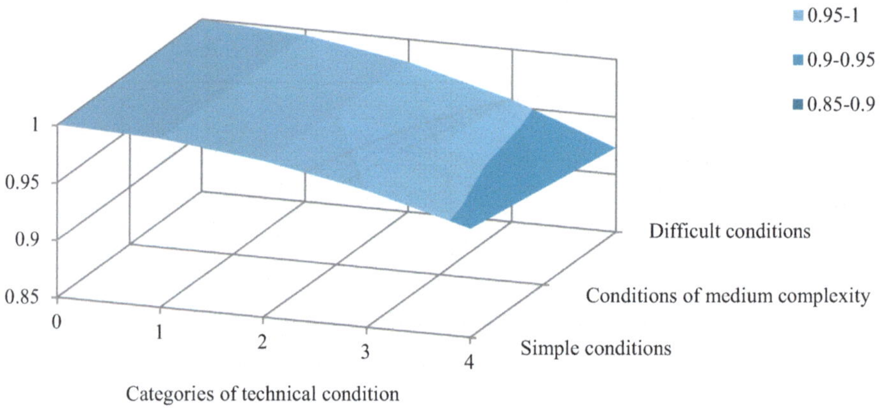

Fig. 6. The surface, which reflects the dependencies between system reliability, operating conditions and technical state categories

Calculation Example 1 - Library Building. We will calculate the system reliability for the time of the examination (t = 45 years, remembering that this system is operated in geotechnical conditions of medium complexity) by the formula (3):

$$P(t) = 0,99986\left(1 - 0,0001 \cdot 1,2 \cdot 1,05 \cdot 0,05983 \cdot 0,99986^{-1} \cdot 45^2\right) = 0,98459.$$

Calculation Example 2 - Public Building. We will calculate the system reliability for the time of the examination (t = 35 years, remembering that this system is operated in geotechnical conditions of medium complexity) by the formula (3):

$$P(t) = 0,99986\left(1 - 0,0001 \cdot 1,2 \cdot 1,0 \cdot 0,06282 \cdot 0,99986^{-1} \cdot 35^2\right) = 0,99.$$

3 Conclusions

In the present study, a life-cycle model of the "building - base" system has been developed. In the suggested model, the numerical reliability parameter is taken as the criterion of the system technical state, which serves as a quantitative integral estimation value of the system technical state.

References

Bobalo T, Blikharskyy Y, Vashkevich R, Volynets M (2018) Bearing capacity of RC beams reinforced with high strength rebars and steel plate. MATEC Web Conf 230:02003. https://doi.org/10.1051/matecconf/201823002003

Bogdanoff J, Kozin F (1989) Probabilistic damage accumulation models/trans. From English Timasheva SA/under. Timasheva SA (ed). World, Moscow (in Russian)

DBN V.1.2-14: 2009 (2009) General principles of ensuring the reliability and constructive safety of buildings, structures, constructions and foundations. Kyiv (in Ukrainian)

DSTU-N B V.1.2-18: 2016 (2017) Guidelines for the buildings and structures inspection for their technical condition determination and assessment. Kyiv (in Ukrainian)

Eurocod 7 (2004) Geotechnical design. General rules

Faber MH (1988) Basics of Structural Reliability Federal Highway Administration (FHWA). Recording and coding guide for the structure in venture and appraisal of the nation's bridges. U.S. Department of Transportation Washington D. C

Frangopol DM (1999) Life-cycle cost analysis for bridges. Bridge Safety and Reliability. ASCE, Reston, Virginia, pp 210–236

Frangopol DM, Kallen MJ, van Noortwijk JM (2004) Probabilistic models for life-cycle performance of deteriorating structures: review and future directions. Prog Struct Mat Eng 6:197–212

Frangopol DM, Okasha NM (2008) Life-cycle performance and redundancy of structures. In: Proceedings of the sixth international probabilistic workshop, Darmstadt, Germany, 26–27 November, 2008 (Keynote Lecture), pp 1–14

Khmil R, Tytarenko R, Blikharskyy Y, Vegera P (2018) Development of the procedure for the estimation of reliability of reinforced concrete beams, strengthened by building up the stretched reinforcing bars under load. Eastern-Eur J Enterp Technol 5/7(95). https://doi.org/10.15587/1729-4061.2018.142750

Lantoukh-Liachtchenko AI (2006) Reliability based service life prediction of concrete bridge superstructures. In: Proceeding EKO MOST 2006. Durable bridge structures in the environment, Kielce, 16–17 May. Warszawa, pp 255–261

Lantoukh-Liachtchenko AI (2008) The bridges operational status assessment clarification. Collection "Roads and Bridges", 9th edn. DerzhdorNDI, Kyiv, pp 12–18 (in Ukrainian)

Maystrenko IY (2008) The supporting structures of metal bridges reliability assessment by statistical modeling method. Kazan State University of Architecture and Construction proceedings "Theory of engineering structures and building structures", 1st edn. KazSACU, Kazan, pp 295–301 (in Russian)

Maystrenko IY, Manapov AZ (2010) The modeling of the structural system reliability level variation process in time. Kazan State University of Architecture and Construction proceedings "Building structures, buildings and structures", 1st edn. KazSACU, Kazan, pp 132–140 (in Russian)

Melchers RE (1999) Structural reliability analyses and prediction, 2nd edn. Wiley, New York

Probabilistic Model Code. 12-th draft (2000) Joint Committee on Structural Safety. PART I. BASIS OF DESIGN. JCSS-OSTI/DIA/VROC–10.11.2000, ETH Zurich

Rzhanitsyn AR (1978) The theory of building structures calculation for reliability. Building Edition, Moscow, p 239 (in Russian)

Rzhanitsyn AR (1952) The use of statistical methods in the structure calculations for strength and safety. Construction industry, № 6, pp 22–25

Thompson P, Small EP, Johnson M, Marshall AR (1998) The pontis bridge management system. Struct Eng Int 8(4):303–308

Thompson PD, Shepard RW (1994) Pontis. Transportation Research Record, vol 324. Transportation Research Board, Washington, D.C., pp 35–42

The Method of Calculating the Bearing Capacity of Compressed Stone Pillars

Ye. Klymenko[1(✉)], I. Grynyova[1], and Z. Kos[2]

[1] Department of Reinforced Concrete Structures and Transport Facilities, Odessa State Academy of Civil Engineering and Architecture, Didrihson, 4, Odesa 65029, Ukraine
klimenkoew57@gmail.com

[2] Department of Civil Engineering, 104. brigade 3, 42 000 Varaždin, Croatia

Abstract. The main goal of the research is to develop a calculation model for determining the residual bearing capacity of compressed stone elements of damaged cross-section. In the laboratory, 15 prototypes with pre-modeled damage were manufactured and tested. From the selected variation factors most affect the carrying capacity proved parameters characterizing element damage eccentricity affects secondary. According to the estimates of the experimental statistical model and single-factor local fields, the depth of damage in the cross section of the column has the greatest impact on the bearing capacity. A calculation model was created to determine the residual bearing capacity of damaged structures of rectangular cross section and the height of the compressed zone (x) and the area of the compressed zone (A_c). At its core, the model is based on the basic assumptions of existing building codes. The article presents two generalized types of damages according to the type of shape: direct damage - when the inclination angle of the damage front is zero and oblique damage - when the inclination angle of the damage front is not zero.

Keywords: Field tests · Eccentric compression · Damage · Operation · Calculation · Residual carrying capacity

1 Introduction

The main problem in determining the technical condition of damaged structural elements of buildings is the reliable determination of their residual carrying capacity.

In current regulatory documents there are no specific recommendations for the calculation of damaged items. In this connection, a number of problems arise that need to be solved. One of the problems is the creation of methods for calculating and estimating the bearing capacity of damaged eccentrically compressed stone pillars; on the basis of this problem, one or another reconstruction or reinforcement scheme is adopted.

It is known that the existing method of calculating stone structures for the first group of limiting states has a sufficiently high reliability and is based on experimentally theoretical studies conducted for centrally and eccentrically compressed elements.

Z. Blikharskyy et al. (Eds.): CEE 2019, LNCE 47, pp. 161–167, 2020.
https://doi.org/10.1007/978-3-030-27011-7_20

Recommendations for the calculation of the bearing capacity of eccentrically compressed elements in existing standards are not available.

2 Analysis of Previous Studies

The complex stress-strain state of masonry does not allow the use of existing theories of strength, so almost all researchers based on the results of processing a large amount of experimental data [] obtained only empirical dependencies of masonry strength.

Most of them take into account the most influential factors (strength of stone and mortar, type and size of stone) [1–3], on which the resistance of masonry depends. But the existing dependences of the strength of the masonry have a partial character, correspond well to certain types of masonry and materials and were obtained in studies on them.

The most successful of them are the empirical dependences of professor Onishchik [4], which reflect the peculiarities of the masonry of stone and mortar, had the greatest application in the 30–50s of the last century. These dependencies form the basis of modern standards and allow one to obtain fairly accurate results for traditional masonry on solutions of medium strength.

The emergence of new types of materials (stone and mortar), an increase in their strength characteristics, the use of industrial masonry manufacturing methods has showed that the dependencies of professor L.I. Onishchik can not take into account all the above features.

Conducted in recent years in the Odessa State Academy of Civil Engineering and Architecture, studies of eccentrically compressed and damaged during the operation of stone structures, allowed us to obtain data for further description of their stress-strain state and develop a methodology for calculating the residual carrying capacity [5].

3 Materials and Methods of Research

To achieve the goals of research as a basic object of study model was developed poles.

For the manufacture of samples used ceramic brick full of corpulent with a brand strength of 100 and a brand of frost resistance F-100 Tatarbunary brick factory. The solution was made of cement PC - II - B - W - 400 at the Odessa Cement Plant and sand with a grain size module - 1.4. To determine the characteristics of the strength of bricks, samples were taken from the party, which were intended for the manufacture of the structure, 10 bricks of the grade M100 - to determine the compressive strength; and 5 bricks to determine the flexural strength.

The studies were conducted in the laboratory for testing building materials and products of the Odessa State Academy of Civil Engineering and Architecture in accordance with the requirements of GOST-8462-85. As already noted in [5], the variation of the experimental design parameters was carried out in a fairly wide (in terms of operation cases) range: the inclination angle of the damage front in one of the main axes of the cross section was $\theta = 0°$; 22.5°; 45° damage depth a = 0 mm; 80 mm; 160 mm and relative eccentricity e0 = 0 mm; 80 mm; 160 mm (Fig. 1).

Fig. 1. Tests of experimental sample pillar

With known cross-sectional dimensions (a, b), as well as damage dimensions (a1, b1), eccentricity (e0), angle of inclination of the damage front line ($\theta = 0°$), the residual carrying capacity (N), height of the compressed zone remains unknown (x), the angle of inclination of the neutral line to one of the main axes of the intact cross section (γ).

4 The Results of Research

In this paper, we developed a method for determining the residual bearing capacity of damaged eccentrically compressed stone pillars, that is, elements that were damaged during operation. Such a calculation will be a priori verifiable and the application of the above premise on the uniform distribution of stresses over the area of a compressed zone should be considered justified.

During the calculation of the bearing capacity of stone elements, the following prerequisites were taken:

1. The tension in the compressed zone is evenly distributed.
2. Damage front has a rectilinear contour.
3. The hypothesis of flat sections is accepted, that is, after the deformation of the section remains flat, and along the height of the cross section, the deformations vary linearly.
4. The center of mass of the compressed masonry zone coincides with the point of application of the calculated longitudinal force N.
5. The work of the stretched zone in the perception of external forces was not taken into account.

The rationale for the first prerequisites is given above. The second prerequisites is justified by the results of field inspections of damaged elements, as the most common type of destruction. The third, fourth and fifth prerequisites were taken from the works of Popov and Zabegaev [8]. To determine the bearing capacity, we use 2 equilibrium Eqs. (1), (2), where f_u is the calculated resistance of masonry to compression, kH/sq.m. In this case, the unknown height of the compressed zone (x) and the area of the compressed zone (A_c) are determined on the basis of prerequisite 3.

While solving the problem, we use the following equations and dependencies (Fig. 2).

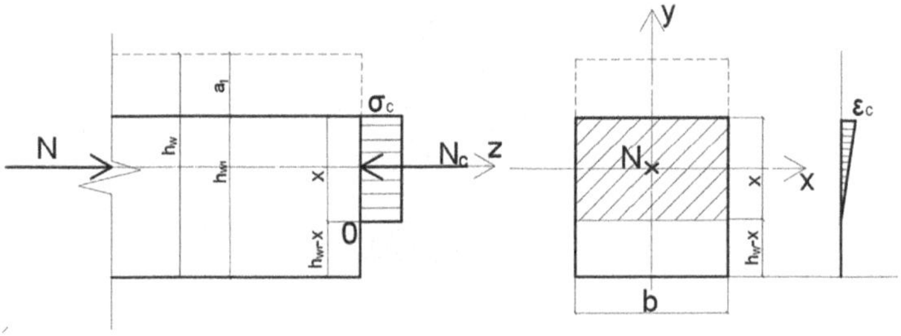

Fig. 2. The scheme for calculating cross-sectional eccentrically compressed element

The equilibrium condition:

$$\sum Z = 0; N - f_u \cdot A_c = 0 \tag{1}$$

$$\sum M_0 = 0; N_c \cdot \frac{x}{2} = N \cdot \frac{x}{2} \tag{2}$$

As a result, from the fourth prerequisite, the coordinates of the center of mass of the compressed zone relative to the main axes with flat damaged sections are accepted (Fig. 3b, c):

$$x_1 = 0; y_1 = e_0. \tag{3}$$

The center of mass of the compressed zone of the pillars coincides with the point of application of the external compressive force.

In this work, we will call flat damage when its front is parallel to one of the main axes of the section, that is, the inclination angle of damage is zero (Fig. 3a).

Oblique damage occurs when the damage front (in this case, also a straight line) is not parallel to one of the main axes of the section, that is, the slope of the damage front line is not zero ($\theta \neq 0$) (Fig. 3b, c).

Depending on the depth of damage (a_1, b_1) and the angle of inclination of the damage front (θ) to the axis (x; y), the calculation can be reduced to several cases. Case I occurs when $b_1 < b/2$ (Fig. 3b). If $b/2 < b_1 < b$, then this is case II (Fig. 3c).

In this case, taking into account the designation of the height (depth) of the damage (a_1) and the distance at which the force N (e_0) was applied, the height of the compressed zone of the pillars is traditionally designated x. Compressed area of complex shape, if necessary, is divided into simple flat sections. We determine the auxiliary geometric parameters of the cross section (Fig. 3a, 3b, 3c).

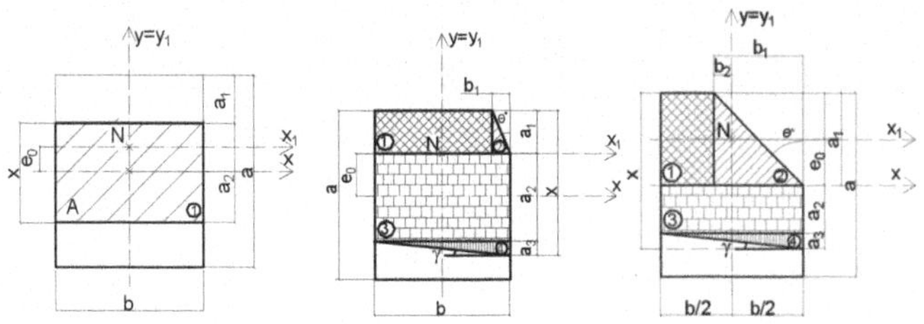

a) b) c)

Fig. 3. Cross section in case of flat damage: (a) - with $a_1 < a/2$, $\theta = 0$; (b) - with $b_1 < b/2$, $\theta \neq 0$; (c) - at $b/2 < b_1 \leq b$, $\theta \neq 0$

Based on conditions (1–2), for the implementation of the fourth prerequisite, we write the equation of static moments and the area of the compressed zone Ac relative to the point of application of force, in general, for plane damage:

$$S_x = A_1 \cdot y_1 + A_2 \cdot y_2 = (x - a_1 - a_3) \cdot b \cdot \left(\frac{a}{2} - a_1 - \frac{a_2}{2}\right) + \frac{a_3 b}{2} \cdot \left(-\left(\frac{a}{2} - (a - x) + e_0 - \frac{2a_3}{3}\right)\right) = 0; \tag{4}$$

$$S_y = A_1 \cdot y_1 + A_2 \cdot y_2 = \frac{a_3 b}{2} \cdot \left(\frac{b}{2} - \frac{b}{3}\right) = 0; \tag{5}$$

$$A_c = A_1 + A_2 = (x - a_1 - a_3) \cdot b + \frac{a_3 b}{2} \tag{6}$$

Based on conditions (1–2), for the implementation of the fourth prerequisite, we write the equation of static moments and the area of the compressed zone Ac relative to the point of application of force, in general, with oblique damage:

– at b1 < b/2, $\theta \neq 0$;

$$S_{x_1} = (b - b_1) \cdot a_1 \cdot \left(\frac{a_1}{2} - e_0\right) + \left(\frac{a_1 b_1}{2} \cdot \left(\frac{a_1}{3} - e_0\right)\right) + (a_2 b) \cdot (a_2 + e_0) + \left(\frac{a_3 b}{2} \cdot \left(x - a_1 + e_0 - \frac{2a_3}{3}\right)\right) = 0; \tag{7}$$

$$S_{y_1} = (b - b_1) \cdot a_1 \cdot \frac{b - (b - b_1)}{2} + \frac{a_1 b_1}{2} \cdot \left(\frac{b}{2} - \frac{2b_1}{3}\right) + \frac{a_3 b}{2} \cdot \left(\frac{b}{2} - \frac{b}{3}\right) = 0; \tag{8}$$

– at $b/2 < b_1 \leq b$, $\theta \neq 0$;

$$S_{x_1} = (b - b_1) \cdot a_1 \cdot \left(\frac{a_1}{2} - e_0\right) + \frac{a_1 b_1}{2} \cdot \left(\frac{a_1}{3} - e_0\right) + a_2 b \cdot \left(x - \frac{a_2}{2} - a_3 - (a_1 - e_0)\right) + \frac{a_3 b}{2} \cdot \left(x - (a_1 - e_0) - \frac{2a_3}{3}\right) = 0; \tag{9}$$

$$S_{x_1} = (b - b_1) \cdot a_1 \cdot \frac{b - (b - b_1)}{2} + \frac{a_1 b_1}{2} \cdot \left(b_1 - b_2 - \frac{2b_1}{3}\right) + \frac{a_3 b}{2} \cdot \left(\frac{b}{2} - \frac{b}{3}\right) = 0; \tag{10}$$

The area of the compressed zone in both cases is given by the expression:

$$A_c = (b - b_1) \cdot a_1 + \frac{a_1 b_1}{2} \cdot \left(b_1 - b_2 - \frac{2b_1}{3}\right) + \frac{a_3 b}{2} \cdot \left(\frac{b}{2} - \frac{b}{3}\right). \tag{11}$$

Solving the three equations, we obtain the compressed area Ac zone. According to the first equilibrium Eq. (1) we find N, the carrying capacity of damaged stone pillars.

5 Conclusions

The basic prerequisites for the calculation of damaged compressed elements with various types of damage and the application of external force were formulated. There are suggestions for calculating the carrying capacity, which will bring closer to the true operation of the structure. Thus, it is necessary to take into account the depth of damage to the element, as well as the eccentricity to which the force is applied. This concerns the determination of the position of the neutral line and, therefore, the height of the compressed zone of concrete. The main settlement cases are highlighted and their differences are described. The calculation by the developed method showed a variation coefficient of 8.9%. Thus, the method for determining the bearing capacity of damaged pillars, described in this paper, is sufficiently accurate and can be based on verification calculations in case of such situations.

References

1. Binda L, Lualdi M, Saisi A (2007) Non-destructive testing techniques applied for diagnostic investigation: syracuse cathedral in Sicily, Italy. Int J Arch Herit 1:380–402. https://doi.org/10.1080/15583050701386029
2. Derkach VN (2016) Stone filling of frame buildings: strength, rigidity and force interaction with the frame: dis. Dr. those. Sciences: 05.23.01. Brest (in Russian)
3. Onishchik LI (1939) Stone structures of industrial and civil buildings. State publishing house of construction literature, Moscow

4. Klymenko EV, Shapoval SL (2001) Experimental studies of masonry with its local compression. Sect Eng Constr 7:58–64
5. Grynyova II (2017) Method of conducting an experimental study of the strained-deformed state of damaged stone pillars. Bull Odessa State Acad Civ Eng Arch 67:20–26
6. Ie Klymenko, Grynyova I, Pents V (2018) The work of stone pillars damaged during operation. Int J Eng Technol 7(4.8):174–179. https://doi.org/10.14419/ijet.v7i4.8.27235
7. Zimin SS, Bespalov VV, Kazimirova AS (2015) The design model of a stone arch design. Bull Donbass Natl Acad Civ Eng Arch 3(113):33–37
8. Milani G (2015) Upper bound sequential linear programming mesh adaptation scheme for collapse analysis of masonry vaults. Adv Eng Softw 79:91–110
9. Surianinov M, Shyliaiev O (2018) Calculation of plate-beam systems by method of boundary elements. Int J Eng Technol (UAE) 7(2):238–241
10. Valente M, Milani G (2016) Non-linear dynamic and static analyses on eight historical masonry towers in the North-East of Italy. Eng Struct 114:241–270. https://doi.org/10.1016/j.engstruct.2016.02.004
11. Lacidogna G, Manuello A, Niccolini G, Carpinteri A (2015) Acoustic emission monitoring of Italian historical buildings and the case study of the Athena temple in Syracuse. Arch Sci Rev 58(4):290–299. https://doi.org/10.1080/00038628.2012.720246
12. Li Z, Chen L, Fang Q, Chen WS, Hao H, Zhu R, Zheng K (2019) Experimental and numerical study on CFRP strip strengthened clay brick masonry walls subjected to vented gas explosions. Int J Impact Eng 129:66–79. https://doi.org/10.1016/j.ijimpeng.2019.02.013
13. Albert ML, Elwi AE, Cheng JJR (2001) Strengthening of unreinforced masonry walls using FRP. J Compos Constr 5(2):76–84. https://doi.org/10.1061/(ASCE)1090-0268(2001)5:2(76)
14. Krainskyi P, Blikharskyy Y, Khmil R, Vegera P (2018) Influence of loading level on the bearing capacity of RC columns strengthened by jacketing. MATEC Web Conf. https://doi.org/10.1051/matecconf/201823002013
15. Vasconcelos G, Lourenco PB (2009) Experimental characterization of stone masonry in shear and compression. Constr Build Mater 23(11):3337–3345. https://doi.org/10.1016/j.conbuildmat.2009.06.045
16. Popov NN, Zabegaev AV (1989) Design and calculation of reinforced concrete and masonry structures. Higher school, Moscow

Investigation of Indoor Air Quality in the Selected Ukraine Classroom – Case Study

P. Kapalo[1], H. Klymenko[2(✉)], V. Zhelykh[2], and M. Adamski[3]

[1] Institute of Architectural Engineering, Faculty of Civil Engineering, Technical University of Kosice, Vysokoskolska 4, 042 00 Kosice, Slovakia
[2] Department Heat and Gas Supply and Ventilation, Lviv Polytechnic National University, St. Bandery 12, 79013 Lviv, Ukraine
anett.lviv@gmail.com
[3] Faculty of Civil and Environmental Engineering, Bialystok University of Technology, Wiejska 45A, 153 51 Bialystok, Poland

Abstract. When reconstructing existing school buildings and building new buildings, great emphasis is placed on increasing their thermal resistance and reducing air tightness. The consequence of these modifications is the reduction of heat flow through the building structure, as well as minimization of uncontrolled ventilation by infiltration through building leaks. According to various studies conducted in the world, it can be said that the exchange of used air for fresh air has a huge impact on the mood of students in the classroom. As part of a research stay at the Lviv Polytechnic, measurements were made in the classroom to determine the air quality during the teaching. The classroom was selected in which measuring instruments were installed for measuring carbon dioxide (CO_2) concentration and air temperature. During the measurements didactic classes took place in the classroom. In addition to the measurement of indoor air parameters, a subjective assessment of air quality was carried out using questionnaires. The students filled out the indoor air quality questionnaire at the beginning of their stay in the auditorium at the end of their stay. An assessment of air quality was carried out based on the measured air parameters and questionnaire ratings in the selected classroom.

Keywords: Classroom · Measurement · Carbon dioxide (CO_2) · Temperature · Ventilation · Questionnaire

1 Introduction

In recent years, there has been a growing consciousness in energy-efficient school buildings. Increasing the thermal resistance of the building envelope cause decrease them heat loss. By Hooff et al. (2014) increasing the thermal resistance of the building envelope increases the number of overheating hours, therefore, in well-insulated buildings shading or additional natural ventilation should be provided to limit the number of overheating hours. The windows are usually the first component to be replaced by new and more efficient ones. Replacing old windows with new ones we

Z. Blikharskyy et al. (Eds.): CEE 2019, LNCE 47, pp. 168–173, 2020.
https://doi.org/10.1007/978-3-030-27011-7_21

achieve a reduction in air exchange by infiltration. The Carlos (2017) in paper presents the obtained airflow and solar gains. The findings are essential for the assessment of passive building energy-saving techniques on a hypothetical building refurbishment by replacing old windows. However, specific adaptive actions to improve indoor environmental quality are also needed.

Therefore is necessary in most buildings ensure additional ventilation. According to Hou et al. (2015) for insure good indoor air quality in primary schools, mechanical ventilation devices with air purification function should be introduced. Research results (Becker et al. 2007) indicate that implementation of improved ventilation schemes in an otherwise well designed energy-conscious building result in savings of classroom orientations. One option is using Demand Controlled Ventilation system. The measurement results (Merema et al., 2018) show that in the case studies the Demand Controlled Ventilation system was able to deliver and maintain a good indoor air quality, even at reduced air flow rates. This shows that demand controlled ventilation is effective in distributing the air even at reduced air flow rates.

Natural ventilation that reduces building energy consumption and improves indoor environment has become a key solution to achieving sustainability in the building industry. According to Chen et al. (2017) the potential for utilizing natiral ventilation strategies depends greatly on the local climate, which varies widely from region to region in the world. According to the results (Han et al. 2018), too can he natural ventilation of open staircase very much influence (negative or positive) the ventilation of rooms. In his research Schibuola et al. (2016) considers the topic of natural ventilation in school buildings that is faced not only for energy saving, but also for the fundamental exigency of the indoor comfort. This analysis is developed by measuring the CO_2 concentration as a significant indicator of indoor air quality when pollution is mainly due to the presence of people.

The method of distributing the outdoor air in classrooms has a major impact on indoor air quality and thermal comfort of pupils. Therefore Karimipanah et al. (2007) to examined the indoor environment in the classroom using confluent jet ventilation. A more general look at the results (Karimipanah et al. 2007) show that ventilation of a school environment is a difficult task and is depended on many parameters that need to be adjusted and controlled. This may mean more responsibility for ventilation engineers and designers to be aware of the fact that a system that works well in a certain environment does not mean it can be adapted for general use.

Information about indoor conditions of classrooms in schools is very scarce and availability of indoor air quality data is important as children are vulnerable to health hazards and spend long times in classrooms. Abdel-Salam (2019) was investigated in 16 mechanically ventilated schools in Qatar during the winter season. Parameters such as temperature, relative humidity and CO_2. High indoor CO_2 concentrations were found in many classrooms. The mean indoor CO_2 concentration in all classrooms was 1776 ± 887 ppm. Teleszewski and Gładyszewska-Fiedoruk (2018) in the paper presents a simplified model of CO_2 concentration in classrooms equipped with stack ventilation systems, based on the experimental research. The test was conducted in six classrooms in the building of the Faculty of Civil and Environmental Engineering of the Białystok University of Technology in north-eastern Poland. In all classrooms, a linear increase in the CO_2 concentration during the classes was observed. According to

Krawczyk et al. (2016) in the school buildings located in two different climates: Białystok (Poland) and Belmez/Córdoba (Spain) the CO_2 concentration in first 45 min met requirements of regulations, but with medium occupation of places in classrooms. The relationship between indoor and outdoor air pollution within classrooms researched Jovanovic et al. (2014), who examined air quality at a primary school located in a city in eastern Serbia. Meciarova et al. (2018) carried out the monitoring two classrooms in primary school and two classrooms in art school were chosen. Very high CO_2 levels were determined in both investigated schools. This study showed a good consistency between data obtained from the measurement and data from the questionnaire study and thus, pointed to the good ability of pupils to evaluate the IEQ in their classrooms.

The perception of air quality by persons it also affects their clothes. The aim of study (Ghaddar et al. (2011) was to develop a modeling methodology to assess thermal comfort and sensation of active people in transitional spaces and consider how comfort can be achieved by air movement while changing upper body clothing properties. Pupil activity in the classroom influence CO_2 productivity, heat and humidity. Kapalo et al. (2019) carried out measurements of indoor air temperature, relative humidity and CO_2 concentrations as well as to determine CO_2 production by students and teacher during various physical activities. Results of objective measurements confirmed strong correlation between CO_2 concentration occupancy for all measurements. Results of this study showed the insufficient ventilation intensity in classrooms as well as obvious rise of CO_2 concentration during the exams. The highest increase of CO_2 was recorded during harder physical activity (run on the spot, squats, right and left side lunges, and rotating of the hips). Regarding CO_2 production by respondents it can be see that it is visibly increased with increasing physical activity.

As part of a research stay at the Lviv Polytechnic, measurements were made in the classroom to determine the air quality during the teaching. The classroom was selected in which measuring instruments were installed for measuring CO_2 concentration and air temperature. During the measurements didactic classes took place in the auditorium. In addition to the measurement of indoor air parameters, a subjective assessment of air quality was carried out using questionnaires.

2 Aim of Research

The aim of the research is to investigate air quality (CO_2 concentration and air temperature) in a selected class during the stay of a person and its impact on people.

3 Materials and Methods

The measurement of indoor air parameters was realized during more days and in the same room. The experimental room is located on the fourth floor of a six storey building. The sizes of the classroom are: length: 5.90 m, width: 6.30 m and height: 3.30 m. The volume of the room is 123 m^3. Windows are northwest oriented. During the stay of people, the room was not exposed to direct sunlight. For measurement indoor air parameters was used the temperature and humidity sensor S3541 and CO_2 concentration sensor C-AQ-0001R.

The research was realized in two days. In the first day (20.3.2019) of the indoor air measurement ventilation was not provided during the lesson and neither during the break. In the second day of the measurement (25.3.2019) ventilation was not provided during the lesson, but room was ventilated during the break. During the research, the selected room was occupied by 7 till 16 persons (Table 1).

Table 1. Number of person in the classroom

	Number of person in the classroom			
	First lesson	Second lesson	Third lesson	Fourth lesson
20.3.2019	9	15	8	14
25.3.2019	5	16	0	10

4 Results and Discussions

The measured indoor air parameters: temperature and carbon dioxide concentration are documented in Fig. 1.

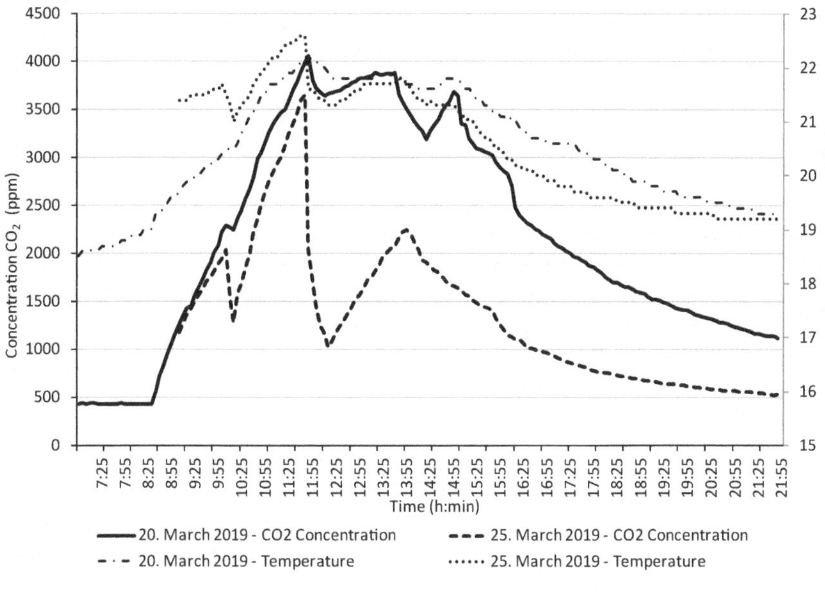

Fig. 1. Measured data for days 20. March 2019 and 25. March 2019

Based on the measured parameters of the indoor air on March 20, it can be seen that in the first day, the temperature increased during the two lessons from 18.6 °C to 22.1 °C. The CO_2 concentration ranged from 430 ppm to 4,061 ppm, which is a value too large. On the basis of the measured parameters of the indoor air in the on March 25, it can be seen that the maximum air temperature was 22.6 °C. CO_2 concentration reached maximum of 3,649 ppm - it is also too large value.

In Fig. 2, we can observe differences in changes of the CO_2 concentration during the day, when the classroom was not ventilated during the break and when the classroom was ventilated during the break. From the Fig. 2, we can see, that when the classroom was ventilated during the break, so the average concentration of CO_2 in the classroom was significantly reduced.

5 Conclusions

By comparing of measured values indoor air parameters on March 20, 2019, when the room was not ventilated during stay of people in the room nor yet during breaks with measured values on March 25, 2019, when the room was not ventilated during stay of people in the room but was ventilated during breaks, it can be stated that values of mean concentration during stay of people in class are 2,845 ppm and 1,896 ppm in time of measurements. The calculated drop in the CO_2 concentration was 33.4%.

It is possible to state that even a short 10-min of class ventilation, has a significant impact on classroom air quality. In conclusion, even a short 10-min ventilation has a significant impact on the air quality in the classroom.

Acknowledgements. This article was elaborated in the framework of the project VEGA 1/0697/17. The article was written during a research stay at the National University "Lviv Polytechnic".

References

Abdel-Salam MMM (2019) Investigation of indoor air quality at urban schools in Qatar. Indoor Built Environ 28(2):278–288. https://doi.org/10.1177/1420326x17700948 ISSN: 1420326X

Becker R, Goldberger I, Paciuk M (2007) Improving energy performance of school buildings while ensuring indoor air quality ventilation. Build Environ 42(9):3261–3276. https://doi.org/10.1016/j.buildenv.2006.08.016 ISSN: 0360-1323

Carlos JS (2017) The impact of refurbished windows on Portuguese old school buildings. Architect Eng Des Manag 13(3):185–201. https://doi.org/10.1080/17452007.2016.1274252

Ghaddar N, Ghali K, Chehaitly S (2011) Assessing thermal comfort of active people in transitional spaces in presence of air movement. Energy Build 43(10):2832–2842. https://doi.org/10.1016/j.enbuild.2011.06.040

Han DH, Kim S, Choi JH, Kim YS, Chung HS, Jeong H, Choi SO, Watjanatepin N, Ruangpattanawiwat Ch (2018) Experimental study on thermal buoyancy-induced natural ventilation. Energy Build 177:1–11. https://doi.org/10.1016/j.enbuild.2018.07.046 ISSN: 0378-7788

Hooff T, Blocken B, Hensen JLM, Timmermans HJP (2014) On the predicted effectiveness of climate adaptation measures for residential buildings. Build Environ 82:300–316. https://doi.org/10.1016/j.buildenv.2014.08.027

Hou YF, Liu JJ, Li JY (2015) Investigation of indoor air quality in primary school classrooms. Procedia Eng 121(2015):830–837. https://doi.org/10.1016/j.proeng.2015.09.037

Chen Y, Tong Z, Malkawi A (2017) Investigating natural ventilation potentials across the globe: regional and climatic variations. Build Environ 122:386–396. https://doi.org/10.1016/j.buildenv.2017.06.026

Jovanovic M, Vucicevic B, Turanjanin V, Zivkovic M, Spasojevic V (2014) Investigation of indoor and outdoor air quality of the classrooms at a school in Serbia. Energy 77:42–48. https://doi.org/10.1016/j.energy.2014.03.080

Kapalo P, Meciarova L, Vilcekova S, Burdova EK, Domnita F, Bacotiu C, Peterfi KE (2019) Investigation of CO2 production depending on physical activity of students. Int J Environ Health Res 29(1):31–44. https://doi.org/10.1080/09603123.2018.1506570 ISSN: 09603123

Karimipanah T, Awbi HB, Sandberg M, Blomqvist C (2007) Investigation of air quality, comfort parameters and effectiveness for two floor-level air supply systems in classrooms. Build Environ 42(2):647–655. https://doi.org/10.1016/j.buildenv.2005.10.016

Krawczyk DA, Rodero A, Gładyszewska-Fiedoruk K, Gajewski A (2016) CO2 concentration in naturally ventilated classrooms located in different climates – measurements and simulations. Energy Build 129:491–498

Meciarova L, Vilcekova S, Kridlova Burdova E, Kapalo P, Mihalova N (2018) The real and subjective indoor environmental quality in schools. Int J Environ Health Res 28(1):102–123. https://doi.org/10.1080/09603123.2018.1429579 ISSN: 09603123

Merema B, Delwati M, Sourbron M, Breesch H (2018) Demand controlled ventilation (DCV) in school and office buildings: lessons learnt from case studies. Energy Build 172:349–360. https://doi.org/10.1016/j.enbuild.2018.04.065

Schibuola L, Scarpa M, Tambani Ch (2016) Natural ventilation level assessment in a school building by CO2 concentration measures. Energy Procedia 101:257–264. https://doi.org/10.1016/j.egypro.2016.11.033

Teleszewski T, Gładyszewska-Fiedoruk K (2018) Changes of carbon dioxide concentration in classrooms - simplified model and experimental verification. Pol J Environ Stud 27(5):2397–2403. https://doi.org/10.15244/pjoes/77074

Technological Approaches to Design of Artificial Lighting Approximated to the Properties of Natural Lighting

L. Koval(✉), V. Yehorchenkov, and O. Sergeychuk

Department of Architectural Designs, Kyiv National University of Construction and Architecture, Povitroflotsky Avenue 31, Kiev 03037, Ukraine
likocolor@gmail.com

Abstract. In the article the characteristic features of natural light are analyzed and on their basis technological methods of designing functional artificial lighting, approximated to the properties of natural light, are determined. In the process of research such methods as analysis, synthesis, generalization, and scientific observation were applied. The consideration of natural light properties was based on its following characteristic features – light intensity, colour temperature, peculiarities of light distribution, dynamics, cyclicity. The research has both theoretical and practical value when forming design proposals and creating appropriate light scenes and scenarios. Consequently, a conceptual list of technological methods of functional artificial lighting has been formed as a part of the design of light environment approximated to natural lighting. These methods include imitation only of that part of the diurnal daylight lighting scenario when the intensity of light is sufficient, and the light distribution has a positive emotional and psychological influence; combination of diffused and direct light and the use of different colour temperature sources for them; coordination of the dynamics of the functional artificial lighting system with a 24-h diurnal cycle, accompanied by transitions between light scenes that are inconspicuous in the course of short-time observation.

Keywords: Light design · Technological methods · Functional artificial lighting · Human centric lighting · Natural lighting

1 Introduction

Studies of the last decades have proved the existence of a correlation between room illuminance and work capacity, health, psycho-physiological state of people. For example, nowadays, it is recognized that the best way to treat seasonal affective disorder (SAD) is the bright light of artificial lighting systems that are used to compensate for insufficient natural lighting (Green Schools 2007). This fact brought the quality of lighting to the forefront of ecological problems of light design of premises. Another tendency is that at the level of modern standards one of the significant requirements for high-quality lighting is the importance of psychological connection of users with the outer space or the possibility of looking at distance for visual comfort and prevention of eye fatigue (Advice on Standards for School Premises 2015).

Z. Blikharskyy et al. (Eds.): CEE 2019, LNCE 47, pp. 174–179, 2020.
https://doi.org/10.1007/978-3-030-27011-7_22

As a result, more and more researchers and promoters of academic and practical studies in light design (Routledge 2015) address the topic of human centric lighting. Such an approach to lighting design was initiated in the late 1990s by the development and production of systems that simulated natural daylight fluctuations (Fiell and Fiell 2013). Today, human centric lighting is regarded as a particularly relevant trend.

The unexpectedly rapid advancement of LED lighting devices since 2000 (Oksanen 2017), new developments in lighting materials, an unprecedented wave of innovations in the field of lighting management continue to transform the possibilities of light design. The consequence was the emergence of a number of new ways of simulating the properties of natural light by means of artificial one, which do not ensure the absolutely accurate reproduction of natural light environment, but extend the qualitative characteristics of artificial light.

Summing up, one can state that natural light is most beneficial to people. Therefore, all the above-mentioned tendencies actualize the search for adequate technological methods of approaching the functional artificial lighting properties to those of natural light.

2 Purpose

The purpose of this study is to determine the technological methods of designing functional artificial lighting, which approximates to the properties of natural light, on the basis of the analysis of its characteristic features.

3 Methods

In the course of this research such widely used general scientific methods as analysis, synthesis and generalization were applied. Furthermore, the formation of judgments and conclusions was based on scientific observation, supported by the ideas of well-known researchers of light and colour in the nature (Minnaert 1969) and the perception of light and colour (Padgham and Saunders 1978).

Consideration of technological methods of functional artificial lighting design for the approximation of its properties to natural lighting was carried out on the basis of such features of the latter as light intensity, colour temperature, peculiarities of light distribution, dynamics, cyclicity.

4 Results

The decisive peculiarity of changes in the natural light intensity during the day is the onset of the moment when its illuminance is insufficient to ensure the process of vision. Thus, the illuminance of the horizontal plane by the cloudless sky after sunset decreases significantly. At solar altitude of 0° it is 400 lx, but at −4° it is already 13 lx (Minnaert 1969), which is even less then illuminance of 20 lx which is necessary for the identification of faces (Perry and Fennelly 2014). On top of that, together with the changes in the illuminance, the observer's vision is switched from photopic to mesopic, from mostly one kind light-sensitive photoreceptors (cones) to mainly the other (rods).

Therefore, in view of these factors, in order to provide functional working lighting, it is appropriate to simulate changes in the colour temperature of light during the day for those intervals of the diurnal cycle when the intensity of natural daylight and the illuminance created by it are sufficient to ensure normal photopic vision. This assertion does not contradict the fact that according to the current lighting standards the average illumination must be significantly increased with the increase in the colour temperature of a light source (Natural and Artificial Lighting 2018). For example, the optimal indices of the illuminance levels dependence on the colour temperature are 200–800 lx for 3500–4000 K respectively (Martirosova 2017).

Moreover, changes in the intensity of natural light during the day are determined by the current cloudiness and clarity of the atmosphere, which, accordingly, lead to changes in light distribution. Through their own experience everyone can feel the oppressive effect of lighting in the dark weather. This observation suggests that it is sensible to imitate only that part of the diurnal daylight-lighting scenario when the intensity of light is sufficient, and the light distribution has a positive emotional and psychological impact.

On a sunny day everything looks brighter and more vibrant. When the sky is clear, 80% of light comes directly from the sun, 20% is formed by the sunlight diffused by the sky (Minnaert 1969). Therefore, the characteristic combination of diffused and direct light is an unchanging attribute of natural daylight, which must be taken into account when attempting to approximate the properties of artificial lighting to natural one.

Continuing the analysis of the interaction between the sunlight and the sky in natural light, one can notice that the true brightness and saturation of the colours are not completely perceived until the sun appears from clouds (Padgham and Saunders 1978). Where the sun shines, its bright yellowish rays predominate over the light emitted by the sky, and only light from the blue or gray sky reaches a shadow, so shadows are more bluish than the surroundings. This difference is further emphasized by light contrast (Minnaert 1969).

All these observations prove the importance of another characteristic of natural light for the formation of artificial light environment approximated in its properties to the natural one. It is the different colour temperature and different intensity of its two components – diffused and direct light. Accordingly, the scheme for the general overhead lighting of the premises can also include two components: a light ceiling that is uniformly lit by a cold white-blue light (close in its colour temperature to the skylight) and, on the ceiling, lamps with more intense direct warm white-yellow light (close in its colour temperature to the sunlight). The lighting obtained as a result of such a method follows the aforementioned characteristics of the emotionally pleasant natural lighting on a sunny day.

Furthermore, thanks to this method, it is possible to get rid of the monotony of uniformly diffused lighting by adding to it some of the direct rays that enrich the overall white lighting by the optical mixing of cold and warm shades. With the help of light of fixtures directed on certain working zones, it is possible to increase their general illuminance, and by changing the orientation of the directed light of a fixture it is possible to accentuate individual objects of the interior. At the same time, their visual perception will be expressed in a way inherent in natural light on a sunny day, including warm illuminated areas, which are subject to direct light rays of a fixture, and cold shadow areas, which are subject mainly to diffused light of ceilings.

According to this lighting scheme, the minimum normalized levels of illuminance for premises of a certain purpose can be provided by diffused light of a ceiling, and by means of the deliberate placement of direct light fixtures over certain working areas of the interior their level of illuminance may increase to optimal.

Kazakov (2010) emphasizes the importance of time as the fourth dimension of architectural compositions and notes that at the level of premises the category of time-movement is primarily represented by the natural dynamics of interior lighting, which is associated with the movement of the sun in the sky and with the transition from dark to light part of the day and vice versa. He also expresses the opinion that to establish a psychological connection between a closed room and open natural environment it is important for a person not only or not so much to see the external environment, but to feel the course of time, which is expressed in the change of natural illuminance.

The dynamics of natural light is determined by a 24-h diurnal cycle and is accompanied by regular alternation of light and darkness. However, the imitation of the daily reduction in the intensity of natural light is meaningless, as the insufficient amount of light is what forms the need for artificial lighting. Consequently, ensuring the natural cycle of light and darkness alternation by means of artificial lighting is inappropriate, but the coordination of the dynamics of any artificial lighting system with a 24-h diurnal cycle is desirable.

Natural lighting, just like any other dynamic lighting, helps people to see (feel) time even in absolutely static objects and in a state of permanent rest. When illuminated by the invariable light of immutable space, a sense of the flow of time is disturbed. Therefore, in order to reflect the cyclicity of the dynamics of artificial lighting, it is important not to precisely imitate the properties of natural light, but the presence of light dynamics as such, since any changes in light help to fix and track time. It is also important that each light scene, being a part of the overall dynamic scenario, provides the quality of lighting related to the visual tasks performed indoors, and has a positive effect on the psychological state of a user.

To the change in overall colour temperature of light associated with the passage of time, it is possible to add changes in the direction of light from the fixtures, as well as coordinated changes in various colour temperatures of the diffused and directed light. The important aspect is not the wide variety of light scenes, but their 24-h cycle and the gradual transition between them (it being unnoticed when observed for a short time). The construction of the indoor light environment in such a way that an indoor user would be able to determine the period (time) of the day only by the state of the light scene, without using any additional equipment, is an important step towards the creation of emotionally and physiologically favourable lighting.

It is possible to facilitate the correlation of some of the light scenes with a certain time interval in a diurnal cycle by means of "light markers", which are predefined in the design of the light scheme. According to Kazakov (2010), in natural lighting the role of such a marker can be performed by the passage of a sunspot on a wall or floor surface located near a window or under a lantern, which in the graphic image has the form of an hourly solar envelope. In artificial lighting the function of such markers can be acquired by light spots from direct light luminaires, whose direction of light rays changes cyclically over time.

Therefore, the analysis of characteristic features of natural light was used as a basis for distinguishing appropriate technological approaches to the design of functional artificial lighting.

5 Scientific Novelty

The subject matter of the article is based on such modern trends in light design as recognition of the correlation between room lighting and human work efficiency, the use of artificial lighting systems to compensate for the lack of natural lighting and, as a consequence, actualization of the problem of approximation of the artificial lighting properties to those of natural light. For the first time in this article the characteristic features of natural light are analyzed with the aim of their further use in the formation of a conceptual list of technological methods of functional artificial lighting as components of the emotional and physiologically favourable light environment.

6 Practical Significance

In addition to the theoretical significance, the results of the study have a specific practical value and can be used by specialists in the field of light design, as well as by interior designers and architects in the design of their project proposals and the creation of appropriate light scenes or scenarios.

7 Conclusions

In the course of the research it was established that for the formation of an emotionally and physiologically favourable artificial light environment, its properties should be as close as possible to the properties of the natural light environment.

Subsequently, basing on the analysis of the characteristic features of natural light, we identified the following technological methods of the functional artificial lighting design approximated to the natural properties:

- imitation only of that part of the diurnal daylight lighting scenario when the intensity of light is sufficient and the light distribution has a positive emotional and psychological influence;
- combination of diffused and direct light and the use of different colour temperature sources for them;
- coordination of the dynamics of a functional artificial lighting system with a 24-h diurnal cycle, accompanied by transitions between light scenes that are inconspicuous in the course of short-time observation.

References

Advice on Standards for School Premises (2015) For local authorities, proprietors, school leaders, school staff and governing bodies. Crown copyright, London. https://assets.publishing.service.gov.uk/government/uploads/system/uploads/attachment_data/file/410294/Advice_on_standards_for_school_premises.pdf

Fiell C, Fiell P (2013) 1000 lights. TASCHEN GmbH, Köln

Green Schools (2007) Attributes for health and learning. National Academy of Sciences USA, Washington

Kazakov HV (2010) Contemporary lighting architecture. «Rastr-7» Publishing, Lviv (in Ukraine)

Martirosova VG (2017) Ophthalmological and lighting engineering prospects for the introduction of LED light sources at industrial enterprises of Ukraine. In: Conference LED progress, Kyiv, Ukraine. http://ualedlight.org/images/LedExpo2017/1/Martirosova.pdf

Minnaert MGJ (1969) Light and color in the outdoors. Nauka, Moscow (translation into Russian)

Natural and Artificial Lighting (2018) DBN V.2.5-28:2018. State Building Codes of Ukraine. Ukrarkhbudinform, Kyiv (in Ukraine)

Oksanen J (2017) Design concepts in architectural outdoor lighting design based on metaphors as a heuristic tool (Doctoral dissertation). Aalto University School of Arts, Design and Architecture Department, Helsinki. https://shop.aalto.fi/media/filer_public/8b/25/8b253f49-c052-4249-b518-5f754dd199b5/oksanen_verkkoversio.pdf

Padgham CA, Saunders JE (1978) The perception of light and colour. «Mir» Publishing, Moscow (translation into Russian)

Perry M, Fennelly L (2014) The handbook for school safety and security. Butterworth-Heinemann is an imprint of Elsevier, Waltham

Routledge G (2015) Is human centric lighting witchcraft? LuxLive. http://luxreview.com/article/2015/10/is-human-centric-lighting-witchcraft-

Application of Thermosiphon Solar Collectors for Ventilation of Premises

M. Ulewicz[1(✉)], V. Zhelykh[1], Kh. Kozak[2], and Y. Furdas[2]

[1] Department of Construction Organization and Technology, Czestochowa University of Technology, Generała Jana Henryka Dabrowskiego 69, 42-201 Czestochowa, Poland
ulewicz@onet.eu

[2] Department Heat and Gas Supply and Ventilation, Lviv Polytechnic National University, St. Bandery 12, Lviv 79013, Ukraine
cr__i@ukr.net

Abstract. The article deals with an analysis of the popular existing solar heating systems and the main directions of modern research on air heating solar collectors. The analysis has made it possible to outline the purpose of the research, namely, to study the nature of the distribution, as well as the rate of distribution of heated air mass in the room heated by the presented solar air heater with turbulators flow. Based on computer simulation and experimental research, it has been found that a thermosiphon air solar collector is capable of providing certain comfortable indoor climate parameters for a person, namely temperature and air flow velocity. In addition, the air solar heating system with a phase-change battery has been presented. The main difference of this system is the availability of the heat-accumulating coating installed in the air channel in the form of a matrix with segments filled with saturated solution of Glauber salt.

Keywords: Solar system · Thermosiphon collector · Heat-accumulating coating · Glauber salt

1 Introduction

Solar energy is considered one of the most popular types of alternative energy worldwide. According to new analytical research by Bloomberg New Energy Finance in 2018, about 50% of the world's energy will be produced by wind and solar energy by 2050 (Bullard 2018). As of 2015, 7% of electricity was produced due to the transformation of solar energy into electricity in the US. Although the transformation of solar energy into electricity confidently dominates, the preparation of hot water, heating and cooling of buildings, drying materials and food products using solar energy is developing in many countries.

However, the installation of solar energy conversion equipment is still expensive enough to restrict its popularization.

That is why the popularity of Heating Homes, where the heating or cooling is carried out with the help of passive cheap solar air collectors has grown significantly in

Z. Blikharskyy et al. (Eds.): CEE 2019, LNCE 47, pp. 180–187, 2020.
https://doi.org/10.1007/978-3-030-27011-7_23

the last decade. Air solar collectors, compared with water, are simpler and therefore cheaper construction. Main advantages: absence of freezing and boiling of coolant; corrosion resistance, higher solar energy usage; easier management, longer lifetime. However, the air also has substantial disadvantages - low heat capacity and low density. In order to increase the efficiency factor, such devices require careful selection of the solar collector design, the shape of heat absorbing plate and heat-accumulative material that will be suitable for low-temperature air heating (Zhelykh et al. 2016).

In addition to heating the premises, solar collectors are able to provide certain air mobility, which is one of the mandatory conditions for a comfortable stay of the person in the room. However, this feature of air heating systems is poorly researched. Therefore, the authors of the article have paid particular attention to determining the speed and nature of movement of air in a room heated by air solar collectors.

2 Analysis of Existing Data

There are many versions of the designs of air solar collectors and air solar heating/cooling systems. From the simplest passive structures that operate due to the difference in temperature of the coolant (Fig. 1a) (Rey 2015), with a useful efficiency of no more than 30% to complex active systems using the fans for transportation of the coolant and accumulation of heat (Fig. 2b) (Matuschka 2014). Passive solar collectors are often called thermosiphon solar collectors because of the nature of their work.

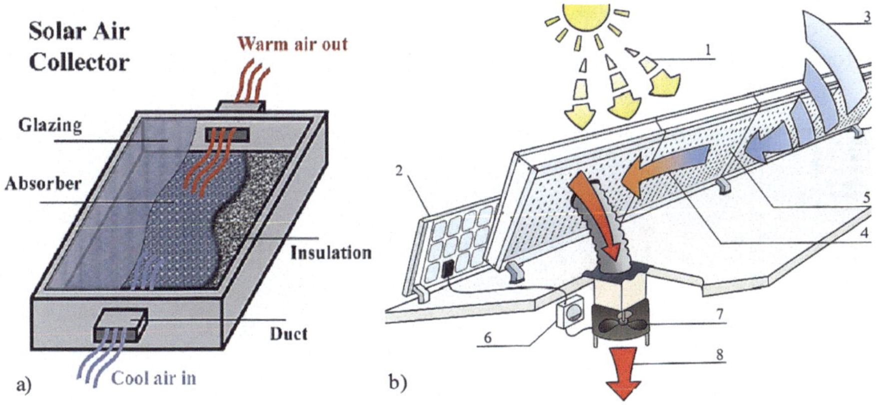

Fig. 1. The design of the passive air solar collector (a) and the diagram of the active solar air heating system (b): 1 – solar radiation; 2 – solar air collector; 3 – cold air; 4 – warm air; 5 – fresh air entry holes; 6 – room thermostat; 7 – fan; 8 – hot air warms the room

What is more, the combined systems of solar heat supply are quite popular in the countries with warm climate. In this system, one part of the converted heat is used for heating or cooling buildings, and the other - for preparing hot water (Fig. 2a) (Kasynets et al. 2018).

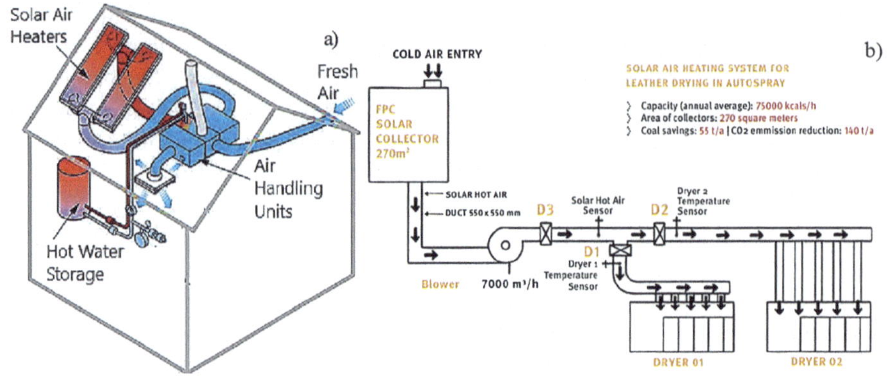

Fig. 2. The design of the passive air solar collector (a) and the diagram of the active solar air heating system (b)

In addition to heat supply and refrigeration, air heaters are used in agrarian industry, construction, etc. Fig. 2b shows a system for drying leather with the main element - the air solar collector (Knoedler 2018).

At present, China, the United States, Canada, and India show a particular interest in research, improvement and introduction of air collectors. 3-D simulation of physical processes occurring in a solar air heater with one/two air channels and a heat-insulated corpus when changing the shape of the heat-absorbing plate is presented in (Jassim and Shbailat 2018).

The studies were carried out for the flat heat absorber (Fig. 3a), as well as for perforated plates with fins of a triangular (Fig. 3b) and a rectangular configuration.

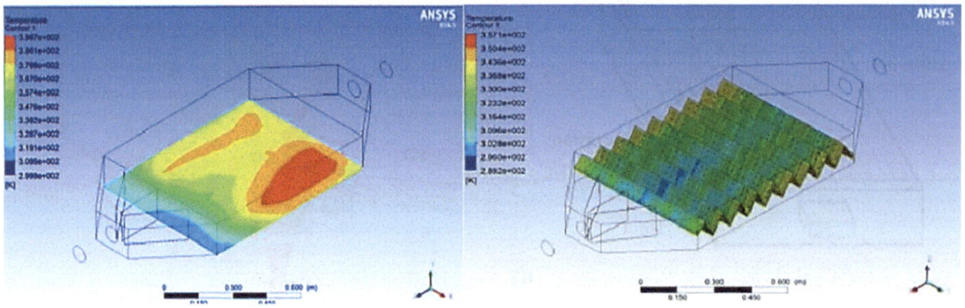

a) b)

Fig. 3. The design of the passive air solar collector (a) and the diagram of the active solar air heating system (b)

It has been found that the highest efficiency of performance has a solar collector with a perforated plate with triangular fins, the efficiency of which is 66.4% as opposed to 35.9% in a solar panel with a flat plate and one air channel, 45.3% with two air channels and 58.1% with a rectangular fin plate.

However, this design does not allow the installation of a solar air heater in the facade or roof of the building without the use of air ducts, the presence of which contributes to additional heat loss. Besides, only the distribution of temperature and velocity of the coolant in the air heater has been considered in the paper, and the nature of the air mass distribution in the heating room has not been studied.

The analysis of literary sources has shown that major work is aimed at studying the distribution of heated air masses in the collector housing, the choice of the effective shape of the heat-absorbing plate, the number of air channels, etc. (Salman and Hamdi 2018; Saha and Sharma 2017). However, the impact of the solar air heater operation on the microclimate in the room needs further research.

3 Experiments and Their Analysis

On the basis of a detailed analysis of literary sources, computer simulation and experimental studies, the design of the air thermosiphon solar collector with an improved shape of heat-absorbing surface has been developed by the authors (Fig. 4). In the solar collector, in addition to the flat heat-absorbing plate in the air channel along the motion of the coolant, air turbulators have been installed. The turbulators are made in a fixed form as a screw made of a sheet steel with the applied selective coating.

With the help of computer simulation, it has been determined that the optimal height of the air channel, taking into account preservation of the integrity of the translucent coating, is 0.08 m with an outer diameter of the screw of 0.05 m (Zhelykh et al. 2018). Moreover, it has been found that when the location of the input and output holes is at the bottom of the collector, and not in the sidewalls, the best heating of the heat transfer agent is obtained.

The bottom and sidewalls of the collector housing are covered with a layer of thermal insulation to minimize heat loss in the environment. The developed design can be installed on the outer wall or roof of the building or mounted in an external constructions.

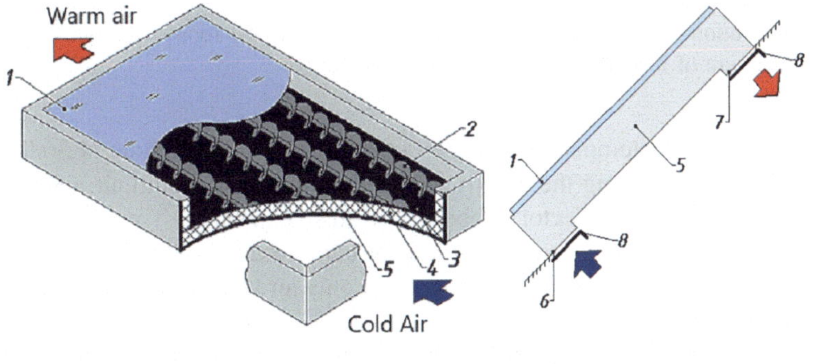

Fig. 4. Structural appearance of the air solar collector with flow turbulators: 1 – glassing; 2 – flow turbulators; 3 – heat-absorbing plate; 4 – thermal insulation; 5 – collector corps; 6 – entrance hole; 7 – outlet hole; 8 – control valve

The next important step was to study the influence of the air thermosiphon solar collector on the parameters of the microclimate in the room, namely temperature and air flow. To begin with, computer simulation has been created using software. A model of the room with its built-in air solar collector with turbulators of airflow has been developed. Output data for computer simulation has been summarized in Table 1.

Table 1. Output data for visualizing the thermal state in an accommodation with an installed air solar collector.

Parameter name	Units	Definition range	
		Minimum	Maximum
Area of the room	F_{room}, m^2	12	
Temperature of the absorber plate	t_{abs}, °C	50	80
Average temperature of the inner surface of the room walls	t_w, °C	17	
Intensity of solar irradiance	I, W/m^2	400	800

The characteristic results of the temperature distribution and the air velocity in the room are given on Fig. 5.

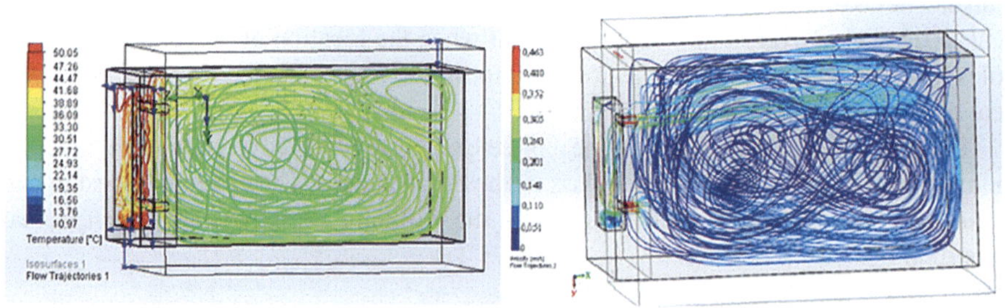

a) b)

Fig. 5. Distribution of temperature (a) and air velocity (b) in the room with the solar collector with a temperature of heat absorber plate tabs equals 80 °C

The maximum air temperature at the outlet from the supply hole is 36.0 °C, followed by its decrease along the X axis (Fig. 6a), and the maximum air velocity at the outlet from the air solar collector is about 0.35 m/s (Fig. 6b).

The maximum air speeds are on the axis of concentrated airflow. According to the data, the air solar collector is able to provide not only air heating but also a comfortable airflow for people.

However, to check the adequacy of the results, and to test the impact of external factors, it was necessary to conduct a series of experimental studies. For this purpose, room with parameters similar to the previous ones and with the built-in air solar collector has been assembled. Since such solar collector can be installed into the roof structure or placed on the outer wall of the building the angle of its inclination varied from 45 to 90° during the experiment. Consumption of the heat transfer agent in the air channel of the collector was regulated with the help of the flaps.

The nature of the distribution of isotherms in the volume of the room was investigated (Zhelykh 2016). It should be noted that the most effective for accepted conditions of operation of the passive air solar collector is its vertical installation within the working area of the room. This happens because this solar collector belongs to passive and works on the principle of free convection.

According to the regulatory documents of Ukraine (Ministry of Regional Development, Building and Housing and Communal Services of Ukraine 2013), the air velocity in the working zone for a non-heating period must not exceed 0.5 m/s, and for heating period – 0.2 m/s. The mobility of the air that the solar collector is able to provide depends on the intensity of the heat flow, the area and the distance between the inlet and outlet of the collector, its inclination angle. Pursuant to International Organization for Standardization ISO 7730-2011 (ISO 7730:2011 2011) the maximum air speed in residential areas must not exceed 0.12 m/s during the non-heating period and 0.1 m/s during the heating period.

The nature of the distribution and air velocity in the experimental module are shown in Fig. 6.

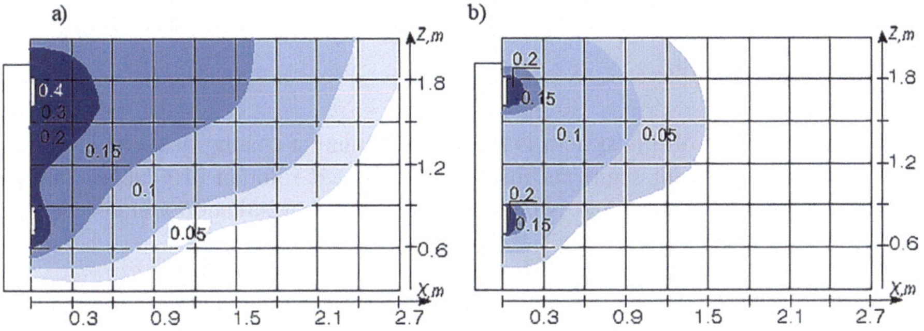

Fig. 6. Speed fields in the experimental module with a passive solar collector with turbulators flow at (a) β = 90 deg., I = 850 W/m^2; (b) β = 90 deg., I = 150 W/m^2

We can see from the figure that given the intensity of radiation of 850 W/m^2, which corresponds to the intensity of radiation during a clear sunny day, and the angle of solar collector of β = 90° the long-range propulsion of the jet is 2.7 m, and the maximum air velocity is observed at the outlet from the solar collector and is 0.4 m/s. Given the intensity of 150 W/m^2, just as in the cloudy period, and vertical installation of solar collector, the long-range air jet is 1.6 m, and the air velocity at the outlet from the solar collector is 0.3 m/s.

It is important to note that the passive solar collector is able to some extent to provide a thermal regime in the room with non-isothermal air jets not only during high solar activity but also during cloudy periods.

However, passive systems can efficiently perform the function of only background heating and have little effect during the heating period, during high clouds or at night. In order to solve this problem a system of solar air heating with a thermal accumulator with phase-change material from Glauber salt has been developed (Fig. 7).

a) b)

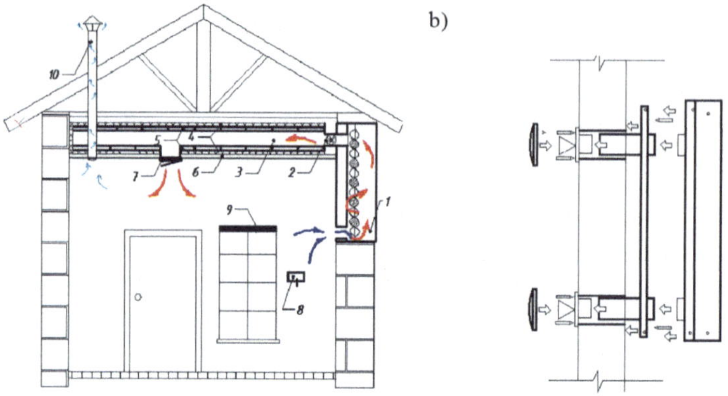

Fig. 7. Diagram of the solar air heating system with a thermal battery based on the air solar collectors installed in the design of the outer wall (a) and on the outside of the house (b): 1 – solar air collector; 2 – air duct; 3 – heat-absorbing coating with Glauber salt; 4 – fan; 5 – heat insulation; 6 – suspended ceiling; 7 – ventilation grate; 8 – temperature sensor; 9 – window vent valve; 10 – ventilation duct

The main feature of this system is a heat-accumulating coating installed in the air channel as a matrix with segments filled with saturated solution of Glauber salt. In addition, ventilation of the room is provided by installing ventilation window valves.

4 Conclusions

This article describes the basic types of solar air heating systems. The main directions of research, which were carried out to determine the effective design of the air heater, have been presented. However, the aspects concerning the influence of the air solar collector on the basic parameters of the microclimate, such as the temperature and airflow in the room, have not been adequately investigated. The results of computer simulation and experimental studies that are related to the study of the nature, features and speed of the heated air mass spreading in the room heated by the air thermosiphon solar collector, have been presented in this paper. The experimentally obtained data showed that for the vertically installed passive solar air collector, given an intensity of thermal radiation of 150 W/m^2, the long-range propulsion of the stream is equal to 1.6 m, and the air velocity at the outlet from the solar collector is 0.3 m/s. At the same time, given the intensity of 850 W/m^2, which relates to the intensity of radiation during a clear sunny day, the long-range jet reaches 2.7 m, while the maximum air velocity is observed at the exit from the solar collector and is 0.4 m/s. Based on the obtained data, it has been concluded that the thermosiphon passive solar collector can be used in a moderate climate or in the cloudy period for partial background heating of the room. To increase the efficiency of use of such devices during the heating period and at night, an active system of solar heating with a phase-change heat accumulator has been presented.

References

Bullard N (2018) Approaches for using scenarios in strategic decision-making. https://about.bnef.com/new-energy-outlook/

International Organization for Standardization (ISO) (2011) ISO 7730:2011. Ergonomics of the thermal environment - Analytical determination and interpretation of thermal comfort using calculation of the PMV and PPD indices and local thermal comfort criteria, Switzerland

Jassim NA, Shbailat SJ (2018) Three-dimensional CFD analysis for simulating dual channel solar collector with different absorbing media. Int J Therm Technol. https://www.researchgate.net/publication/324992242_Three-dimensional_CFD_Analysis_for_Simulating_Dual_Channel_Solar_Collector_with_Different_Absorbing_Media

Kasynets M, St Shapoval, Kozak Kh, Hulai B (2018) Analytical studies of coolant temperature in solar panel. JEECS 4(1):37–44

Knoedler J (2018) Solar air heating system. UNIDO. Kanpur Leather Development Project. https://leatherpanel.org/sites/default/files/publications-attachments/factsheet_4_solar_air_heating_system_p.pdf

Matuschka M (2014) Solar heating. https://www.solamate.com.au/solar-heating/

Ministry of Regional Development, Building and Housing and Communal Services of Ukraine (2013) State construction norms of Ukraine V.2.5-67:2013. Heating, Ventilation And Air Conditioning, Kyiv

Rey JF (2015) Project of application the cogeneration systems in hospitals in Colombia. https://www.monografias.com/usuario/perfiles/javier_fernandez_rey/monografias. (in Spanish)

Saha SN, Sharma SP (2017) Performance evaluation of corrugated absorber double flow solar air heater based on energy, effective and exergy efficiencies. Int J Mech Mechatron Eng 17 (1):63–76

Salman Ab Z, Hamdi RT (2018) Theoretical technique for studying the effecting factors for loss coefficients in solar collectors. Int J Trend Res Dev 5(2):359–363

Zhelykh V, Kozak Ch, Savchenko O (2016) Using of thermosiphon solar collector in an air heating system of passive house. Pollack Period 11(2):125–133

Zhelykh V, Kozak Kh, Dzeryn O, Pashkevych V (2018) Physical modeling of thermal processes of the air solar collector with flow turbulators. JEECS 4(1):9–16

Application of Theory of Experimental Design and FEA to Assessment of Rotation Capacity of Steel Joints

Aleksander Kozłowski[1(✉)] and Krzysztof Ostrowski[2]

[1] Department of Building Structures, Faculty of Civil and Environmental Engineering and Architecture, Rzeszow University of Technology, Poznanska 2, 35-084 Rzeszow, Poland
kozlowsk@prz.edu.pl
[2] Design Office, MTA Engineering Ltd., Rzeszow, Poland
krzysztof.ostrowski@mta-online.net

Abstract. Rotation capacity plays a very important role in an advanced analysis of steel frames, especially in plastic design of structures. In current version of Eurocode 1993-1-8 there is a lack of design procedure to calculate rotation capacity of steel structure joints. In the paper the calculation procedure of the rotation capacity for stiffened bolted end-plate beam-to-column connections has been proposed. Because of many variables influencing rotation capacity of such joint, theory of experimental design was used. The analysis was performed with the use of finite element method, based on the numerical experiment plan. The determination of maximal rotation angle was carried out with the use of regression analysis. The main variables analyzed in parametric study were: pitch of the bolt "w" (120–180 mm), the distance between the bolt axis and the beam upper edge c_{g1} (50–90 mm) and the thickness of the end-plate t_p (10–20 mm). Obtained simple formula can be used in everyday design practice to estimate rotation capacity of the joint.

Keywords: Steel joint · Rotation capacity · FEM analysis · Experimental design · Regression analysis

1 Introduction

The design procedures given in (EN 1993-1-8. 2005. Eurocode 3: Design of steel structures – Part 1-8: Design of joints, 2005) allow estimating the moment resistance $M_{j,Rd}$ and initial stiffness $S_{j,ini}$ of the bolted joints, but there is not enough information for the calculation of a connection rotation capacity φ_{cu}. There are only qualitative descriptions of rotation capacity of selected connection types. The previous analysis of rotation capacity was carried out with the use of component method (Kuhlmann and Fürch 1997; Silva and Girao 2001; Beg et al. 2004), probabilistic analysis (Beg and Zupancic 2004; Silva et al. 2004) and finite element method (Ostrowski 2014; Ostrowski and Kozlowski 2014), but without a practical guidelines how to use them in the design.

Z. Blikharskyy et al. (Eds.): CEE 2019, LNCE 47, pp. 188–194, 2020.
https://doi.org/10.1007/978-3-030-27011-7_24

Rotation capacity is influenced by many variables, as:

- material (steel) used,
- type of joint,
- geometry of the joints (configuration, bolts arrangement, methods of joint stiffening),
- profiles used for beam and column section,
- joint location (external or internal joint).

Large number of factors influencing rotation capacity forced researchers to find simple, alternative to component method, design procedure. Such procedure is presented in the paper based on theory of experimental design and fine element analysis.

2 Theory of Experimental Design

Theory of experimental design (Polanski, 1984) gives researches a very effective and efficient tool to analyse multi-factors processes in the technology and science. It allows to remarkable reduction of needed experiments which must be conducted to obtain analytical model of investigated object.

Creating a mathematical model should follow as below:

- The characterization stage - compilation, based on general substantive analysis the set of factors affecting the research object,
- parametric study to find factors having the strongest impact on object,
- classification of these factors into:
  - research factors X (controlled, independent) feature of them is the ability to change during the implementation of tests,
  - resultant factors Z (dependent); have a basic effect on the object, the values are obtained from the measurements,
  - constant factors C, so-called unchanged test conditions, although they have an impact on the object, for other reasons, their impact is not the purpose of the research, e.g. the effect is known, or goes beyond the scope of technical possibilities,
  - disturbance factors H, measurement inaccuracies, errors - they are the basis for the spread of results.
- The concept of a mathematical model, that is, a preliminary description of the type of the relationship (function) including the factors Z and X, e.g. polynomials, logarithmic, exponent.

The following types of models can be used in research:

- deterministic,
- random:
  - Probabilistic,
  - Statistical,
  - Stochastic,
  - Strategic.

Models can be continuous or discreet.

3 Methodology for Rotation Capacity

In the paper theory of experimental design was adopted to numerical simulation by FEM, used to obtain rotation capacity of selected steel joint (Fig. 1).

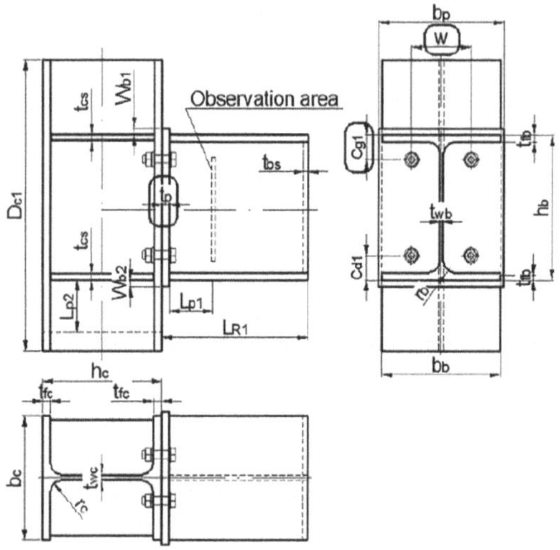

Fig. 1. View of investigated join and designation of variable factors: „t_p", „c_{g1}" and „w".

Rotation capacity was assess by using the following procedure:

- experimental tests of materials used to make the joints were carried out, i.e. steel used to make beams, columns and end plates. Similar tests were carried out for the steel from which the bolts were made,
- experimental tests of bolt assemblies (bolt-washers –nut) were carried out, with the measurement of deformation of individual bolt sections along its length,
- experimental tests of T-stubs, of various configurations, designed to obtain all models of failure proposed by (EN 1993-1-8. 2005. Eurocode 3: Design of steel structures – Part 1–8: Design of joints, 2005),
- hierarchical validation of FEM models was performed by comparing the results obtained from the above-described experimental tests with the results of FEM models, which enabled the calibration of models (Ostrowski and Kozlowski, 2015),
- a parametric study of factors influencing the joint behavior, in particular the rotation capacity. As a result of the analysis, the factors that have the strongest influence on the rotation of the joint were selected, i.e. thickness of the end-plate t_p, horizontal bolt spacing w, distance from the bolt axis to the upper edge c_{g1}.
- on the basis of the substantive analysis of the most frequently used in practice solutions, the range of variability of the tested factors was determined as:
  $t_p = 10 \div 20$ mm,
  $w = 120 \div 180$ mm,
  $c_{g1} = 50 \div 90$ mm,

- constant factors have been determined as:
  Beam section HEA 360, column section HEB 300,
  S235 steel grade, from which joint will be made,
  M20 bolts, class 10.9, according to ISO 4014.
- in order to determine the function of the maximum rotation angle of the joint, the theory of experimental design (Polanski 1984) adapted to numerical simulations was used. Hartley's PSIS, PS/DS-P: Ha3 (Table 1), based on the hyper sphere (hK) with 11 variable factor systems was chosen,
- using a validated FEM model, calculations of joints with dimensions resulting from the experiment plan were made in order to obtain a rotation angle at which there is no increase of the joint capacity (Fig. 2),
- the resulting values from the above calculations were used in the regression analysis to determine the function of the achievable angle of rotation.

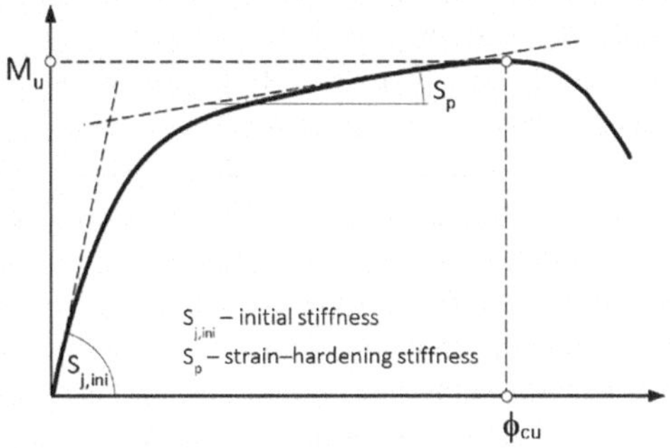

Fig. 2. Graphical representation of rotation capacity

Table 1. Experimental plan Hartley's PSIS

DPi	$\breve{x}_k$			x_k			Rotation [mrad]
	$\breve{x}_1$	$\breve{x}_1$	$\breve{x}_2$	t_p	c_{g1}	w	f_{cu}
DP1	−1	−1	1	12,11	58,45	167,32	83.07
DP2	1	−1	−1	17,89	58,45	132,68	13.98
DP3	−1	1	−1	12,11	81,55	132,68	106.69
DP4	1	1	1	17,89	81,55	167,32	43,75
DP5	−1,732	0	0	10,00	70,00	150,00	106,84
DP6	1,732	0	0	20,00	70,00	150,00	15,85
DP7	0	−1,732	0	15,00	50,00	150,00	29,38
DP8	0	1,732	0	15,00	90,00	150,00	86,58
DP9	0	0	−1,732	15,00	70,00	120,00	45,37
DP0	0	0	1,732	15,00	70,00	180,00	72,59
DP1	0	0	0	15,00	70,00	150,00	59,81

The function of the maximal rotation angle was defined as a polynomial function as below:

$$\varphi_{cu} = t_p^{\alpha} \cdot c_{g1}^{\beta} \cdot w^{\gamma} - A \tag{1}$$

where: α, β, γ – are the coefficients of polynomial,
t_p [mm] – is the thickness of end-plate,
c_{g1} [mm] – is the distance from the top edge of upper flange to the axis of bolt,
w [mm] – is the spacing between the centers of bolts,
A – is the constant of the polynomial.

4 Hierarchic Validation of FEM Models

The use of correct and actual material characteristics in FEM analysis is the basic factor affecting the accuracy of the results obtained. The reliability of the finite element method results can be considered satisfactory if results coincide with the results of laboratory tests. In order to obtain convergent results, it is necessary to tune the MES models. Tuning of models can be obtained by hierarchical validation of joint components. The hierarchical validation process of FEM models is based on a comparative analysis of models with a different degree of complexity. Four stages of hierarchical validation of FEM models was applied in the present analysis.

In the first validation stage, the characteristics σ-e of the steel specimen used for the execution of the joint elements and the steel of the bolt subjected to tension were determined (Fig. 3a). In the second stage of the test, isolated fasteners in a bolt-washer-nut system (Fig. 3b) were tested. The third stage consists in determining a comparative characteristic F-Δ describing the deformation state of the end-plate joint defined in the form of extended T-stubs (Fig. 3c). In the last stage of hierarchical validation, the model of beam-to-the column joint was analyzed (Fig. 3d). The measure of validation

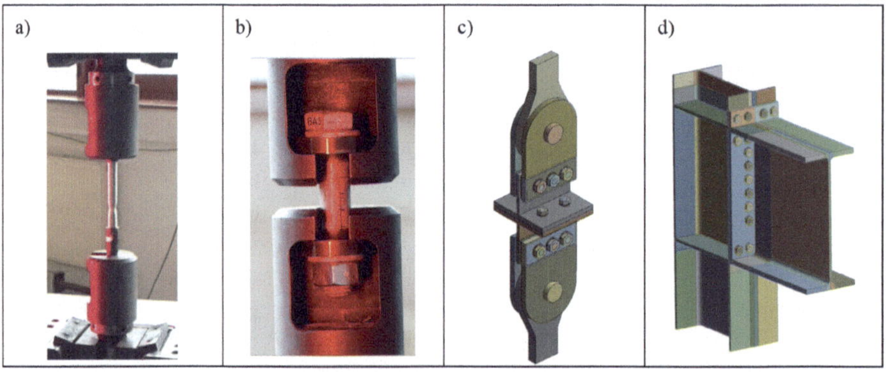

Fig. 3. Stages of hierarchical validation: (a) Stage I – tension test of a steel sample, (b) Stage II - tension of a bolt in a bolt-washer-nut set, (c) Stage III - tension of T-stubs, (d) Stage IV - bending of the beam-to-column joint.

of this stage is to obtain convergent rotation angle results based on the obtained in test characteristics M- φ. The tests were carried out in the Structural Research Laboratory of Rzeszow University of Technology, Faculty of Civil and Environmental Engineering and Architecture.

In the FEM analysis validated material models have been applied.

5 Results of the Analysis

Solving all configurations of the experiment's plan allowed elaborating the response surface of the rotation angle. Results of this analysis are presented in Fig. 4.

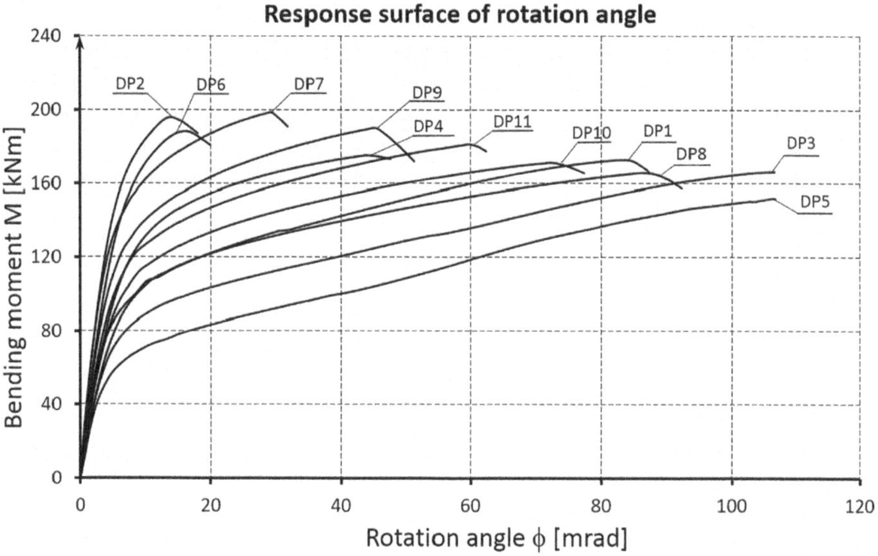

Fig. 4. Response surface of rotation angle of experiment plan.

Based on the calculations carried out, a large variation in the received angles of rotation with respect to individual layouts of the experimental plan is observed. The values of the angles of rotation obtained are within the following limits: φ_{cu} = 14 ÷ 106 [mrad].

The coefficients of polynomial described by the Eq. (1) were calculated by the program Statistica 12, with the use of the nonlinear regression anlysis. Inserting the calculated coefficients one can receive the function of available rotation capacity of joint in the following form:

$$\varphi_{cu} = t_p^{-1,267} \cdot c_{g1}^{1,044} \cdot w^{0,714} - 42,48 \quad (2)$$

6 Conclusion

The component method proposed in the standard (EN 1993-1-8. 2005. Eurocode 3: Design of steel structures – Part 1–8: Design of joints, 2005) in the field of forecasting the rotation capability has not yet been fully developed, which leads to the search for other, alternative but possibly reliable methods for estimating the rotation capacity of joints in steel structures. Such attempt with the application of theory experimental design is presented in the paper.

The presented analysis show, a clear tendency of reaching the maximum rotation angle φ_{cu} obtained as a result of the bolt capacity being exhausted, regardless of the mode of failure presented in the code. In connections in which the first mode of failure occurs force redistribution took place, which results in the "membrane" effect. The end plate following the progressive degradation of stiffness passes from the flexion state into the membrane state, which has not yet analytical description.

Introduction in FEA material characteristics which are the result of a multi-stage process of fine-tuning FEM models implemented as part of hierarchical model validation is a prerequisite for obtaining reliable results of FEA used in simulation analysis.

The relationship proposed by formula (2) allows estimation of the achievable rotational capability of stiffened joints. Analyzes of other types of joints is the subject of a dissertation of the second author of the work.

References

EN 1993-1-8 (2005): Eurocode 3: Design of steel structures – Part 1-8: Design of joints

Kuhlmann, U, Fürch, A (1997): Rotation Capacity of Steel Joints. COST Project C1 Meeting

Silva S, Girao A (2001) A ductility model for steel connections. J Constr Steel Res 57:45–70

Beg D, Zupancic E, Vayas I (2004) On the rotation capacity of moment connections. J Constr Steel Res 60:601–620

Beg D, Zupancic E (2004) Statistical evaluation of rotation capacity of moment connections. Connections in Steel Structures V, Amsterdam

Silva S, Borges L, Garvasio H (2004) A probabilistic evaluation of the rotation capacity of end-plate beam-to-column steel joints. Connections in Steel Structures V, Amsterdam

Ostrowski, K (2014): Finite element analysis of the rotation capacity of beam-to-column end-plate bolted joints. Eurosteel 2014, 10–12 September, Naples

Ostrowski K, Kozlowski A (2014) Parametric study on the rotation capacity of stiffened end-plate beam-to-column connection (In Polish). Build Architect 13(3):2014

Polanski Z (1984) Experimental design in technique (In Polish). PWN, Warsaw

Ostrowski, K, Kozlowski, A (2015): Hierarchical validation of FEM models. In: XIV International Scientific Conference Rzeszow – Lwow – Koszyce, Rzeszow

Crack Resistance of RC Columns Strengthened by Jacketing

Pavlo Krainskyi[1], Yaroslav Blikharskyy[2(✉)], Roman Khmil[1], and Pavlo Vegera[1]

[1] Department of Building Constructions and Bridges, Lviv Politechnic National University, Karpinskogo 6, Lviv 79013, Ukraine
[2] Department of Highways and Bridges, Lviv Politechnic National University, Karpinskogo 6, Lviv 79013, Ukraine
yaroslav.z.blikharskyy@lpnu.ua

Abstract. The results of an experimental study on crack-resistance of large scale compressed-bent reinforced concrete columns strengthened under initial load are represented in this paper. Structural performance of unstrengthened reference samples and jacketed samples under initial load of 30%, 50%, 70% and 90% of the bearing capacity of the reference samples was investigated. During the tests of the columns cracks pattern along the length of the columns, cracks length and width development were observed and recorded. Crack width vs Load graphs were plotted for the analysis of cracks development. Crack width at each loading stage was recorded to determine the critical crack. The load N_w when the critical crack reached the width of w_{max} = 0.3 mm (according to Eurocode-2: Design of Concrete Structures) was considered as a loss of serviceability of the column. Analyzing the increase of columns serviceability after reinforced concrete jacketing the conclusion was made that higher load level during strengthening results in the lover strengthening effect. The maximum increase in serviceability of the columns considering the ultimate crack width was achieved when the sample was jacketed without any initial loading and the minimum - for the samples jacketed under the load of 90% of their bearing capacity.

Keywords: Crack resistance · RC columns · Strengthening · RC jacket · Large scale · Strengthened under initial load

1 Introduction

Reinforced concrete constructions occupy a large share in the sphere of construction products for the last 50–70 years. Recent studies are devoted to the use of up-to-date materials for improving the properties of reinforced concrete, structural solutions using high-strength steel in the form of steel concrete structures and the use of high-strength concrete (Bobalo et al. 2018; Al Saffar et al. 2019; Fadhil et al. 2018; Sobol et al. 2014; Tayeh et al. 2013). However, the costs of disassembling, replacing and disposing of individual structural elements or entire buildings are usually extremely large. In some cases, it is simply impossible to replace a particular structural element. Therefore, it is

Z. Blikharskyy et al. (Eds.): CEE 2019, LNCE 47, pp. 195–201, 2020.
https://doi.org/10.1007/978-3-030-27011-7_25

reasonable to strengthen and restore the bearing and enclosing structures of existing buildings in order to extend their operational suitability (Kramarchuk et al. 2019).

At the same time, the aggressive production environment of many enterprises, mechanical damage and untimely removal of acquired defects cause an urgent need to reinforce individual elements or structures in general (Al Sherravi et al. 2018; Blikhars'kyi and Obukh 2018; Khmil et al. 2018; Blikharskyy et al. 2018a, b, c; Selejdak et al. 2018; Kos et al. 2017). Along with physical deterioration, the need for reconstruction or reinforcement is dictated by the introduction of new design codes and standards. Considering the global use of reinforced concrete as a structural material, the research of reinforced concrete structures is relevant. Among the newest strengthening materials for reinforced concrete structures, modern composite materials are now widely discussed and investigated as both bending (Blikharskyy et al. 2018a, b, Brozda et al. 2017) and shear reinforcement (Blikharskyy et al. 2019; Vegera et al. 2018, Blikharskyy et al. 2018a, b, c). However, these materials have a number of disadvantages: the negative effects of ultraviolet radiation, the necessary use of additional protection for fire resistance and the high cost of materials that use different fibers. The method of strengthening using reinforced concrete jacketing is quite known and widespread. However, in the common practice of structural retrofitting and strengthening, it is rarely possible to completely remove the load on the structure, thus leaving at least self weight. That is why there is a need to find reliable methods for calculating and predicting the stress-strain state of strengthened reinforced concrete structures, taking into account the load on the element during strengthening. Therefore, in this paper we investigate the crack resistance of reinforced concrete columns strengthened by jacketing under the action of loads of different levels.

2 Experimental Program

In order to achieve the set goal and fulfill the set tasks of the research, 12 full-size research samples of reinforced concrete columns were designed and manufactured (Table 1).

Table 1. Test program

№	Specimens	Test type
1	C-1 and C-2	Without jacketing (reference samples)
2	C-3-0 and C-4-0	Jacketed without initial loading
3	C-5-0.3 and C-6-0.3	Jacketed with initial loading (30% bearing capacity of the reference samples)
4	C-7-0.5 and C-8-0.5	Jacketed with initial loading (50% bearing capacity of the reference samples)
5	C-9-0.7 and C-10-0.7	Jacketed with initial loading (70% bearing capacity of the reference samples)
6	C-11-0.9 and C-12-0.9	Jacketed with initial loading (70% bearing capacity of the reference samples)

Two reference columns were tested without jacketing to determine the actual parameters of strength, deformability and crack resistance of the samples. The remaining 10 columns were tested with reinforced concrete jacketing. The jacketing was performed without the any load and at a constant initial load of 30, 50, 70, and 90% of the bearing capacity of the reference samples.

3 Research Methodology

All specimens were tested as pinned columns on a compression test stand (Fig. 1). This stand was designed to conduct compression tests in horizontal position. The load was applied with eccentricity to create both axial compression and bending. For every column eccentricity equaled 150 mm. Load to the columns was applied incrementally by hydraulic jack. Loading level was determined from the readings of a ring dynamometer. The length of the columns was 2.2 m and a cross section dimensions before jacketing was 0.14 m × 0.18 m. Cross section dimensions of the jacketed columns were increased to 0.2 m × 0.26 m. All 12 columns were manufactured from the same materials. Reinforcement ratio of unstrengthened column equalled 1.8%. Four 12 mm rebar were used as longitudinal reinforcement with 6 mm rebar used for ties. The rupture stress of the rebar was 720 MPa; yield stress – 630 MPa; Young's modulus – 210 GPa. Average concrete cube strength equalled 37 MPa which corresponds to C25/30 concrete class.

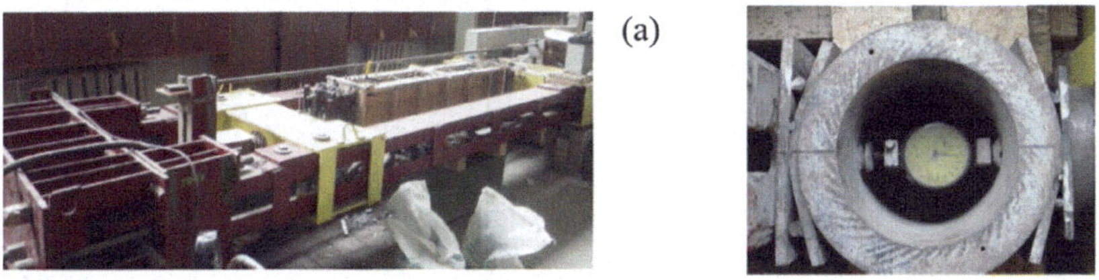

(a) (b)

Fig. 1. Jacketing of the test sample under constant initial load (a) and mechanical dynamometer (b)

Reinforced concrete jacketing length equalled 1.7 m. Reinforcement ratio equalled 1.17%. Four 10 mm rebar were used as jacketing longitudinal reinforcement with 6 mm rebar used for ties. Mechanical properties of concrete and rebar were the same as previous. More detail methodology are showed in previous articles (Krainskyi et al 2018a, b; Zhang et al. 2015).

4 Results and Discussion

4.1 Crack Resistance of Reference Columns

During the tests of the columns cracks pattern along the length of the columns, cracks length and width development were observed and recorded. Crack width at each loading stage was recorded to determine the critical crack. Crack width vs Load graphs were plotted for the analysis of cracks development.

The ultimate crack width of $w_{max} = 0.3$ mm was adopted according to Eurocode-2: Design of Concrete Structures. Column C-1 reached the ultimate crack width under the load of $N_w = 166.71$ kN, for C-2 - $N_w = 164.75$ kN. The difference between the results equals 1.2%. The difference between N_w and the ultimate load at failure N_u for the reference columns equals 5.6% and 4.5% for columns C-1 and C-2 respectively. Approximately $N_w = 0.95N_u$.

4.2 Crack Resistance of Jacketed Columns

In case of the rest tested samples the pattern of crack development before jacketing was the same as for reference samples. The ultimate crack width of $w_{max} = 0.3$ mm was adopted for jacketed columns. Crack width vs Load graphs were plotted for the analysis of cracks development. Strengthening effect diagram and crack width development graphs for all strengthened samples are presented on Fig. 2.

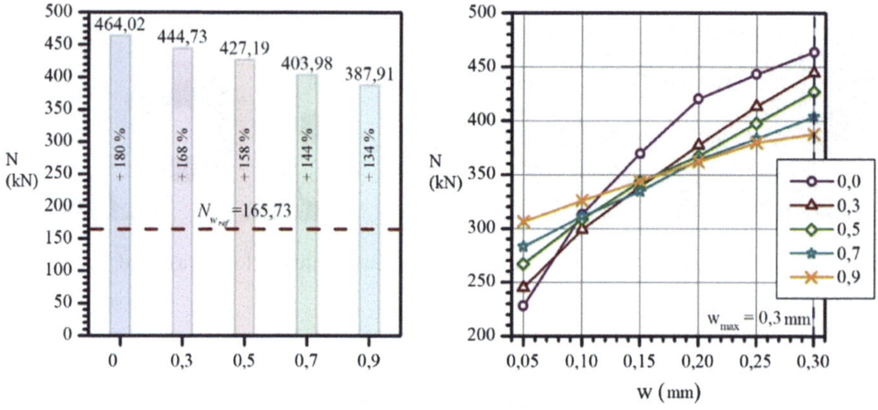

Fig. 2. Test results of serviceability of jacketed reinforced concrete columns.

Columns C-3-0 and C-4-0 reached the ultimate crack width at the load N_w = 471.53 kN and N_w = 456.50 kN respectively. C-5-0.3 and C-6-0.3 at the load N_w = 438.16 kN and N_w = 451.29 kN respectively. C-7-0.5 and C-8-0.5 at the load N_w = 443.21 kN and N_w = 411.16 kN respectively. C-9-0.7 and C-10-0.7 at the load N_w = 402.07 kN and N_w = 405.89 kN respectively. C-11-0.9 and C-12-0.9 at the load N_w = 387.45 kN and N_w = 397.37 kN respectively.

The crack pattern on the column before strengthening was recorded for comparison with the crack pattern of jacketed columns (Fig. 3). No clear match was detected between the cracks patterns before and after jacketing.

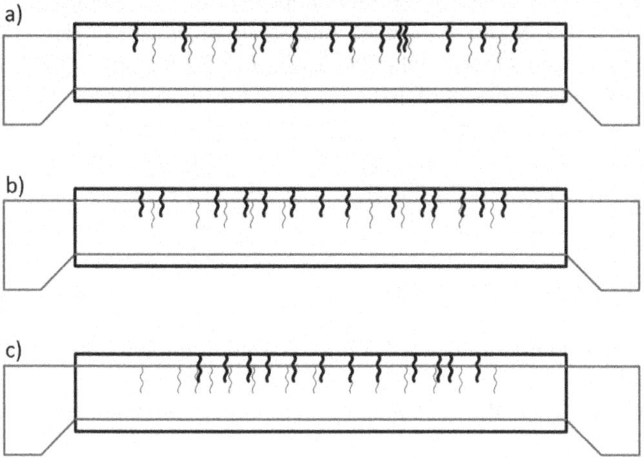

Fig. 3. Crack patterns for columns C-7-0.5 (a), C-10-0.7 (b) and C-11-0.9 (c)

The main test results are summarized in the Table 2.

Table 2. Serviceability of jacketed columns considering ultimate crack width

Specimens	Bearing capacity load, N_u, kN	Ultimate crack width load, N_w, kN		N_w/N_u	Strengthening effect, %
		Specimen	Average		
C-1	174.56	166.71	167.73	0.95	-
C-2		164.75			
C-3-0	474.93	471.53	464.02	0.98	180
C-4-0		456.50			
C-5-0.3	465.62	438.16	444.73	0.96	168
C-6-0.3		451.29			
C-7-0.5	442.52	443.21	427.19	0.96	158
C-8-0.5		411.16			
C-9-0.7	430.28	402.07	403.98	0.94	144
C-10-0.7		405.89			
C-11-0.9	397.51	378.45	387.91	0.98	134
C-12-0.9		397.37			

The highest strengthening effect was reached by the samples jacketed without initial loading and equaled 180%. The lowest strengthening effect was reached by the samples jacketed under initial loading of 90% of reference column capacity and equaled 134%.

First cracks with the width of 0.05 mm appear faster on samples C-3-0 and C-4-0 but the ultimate crack width is reached later than in other samples. The following observations were made: (1) the higher the level of initial loading before strengthening, the later comes the development of first cracks; (2) the higher the level of initial loading before strengthening, the sooner ultimate crack width is reached.

N_w/N_u for all test samples is in range 0,94–0,98. This shows that the ultimate crack width was reached almost before the failure of all samples.

5 Conclusions

The ultimate crack width in all samples was reached at the load N_w 2…6% lover then the failure load N_u.

The higher the level of initial loading before strengthening, the sooner ultimate crack width is reached.

The higher the level of initial loading before strengthening, the later comes the development of first cracks.

The maximum achieved strengthening effect was 180% for columns jacketed without initial loading. The lowest (134%) - for columns strengthened under load close to failure.

References

Al Saffar D, Alsaad A, Tayeh B (2019) Effect of internal curing on behavior of high performance concrete: an overview. Case Stud Constr Mater. https://doi.org/10.1016/j.cscm.2019.e00229

Al Sherrawi M, Lyashenko V, Edaan E, Sotnik L (2018) Corrosion as a source of destruction in construction. Int J Civ Eng Technol 9:306–314

Blikhars'kyi Z.Ya, Obukh Y (2018) Influence of the mechanical and corrosion defects on the strength of thermally hardened reinforcement of 35GS steel. Mater Sci 54. https://doi.org/10.1007/s11003-018-0183-2

Blikharskyy Y, Khmil R, Blikharskyy Z (2018a) Research of RC columns strengthened by carbon FRP under loading. Matec Web Conf 174:1–8. https://doi.org/10.1051/matecconf/201817404017

Blikharskyy Z, Brózda Kinga, Selejdak J (2018b) Effectivenes of strengthening loaded RC beams with FRCM system. Arch Civ Eng 64:3–13. https://doi.org/10.2478/ace-2018-0025

Blikharskyy Z, Selejdak J, Blikharskyy Y, Khmil Roman (2019) Corrosion of reinforce bars in RC constructions. Syst Saf Hum Tech Facil Environ 1:277–283. https://doi.org/10.2478/czoto-2019-0036

Blikharskyy Z, Vegera P, Vashkevych R, Shnal T (2018c) Fracture toughness of RC beams on the shear, strengthening by FRCM system. Matec Web Conf 183:02009. https://doi.org/10.1051/matecconf/201818302009

Bobalo T, Blikharskyy Y, Vashkevich R, Volynets M (2018) Bearing capacity of RC beams reinforced with high strength rebars and steel plate. Matec Web Conf 230:02003. https://doi.org/10.1051/matecconf/201823002003

Brózda K, Selejdak J, Koteš P (2017) The analysis of beam reinforced with FRP bars in bending. Procedia Eng 192:64–68. https://doi.org/10.1016/j.proeng.2017.06.011

Fadhil B, Al-Rumaithi A, Al-Sherrawi M (2018) Properties of reactive powder concrete with different types of cement. Int J Civ Eng Technol 9:1313–1321

Khmil R, Tytarenko R, Blikharskyy Y, Vegera P (2018) Development of the procedure for the estimation of reliability of reinforced concrete beams, strengthened by building up the stretched reinforcing bars under load. East Eur J Enterp Technol 5/7(95):32–42. https://doi.org/10.15587/1729-4061.2018.142750

Kos Ž, Gotal Dmitrović L, Klimenko E (2017) Developing a model of a strain (deformation) of a damaged reinforced concrete pillar in relation to a linear load capacity. Tech J 11(4):150–154. https://hrcak.srce.hr/190990

Krainskyi P, Blikharskyy Y, Khmil R, Blikharskyy Z (2018a) Experimental study of the srengthening effect of reinforced concrete columns jacketed under service load level. Matec Web Conf 183:1–5. https://doi.org/10.1051/matecconf/201818302008

Krainskyi P, Blikharskyy Y, Khmil R, Vegera P (2018b) Influence of loading level on the bearing capacity of RC columns strengthened by jacketing. Matec Web Conf 230:02013. https://doi.org/10.1051/matecconf/201823002013

Kramarchuk A, Ilnytskyy B, Lytvyniak O, Grabowski A (2019) The increase of seismic stability for existing industrial buildings. AIP Conf Proc 2077:020029. https://doi.org/10.1063/1.5091890

Selejdak J, Khmil R, Blikharskyy Z (2018) The influence of simultaneous action of the aggressive environment and loading on strength of rc beams. Matec Web of Conf 183:1–6. https://doi.org/10.1051/matecconf/201818302002

Sobol K, Blikharskyy Z, Petrovska N, Terlyha V (2014) Analysis of structure formation peculiarities during hydration of oil-well cement with zeolitic tuff and metakaolin additives. Chem Chem Technol 8(4):461–465. https://doi.org/10.23939/chcht08.04.461

Tayeh B, Bakar BH, Megat J, Megat A, Voo Y (2013) Evaluation of bond strength between normal concrete substrate and ultra high performance fiber concrete as a repair material. Procedia Eng 54:554–563. https://doi.org/10.1016/j.proeng.2013.03.050

Vegera P, Vashkevych R, Blikharskyy Z (2018) Fracture toughness of RC beams with different shear span. Matec Web Conf 174:1–8. https://doi.org/10.1051/matecconf/201817402021

Zhang Q, Mol'kov YV, Sobko YM, Blikhars'kyi YZ, Khmil RE (2015) Specific fracture energy of thermally hardened reinforcement. Mater Sci 50(6):824–829. https://doi.org/10.1007/s11003-015-9789-9

The Research Bearing Capacity of Crane Beams for Possible Establishment of Bridge Crane on Them

A. Kramarchuk[1], B. Ilnytskyy[1], T. Bobalo[1], and O. Lytvyniak[2(✉)]

[1] Department of Building Constructions and Structures, Lviv Polytechnic National University, 12 St. Bandery Str., 79013 Lviv, Ukraine
[2] Department of Civil Safety, Lviv Polytechnic National University, 12 St. Bandery Str., 79013 Lviv, Ukraine
lytvyniak.oksana@gmail.com

Abstract. At the time of reconstruction of edifice, it often appears a need in researches of bearing capacity for prefabricated reinforced concrete crane beams for the purpose of their exploitation in future. And also the designers look into possibility of installation modern bridge cranes with technical characteristics, which are different from foreseen or existing bridge cranes for establishment on these constructions. Different mark of cranes by equal weight-bearing capacity during the exploitation cause appearance of bending moments and shear in reinforced concrete crane beams, the values of which differ depending on technical characteristics of cranes. The maximum load of the wheel on the crane and the distance between the wheel's axes are important characteristics of hoisting cranes during researches of bearing capacity for prefabricated reinforced concrete crane beams. Accordingly, the exploitation of reinforced concrete crane beams is possible only after research of bearing capacity for these constructions by means of plotting of bending moment diagram and shear diagram, and moreover analysis of changing for deflected mode of these constructions.

Keywords: Crane beams · Hoisting crane · Crane loading · Reinforced concrete constructions

1 Introduction

Crane beams sustain vertical and horizontal loads, longitudinal and lateral loads, which originate from load lifting and load-lowering and also from braking of crane. In industrial buildings, depending on manner of manufacture, we can use cranes of light, medium, heavy and special operating mode. The difference between these cranes consists in character and duration of bridge cranes' work during day, speed of load lifting by cranes etc.

The loading from bridge cranes have dynamic character. Main characteristics of bridge cranes, which influence on loadings for crane beams, are their weight-bearing capacity, dimensions and crane span, own weight with carriage, operating mode and position of load on bridge crane. Limit vertical crane loads and limit transversal

Z. Blikharskyy et al. (Eds.): CEE 2019, LNCE 47, pp. 202–210, 2020.
https://doi.org/10.1007/978-3-030-27011-7_26

horizontal crane loads are taken into consideration from action two approached cranes on one track, however, at the same time, operational load is taken into consider from only one crane on track by (DBNV.1.2-2:2006 2006).

Reinforced concrete crane beams are used for frame industrial buildings with baulks (their widths are to 24 m) for bridge cranes with weight-bearing capacity to 20 t inclusive. Reinforced concrete crane beams are predominantly designed divided and they are reinforced by prestressed reinforcement in lower stretched zone.

When spacing of columns is 6 m, then reinforced concrete crane beams have T-section. Their heights are equal to 800 mm (for height of columns - 8.4 m) and are equal to 1000 mm (for columns with bigger height). If the spacing of columns is 12 m, then crane beams have I-section and their heights are 1400 mm.

This renovation was carried out according to the researches and propositions of the scientists such as (Blikharskyy et al. 2018; Kramarchuk et al. 2019; Kramarchuk et al. 2018; Krainskyi et al. 2018 and Khmil et al. 2018).

2 Materials and Methods

This research consisted in determination of bearing capacity for reinforced concrete crane beams BKNB6-2c and BKNB12-1c, which were reinforced by steel reinforcement A500 in conjunction with steel reinforcement A400C. Herewith, we had to determine if prefabricated reinforced concrete crane beams will have sufficient strength that we could change old crane by (GOST 3332-54, 1954) on crane KMO-10.0-16,5-5,9GCH with weight-bearing capacity 10 t. Moreover, we additionally determined utilization percentage of beam's strength BKNB6-2c and BKNB12-1c by bending moment, and also utilization percentage of compression area concrete strength through maximum deflection of beam in their loading by operational load.

During the inspection of reinforced concrete crane beams, there were executed their in-situ measurement for series identification, according to which they were manufactured. According to the results of measurement were determined that reinforced concrete crane beams, which were installed on outer axis of production department with their overall dimensions, meet (Series KE-01-50 edition 1, 1950) and they have length, which is equal to 6 m, and crane beams, which were installed on inner axis of production department, meet (Series KE-01-50 edition 2, 1950) and they have length that is equal to 12 m (Fig. 1).

Fig. 1. Crane beams BKNB6-2c (on the left) and BKNB12-1c (on the right)

After the uncovering of crane beam principal reinforcement (length 6 m) was determined that this beam was reinforced by 2∅32 A500. The place of re-bars arrangement, protective layer thickness and rod spacing were coincided with beam BKNB6–2c of (Series KE-01-50 by edition 1, 1950) (Fig. 2). Strength quality of concrete was determined by SILVER SCHMIDT HAMMER and it was responsible for concrete strength M350 that was foreseen for beams BKNB6-2c of (Series KE-01-50 by edition 1, 1950). Maximum bending moment for these beams by (Series KE-01-50, 1950) was M_{max} = 49.8 t·m, and maximum shear was equal to Q_{max} = 38.5 t.

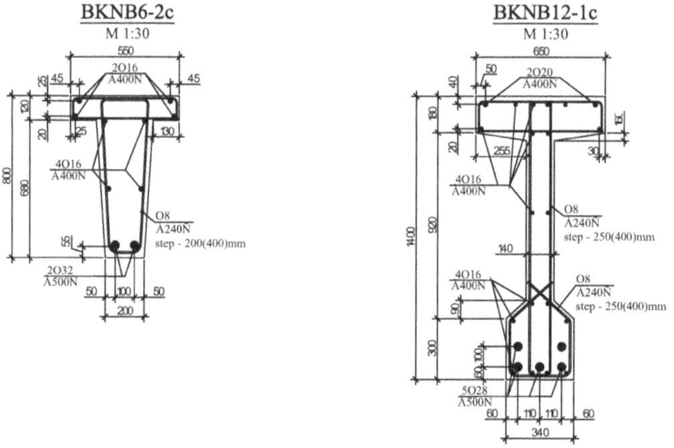

Fig. 2. Scheme of frameworks of crane beams BKNB6-2c and BKNB12-1c

After the uncovering of crane beam principal reinforcement (length 12 m) was fixed that this beam was reinforced by 5∅28 A500 (area of reinforcement was 3079 mm^2) that agree to BKNB12-1c of (Series KE-01-50 by edition 2, 1950) (Fig. 2). This beam had 3∅36 A500 (area of reinforcement was 3054 mm^2) by the project. Strength quality of concrete was determined by SILVER SCHMIDT HAMMER. It agreed with concrete strength M350 that was foreseen for beam BKNB12-1c of (Series KE-01-50 by edition 2, 1950). Maximum bending moment for these beams according to series (KE-01-50, 1950) was M_{max} = 138.3 t·m, and maximum shear was equal to Q_{max} = 54.1 t.

Hence, there were determined that the series of crane beams was assigned for establishment of cranes with weight-bearing capacity 10 t. According to (Series KE–01–50 by edition 1 and 2, 1950), the beams were calculated by divided scheme from loading of two approached cranes in one span, own weight of beam and crane rail. The loading of cranes for this series was accepted in accordance with (GOST 3332-54, 1954). The dynamic factor for crane loading was 1.2; overload factor was 1.3 and overload factor for own weight of beam and crane rails was 1.1.

According to (GOST 3332-54, 1954) maximum loading for one wheel from crane with medium operating mode was 12.5 t (weight-bearing capacity of crane was 10 t; crane span was 17 m and own weight of crane was 21 t). The distance in axes between wheels of crane was 4400 mm by width – 6300 mm according to (GOST 3332-54, 1954).

On the basis of above-named facts was executed calculation of maximum bending moment and maximum shear from action two cranes in the most inconvenient location by means of software package 'LIRA'. The calculation confirmed that the efforts, which appeared in beams BKNB6-2c and BKNB12-1c from the most inconvenient location of cranes, were similar to maximum efforts, which are presented in (series KE–01-50 by edition 1 and edition 2, 1950).

A crane KMO-10.0-16,5-5,9GCH, which we had to install on existing investigated reinforced crane beams, has following technical characteristics: maximum load of wheel on crane way is 70 kN; maximum proper weight of crane is 50 kN; distance between wheels (in axis) is 2650 mm; width of crane is 3150 mm.

As long as technical characteristics of crane KMO-10.0-16,5-5,9GCH are different from characteristics of cranes by (GOST 3332-54, 1954), therefore we executed a calculation of maximum bending moments and maximum shears from action of two near standing cranes KMO-10.0-16,5-5,9GCH with maximum load of wheel on crane way in places of its unfavourable location on crane beam.

There were accepted for calculation: dynamic factor for crane loads was 1.2; overload factor was 1.3; ratio for own weight of beam and ratio for crane ways were 1.1, which were determined according to (DBN V.1-1-2002, 2002), (DBN V.2.6-98:2009, 2009) and (DSTU B V.2.6–156:2010, 2010).

Calculated load scheme of beams BKNB6-2c and BKNB12-1c for finding maximum bending moment and maximum shear from weight of two near standing cranes KMO-10.0-16,5-5,9GCH with load 10 t are shown in Figs. 3, 4, 5 and 6.

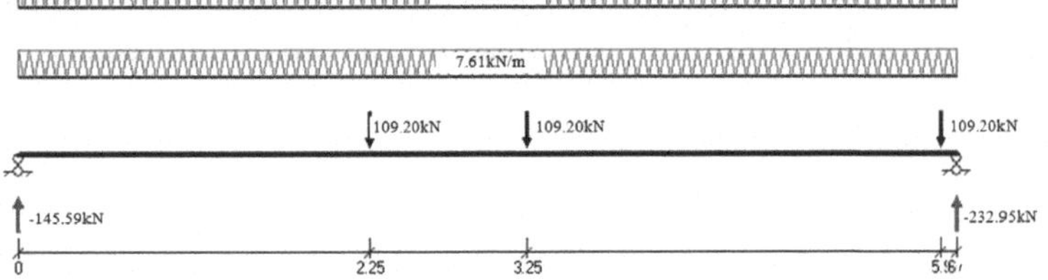

Fig. 3. Calculated load scheme of beam BKNB-6-2c for finding maximum bending moment from weight of two near standing cranes KMO-10.0-16,5-5,9GCH with load 10 t

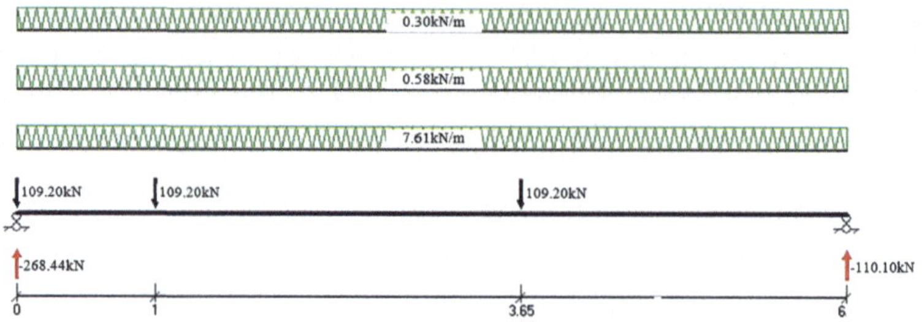

Fig. 4. Calculated load scheme of beam BKNB-6-2c for finding maximum shear from weight of two near standing cranes KMO-10.0-16,5-5,9GCH with load 10 t

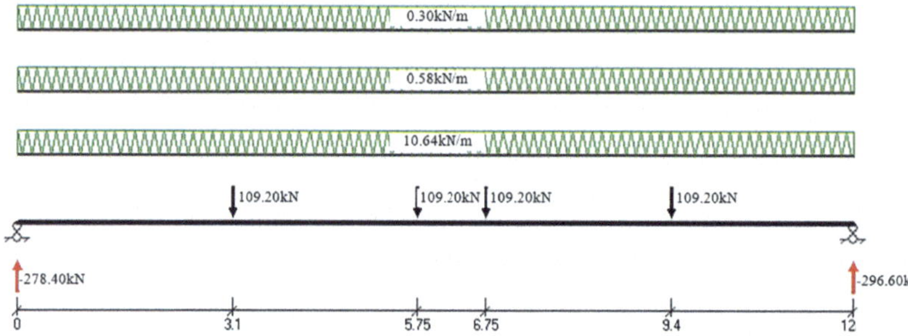

Fig. 5. Calculated load scheme of beam BKNB-12-1c for finding maximum bending moment from weight of two near standing cranes KMO-10.0-16,5-5,9GCH with load 10 t

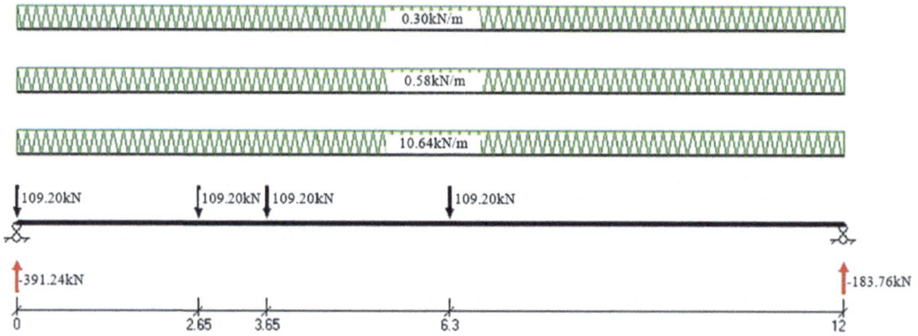

Fig. 6. Calculated load scheme of beam BKNB-12-1c for finding maximum shear from weight of two near standing cranes KMO-10.0-16,5-5,9GCH with load 10 t

3 Results and Discussions

The results of analysis by deflected mode of experimental crane beams BKNB6-2c and BKNB12-1c under loading are shown in Table 1.

Deflected mode analysis of investigated crane beams was executed by program 'LIRA'. For that was minutely modeled a construction of beams BKNB6-2c and

BKNB12-1c with accounting crowded reinforcement by (Series KE-01-50, 1950). For maximum investigation accuracy, besides principal reinforcement, which was constructed for perception of maximum bending moments and maximum shears, in calculated model was also included structural reinforcement in compression area and tension area, which was installed with spatial framework in factory.

Table 1. The comparison maximum bending moments and maximum shears

Crane beams	Maximum bending moment from action of two cranes KMФ10,0-16,5-5,9GCH M_p kN·m	Maximum calculated moment by (Series KEKE-01-50, 1950), M_A kN·m	$100-(M_A-M_p)/M_A$, %	Maximum shear from action of two cranes KMФ10,0-16,5-5,9GCH Qp, kN	Maximum calculated shear by (Series KE-01-50, 1950), Q_A kN	$100-(Q_A-Q_P)/Q_A$, %
BKNB-6-2c	319.1	488.5	65	268.4	377.7	71
BKNB-12-1c	1121.0	1356.7	83	391.2	530.7	74

The simulation was determined in spacing out geometrical scheme of crane beams on finite tetragonal plane elements by sizes 30 x 100 mm, 50 x 100 mm and 60 x 100 mm. These types of sizes were sufficiently for basic geometry simulation of beams. Principal and structural reinforcements were preset as finite bar elements, which were joined between them in knots by finite tetragonal plane elements (Figs. 7, 8, 9 and 10).

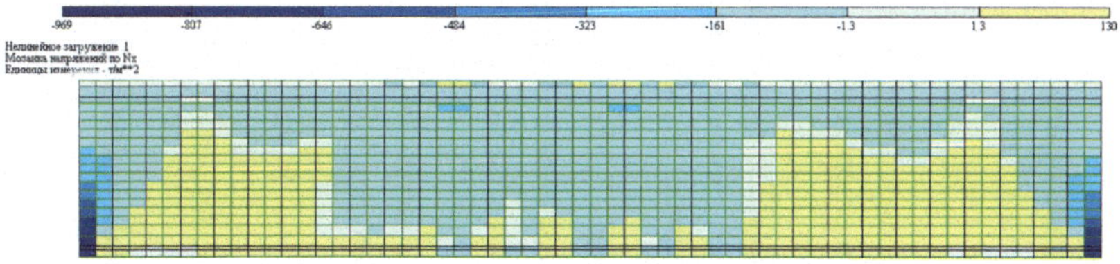

Fig. 7. The stress of compression and the tensile stress in direction of axis X of beam BKNB-6-2c (maximum value of compression – 9.69 MPa)

Calculation was executed with accounting physical nonlinearity of concrete by nonlinear deformation law (exponential law). The characteristics of selected strength quality of concrete C20/25 coincided with characteristics of strength and deformability of concrete of researched models. Chosen type of finite elements was physical nonlinearity finite tetragonal plane element for two-dimensional problem.

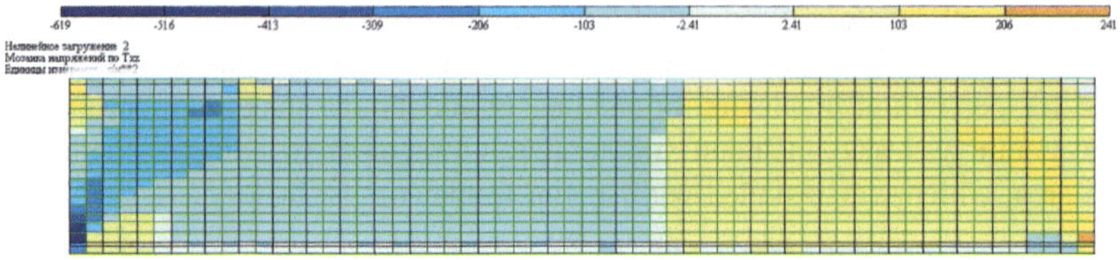

Fig. 8. Tangential stress in beam BKNB-6-2c (maximum value of shear – 6.19 MPa)

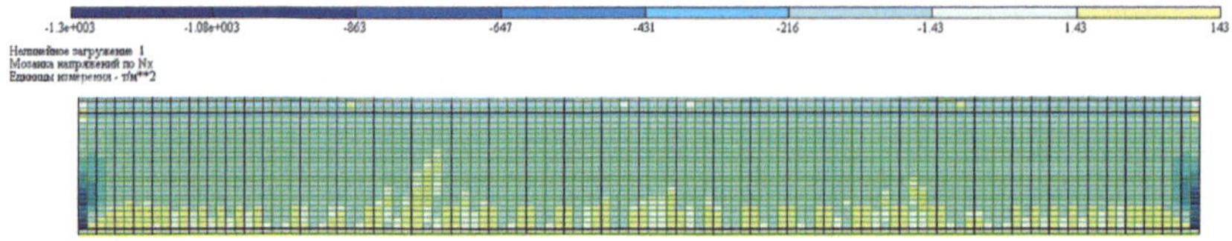

Fig. 9. The stress of compression and the tensile stress in direction of axis X of beam BKNB-12-1c (maximum value of compression – 13.00 MPa)

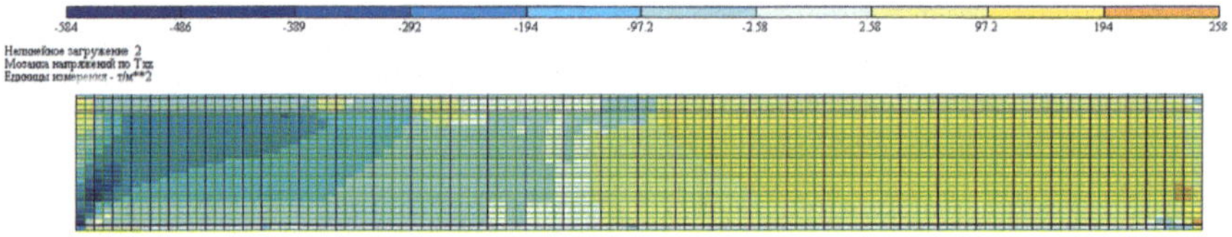

Fig. 10. Tangential stress in beam BKNB-12-1c (maximum value of shear – 5.84 MPa)

Table 2. The results of investigations by experimental crane beams

Crane beams	Maximum tensile force in tensile reinforcement, Fp, kN	Maximum calculated tensile force, which can receive reinforcement, F_A, kN	$100\cdot(F_A-F_P)/F_A$, %	Maximum concrete compression stress of compression area, σ, MPa	Calculated concrete compression area, f_{ud}, MPa	$100\cdot(f_{ud}-\sigma)/f_{ud}$, %	Maximum deflection, f, mm	Maximum allowable deflection $(1/250)\cdot L$, f_n, mm	$100\cdot(f_n-f)/f_n$, %
BKNB-6-2c	322.0	736.7	44	9.69	14.5	68	5.18	24	22
BKNB-12-1c	1024.2	1453.9	71	13.0	14.5	89	26.2	48	55

For principal and structural reinforcements was set linear law of deformation on bases elastic coefficient by steel, inasmuch as the reinforcement works only in elastic stage in real constructions. Selected type of finite linear elements was universal spatial bar finite element.

The executed analysis results of deflected mode for experimental crane beams BKNB-6-2c and BKNB-12-1c under loading are shown in Table 2.

Taking into account executed analysis of deflected mode by experimental beams BKNB-6-2c and BKNB-12-1c was determined possibility of their exploitation under loading from two cranes KMO-10.0-16,5-5,9GCH with full load.

4 Conclusions

The renovation of industrial building demanded the determination of bearing capacity of existent prestressed reinforced concrete structures for further exploitation of all building and replacement of old crane to new crane KMO-10.0-16,5-5,9GCH.

1. Two cranes KMO-10.0-16,5-5,9GCH can install on reinforced concrete crane beams BKNB-6-2c and BKNB-12-1c according to the realized calculation and executed analysis of deflected mode. The bearing capacity of these cranes is 10 t. The technical characteristics of cranes: maximum load of wheel on crane way is 70 kN; maximum proper weight of crane is 50 kN; distance between wheels (in axis) is 2650 mm; width of crane is 3150 mm.
2. During exploitation of two cranes KMO-10.0-16,5-5,9GCH with full load, utilization percentage of beam strength BKNB-6-2c by bending moment was 44%; utilization percentage of beam strength by compression area was 68%; percentage of maximum deflection by maximum allowable deflection was 22%.
3. Accordingly, utilization percentage of beam strength BKNB-12-1c by bending moment was 71%; utilization percentage of beam strength by compression area was 89%; percentage of maximum deflection by maximum allowable deflection was 55% by exploitation of two cranes KMO-10.0-16,5-5,9GCH with full load.

References

DBN V.1.2-2:2006 (2006) Loads and impacts. National standard of Ukraine [in Ukrainian]

Series KE-01-50 edition 1; edition 2 (1950). Prefabricated reinforced prestressed crane-rail beams (in [Russian])

GOST 3332-54 (1954) General-purpose electrical bridge cranes with carrying capacity from 5 t to 50 t of medium duty and heavy duty (in [Russian])

DBN V.1-1-2002 (2002) Repairs and strengthenings of bearing and fencing constructions and bases of industrial buildings and structures. National standard of Ukraine [in Ukrainian]

DBN V.2.6-98:2009 (2009) Concrete and reinforced concrete structures. National standard of Ukraine [in Ukrainian]

DSTU B V.2.6-156:2010 (2010) Construction of houses and buildings. Concrete and reinforced concrete structures made of heavy concrete. Designing rules. National standard of Ukraine [in Ukrainian]

Blikharskyy Z, Vegera P, Vashkevych R, Shnal T (2018) Fracture toughness of RC beams on the shear, strengthening by FRCM system. MATEC web of conferences, vol 183, p 02009. https://doi.org/10.1051/matecconf/201818302009

Kramarchuk A, Ilnytskyy B, Lytvyniak O, Grabowski A (2019) The increase of seismic stability for existing industrial buildings. In: Scientific session on applied mechanics X. AIP conference proceedings 2077, 020029. Published Online: 21 February 2019

Kramarchuk A, Ilnytskyy B, Lytvyniak O (2018) Arrangement of the foundations under the new hotel in Lviv. In: MATEC web of conferences. 183. 02007. https://doi.org/10.1051/matecconf/201818302007

Krainskyi P, Blikharskyy Y, Khmil R, and Vegera P (2018) Influence of loading level on the bearing capacity of RC columns strengthened by jacketing. In: MATEC web of conferences, vol 230, p 02013. https://doi.org/10.1051/matecconf/201823002013

Khmil R, Tytarenko R, Blikharskyy Y, Vegera P (2018) Development of the procedure for the estimation of reliability of reinforced concrete beams, strengthened by building up the stretched reinforcing bars under load. Eastern-Eur J Enterp Technol 5/7(95), https://doi.org/10.15587/1729-4061.2018.142750

Development of Eco-Efficient Composite Cements with High Early Strength

Tetiana Kropyvnytska[1(✉)], Teresa Rucinska[2], Hanna Ivashchyshyn[1], and Roman Kotiv[3]

[1] Department of Building Production (DB), Lviv Polytechnic National University, S. Bandera Str. 12, Lviv 79013, Ukraine
tkropyvnytska@ukr.net

[2] Faculty of Civil Engineering and Architecture, West Pomeranian University of Technology Szczecin, Piastów Str. 50, 70-311 Szczecin, Poland

[3] Department of Architectural Constructions (DA), Lviv Polytechnic National University, S. Bandera Str. 12, Lviv 79013, Ukraine

Abstract. It was shown that a significant reduction of CO_2 emissions in construction is achieved through the using of composite cements with a lower clinker factor. A quaternary composite cements (clinker factor - 50%) containing granulated blast furnace slag, natural zeolite and limestone has been presented. The composition and particle size distribution of the constituents are optimized by the incremental coefficient of the surface activity. Synergistic combination of fine mineral additives and fillers in composite cements ensures the formation of dense microstructure of paste. It is shown that the addition of microsilica (MS) and polycarboxylate ether (PCE) admixtures ensures the production of higher strength cements. In order to accelerate the process of hydration of cement at an early age, the suspension of the C-S-H (X-Seed) nanoparticles was introduced. The effectiveness of a MS – C-S-H – PCE admixture in the early strength development of composite cement with zeolite was investigated. It has been found that the addition of MS – C-S-H – PCE significantly increases the compressive strength of cement mortar, particularly at 12–24 h of hydration. XRD and SEM measurements confirm that the hydration of composite cement is greatly accelerated. Based on the obtained results, high quality eco-efficient composite cements with high early strength can be produced.

Keywords: Composite cements · Microsilica · Polycarboxylate ether · C-S-H nanoparticles · High early strength · Clinker factor

1 Introduction

Cement, as the main component of concrete, will continue to play an important role in resources management and solving problems related to population growth and progressive urbanization (Schneider 2018). Significant potential for reducing of CO_2 emissions in the construction sector is achieved by the introduction of new types of cements. One of these possibilities is the wider use of composite cements, which provide increased durability, economic and ecological benefits (Miller et al. 2018; Krivenko et al. 2018b).

Z. Blikharskyy et al. (Eds.): CEE 2019, LNCE 47, pp. 211–218, 2020.
https://doi.org/10.1007/978-3-030-27011-7_27

New low-carbon eco-efficient cements in the direction of sustainable development will become increasingly important (Limbachiya et al. 2014). The most widely used additives in the production of cement are technological waste, fly ash, granulated blast furnace slag or microsilica (Lothenbach et al. 2011; Zhang et al. 2016; Pluhin et al. 2017; Solodkyy et al. 2017). As a natural pozzolanic material, the practical interest is the use of zeolite tuff, the main mineral of which is clinoptilolite that has unique characteristic such as the ability to cation exchange (Ahmadi and Shekarchi 2010; Krivenko et al. 2018a).

Limestone is one of the most common components of composite cements in the EU (Scrivener et al. 2016; Schneider 2018). The peculiarities of its application are caused by the chemical interaction of calcite with the aluminate phases of Portland cement clinker with the formation of stable AF_m-phases, which provides increased kinetics of solidification of composite cements at an early age (Sanytsky et al. 2018; Kropyvnytska et al. 2017). At present, alternative solutions are the expansion of the second group of Portland composite cements by the subgroup CEM II/C by reducing the clinker factor to 64-50% and increasing the content of the main components (S - granulated blast furnace slag, V - fly ash, P - natural pozzolana, L/LL - limestone) in their three-component composition (K-S-L/LL, K-S-V and K-V-L/LL) to 36–50% by weight (Kuteranska and Krol 2016). The new direction in cement production is the use of low-carbon cement composites with additives of calcined clay and ground limestone (LC^3 - Limestone calcined clay cement) (Scrivener et al. 2016).

At the same time, the replacement of clinker with mineral additives in the Portland composite cements is limited due to the slow kinetics of their early strength caused by their low reactivity compared to the clinker phases. Therefore, in the case of composite cements with mineral additives, especially the CEM V/A, B type, it is extremely important to find ways to increase the kinetics of their early strength increase (Smrckova et al. 2014; Ghiasvand and Ramezanianpour 2017). Nanopowders have an important role in the production of cements and concrete (Sobolev and Ferrada-Gutiérrez 2005; Sikora et al. 2018; Kropyvnytska et al. 2018). Increased strength can be achieved by the use of micro- and nanosilica, which accelerates the early pozzolanic reaction (Ajileye 2012; Sikora et al. 2015). A significant acceleration of the hardening of eco-efficient composite cements can be achieved through alkaline activation, which provides the transition to new alternative cements (Krivenko et al. 2017; Krivenko et al. 2018a, 2018b; Rudenko et al. 2018).

In order to accelerate the process of hydration of cement at an early age (6–12 h), the new Crystal Speed Hardening ™ technology is applied. When introducing a suspension of active X-Seed nanoparticles, additional crystallization centers of hydrocarbonate calcium C-S-H are formed between the cement grains, which allows to accelerate the growth of early strength. At the same time, zeolite and microsilica, as components of low-carbon composite cements, lead to the decrease of the mixture workability. Therefore, one of the basic principles is the use of superplasticizers from the polycarboxylate group (Plank et al. 2009, Krivenko et al. 2018b).

The combination of microsilica and active X-Seed nanoparticles in composite cements with high content of mineral additives, as well as PCE will increase the early strength and provide increased durability of concrete, including corrosion resistance (Kanchanason and Plank 2015, Marushchak et al. 2016). That is why, it is very relevant to introduce cost-effective, balanced solutions for the low carbon industry of materials based on eco-efficient composite cements with a low clinker factor which are characterized by high early strength.

2 Experimental Program

2.1 Raw Materials

Ordinary Portland Cement (OPC) CEM I 42.5R JSC "Ivano-Frankivsk Cement" (Manufacturer - JSC "Ivano-Frankivsk Cement", Ukraine) was used in experiments for the development of composite cements. The contents of the main clinker phases were as follows, mass. %: C_3S – 60.2; C_2S – 4.25; C_3A – 7.20; C_4AF – 11.85. Ground granulated blast furnace slag (S) and zeolite tuff (P) were used as SCMs. The content of clinoptilolite $[(Na_4K_4)(Al_8Si_{40}O_{96})^{\cdot}24H_2O]$ in natural zeolite is 58%. Limestone powder with 95 wt. % $CaCO_3$ was used as micro-filler. Limestone belongs to the L type according to the EN 197-1 in terms of total content of organic carbon. Elkem Microsilica Grade 940-U was used to increase the early strength and acceleration of the hydration process of composite cement in the early age.

Composite cements were obtained by mixing of OPC, GGBFS, zeolitic tuff and limestone powder. Master X-Seed 100 (BASF) was used to accelerate the hydration process of cement in the early stages. Master X-Seed 100 forms additional centers of crystallization of calcium hydrosilicates between cement grains, which greatly accelerates the development of early strength. MasterGlenium ACE 430 (BASF) based on modified polycarboxylic ether (PCE) was used as a superplasticizer.

2.2 Experimental Process

The use of the chemical composition of raw materials was determined on X-ray spectrometer ARL 9800 XP. A particle size distribution (PSD) was studied by a laser granulometer Mastersizer 3000. Hydration processes of the cements were examined by means of X-ray diffractometry (XRD), scanning electron microscopy (SEM) and energy-dispersive X-ray spectroscopy (EDX).

The compressive and flexural strengths of the tested cements under study were determined on mortar prisms (40 × 40 × 160 mm) in accordance with in EN 196-1. Workability of fresh mortars (consistency measured on a flow table) was measured in accordance with EN 1015-3. The cement mortars were adjusted to obtain a constant workability of approximately 180 ± 2 mm (flow).

3 Results and Discussion

There were studied PSD of materials that had been used for the production of blended and composite cements (Table 1). It was shown that values of the volume mean diameter D[4;3] for the OPC corresponds to 24.8 μm, and for the SCMs it changes from 18.1 to 46.4 μm. The surface area mean diameter D[3;2] for the OPC is 5.21 μm,

Table 1. Particle size distribution for the OPC and SCMs

Material	SSA, m^2/kg	Ø < 5 μm, %	Ø < 10 μm, %	Ø < 20 μm, %	Ø < 60 μm, %	D[3;2] μm	D[4;3] μm	d10, μm	d50, μm	d90, μm
OPC	360	21.86	37.46	60.41	97.52	5.21	24.8	2.76	18.1	56.5
GGBFS	650	51.54	60.43	80.32	94.37	3.95	18.1	1.64	7.87	43.5
P	1100	40.63	56.88	72.67	94.39	3.81	19.6	1.40	9.31	51.1
L	950	16.67	34.59	50.85	75.24	8.29	46.4	3.92	23.3	125

and for the SCMs it changes from 3.81 to 8.29 μm. This indicates that the fine fractions of SCMs determine their surface area.

There was investigated the influence of additives GGBFS, zeolite and limestone on the properties of mixed cements (clinker factor - 50%). The content of SCMs up to 50 wt.% as part of cements leads to a reduction in their strength especially at an early age (Table 2). The addition of fine zeolite significantly reduces the workability of blended cement but small particles of calcium carbonate provide to an increase in their workability.

Table 2. Influence of SCMs on physical and mechanical properties of blended cements

Cement	SSA, [m^2/kg]	Water demand [%]	Setting time [min]		Flow, mm	Compressive strength [MPa]		
			Initial	Final		2	7	28
100% OPC	346	29.5	235	340	172	21.5	30.1	44.6
50% OPC + 50% S	370	27.5	250	355	180	5.7	17.0	27.7
50% OPC + 50% P	590	38.5	195	220	160	5.6	12.4	24.5
50% OPC + 50% L	620	26.0	155	220	185	6.9	11.3	21.6

The composite cement CEM V/A was made by mixing 50 wt.% OPC, 20.0 wt.% GGBFS, 25.0 wt.% zeolitic tuff and 5 wt.% limestone powder. The combination of fine particles of various types with high "superficial energy" has an even more significant effect on the strength of composite cements. The microsized level of composite cements CEM V/A was enhanced by the addition of microsilica (MS). According to the particle size distribution of composite cement CEM V/A fractions Ø1, Ø5 and Ø20 μm are respectively 9.41, 34.85 and 52.76%, and the grain size d10, d50 and d90 corresponds to 1.14, 9.15 and 49.6 μm. A volume mean diameter D[4;3] for the CEM V/A corresponds to 20.5 μm, and for the MS – to 1.36 μm. At the same time, the surface area mean diameter D[3;2] for the CEM V/A corresponds to 2.72 μm, whereas for the MS – to 1.06 μm.

The calculated specific surface area for the CEM V/A corresponds to 650 m^2/kg, and for the MS –16750 m^2/kg. In this case, the ratio of the value for the MS to the value for the CEM V/A is 25.7. The degree of additional interfacial active surface area of the CEM V/A and the MS have increased and can be determined by calculating of the ratio of surface area of particles to their volume (coefficient A/V) according to special methodology (Sanytsky et al. 2018). To assess the contribution of individual particles to total specific surface area, it was calculated particle size distribution by surface area and suggested an incremental coefficient of surface area (K_{isa}), which is determined by multiplying the coefficient A/V by the incremental volume of each fraction of the material. As seen in Fig. 1a, the maximum value of K_{isa} (8.72 $\mu m^{-1} \cdot$vol.%) for the CEM V/A was achieved for the fraction of 0.315 μm and for the fraction of 1.0 μm this coefficient was 6.34 and for the fraction of 10 μm decreased by 3.2 times and with further increase in particle size was significantly lower. The maximum value of K_{isa} for the MS is 59.2 $\mu m^{-1} \cdot$vol. % at 0.786 μm, for the most reactive fractions the incremental coefficient of surface area of the MS was by 6.8 times higher than the same of the CEM V/A (Fig. 1b).

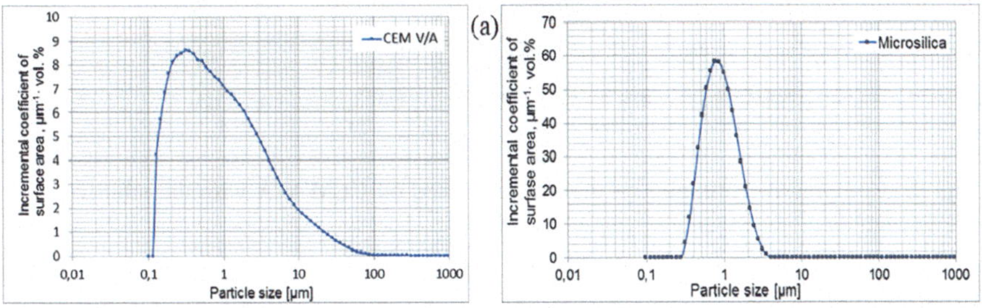

Fig. 1. PSD by surface area of composite cement CEM V/A (a) and microsilica (b)

Compressive strengths of the mortars were measured at the following early ages: 12 h, 24 h and 2, 7 and 28 days. The results are shown in Fig. 2 and demonstrate that the CEM V/A is characterized by a low strength at the early age, but the addition of 3% MS by weight (W/C = 0.5, flow = 180 mm) promotes strength development of the mortar. However, a greater increase in strength is observed in case by the addition of MS and PCE. In this case, the early strength after 12 and 24 h has increased by 2.7 and 1.8 times, respectively. The additives of the MS (3.0% by wt.), C-S-H (X-Seed) (2.0% by wt.) and PCE (1.0% by wt.) has increased the strength of the mortar after 12 and 24 h to 11.5 and 19.7 MPa (or by 4.6 and 2.6 times, respectively). The 2 and 7 days compressive strength values of 32.6 and 63.9 MPa were obtained in case of the MS – C-S-H – PCE were 24.0 and 3.7% higher than in the case of MS-PCE. In this case, at an age of 28 days, the compressive strength of the cement was by 1.67 times higher than that of the reference cement (plain CEM V/A).

According to the XRD, the lines of calcium hydroxide (d/n = 0.493; 0.263 nm), calcite (d/n = 0.303; 0.191 nm) and ettringite (d/n = 0.973; 0.561 nm) are identified on diffractograms of the CEM V/A cement paste the MS – C-S-H – PCE after 1 day of curing.

After 1 day of hardening an intensity of the calcium hydroxide lines decreases by 3.5 times for the paste based on the CEM V/A with the MS – C-S-H – PCE addition compared to the CEM I paste due the pozzolanic reaction with binding of calcium hydroxide in calcium silicate hydrates that significantly effects the ultimate compressive strength. The microsilica as a part of the complex three-component additive MS – C-S-H – PCE leads to a significant decrease in the intensity of $Ca(OH)_2$ lines, with slightly increased ettringite lines.

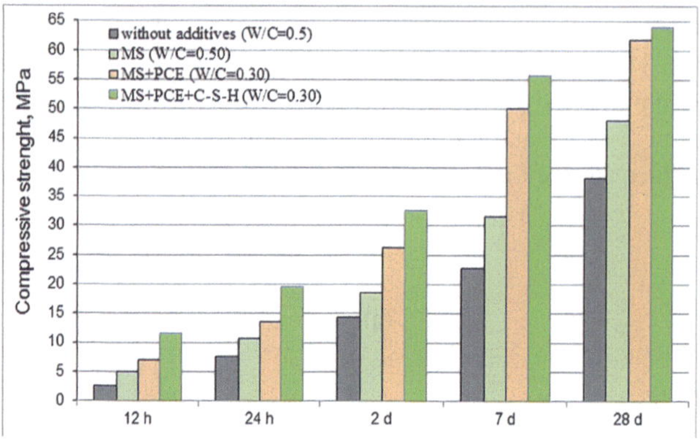

Fig. 2. Compressive strengths of the mortars based on CEM V/A with the additives

Microstructure of the CEM V/A with the complex three-component additive MS – C-S-H – PCE at the age of 1 day is dense (Fig. 3) as a result of the formation of amorphous C-S-H phases, reinforced with needle-like ettringite crystals. Synergetic combination of ultrafine mineral additives and fillers in composite cement CEM V/A ensures the formation of dense microstructure of the paste and increase its strength. In addition to the environmental benefits generated by the use of high content of SCMs as substitute of cement, the obtained results reveal a significant long-term improvement in strength properties and durability performance of composite cements.

Fig. 3. SEM images of the CEM V/A paste with the MS – C-S-H – PCE at an age of 1 day with various magnification

4 Conclusions

It was established that the fine particles are characterized by a very high surface area to volume ratio and possess high reactivity compared to the cement particles. The C-S-H (X-Seed) + PCE nanomodifier also stimulates the pozzolanic reaction of fine zeolite and microsilica, thus explaining the high early strength values. The strength of CEM V/A with MS – C-S-H – PCE supplements at an age of 1 day has increased by 1.4 time than CEM V/A without additives at the age of 2 days. Due to the synergistic effect of ultrafine particles of nanomodifier C-S-H + PCE, a significant acceleration of the cement hydration processes and improvement of its mechanical properties at an early age of hardening (12 and 24 h) and at standard age is provided. The MS – C-S-H – PCE combination can successfully compensate for the low early strength of large quantities of SCMs and opens up new areas of application for eco-efficient composite cements.

References

Ahmadi B, Shekarchi M (2010) Use of natural zeolite as supplementary cementitious material. Cement Concr Compos 32(2):134–141

Ajileye EV (2012) Investigations on microsilica (silica fume) as partial cement replacement in concrete. Glob J Res 12(1):17–23

Ghiasvand E, Ramezanianpour A (2017) Effect of grinding method and particle size distribution on long term properties of binary and ternary cements. Constr Build Mater 134:75–82

Kanchanason V, Plank J (2015) C-S-H−PCE nanocomposites for enhancement of early strength of cement. In: 19th international conference on building materials, Weimar, Germany, 16–18 September

Krivenko P, Petropavlovskiy O, Kovalchuk O (2018a) A comparative study on the influence of metakaolin and kaolin additives on properties and structure of the alkali activated slag cement and concrete. East Eur J Enterp Technol 1(6–91):33–39

Krivenko P, Runova R, Rudenko I, Skorik V, Omelchuk V (2017) Analysis of plasticizer effectiveness during alkaline cement structure formation. East Eur J Enterp Technol 4(6 (88)):35–41

Krivenko P, Sanytsky M, Kropyvnytska T (2018b) Alkali-sulfate activated blended portland cements. Solid State Phenom 276:9–14

Kropyvnytska T, Semeniv R, Ivashchyshyn H (2017) Increase of brick masonry durability for external walls of buildings and structures. In: MATEC web of conferences, vol 116

Kropyvnytska T, Semeniv R, Kotiv R, Kaminskyy A, Hots V (2018) Studying the effect of nano-liquids on the operational properties of brick building structures. East Eur J Enterp Technol 5 (6(95)):27–32

Kuteranska J, Krol A (2016) New types of low-carbon cements with reduced Portland clinker content as a result of ecological actions of cement industry towards sustainable development. Econ Environ Stud 3(39):403–419

Limbachiya M, Bostanci SC, Kew H (2014) Suitability of BS EN 197-1 CEM II and CEM V cement for production of low carbon concrete. Constr Build Mater 71:397–405

Lothenbach B, Scrivener K, Hooton R (2011) Supplementary cementitious materials. Cem Concr Res 41(12):1244–1256

Marushchak U, Sanytsky M, Mazurak T, Olevych Y (2016) Research of nanomodified portland cement compositions with high early age strength. East Eur J Enterp Technol 6/6(84):50–57

Miller SA, John VM, Pacca SA, Horvath A (2018) Carbon dioxide reduction potential in the global cement industry by 2050. Cem Concr Res 114:115–124

Plank J, Schroefl C, Gruber M, Lesti N, Sieber R (2009) Effectiveness of polycarboxylate superplasticizers in ultra-high strength concrete: the importance of PEC compatibility with silica flume. J Adv Concr Technol 7(1):5–12

Pluhin O, Plugin A, Plugin D, Borziak O, Dudin O (2017) The effect of structural characteristics on electrical and physical properties of electrically conductive compositions based on mineral binders. In: MATEC web of conferences, vol 116, p 01013

Rudenko II, Konstantynovskyi OP, Kovalchuk AV, Nikolainko MV, Obremsky DV (2018) Efficiency of redispersible polymer powders in mortars for anchoring application based on alkali activated Portland cements. Key Eng Mater 761:27–30

Sanytsky M, Kropyvnytska T, Kruts T, Horpynko O, Geviuk I (2018) Design of rapid hardening quaternary zeolite-containing Portland-composite cements. Key Eng Mater 761:193–196

Schneider M (2018) Innovation and technical trends in cement production. 20. Internationale Baustofftagung, Weimar, Germany, 12–14 September

Scrivener KL, John VM, Gartner EM (2016) Eco-efficient cements: potential, economically viable solutions for a low-CO_2, cement based materials industry (Rep.). United Nations Environment Programme, Paris, France

Sikora P, Abd Elrahman M, Stephan D (2018) The influence of nanomaterials on the thermal resistance of cement-based composites - a review. Nanomaterials 8(7):465

Sikora P, Horszczaruk E, Rucinska T (2015) The effect of nanosilica and titanium dioxide on the mechanical and self-cleaning properties of waste-glass cement mortar. Procedia Eng 108:146–153

Smrckova E, Bacuvcík M, Janotka I (2014) Basic characteristics of green cements of CEM V/A and CEM V/B kind. Adv Mater Res 897:196–199

Sobolev K, Ferrada-Gutiérrez M (2005) How nanotechnology can change the concrete world. Am Ceram Soc Bull 10:14–17

Solodkyy S, Markiv T, Sobol K, Hunyak O (2017) Fracture properties of high-strength concrete obtained by direct modification of structure. In: MATEC web of conferences, vol 116, p 01016

Zhang J, Wang Q, Wang Z (2016) Optimizing design of high strength cement matrix with supplementary cementitious materials. Constr Build Mater 120:123–136

FE Analysis Versus Experimental Test Results of FRP Deck System

Maciej Kulpa(✉) and Tomasz Siwowski

The Faculty of Civil and Environmental Engineering and Architecture,
Rzeszow University of Technology,
Powstancow Warszawy 12, 35-959 Rzeszow, Poland
kulpa@prz.edu.pl

Abstract. The growing need of durability enhancement for road bridge decks has recently caused the big impulse for research on a new, durable, lightweight and easy to handle bridge decks, made of advanced materials, f.e. FRP (fibre reinforced polymers). Therefore, in the frame of UE 7FP project PANTURA, three structural solutions of sandwich FRP bridge deck have been elaborated, fabricated and tested under static load. The panels are planned to form a deck slab in a vehicular girder bridge. The span of 2.4 m in a simply supported system was adopted as a typical span length. After the panels were manufactured by the vacuum assisted resin transfer molding method (VARTM) they were tested in the laboratory under static load. At the same time, a numerical model based on the finite element method was created to analyse panels' behaviour under the test load. The paper contains a short description of the experiment procedure, FE analysis and a comparison of results obtained from tests with results obtained by means of numerical simulations.

Keywords: FRP composite · Bridge deck · Sandwich structures · FE analysis · Static test

1 Introduction

The growing need of durability enhancement for road bridge decks has recently caused the big impulse for research on durable, lightweight and easy to handle bridge decks, made of advanced materials, f.e. FRP (fibre reinforced polymers) (Williams et al. 2003; Hollaway 2010; Manalo et al. 2016; Chróścielewski et al. 2017, 2019). Therefore, in the frame of UE 7FP project PANTURA, three structural solutions of sandwich FRP bridge deck have been elaborated, fabricated and tested under static load. On the base of initial test results the stiffness and load carrying capacity of panels have been estimated and the best solution for further research has been chosen (Kulpa and Siwowski 2015). The paper contains a description of the procedure applied to analyse the behaviour of the FRP panels. The panels were manufactured by the vacuum assisted resin transfer molding method (VARTM) and statically tested in a simply supported scheme with a typical span length of 2.4 m. The experiment and FE analysis results were compared to validate the numerical model to be applied in further design works.

Z. Blikharskyy et al. (Eds.): CEE 2019, LNCE 47, pp. 219–226, 2020.
https://doi.org/10.1007/978-3-030-27011-7_28

2 Deck Structures

In general the FRP sandwich panels are made of face laminates and a core. Before starting work on shaping the structure of a panel, it was necessary to choose the basic materials that make up the composite (i.e. reinforcing fibres, matrix) and the core material. On the basis of the state-of-art review and own previous work it was decided to use a glass composite (GFRP) based on epoxy resin and PVC foam core due to their very good strength and durability parameters and optimal cost. The vacuum assisted resin transfer molding method (VARTM) was chosen by the manufacturer to produce panels for testing. On the basis of the analysis of the previous applications of FRP sandwich decks (Bakis et al. 2002; Keller 2003; Friberg and Olsson 2014) and own preliminary calculations, three different cross-sections of the panel were designed (Fig. 1). In the case of panel III a two-component epoxy adhesive was applied as a bondline.

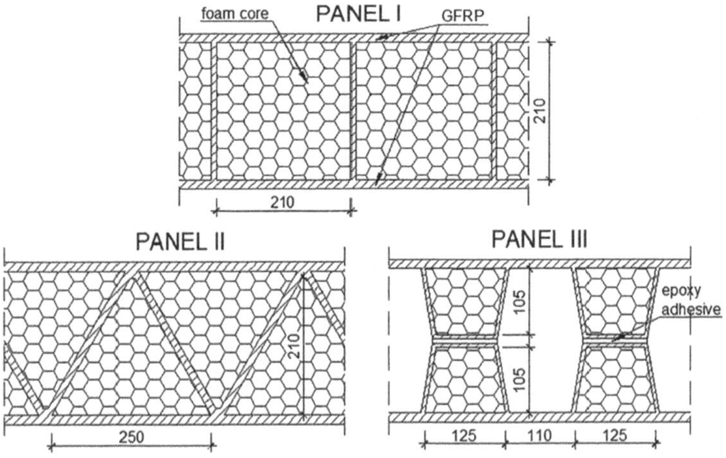

Fig. 1. Three cross-sections of tested FRP panels

3 Numerical Analysis

3.1 Lamina Level

The first stage of designing involved the selection of the number and orientation of the fibres. Due to the orthotropic nature of work of the deck panel, fabrics with orthogonally arranged glass fibres were used. The main type of reinforcement was woven fabric with fibres longitudinally and perpendicularly to the panel's axis (0/90). There were also interleaving layers with sewn fabrics in which the fibres were oriented at an angle of ± 45° to panel's axis (±45). In addition, the fabrics also differed in grammage: ρ = 800 g/m^2 for 0/90 fabrics and ρ = 1200 g/m^2 for ±45 fabrics. Grammage of applied fabrics affects the thickness of the final lamina after saturation with resin.

Basing on previous experience gained with production of similar GFRP structures (Siwowski et al. 2018) the assumption was made that each 1000 g/m^2 of fabrics grammage results in the laminae thickness of 0.7 mm.

The mechanical properties of a laminae were determined experimentally. Material tests were carried out on coupons with a thickness of approx. 2 mm, containing 3 or 4 layers of fabrics (laminas). The values obtained in the test were presented in Table 1. The engineering constants for foam core were assumed according to the manufacturer's data.

Table 1. Engineering constants of panel's materials

	Longitudinal modulus of elasticity $E_x = E_y$ [GPa]	Transverse modulus of elasticity G_{xy} [GPa]	Poisson's ratio ν_{xy} [-]
Laminae	24.0	4.37	0.14
Foam	0.085	0.027	0.40

3.2 Laminate Level

In the next step, number, orientation and stacking sequence of fibres in the mould were precisely determined (Kulpa and Siwowski 2019). After that it was possible to determine the appropriate thicknesses of composite parts of the panel: external and internal.

The numerical analysis were carried out using the finite element method (FEM) in SOFiSTiK environment. The software made possible to describe a single material as a laminate of up to 10 individual layers (laminae). That was enough to describe the adopted solution. The engineering constants for whole laminate in the analysis were then determined automatically based on the Classical Lamination Theory (CLT). An additional benefit was obtaining the results in the form of stresses in each layer within one laminate, which allowed direct application of stress criteria for each single laminea.

3.3 Structure Level

Static analysis was performed on global models of the whole panel (Fig. 2). For each of the three construction variants, a deck panel with dimensions of 2.70 × 1.10 × 0.24 m was analysed. Load was applied as single patch load with dimension of 0.40 m 0.40 m in panel's centre according to load path in Eurocode 1 (EN 1991-2 2003).

The laminates were modelled using shell elements. The dimensions of single element were about 3 × 4 cm. The element remains plane, so the bending and the membrane structural behaviour are decoupled. The plate bending properties are evaluated according to the Reissner-Mindlin theory. Due to the usage of discrete Kirchhoff conditions it is applicable for thin plates such as laminate in this case. The cross sections remain plane also according to Mindlin's theory, however, they are not perpendicular anymore to the neutral axis. The same shape functions as for the displacements are used for the additional shear rotations. The total rotation is then the sum of the shear deformation and the bending rotation (Kreja 2011).

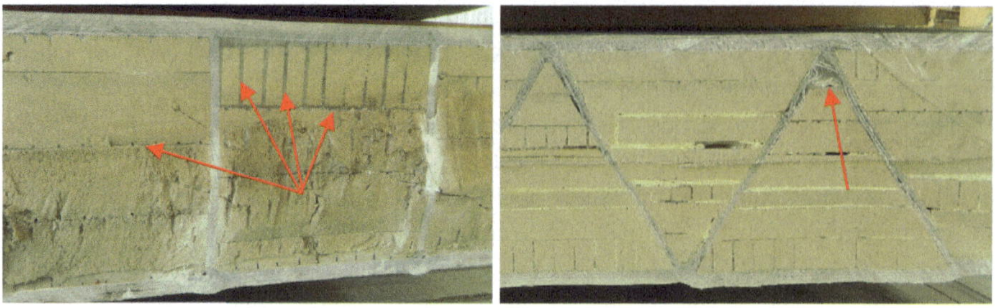

Fig. 2. The resin between foam blocks (a) and examples of executive imperfections (b)

The element formulation of the membrane stress state occurs via a classical non-conforming formulation written by Taylor et al. (1976). These functions lead to a substantial improvement of the results, however, they violate the continuity of displacements between elements (thus they are called non-conforming elements). These elements can describe a linear membrane forces variation and in general approximately of moment variation inside them.

The rotational degree of freedom around the shell normal is not contained in both load bearing behaviours. In order to prevent numerical difficulties for three-dimensional structures, the in-plane-rotation of the nodes is coupled via a weak torsional spring at the displacements of the corner nodes in an internway.

A separate and quite important issue was the approach to modelling the foam forming the core of the panel. It should be emphasized here that the panels were not completely traditional sandwich constructions. In this case, besides the outer laminates and the inner core, there are also internal longitudinal ribs. As a result, the foam core is less involved in the transmission of loads. The mere comparison of the foam and composite stiffness material parameters (Table 1) shows that the stiffness of the rib material is more than 250 times greater. However, in the transverse direction, the rib does not exist and the composite system begins to resemble the Vierendeel system.

To solve engineering problems, the most sensible approach would be to skip (to the safe side) the foam core, treating it as a technological element separating external laminates and giving the panel height. However, from the scientific point of view, the core participation in the obtained results should also be taken into account. This is an extremely difficult issue to quantify.

The starting point is the parameters of the foam itself given by the manufacturer (Table 1). The foam has closed pores, so that absorption of the resin is limited and takes place mainly in the surface zone of the foam block. The foam, however, has technological grooves and cuts to improve the flow of resin during production by infusion. In addition, the core with a nominal height of 21 cm consists of blocks that do not have to be strictly in contact on the whole surface. The resin penetrate all these voids, which then forms a specific "skeletal reinforcement" of such a foam core (Fig. 2a). Because the models of panels were prototypes, there were also a local executive failures (Fig. 2b).

In order to quantify this problem, the analysis of the impact of the core modelling method on the was carried out. The static analysis was carried out with the omission of the foam (f0) and with the core described with the material parameters of the foam itself from Table 1 (f1). Then the possibility of increasing foam parameters after saturation with the resin was considered, increasing the basic parameters twice (f2), three times (f3) and four times (f4). The analysis was carried out for the displacement distribution on width of the panel 1 and are shown in Fig. 3.

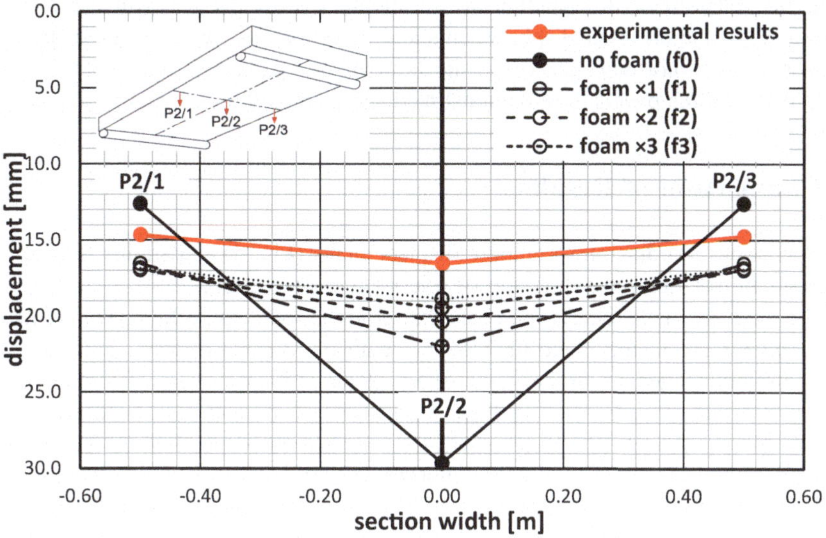

Fig. 3. The impact of core parameters on the strain distribution in the cross-section.

The analysis of Fig. 3 indicates that modelling of the foam core far better reflects the transverse distribution of displacements. However, the difference between the further increase in parameters is not critical (between f2 and f4). The best match between the ratio of displacement in the middle (P2/2) and on the edge (P2/1) occurs for the variant with doubled foam parameters (f2). This assumption was adopted for further analyses.

Foam cores were discretized using solid elements with eight nodes (brick). The dimensions of a single finite element were about 3 × 4 × 3 cm (Fig. 4a).

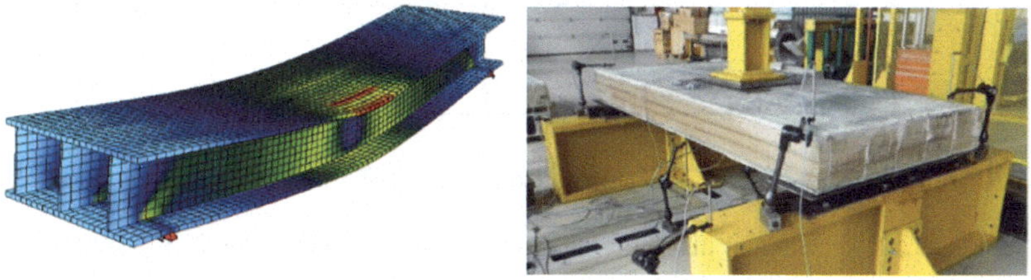

Fig. 4. Numerical model (half-symmetry, foam core hidden) (a) and laboratory test of a panel I (b)

4 Laboratory Tests

For each of the three construction variants, a deck panel with dimensions of 2.70 × 1.10 × 0.24 m was fabricated. The production process included vacuum infusion of dry fabrics by resin, arranged in a non-deformable steel form with a target shape.

In the experiment the patch load was applied on a 0.4 × 0.4 m area. During loading of the panels, their displacements and deformations were measured (Fig. 4b). The panels were not damaged up to the limit of the loading system (630 kN). This meant that all three panels have demonstrated sufficient stiffness and carrying capacity.

5 Comparison of Results

The following two diagrams compare displacements (Fig. 5) and strains (Fig. 6) distributions in the transverse direction of mid-span cross-section.

The analysis of the charts indicates that compliance is quite high. Differences of displacements are in the range of 77–91%, with an average of 85%. Differences of strains in bottom face are in the range of 84–99%, with an average of 88%.

The distributions presented in Figs. 5 and 6 prove correct reproduction of transverse stiffness in models (the corresponding curves are almost parallel). In fact, the only anomaly can be observed for the distribution of deformations in panel 2. The distribution obtained in the calculations is more even, which mean that is in reality the stiffness was lower than resulted from numerical simulations.

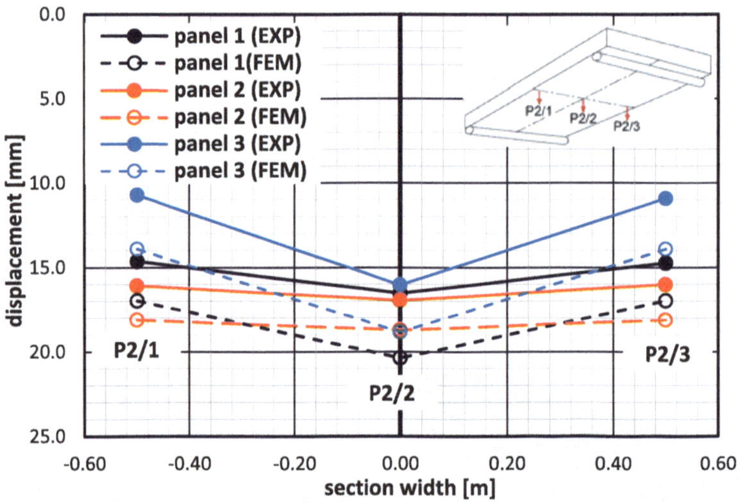

Fig. 5. Displacement distributions at mid-section of panels for P = 630 kN.

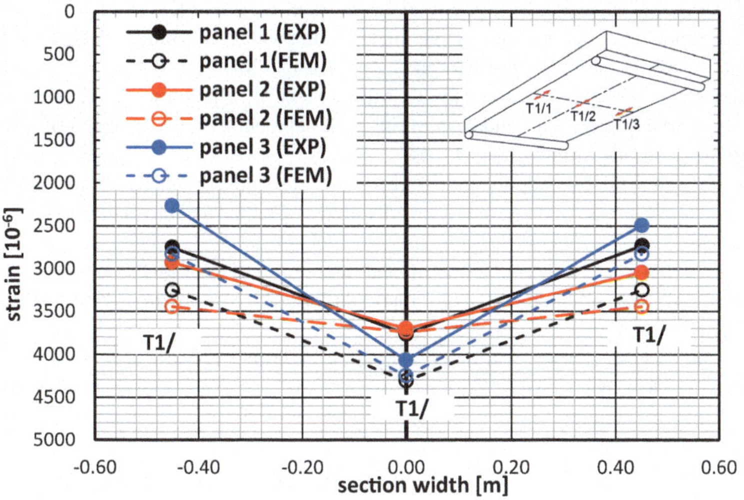

Fig. 6. Strain distributions at mid-section of panels for P = 630 kN.

6 Conclusions

The compatibility between the results of calculations and the experiment on the FRP deck prototypes was at a satisfactory level. The most troublesome was to estimate the parameters of the saturated core foam forming a cross-section. Based on the observation of the distribution of results in cross-sections, it was assumed that the appropriate approach was to double the parameters declared by the manufacturer. This assumption is difficult to prove also because that specific the prototypes was burdened with imperfections caused locally concentration of resin in the foam core. After adopting this assumption, the transverse distributions show satisfactory compatibility.

Despite the correct estimation of transverse stiffness the models still show a global underestimation (about 10%) of vertical displacements resulting from longitudinal stiffness (along the ribs).

Acknowledgements. This work was supported by the European Commission's 7th Framework Programme, the project titled: "*PANTURA: Flexible Processes and Improved Technologies for Urban Infrastructure Construction Sites*", No. CP–IP CP–FP 265172 (www.pantura-project.eu) and Mostostal Warszawa SA.

References

Bakis CE, Bank LC, Brown VL, Cosenza E, Davalos JF, Lesko JJ, Machida A, Rizkalla SH, Triantafillou TC (2002) Fiber-reinforced polymer composites for construction, state-of-the-art review. J Compos Constr 6(2):73–87

Chróścielewski J, Miśkiewicz M, Pyrzowski Ł, Sobczyk B, Wilde K (2017) A novel sandwich footbridge-practical application of laminated composites in bridge design and in situ measurements of static response. Compos Part B Eng 126:153–161

Chróścielewski J, Ferenc T, Mikulski T, Miśkiewicz M, Pyrzowski Ł (2019) Numerical modeling and experimental validation of full-scale segment to support design of novel GFRP footbridge. Compos Struct 213:299–307

Friberg E, Olsson J (2014) Application of fibre reinforced polimer materials in road bridges – general requirements and design considerations. Chalmers University of Technology, Goeteborg

Hollaway LC (2010) A review of the present and future utilisation of FRP composites in the civil infrastructure with reference to their important in-service properties. Constr Build Mater 24(12):2419–2445

Keller T (2003). Use of fibre reinforced polymers in bridge construction. IABSE Structural Engineering Documents, vol 7, Zurich

Kreja I (2011) A literature review on computational models for laminated composite and sandwich panels. Open Eng 1(1):59–80

Kulpa M, Siwowski T (2015) Structural shaping of FRP bridge decks. J Civ Eng Environ Arch 32(62):279–300 (in Polish)

Kulpa M, Siwowski T (2019) Stiffness and strength evaluation of a novel FRP sandwich panel for bridge redecking. Compos Part B Eng 167:207–220

Manalo A, Aravinthan T, Fam A, Benmokrane B (2016) State-of-the-art review on FRP sandwich systems for lightweight civil infrastructure. J Compos Constr 21(1):04016068

EN 1991-2 (2003) Eurocode 1: actions on structures Part 2: traffic loads on bridges. CEN Brussles

Siwowski T, Kulpa M, Rajchel M, Poneta P (2018) Design, manufacturing and structural testing of all-composite FRP bridge girder. Compos Struct 206:814–827

Taylor RL, Beresford PJ, Wilson EL (1976) A non-conforming element for stress analysis. Int J Numer Methods Eng 10:1211–1219

Williams B, Shehata E, Rizkalla SH (2003) Filament-wound glass fiber reinforced polymer bridge deck modules. J Compos Constr 7(3):266–273

Vibration Monitoring of Steel Shaft Headgears

V. M. Kushchenko(✉) and D. O. Khomitskyi

Department of Building Constructions and Bridges (BCB),
Institute of Building and Environmental Engineering (IBEE),
Lviv Politechnic National University (LPNU),
12 Bandera Street, Lviv 79013, Ukraine
Volodymyr.M.Kushchenko@lpnu.ua

Abstract. The structures of shaft headgears belong to the high responsibility class of structures, which have a requirement regulatory - monitoring equipment for technical condition. Due with the dynamic tone of operational loads, for the construction of shaft headgears is relevant to create vibration monitoring systems.

In the experimental researches of dynamic behavior of shaft headgears structures, were applied measuring method of vibration parameters by vibration transducers. Long-term observations of peak values of vibration velocity amplitudes were statistically processed and established by distribution laws of their random variables.

Comparison of the obtained amplitude-frequency characteristics of general vibration structures with the results of modal analysis was made.

In this article a rational scheme of the vibration transducer location in the upper part of the structures is proposed. Based on long-term statistical data analysis observations of the peak values of the vibration amplitudes the criterion for determining the limiting values of the vibration speed amplitude is formulated. But exceeding of that is a diagnostic sign of unsatisfactory technical condition of the structure. That are under condition of the spectrum deviation of own frequencies from the reference.

The concept of vibration monitoring of shaft headgears structures is formulated, which is based on the comparison of constantly measured vibration speed amplitude with boundary values, and also on the control of the spectrum of the own frequencies oscillations.

Keywords: Steel shaft headgear · Hoisting machine mechanism · Vibraition monitoring

1 Introduction

Buildings of mine winding plants (Headgears) in accordance with the normative requirements of Ukraine (DBN, B.1.2-14-2009; 2009; RD, 12.005 94; 1994; DSTU, NB.1.2-18: 2016, 2017) relate to the hign class of responsibility SS-3, which have the requirement of mandatory equipment by monitoring systems of technical condition. Buildings of mine winding plants are a combination of a geometrically invariable mechanical system of the headgear and hoisting machines mechanism (Fig. 1a), which have in the process of operation a dynamic interaction with them related local and

Z. Blikharskyy et al. (Eds.): CEE 2019, LNCE 47, pp. 227–234, 2020.
https://doi.org/10.1007/978-3-030-27011-7_29

general vibrations of headgears (Kushchenko; 2006; Kushchenko and Kostrickiy; 2003; Kushchenko; 2010; Kuschenko; 2007). Therefore, construction of mine headgears is the actual to creation of passive vibration monitoring systems.

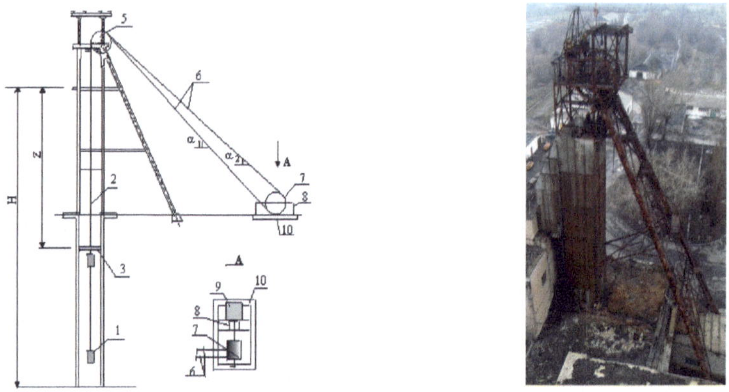

a) b)

Fig. 1. Object of research is steel structures of mine headgears: (a) mine headgear in part of mine winding plants (1 – hoisting vessel; 2 – branches of a hoisting rope; 3 – reinforcement of a shaft trunk; 4 – mine headgear; 5 – pulleys; 6 – strings of hoisting rope; 7 – a drum of hoisting machine; 8 – reducer; 9 – engine; 10 – the foundation); (b) man-load mine winding plants, p. 1, Table 1.

In this work we have considered seven steel headgears with ground-based hoisting machines, one of the characteristic research objects, man-load hoisting headgear is shown in Fig. 1b. General research characteristics of the objects are shown in Table 1.

Table 1. Research characteristics of the objects (steel headgears with ground-based hoisting machines)

№ object	Technological signs of the mine headgear	Hight of headgear (m)	Hoisting hight (m)	Hoisting speed (m/s)	Working lifespan (years)
1	Man-load	35,0	222	5,07	62
2	Man-load	25,2	497	8,92	57
3	Load	44,5	520	9,0	57
4	Man-load	38,5	540	7,16	42
5	Load	45,0	551	7,29	42
6	Man-load	38,12	651	10,1	54
7	Load	50,0	691	4,5	54

2 Analysis of Literary Sources

Vibration monitoring is widespread for technical control of buildings and structures, both at the stage of building and at the stage of operation (Wenzel; 2009; Brownjohn et al.; 2011; Wenzel; 2008; Jaishi and Ren, 2005; Ivanovic et al.; 2000). In connection with the fact that the operational technological loads on the structure of mine headgears are dynamic. In article (Kuschenko; 2007) evaluation the technical condition of the headgears was proposed by comparison an experimentally determined and spectrum of eigenfrequencies with the reference spectrum of frequencies, which is fixed when commissioned in the facility or during his planned inspection. However, this technique is imperfect, since it isn't always possible to reliably measure ones, because process of fluctuations in the structures of mine headgears is under the influence of occasional constraints on the side of the hoisting machine (Kushchenko; 2006). For monitoring structures with complex dynamic behavior there are rational methods, which based on comparing the measured amplitude of vibration limit values (Redchenko; 2010). They are determined on the basis statistical analysis of data on parameters of vibration an object in the course in normal operation.

3 The Purpose of this Work

The purpose of this work is a generalization experimental data to measure vibrations of a number of mine headgears with different structural and technological features and formulation concept of vibration monitoring to the structures of mine winding plants.

4 Method of Work Execution

In this work, as primary diagnostic parameters the technical condition of headgears vibration amplitude values and their mean-square values at intervals of implementation, which correspond to the length of the technological cycle of mine hoisting installations were considered.

Table 2. Results rms measurement values of vibration amplitudes overall analyzer vibration spectrum «795 M»

№ object	Structural and technological features	The registered rms amplitudes vibration in the direction of the respective axes of the coordinate system (Fig. 3) (mm/ s)			
		X (forced)	X (own)	Z (forced	Z (own)
1	Man-load	0,6-1,2	0,1-0,5	0,72	–
2	Man-load	0,5-1,5	0,1-0,4	–	0,11
3	Load	0,5-1,6	0,1-0,4	0,23	0,12
4	Man-load	0,4-1,3	0,1-0,3	0,2-0,4	0,01-0,1
5	Load	0,4-2,2	0,1-0,3	0,2-0,5	0,01-0,1
6	Man-load	0,3-1,5	0,1-0,2	0,2	0,1
7	Load	0,5-2,3	0,1-0,4	0,12	–

On steel mine headgears with different structural and technological features (Table 1) vibration analyzer are determined the amplitude-frequency characteristics of their own (Fig. 2a) and forced oscillations of buildings (Fig. 2b). Also the mean square values of the amplitudes of vibrations within the technological cycles of hoisting machines was determined (Table 2).

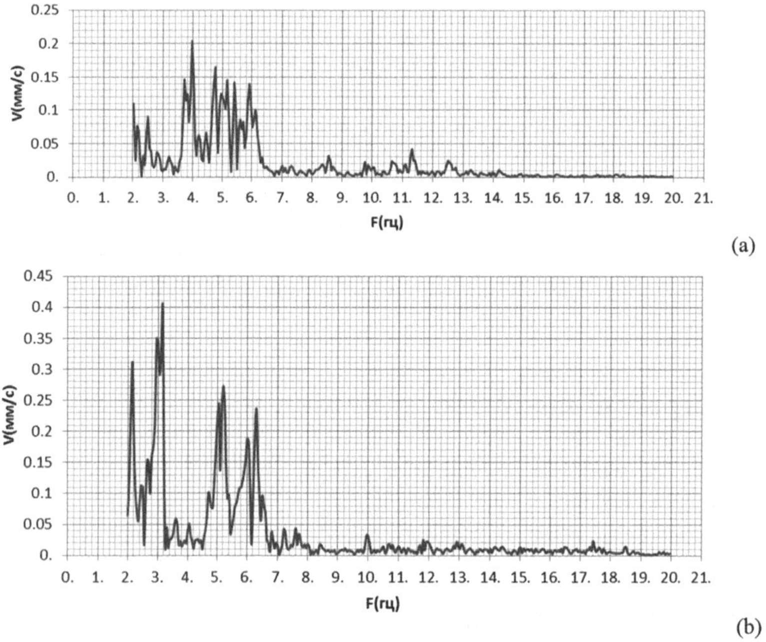

Fig. 2. The amplitude-frequency characteristics of mine construction headgear (p. 3 Table 1), which obtained with the use of the vibration analyzer "795 M": (a) amplitude-frequency characteristic of free fluctuations of the structure; (b) frequency response forced vibrations during hoisting machine

After the analysis and generalizing of these parameters measuring overall vibration, one of the objects (p. 1 Table 1) were performed measuring the amplitudes of vibration transducers VPE-085-T-32 (measurement frequency range 10–1000 Hz; vibration measurement range 0–32 mm/s) which were fixed in the upper part of the building on the pullout structures in two directions of measurement: horizontal «X» and vertical «Z»» according to the scheme in Fig. 3a. Indications vibration sensor registered with intervals 1 s and accumulated in a database, visualization of the results was carried out on the display in the diagrams form, which represent the discrete values of peak vibration velocity values of mine headgear forced fluctuations in time of 5–8 h (Fig. 3b). The obtained experimental data were divided into statistically homogeneous groups, which the following numerical characteristic are determined by mathematical expectation, variance and confidence intervals.

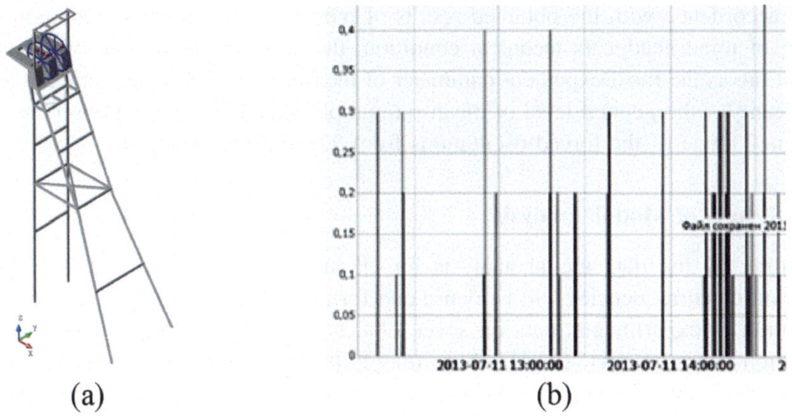

Fig. 3. (a) The scheme of placing vibration sensors on structures of mine headgears; (b) The results of vibration recording velocity amplitudes (mm/s) in the direction of "X" by VPE-085-T-32 transducers in the diagram form for 2 h periods (object p. 1 of Table 1)

Based on these defined thresholds amplitude vibrations, excess of which is the manifestation fact of destructive processes in the system of mine winding plant construction.

For verification amplitude-frequency characteristics of mine headgears structures, which obtained experimentally and for a number of sites was performed modal analysis in accordance with their own frequencies and orms of own oscillations on the calculation models implemented in the software complex «SCAD-11.1»

5 Results of the Work

5.1 Results of Vibrations Analysis with Appliance of the Vibration Analyzer «795 M»

In Table 2 shows the recorded values of the mean-square values of the vibration amplitudes forced oscillations in the horizontal direction, which coincides with direction of the horizontal component of the load from the hoisting ropes (direction of the axis «X», Fig. 3a), the average values rms amplitude values of the vibration velocity in this direction change on the range of: 0.85… 1.4 mm/s. The highest values of mean-square values vibration amplitudes were recorded on load hoisting headgears. The upper limit of confidence intervals is: 2.3 mm/s. The same for the loads of man-loads elevations is: 1.6 mm/s. This fact is explained by the higher height of the freight hoisting headgears, which height is from 45 to 51.6 m. The height headgears of man-loads elevations is from 25.2 to 38.75 m.

The recorded mean square values of the vibration amplitude of direction "X" in the free oscillation mode (excitation by anthropogenic microseismicity) for all investigated structures are: 0.1… 0.5 mm/s.

In accordance with the obtained results of measuring, for purpose vibration monitoring of mine headgears technical condition, the vibration amplitude of the forced oscillations is the most expedient parameter of measurement. Vibration amplitudes are significant for the general level of mean-rpm amplitudes and them are observed in the frequency range of the forced oscillations from 2.0 to 20 Hz (Fig. 2b).

5.2 Results of Modal Analysis

According by the data modal analysis for all investigated structures (Table 1), the Spectrum eigenfrequencies will be dense and forms of oscillations (Fig. 2a). The generalization of experimental data on seven objects e showed a range tone of the own oscillations from 2.0 up to 2.2 Hz. According to the results of modal analysis the same parameter is from 1.5 up to 2.1 Hz. In general, there is a coincidence between calculated and experimental values of own frequency with a confidence probability $p_0 = 0.95$.

5.3 Results of Statistical Analysis Data Experimental, Which Obtained with the Use of Vibration Sensors VPE-085-T-32

In total, the registration data of the vibration amplitudes with a frequency of 1 s from the period of 45 technological cycles of the technological mine hoisting machine operation (Fig. 2b). As a result of the statistical processing of data vibration amplitude in horizontal and vertical directions (Fig. 3a), was found that the data sample represents a probabilistic mixture of two general aggregates. The statistical analysis of these general aggregates for vibrations in the horizontal direction showed that they correspond to the normal distribution law. The sampling from the first general population is limited by the values of the vibration amplitude as: 0.5… 1.6 mm/s; the same sample from the second general population is: 0.1… 0.4 mm/s. The mathematical expectation for the first general population is defined in the following confidence interval: 0.81… 1.15 mm/s; the same for the second general population: 0.2… 0.3 mm/s. The variation coefficient for the first general population is: 0,19; the same for the second general population is: 0.12. Thus, the sample from the first general population is the most informative according to the extreme values of the vibration amplitude. The Similar results were shown by the statistical analysis for measuring the vibration velocity amplitudes in the vertical direction «Z»: mathematical expectations for the sample from the first general population constitute the confidence interval: 0.64… 0.82 mm/s, with variation coefficient is: 0.17 … 0.3. The obtained results were compared with the results of «795 M» vibration analyzer measurements, which showed the convergence of the average values of the vibration amplitude: 13… 21%.

5.4 Concept of Steel Headgears Vibration Monitoring

Because speed vibration amplitude is depending of the one side from the technical state of the headgears structures. And for other side from the level of dynamic component of the load hoisting ropes. Vibromonitoring of the headgears construction are carries information about both the quality of the headgears construction and the quality of the mechanical system of the mine winding plant. Accumulation of information in accordance with the speeds vibration amplitudes under normal operation allows us to statistically determine of the amplitude level, which is unlikely to occur and the exceeding probability of this level is $2.0 \cdot 10^{-5}$. If this level is exceeded, the spectrum of the own frequencies facilities of the structure wouldn't coincided with the benchmark and that it's a sign of damage to the mine headgears structures. If for these situations the spectrum of the own frequencies of the mine shaft structures remains unchanged, it will be fact determines of destructive changes in the system of the mine winding plant.

6 The Practical Significance of the Obtained Results

The practical significance of the obtained results to create the concept of passive vibration monitoring of mine headgears structures, which allows to receive in real time the information about technical condition of the mine headgears structures and elements of the mine hoisting system. That are provides the opportunity to prevent accidents in the construction of mine hoisting facilities.

7 Conclusions

In the vibration monitoring of the mine hoisting installations structures, as the main diagnostic parameters the measure amplitude of vibration in two areas of active dynamic forces of load hoisting ropes: (a) horizontal "X"; (b) vertical "Z" (Fig. 3a) is proposed.

An additional diagnostic sign of the technical condition of the mine shafts structures the spectrum of the own frequency oscillations of the structure should be considered.

Vibrating sensors should be placed rationally in the upper part of the structure on sub-shell structures in the directions of measurement "X" and "Z", as shown in Fig. 3a. Recommended measurement range for frequencies is from 2 up to 20 Hz and measurement range for the amplitudes of vibration is from 0.1 up to 10 mm/s.

Assessment of the technical condition of mine shaft structures on the basis of vibration monitoring data is proposed. We recommend to determine the criterion for exceeding the critical value of the vibration speed amplitudes, which is defined as the quantile of the normal distribution law of the first general population and when the probability of an excess being at $2.0 \cdot 10^{-5}$.

References

DBN B. 1.2-14-2009 (2009): General principles of ensuring the reliabiliny and constructive safety of buildings, structures, constructions and foundations. Kyiv: Minregionbud Ukraine, 30 (in Ukrainian)

RD 12.005_94 (1994): Metal structures of mine headframes. Operation requirements. Kyiv: Gosugleprom of Ukraine, 68 (in Russian)

DSTU NB 1.2-18: 2016 (2017): Guidelints for the inspection of building and structures for the determination. Kyiv: DP «UkrDPC», 44 (in Ukrainian)
Kushchenko, VN (2006): Organizations of safety of building constructions of cut_sample mine headframe. Monograph: Makiivka, 203 (in Russian)
Kushchenko, VN, Kostrickiy, AS (2003): Experimental research of dynamic behavior of mine head frames structures. Compendium Proceeding of the Donbas State Academy of Civil Engineering and Architecture, pp. 46–52 (in Russian)
Kushchenko VN (2010) Analysis of the influence of magnetic components in the dynamic load character from the power of handles. Metal Constructions 16(1):19–30 (in Russian)
Kuschenko VN (2007) Application of dynamic tests for technical diagnostics of building structures of staying head frames. Metal Constructions 2(13):113–122 (in Russian)
Wenzel H (2009) Ambient vibration monitoring, encyclopedia of structural health monitoring. John Wiley & Sons, New Jersey
Brownjohn JM, De Stefano A, Xu YL, Wenzel H, Aktan AE (2011) Vibration-based monitoring of civil infrastructure: challenges and successes. J Civil Structural Health Monit 1(3–4):79–95
Wenzel H (2008) Health monitoring of bridges. John Wiley & Sons, New Jersey
Jaishi B, Ren WX (2005) Structural finite element model updating using ambient vibration test results. J Structural Eng 131(4):617–628
Ivanovic SS, Trifunac MD, Todorovska MI (2000) Ambient vibration tests of structures-a review. ISET J Earthq Technol 37(4):165–197
Redchenko, VP (2010): Vibration diagnostics of large-scale spatial structures, problems and perspectives. Collection of scientific works of the Ukrainian research and design institute of steel structures named after VM Shimanovsky (5), 52–59 (in Ukrainian)

Criterial Equation for the Description of Low-Speed Air Distributor Operation

H. Klymenko, V. Labay(✉), V. Yaroslav, and M. Gensetskyi

Department Heat and Gas Supply and Ventilation, Lviv Polytechnic National University, St. Bandery 12, Lviv 79013, Ukraine
wlabay@i.ua

Abstract. The advantages of Displacement Ventilation systems as an energy-efficient way of air distribution are indicated. The task Displacement Ventilation systems is to provide normative parameters of the microclimate in the working area. For the most part, this task cannot be solved without uniform air distribution directly to the working area, for which source perforated low-speed air distributors are used. The characteristics of air distribution devices of Displacement Ventilation systems is are presented. There is no analytical justification for the design solutions of these devices. The necessity of analytical study of aerodynamic processes in the body of the proposed air distribution device is indicated, and it is also advisable to obtain a criterion equation of similarity for dissemination of the results of experiments on similar, but not yet studied objects. To obtain generalized dependencies, it is suggested to use the theory of similarity. The scheme of the experimental installation, the general scheme of the air distributing device, and the schemes of structures of the investigated air distributor are presented. The results of the research are presented. As a result of mathematical description of aerodynamic processes in the case of a low-speed air distributor, a criterial equation of similarity was obtained.

Keywords: Energy-efficient ventilation · Low-speed air distributor · Air flow separators · Criterion equation

1 Introduction

The energy efficiency of microclimate maintenance systems for production, administrative and public premises is determined by air exchange organization and air distribution patterns. Stricter requirements (Ministry of Regional Development, Building and Housing and Communal Services of Ukraine 2013; Standard ASHRAE 62.1-2013 2013) to the energy efficiency of ventilation and air conditioning systems cause the necessity to develop theoretical basics for planning the operation of ventilation systems and air distribution devices, which form indoor air parameters. Microclimate regulation systems must meet normative requirements and technological process needs.

Z. Blikharskyy et al. (Eds.): CEE 2019, LNCE 47, pp. 235–242, 2020.
https://doi.org/10.1007/978-3-030-27011-7_30

2 Analysis of Recent Studies and Publications

In recent years, Displacement Ventilation Systems, in which special air distributors are used, are becoming more and more common. In the course of analysis of the operation of displacement ventilation systems, a number of their advantages were revealed (Zhivov et al. 2001; Livchak 2001). At the same time, incoming air with a low turbulence level does not contribute to air mixing in the lower zone. Artificial microclimate is maintained at minimum energy expenditure. At uniform air distribution directly into the operation area, uniformly normative microclimate parameters are maintained in it. This helps to protect employees' health and increase productivity. To ensure air distribution, panel distributors with horizontal slits in the air distribution wall, wedge-shaped pressure chambers and a perforated wall between them are used (panel near-wall air distributors (Staroverov et al. 1977). The incoming air mixes less intensively with the ambient indoor air as air distributors (their air distribution walls) become bigger. The local resistance coefficient of a panel near-wall air distributor ξ = 1, 9... 2, 3, pressure loss at $w_{en} = 0{,}6$ m/s within 40–50 Pa. There are no analytical studies of justification of the suggested construction solution of the given air distributor. There are no experimental studies of aerodynamic characteristics, in particular initial velocity change of the formed jet, either.

An improved construction of source cylindrical air distributors with the placing of an axisymmetric conic cavity insert in their body was suggested by the members of the Department of Heat and Gas Supply and Ventilation of Lviv Polytechnic National University (Zhukovskyi and Shcherbatiuk 1990). The drawback of this air distributor construction is its significant pressure loss (ξ = 4, 4), attributed to the dynamic pressure in the inlet nozzle. There are no analytical studies of aerodynamic characteristics of this air distributor available.

In air distributors with a constant cross-section of the body aimed at equalizing the internal static pressure and ensuring uniform initial velocity and expenditure, different construction solutions were suggested.

Among the latest developments to ensure equal distribution, let us mention a technical solution that provides equal distribution of expenditure, uniform velocity by height and a direction of air escape without additional initial excitation of the jet at the same time. This solution was suggested by researchers from Kyiv National University of Construction and Architecture, in particular V. B. Dovhaliuk V. O. Mileikovskyi, and others (Dovgalyuk and Mileikovskyi 2006). There are no results of experimental studies of this air distributor that can be expanded onto similar, but not yet studied objects.

Changing thermal conditions inside a room require changing air expenditure in this room. This necessitates air distribution devices that are able to change the area of air penetration. The aforesaid air distributor constructions are not adapted to systems with a changing air expenditure. In general, the constructions of the aforesaid air distributors do not fully meet displacement ventilation requirements. Moreover, the mathematical description of experimental research in the aforesaid publications is not sufficient.

As a result of numerous experimental studies in Lviv Polytechnic National University, new air distribution devices were developed and patented with low aerodynamic resistance for supplying air to the operation area with low velocities and a uniform and changing area of air escape (Zhukovskyi et al. 2006). Figure 2 shows the assembled representation of one of the constructions of a two-chamber low-velocity panel-sectional air distributor with a constant cross-section of the body.

The conducted experimental and analytical studies (Klymenko 2009) of the uniformity of distribution of excess static pressure in the distribution chamber of the low-velocity panel air distributor with air flow separators showed that a construction solution is possible for this air distributor to achieve highly uniform distribution of the excess static pressure in the distribution chamber. Accepted correlations of factors were determined, at which the distribution uniformity coefficient for the excess static pressure in the distribution chamber of the air distributor exceeds 90% (Zhukovsky and Klymenko 2009). However, analytical research of the experiments conducted is not sufficient. Therefore, an opportunity arises to conduct analytical research of the operation of a low-velocity air distributor with a more detailed mathematical description of experimental studies.

3 Research Topicality

At the present stage of development of construction and civil engineering, experimental and analytical studies of the effectiveness of using optimized constructions of air distribution devices in displaced ventilation systems allow suggesting new solutions for energy-effective microclimate formation inside buildings. This is an urgent task connected with ensuring the energy efficiency of engineering solutions for indoor premises according to (DSTU B EN 15232 2011; DSTU B A.2.2-12 2015).

4 The Purpose of the Research

The aim of the study is to create a mathematical description of aerodynamic processes that take place in the body of the air distributor on the basis of results of experimental research into panel low-velocity air distributors. For this purpose, it is worth using the dimensional method, take into account all the crucial parameters of impact on the uniformity of distribution of excess static pressure along the height of the distribution chamber of a low-velocity panel air distributor and obtain a criterial equation of similarity to expand the results of experiments onto similar, but yet unstudied objects.

5 Description of the Pilot Unit

The pilot unit (Fig. 1) was created in the laboratory of Lviv Polytechnic National University to conduct experimental research into the low-velocity panel-sectional air distributor (Fig. 2).

The air distributor (Fig. 2) has the first distribution chamber with air flow separators and the second one – the stabilization chamber that can be constructed as a separate module.

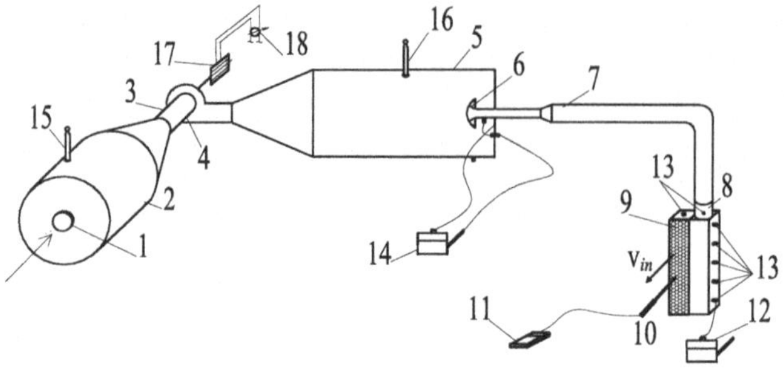

Fig. 1. Design of the pilot unit for researching the low-velocity panel-sectional two-chamber air distributor: 1 – inlet nozzle; 2 – pressure chamber; 3 – air line; 4 – centrifugal blower; 5 – pressure chamber; 6 – flow collector; 7 – airline; 8 – inlet nozzle of AD; 9 – air distributor; 10 – thermal electrical anemometer sensor, 11 – thermal electrical anemometer "TESTO-405"; 12, 14 – micromanometers MMH-240; 13 – check points (connections) for measuring excess static pressure; 15–16 – alcohol thermometers; 17 – direct current electric engine; 18 – rotation regulator

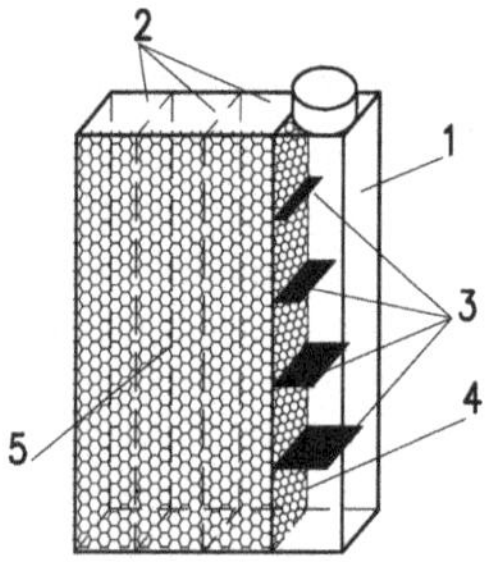

Fig. 2. Assembled representation of a low-velocity panel-sectional air distributor (patent of Ukraine No. 19497): 1 – distribution chamber; 2 – stabilization chamber; 3 – air flow separators; 4 – internal air-permeable wall; 5 – external air-permeable wall

The construction of the low-velocity panel-sectional air distributor ensures initially equal velocity of the air flow-out and the possibility to effectively regulate the expenditure of the incoming air (Klymenko 2009).

Dependence has been obtained to determine the optimum size of shelf separators of the flow in a low-velocity air distributor. This dependence allows ensuring equal air distribution (DSTU B EN 15232 2011; DSTU B A.2.2-12 2015).

6 Obtaining Criterial Equation for the Process of Ensuring Uniform Distribution of Excess Static Pressure in the Distribution Chamber of a Low-Velocity Distributor

With the difficulties of aerodynamic processes and the impact of numerous factors on the change of effectiveness of the suggested means of flow equalization and with the lack of mathematical description of the processes analyzed, we can obtain the criteria of similarity and the general view of the similarity equation (Labay 2004) taking into account all the crucial parameters of impact. For a particular case of the aerodynamic process, the change of excess pressure in the air distributor body depends on the following physical parameters:

$$\Delta p_{st} = f(w_{en}, \rho, d_e, H, h, l, v, g), \text{Pa}. \tag{1}$$

Then the correlation between the pressure forces that affect the elementary volume of air and the inertia forces is expressed by the equation:

$$\overline{\text{Eu}} = A \cdot \text{Re}^{\alpha} \cdot \left(\frac{H}{d_{in}}\right)^{\beta} \cdot \left(\frac{\ell}{h}\right)^{\gamma}, \tag{2}$$

where $\overline{\text{Eu}} = \frac{\Delta p_{st}}{\rho \cdot w_{en}^2}$ – average Euler's criterion at the entrance of the air distributor; Δp_{st} – the difference of static pressures at the entrance of the air distributor, Pa; ρ = 1,2 kg/m^3 – air density at t_{en} = 20 °C; w_{en} – air velocity at the entrance of the air distributor, m/s; $\text{Re} = \frac{w_{en} \cdot d_e}{v} = \frac{w_{en} \cdot d_e}{v}$ – Reynold's criterion at the entrance of the air distributor; $d_e = d_{in}$ – equivalent (internal) diameter of the cross-section at air entrance into the air distributor, m; $v = 15 \cdot 10^{-6}$ m^2/s – coefficient of kinematic viscosity of air

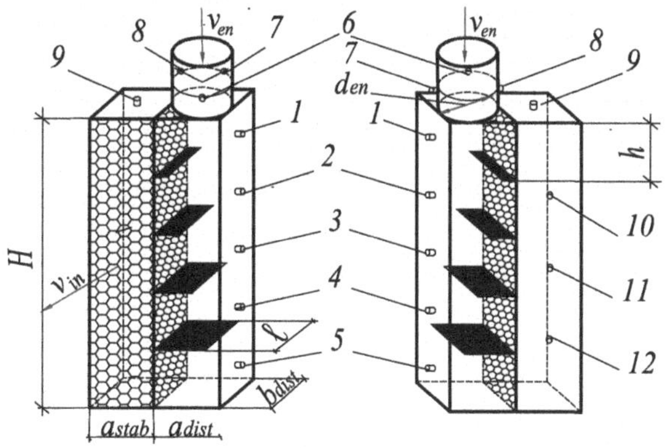

Fig. 3. The air distribution device with checkpoints marked to measure excess static pressure $\Delta p_1 \ldots \Delta p_{12}$ in the body of the air distributor with air flow separators; 1…12 – connections for measuring point excess static pressure in the body of the air distributor H = 0,4 m; $a_{dist} = b_{dist} == a_{stab} = \ell = d_{en}$ = 0,1 m; h = 0,08 m

at t_{in} = 20 °C; H – height of the air distributor, m; ℓ – length of the equalizer (separator) of the air flow in the air distributor (equals the width of the distribution chamber of the air distributor $\ell = b_{dist}$), m; h – distance from air entrance to the equalizer (separator) of the air flow, m. In a particular air distributor $\frac{\ell}{h} = \text{const}$.

To analyze the impact of separators of the air flow on excess static pressure distribution at the entrance to the air distribution device, experimental research was held into the construction with air flow separators (Fig. 3). Points of research: No. 6; 7; 8.

7 Research Results

Excess pressure value at the entrance of the air distributor depends on the velocity of the incoming air. As a result of experimental research, it was established that velocity change affects the distribution of excess pressure at the entrance of the air distributor. The values of Euler's and Reynold's criteria were obtained for the construction with air flow separators (Table 1 and Fig. 3). The length of the stabilization chamber (a_{stab}) of the air distributor equals that of the distribution chamber (a_{dist}).

Table 1. Results of research into excess pressure distribution at the entrance of the air distributor with air flow separators (Fig. 3)

Δp_{st}, Pa	w_{en}, m/s	$\overline{\text{Eu}}$	Re	$\frac{H}{d_{in}}$	$\frac{\ell}{h}$
3.43	1.58	1.145	10533	4	0.8
4.41	1.99	0.928	13267	4	0.8
6.86	2.71	0.778	18067	4	0.8
13.39	3.85	0.753	25667	4	0.8
23.52	5.27	0.706	35133	4	0.8
36.31	6.73	0.668	44867	4	0.8
47.06	7.75	0.653	51667	4	0.8

Therefore, the dependence of Euler's criterion (Eu) on Reynold's criterion (Re) is approximated by the following formula for the air distributor with air flow separators

$$\text{Eu} = 15{,}907 \cdot \text{Re}^{-0{,}2966}, \tag{3}$$

with the maximum error 11.3% (Fig. 4).

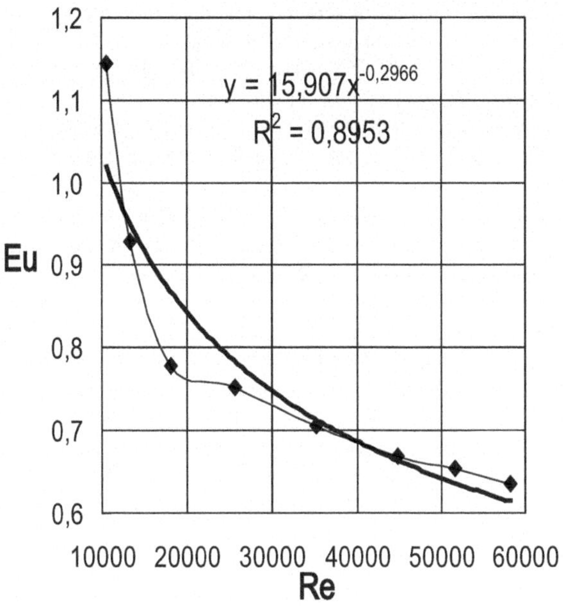

Fig. 4. Dependence of Euler's criterion (Eu) on Reynold's criterion (Re) for the air distributor with air flow separators

8 Conclusions

Constructions of low-velocity panel-sectional air distributors that ensure initially uniform velocity of air flow-out and the possibility of effective regulation of incoming air expenditure are suggested.

The effect of improving the construction of the low-velocity air distributor on the uniformity of excess static pressure distribution in the body of the low-velocity air distributor is analyzed. By analyzing dimensions, the criterial equation of similarity was obtained. This will allow expanding results of the experiments onto similar but not yet studied objects.

References

Ministry of Regional Development, Building and Housing and Communal Services of Ukraine (2013) State construction norms of Ukraine V.2.5-67:2013. Heating, ventilation and air conditioning (in Ukrainian)

Standard ASHRAE 62.1-2013 (2013) Ventilation for acceptable indoor air quality (ANSI approved), p 53

Zhivov A, Nielsen PV, Riskowski G, Shilcroft EO (2001) Exhaust ventilation systems for industrial buildings. Types, field of application, design principles. AVOK. 5:36–47 (in Russian)

Livchak. A, Nall D (2001) Displacement ventilation – application for hot and humid climate. In: Clima 2000/Napoli 2001 world congress, 15–18 September 2001

Staroverov, MG, Bohoslovsky, VN, Shepelev IA (eds) (1977) Internal sanitary devices. Reference book of the 502, pp 191–193 (in Russian)

Zhukovsky SS, Shcherbatyuk BI (1990) Author's certificate 1564478. Moscow, State Committee for Inventions and Discoveries at the National Technical University of the USSR (in Ukrainian)

Dovgalyuk VB, Mileykovsky VO (2006) Air distribution using curvilinear flooring. New Top J Assoc Eng Energy Effic Technol Ukr Sci Tech J 42:25–27 (in Ukrainian)

Zhukovsky SS, Dovbush OM, Klymenko HM (2006) Patent of Ukraine 19497. State Patent Office of Ukraine, Kyiv (in Ukrainian)

Klymenko GM (2009) Pressure influences in the case of a two-chamber panel air distributor and ensuring their uniformity. Theory Pract Constr Visnyk NU LP 655:140–147 (in Ukrainian)

Zhukovsky S, Klymenko H (2009). Experimental and analytical research of pressure effects inside the sectional source air distributor. Zeszyty naukowe Politechniki Rzeszowskiej, № 266. Budownictwo i inżynieria środowiska. Z. 54:151–157 (in Polish)

DSTU B EN 15232 (2011) Energy efficiency of buildings, influence of automation, monitoring and building management (EN 15232: 2007, IDT). Effective from 2012.04.01. Ukrrahbud-inform (in Ukrainian)

DSTU B A.2.2-12 (2015) Energy efficiency of buildings. Method of calculating energy consumption for heating, cooling, ventilation, lighting and hot water supply. Effective from 2016.01.01. Ukrexbudinform (in Ukrainian)

Labay VYo (2004) Heat-mass transfer. In: Textbook for high schools, p 260 (in Ukrainian)

Planning Experiment for Researching Reinforced Concrete Beams with Damages

M. Lobodanov, P. Vegera(✉), and Z. Blikharskyy

Department of Building Constructions and Bridges, Lviv Polytechnic National University, Karpinskoho Str. 6, 79013 Lviv, Ukraine
Pavlo.I.Vehera@lpnu.ua

Abstract. In the article, theoretical researching of the influence of damage in the compressed area in bending reinforced concrete elements is presented. The main attention is paid to the comparison of the impacting two factors on the bearing capacity: level of loading and the loss of the cover in the compressed area. In the first step, carried the formation of a matrix of the dependence impact of each factor on the bearing capacity with the subsequent determination of coefficient of influence and the formation of the regression equation. Analyzing the regression equation, obtained results of the impact of each factors, and their interaction, on the bearing capacity of the element. Second step is modeling of the experiment research using software "Femap with NX Nastran". Thereafter, there was created a table of the distribution factors and their combinations with the indications of their impact on the carrying capacity of the model. It is possible to concluding, that the main factor is the initial level of loading, when the bending reinforced concrete beams received damages. The obtained results show the relevance and expediency of conducting experiments with the aim of determination of the residual bearing capacity in damaged reinforced concrete elements.

Keywords: Planning experiment · Reinforced concrete beam · Damages · Defects · Bearing capacity

1 Introduction

Nowadays the tendence of reconstruction of existing buildings and structures is becoming more and more popular as there are a large number of buildings in a preserved state. However, before the beginning of the reconstruction, it is necessary to research the actual technical conditions level of the elements.

Using reinforced concrete structures has a major impact at the territory of Ukraine, which gives priority to the research of damaged flexural reinforced concrete elements of a rectangular section, since they are most commonly used in the construction of existing buildings and constructions (Blikhars'kyi and Obukh 2018; Zhang et al. 2015).

Nowadays, the issue of research of damaged reinforced concrete elements is a complex and relevant topic. Mostly, this follows from the complicated process of forecasting and modeling of the physical and mechanical characteristics of the constructions materials through the composite properties of the reinforced concrete.

Z. Blikharskyy et al. (Eds.): CEE 2019, LNCE 47, pp. 243–250, 2020.
https://doi.org/10.1007/978-3-030-27011-7_31

A large number of different factors and their possible combinations caused the significant complication of calculations (Klymenko et al. 2013). Influence of the factors of damage of T-beams on the magnitude of their destructive load is considered. In the operation of the experimental obtained data processing by the method (Voskobiinyk et al. 2011), with the removal of insignificant coefficients of the regression equations, an adequate mathematical model has been obtained, which has a sufficient information usefulness and by which the influence of the investigated factors on the initial parameters of beams can be used, like geometric interpretation.

The researches in the creation of universal algorithm of numerical modeling of nonlinear deformation processes of reinforced concrete elements is considered in the paper (Vozniak and Zhelykh 2003), where the results and examples of the implementation of this model are provided. These developments are considered through the use of norms (SNIP 2.03.01.84, 1989), on the territory of Ukraine. Nevertheless, the given research can be used as the basis for the formation of the numerical simulation algorithm for the nonlinear deformation processes of damaged reinforced concrete elements with taking into account current norms (DBN V.2.6-98: 2009, 2011; DSTU: B V.2.6-156 ISO: 2010, 2011).

Damage caused by alkaline-aggregate reactions in the bending reinforced concrete elements can also be considered through numerical simulation, with taking into account the complexity of the increase in the volume of reinforcement. In the research (Capra and Sellier 2003) the simultaneous action of chemical and mechanical processes under the action of the load is considered. The results of this research will be useful for the formation of the numerical model of nonlinear processes with a combination of defects and damage.

In some cases, the influence of the external environment may entail a change in the work of the reinforced concrete element, for example, the formation of oblique bend, for this case we can use the research (Pavlikov et al. 2012; Pavlinkov et al. 2012), where are the mathematical models of calculation according to the current norms in the territory of Ukraine are reviewed (DBN V.2.6-98: 2009, 2011; DSTU: B V.2.6-156 ISO: 2010, 2011).

While considering the issue of residual bearing capacity research, it should be payed attention to the research (Brara and Klepaczko 2006) where the issues of the influence of dynamic tension on the strength of concrete depends on the time changing, are considered. The main attention is paid to the factors of the speed of the development of deformation and changes in the humidity of concrete. Understanding of this problem can make more useful for strengthening construction (Blikhars'kyi and Obukh 2018; Krainskyi et al. 2018)

On the basis of the above, it can be concluded that the research of damaged reinforced concrete elements is an actual task, with significant practical value.

2 Aim of the Researching

The aim of the researching is determining impact of loading level and damages of compressed area which have influence on the stress-strain state of reinforced concrete beams with rectangular section and to determine the priority factor for conducting an experimental testing.

3 Methodology of Planning Researching

For determine the influence of each factor, it must be considered the definition of the concept "factor". A factor is an independent measured variable, what at a certain time acquires a certain value. During operation, every factor, denoted by x_i, may become one of several values. Such values are called levels. So, the totality of combinations of factors levels is the number of experiments (Vozniak and Zhelykh 2003) what we can be determined by the formula:

$$N = p^{(k)} \tag{1}$$

Where p is the number of levels of factors, k is the number of factors. In the analysis of the priority of two main factors, the methodology of planning the experiment 2^2 with the effect of interaction of factors is described in (Vozniak and Zhelykh 2003). Where the regression equation looks like:

$$y = b_0 + b_1x_1 + \ldots + b_kx_k \tag{2}$$

The coefficients are calculated according to the equation:

$$b_j = \frac{\sum_{i=1}^{N} x_{ji}y_{ji}}{N}; j = 0, 1, \ldots, k \tag{3}$$

Particularly, for b_1 and b_2:

$$b_1 = \frac{-y_1 + y_2 - y_3 + y_4}{4} \tag{4}$$

$$b_2 = \frac{-y_1 + = -y_2 + y_3 + y_4}{4} \tag{5}$$

Then the coefficient b_0 is determined by the dependence of:

$$b_0 = \frac{\sum y_1}{4} \tag{6}$$

4 Analyzing Results of Modelling

According to this method, factor x_1 and x_2 are considered, where x_1 is the initial load level which is 0,3 from the bearing capasity, and x_2 is damage which carries the loss of the cover in the compression area in the size of 30 mm. It is considered that these factors as the most widespread in the practice. Consequently, $y = b_0$, where b_0 is the arithmetic mean of the optimization parameter. To perform this operation, we introduce a vector-column of a fictitious variable x_0, which acquires the value of +1 in experiments. In this study the reinforced concrete beam is presented in Fig. 1.

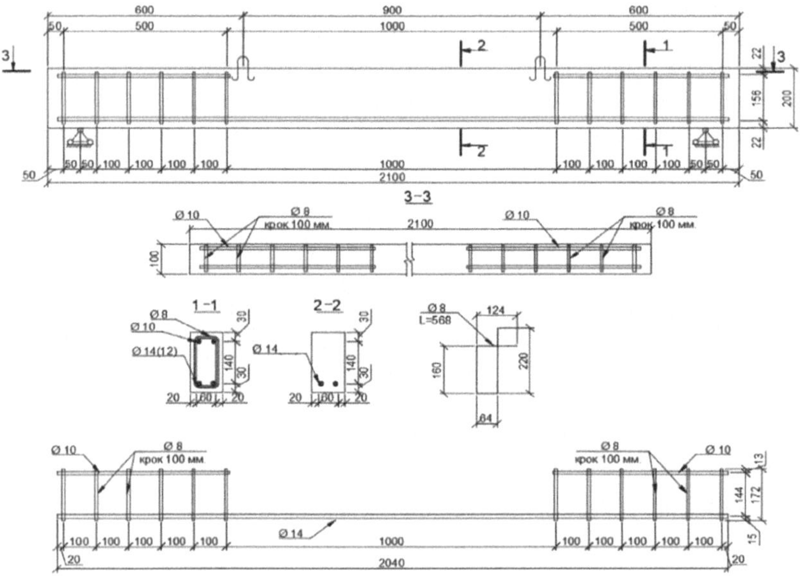

Fig. 1. Reinforcement and dimensions of the testing sample.

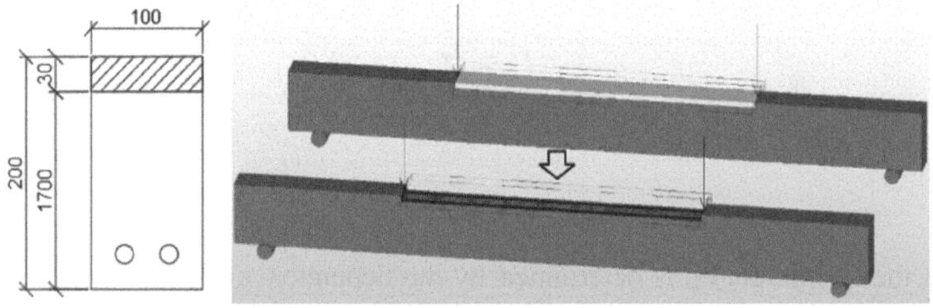

Fig. 2. Scheme of placement damage in compressed area

The reinforcement is made in the form of working tensile rebar Ø14 mm, compressed reinforcement in the area of maximum transverse force - Ø10 mm. Transverse reinforcement is made with a smooth rebar Ø8 mm located in support area. Concrete is used C35/40. Sceme of future damage of researching samples are presented at Fig. 2.

The calculation scheme of the beam Fig. 3 is made according to the scheme of a single-span beam on two supports, loaded by two forces at the same distance from the supports, with the purpose of the formation of a "net bend."

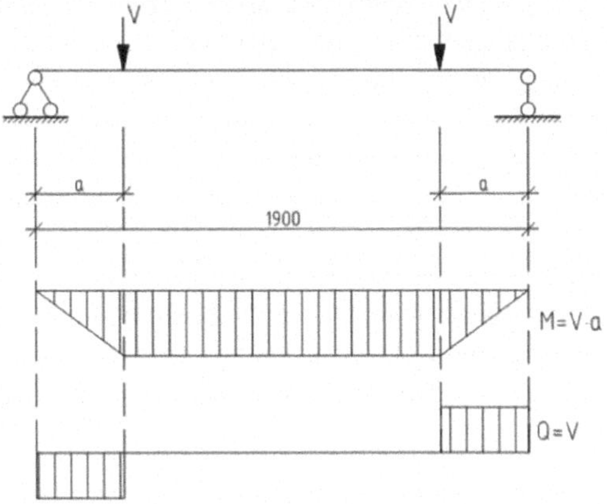

Fig. 3. Design scheme of the beam.

While forming a matrix of the interaction of two factors, we assume that in the presence of the factor x_i or x_j, the factors have a "+" value, in the absence of the "–" value. For example, if there is a load and no damage, get the value in the line $+x_1$ and $-x_2$. The value of y_i is the bearing capacity of the beam at various combinations of factors in Table 1, for example, in case 1, y_i is defined as bearing capacity in the presence of load (loss of protective layer in the size of 30 mm.) and loading equal to 8,786 kN (load level 0,3 from the bearing capacity of regular undamaged samples).

Table 1. The planning experiment's matrix 2^2 with the effect of the interaction of factors

№	x_0	x_1	x_2	x_1 x_2	y
1	+	+	+	+	9,133
2	+	–	+	–	14,756
3	+	+	–	–	13,12
4	+	–	–	+	18,964

The following dimensions of the coefficients for the regression equation are obtained: b_0 = 13,993, b_1 = −2.866, b_2 = −2.049, b_{12} = 0.055. The regression equation will have the following form:

$$y = 13,993 - 2,866x_1 - 2,049x_2 + 0.255x_1x_2 \tag{7}$$

Also, this equation indicates the simultaneously action of two factors. The obtained results indicate a decrease in the bearing capacity with an increase in the factor x_1 and x_2, where the factor x_1 is more influential. Consequently, the key factor is the level of loading.

In order to investigate the priority of factors in the research, an experiment was simulated in the software Femap, which contains the NX Nastran calculation module. This complex gives a significant range of variations, both during modeling and in the analysis.

Modelling reinforced concrete beams, we divided it into separate elements "Solid" with the dimensions of elements 10 × 10 × 10 mm. This size is chosen to facilitate further analysis of damage in the compressed area that was considered in a step of 10 mm. The reinforcement was modeled as a linear element with the Beam subtype. Since the concrete grid does not coincide with the broken reinforcement elements, the "Move Merge Nodes" operation was performed with subsequent copying of the elements.

The modeling results were obtained in different variations with the results of the total displacements of the model obtained: regular beam without any damages (Fig. 4), beam with damages without initial level of loading (Fig. 5) and with damages which was taken with initial loading 0.3 from bearing capacity from regular beam (Fig. 6).

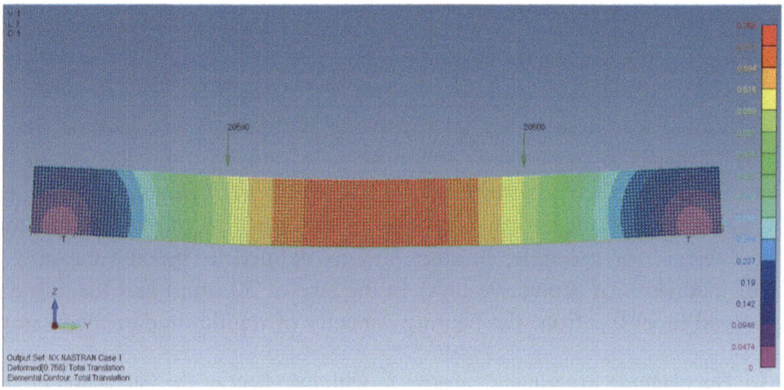

Fig. 4. Total displacement of model of regular sample.

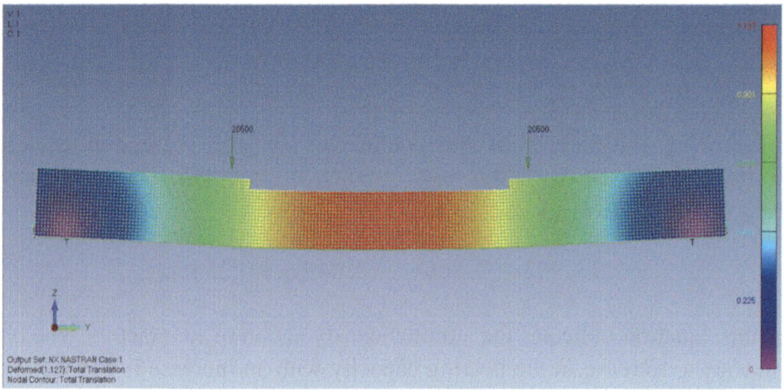

Fig. 5. Total displacement of model of damaged sample in the compressed area.

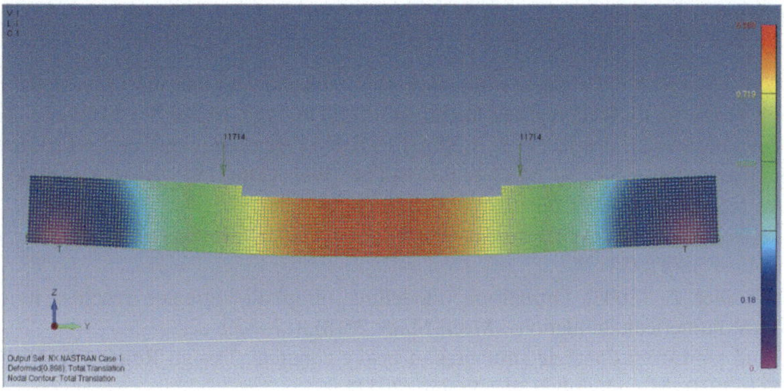

Fig. 6. Total displacement of model of damaged sample in the compressed area which are obtained at the initial level loading.

Table 2. Comparison of simulation results with different combinations of factors

№	Type of test	Maximum total displacement of the model, mm	Divergence with regular sample
1	Regular sample,	0.758	–
2	Sample with damaged compressed cover layer by 30 mm. without initial load	1.127	1.48
3	Sample with damaged compressed cover layer by 30 mm. with initial load of 30% of bearing capacity	0.898	1.18

Using data from Table 2, beam with damages had maximum displacement in 1.48 more than regular beam. But beam with same damages obtained at the initial level loading of 0.3 from bearing capacity, was had displacement only in 1.18 times more.

As the result of the analysis above, we can conclude that the initial level of loading changed stress strain state, for elements which obtained damage during exploitation.

5 Conclusions

On the basis of the above, we can conclude that the initial load level substantially affects the residual bearing capacity. Reinforced concrete beams damaged without an initial load level showed a deformability of 1.48 times greater than the regular. Samples, what received a similar damage at a load level of 0.3 from the bearing capacity of the regular, showed a smaller increase in deformability (1.18 times) than regular beam. This means that for researching the influence of defects and damage on bearing capacity of the bending reinforced concrete elements, it is necessary to take into account the initial stress-strain state.

References

Blikhars'kyi ZY, Obukh YV (2018) Influence of the mechanical and corrosion defects on the strength of thermally hardened reinforcement of 35GS steel. Mater Sci 54(2):273–278

Blikharskyy Z, Vegera P, Vashkevych R, Shnal T (2018) Fracture toughness of RC beams on the shear, strengthening by FRCM system. MATEC Web Conf 183:02009. https://doi.org/10.1051/matecconf/201818302009

Brara A, Klepaczko JR (2006) Experimental characterization of concrete in dynamic tension. Mech Mater 38(3):253–267

Capra B, Sellier A (2003) Orthotropic modelling of alkali-aggregate reaction in concrete structures: numerical simulations. Mech Mater 35(8):817–830

Concrete and reinforced concrete structures of heavy concrete. Design Rules: DSTU: B V.2.6-156 ISO: 2010. Minrehionbud of Ukraine (2011)

Concrete and reinforced concrete structures. Key provisions: DBN V.2.6-98: 2009. Minregionstroy of Ukraine (2011)

Klymenko EV, Cherneva OS, Dovgan OD, Aries Mohammed Ismael (2013) Vplyv faktoriv poshkodzhenykh tavrovykh balok na velychynu yikh ruinivnoho navantazhennia [Influence of the factors of damaged taurian beams on the magnitude of their destructive load]. Intercollegiate collection "SCIENTIFIC NOTES". No 43, pp 94–97

Krainskyi P, Blikharskyy Y, Khmil R, Vegera P (2018) Influence of loading level on the bearing capacity of RC columns strengthened by jacketing. MATEC Web Conf 230:02013. https://doi.org/10.1051/matecconf/201823002013

Pavlikov AM, Harkava OV, Horbun AV, Baliaba AV, Detiuchenko PA (2012) Rozrakhunok mitsnosti zalizobetonnykh elementiv pry kosomu zghyni za DBN V. 2.6-98: 2009. Zbirnyk naukovykh prats [Poltavskoho natsionalnoho tekhnichnoho universytetu im. Yu. Kondratiuka]. Ser.: Haluzeve mashynobuduvannia, budivnytstvo, (5), 111–115

Pavlinkov AM, Boiko OV, Kharchenko MO (2012) Rozmezhuvannia form stysnutoi zony betonu v pererizi koso zihnutykh balok tavrovoho profiliu. Visnyk Odeskoi derzhavnoi akademii budivnytstva ta arkhitektury 47:255–260

SNiP 2.03.01.84. Concrete and Reinforced Concrete Structures [in Russian], TsITP Gosstroi USSR, Moscow (1989)

Voskobiinyk OP, Kitaiev OO, Makarenko YV, Buhaienko YS (2011) Eksperymentalni doslidzhennia zalizobetonnykh balok z defektamy ta poshkodzhenniamy, yaki vyklykaiut kosyi zghyn. Birnyk Nauk. Prats (haluzeve Mashynobud., Bud-vo), 1(29), 87–92

Vozniak OT, Zhelykh VM (2003) Osnovy naukovykh doslidzhen u budivnytstvi. Lviv: Vyd-vo NU" Lvivska politekhnika

Zhang Q, Mol'kov YV, Sobko YM, Blikhars'kyi YZ (2015) Determination of the mechanical characteristics and specific fracture energy of thermally hardened reinforcement. Mater Sci 50 (6):824–829

To the Calculation of the Optimal Level of Reliability by Using Economic Indicators

A. Makhinko[1] and N. Makhinko[2(✉)]

[1] ETUAL LLC, Bortnytska Street 1, Petropavlivske, Boryspil District, Kiev 08341, Ukraine

[2] National Aviation University, Kosmonavta Komarova 1, Kiev 03058, Ukraine

Pasargada1985@gmail.com

Abstract. This paper deals with the study of the minimum required level of reliability of the steel capacities for grain storage and products of its processing, which should be ensured on the terms of economic expediency. This indicator is determined from the stochastic calculation of the construction, considering the stochastic nature of external influences and strength indexes of steel. The storage capacities are considered as constructions with exclusively economic responsibility. The calculation algorithm is based on the definition of the construction risk indicator. Graphically, it is expressed on the basis of the probability of failure-free operation or geometric characteristics of the cross-section. Considering the direct dependence of the element cost and its cross-sectional area, on the basis of a minimum of the risk function, it was obtained the formula of economic loss. It depends on the optimal probability of failure and the coefficient of the load variation. For steel storage capacities the generalized effort is schematized by two distribution laws: the normal one, which is used to describe the pressure of the bulk material on the body walls of the storage capacity, and the double exponential distribution of Humbel, which is used to describe the maximums of the snow and wind loads.

Keywords: Storage capacities · Level of reliability · Stochastic calculation · Construction risk · Economic loss · Cross-section

1 Introduction

Ukraine is part of the global grain market with a positive trend of annual growth. Seasonal character of grain production and constant high consumer demand, create the need for its storage for an extended time. The most of popularity and widespread use obtained the galvanized steel silos which meet the requirements (Fig. 1a) Kachurenko and Bannikov (2016). Steel containers for storage of grain and products of its processing occupy leading positions both in the construction of elevator enterprises and with individual farmers. The demand for such structures not only subsists, but also constantly increases. At the same time, in the development trend, there is an increase in the array of existing capacities and simultaneous increase of overall dimensions of new ones. This of course, represents the need for calculations of these designs. There are several ways to do this - automated calculation by Gallego et al. (2015), Mehretehran

Z. Blikharskyy et al. (Eds.): CEE 2019, LNCE 47, pp. 251–259, 2020.
https://doi.org/10.1007/978-3-030-27011-7_32

and Maleki (2018), Goodey et al. (2019); analytical solution of the task by Gallego et al. (2011), Makhinko (2018) and experimental research (Dooms et al. 2006, Chen et al. 2018).

The use of calculation software systems based on the MKE (Fig. 1b), certainly increases the speed and accuracy of calculations, allows expending the designer's capabilities and achieve the optimal economic solution, ensuring the conditions of strength and stability (Iwicki et al. 2016).

a) b)

Fig. 1. Steel containers for storage of grain: a general view of the vertical silos Lubnymash (a) https://lubnymash.com; calculation scheme of capacitance; calculations of MSC (b)

Nonetheless, the analytical solution is little used at present, due to the complexity of the analysis of the performance of the multi-element design, the large number of computational procedures and obsolete normative calculation methods.

An important contingency in the design of this type of construction is the existing practice of manufacturers who, in search of economic benefits, seek to reduce the metal content of the existing construction. In some cases, it turns out to be justifiable, while in others it leads to catastrophic consequences. Such a situation is provoked by the traditions of designing storage tanks, which involve only the calculation in the context of the method of boundary states using the calculated values of external loads and the characteristics of the strength of the material. A successful solution always shows only the implementation of the marginal inequality with some standard and cannot serve as a characteristic of reliability of the design as a whole. In the process of analyzing the level of reliability, the issue of costs to ensure its minimum value and losses from a possible refusal is important. This direction is the subject of the optimization theory of initial risks. Within the framework of this study, the main focus will be on effective risk, as used most often in technical disciplines by Kumamoto and Henley (2000), Singpurwalla (2006).

In the context of the deterministic calculation of silos, one should note the well-known scientific works (Rotter 2001, Brown and Nielsen 2005). Also, an analytical evaluation of the most significant internal factors of storage capacity was investigated by Lapenko et al. (2018), Pichugin and Makhinko (2018). The issue of the reliability of building structures is fundamental to the construction industry and is represented by a large number of scientific works (Bolotin 1982, Venttsel 2000, Pichugin 2010).

2 The Purposes Formulation of Article

Probabilistic calculation of silos is not limited to determining the actual level of reliability or the search for geometric design parameters for a given reliability index. Before the engineer raises the question of determining minimum required level of reliability, which must be achieved by construction design, in fact, this indicator is a necessary requirement that must be fulfilled in order to achieve safe operation of the facility.

3 General Problem Solving Method

In solving the problem, general methods of the theory of reliability of buildings and structures and the theory of risks were used. Several probabilistic load models were used. The random value of generalized strength was described by the normal distribution law, and the effect of generalized effort was considered in two versions. In the first, the normal distribution was used to describe the pressure of the loose material on the walls of the silo body. In the second, the double exponential distribution of the Humbel was used to describe the maximum snow and wind load. These characteristics were taken into account by the system of ratios that are present in the resulting formula for the function of failure-free operation.

4 Research Results

Probabilistic calculation of storage capacities may be branched out into three variants - determination of the actual level of reliability; definition of geometrical parameters of design with a predetermined index of reliability and determination of the minimum required level of reliability. The latter version is the subject of research of this scientific article.

Steel containers for storage are buildings where people are practically absent, that is, they can be considered as buildings with exclusively economic purpose. This allows us to solve the problem of determining the optimal level of reliability in a simple form.

We introduce to the concept of construction risk R_Q as a value equal to the outcome of the probability of failure Q volume per number of monetary units C, which will be required for the elimination of all the consequences of an accident. Typically, it may include the cost of capacity C_C and the value of the product stored in it C_G. The cost of capacity is estimated by its metal content, which depends on the dimensions of the cross-section elements. The larger the cross-sectional dimensions, the higher is the reliability of the capacity and its cost. From the mathematical point of view, this can be expressed as follows:

$$R_{\Sigma}(Q) = R_C(Q) + R_Q(Q) = \alpha_C \cdot C_C \cdot [1 + Q \cdot (1 + \alpha_Q)], \tag{1}$$

where α_C is the ratio associated with the losses that do not depend on the size of the sections of the container (for example, cleaning space, a line for transporting grains in galleries, paths, automation means that provide control overloading, unloading and storage of bulk material) and α_Q is the dimensionless ratio, which shows how many times the cost of a storage product is greater or less than the value of the container itself:

$$\alpha_Q = C_G / C_C. \tag{2}$$

The geometric interpretation of the described approach is illustrated in Fig. 2 in two variants. In the first case, the argument of total risk is the probability of failure-free operation, in the second case, a certain geometric characteristic of the cross-section.

Constructed dependencies have two important features. Firstly, the function of total risk has a clearly defined minimum, which corresponds to the optimal value - the optimal failure probability or the ideal characteristic of the cross-section of the element. Secondly, with an increase in the likelihood of failure, compared with optimal values, the full expected costs increase much faster than with the reduction of this probability. Therefore, the excess probability costs less than its lack. This dependence is more pronounced, with a greater parameter of economic damage or the greater the responsibility for the construction. This is due to the fact that the cost of increasing the reliability, which is determined by the cost of the construction itself, becomes much less than the cost of the accident, which is determined by the importance of the object.

Let's examine Eq. (1) on the example of the optimal level of reliability of a single element under conditions of the uniaxial stress-strain state.

The actual probability of failure-free operation of the construction y_F, obtained by direct modeling, can be estimated by the formula:

$$y_F = \left[\sqrt{B_K^2 - 4A_K(C_K - 1/m_K)} - B_K\right] \Big/ (2A_K), \tag{3}$$

where A_K, B_K and C_K are the dimensionless ratios depending on the laws of the distribution of random variables of generalized strength and effort.

When expressing the area, we will receive the following as a function of the probability of failure:

$$A_p(Q) = \frac{m_N}{m_R}\left(A_K[\ln(-\ln(1-Q))]^2 - B_K \ln(-\ln(1-Q))\right), \tag{4}$$

where m_N and m_R are the mean values of the longitudinal force in the element and the boundary of the steel flux.

a)

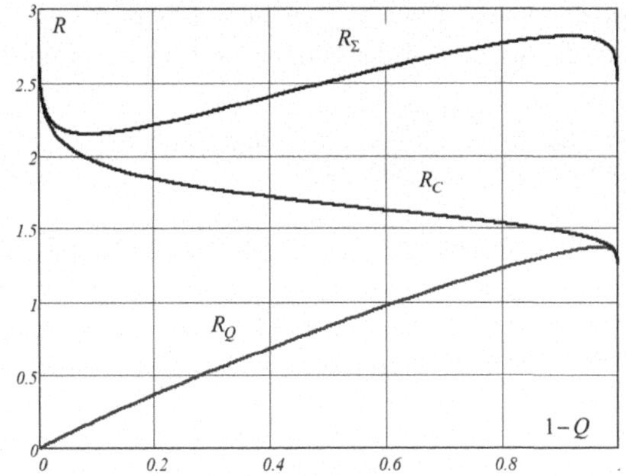

b)

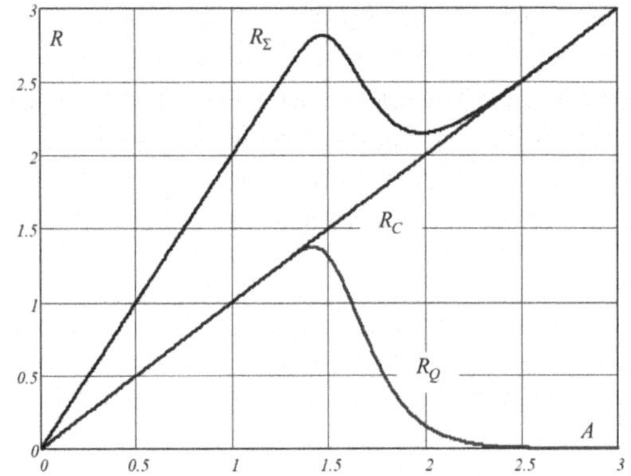

Fig. 2. Representation of the total risk as a function of the probability of failure-free operation (a) and depending on the cross-sectional parameter of the element (b)

Let's suppose that the cost of an element is proportional to the square of its cross-sectional aspect ratio k_C. We will insert Eq. (4) to Eq. (1) and simplify:

$$R_\Sigma(Q) = k_C \frac{m_N}{m_R}(1 + Q\Omega_Q) \cdot \left\{ A_K y_Q^2 - B_K y_Q + C_K \right\}. \quad (5)$$

where $y = \ln[-\ln(1 - Q)]$ is the value entered to shorten the description.

Condition of the minimum risk function $R_\Sigma(Q)$ for differentiation according to the probability of failure Q:

$$\frac{1 + Q \cdot \Omega_Q}{\ln(1 - Q) \cdot (Q - 1) \cdot \Omega_Q} = \frac{A_K y_Q^2 - B_K y_Q + C_K}{B_K - 2A_K y_Q}. \quad (6)$$

For the case of zero value of ratio A_K, characteristic of the load is represented by the double exponential distribution of the Humbel, we have a simpler form of record:

$$\frac{1+Q\cdot\Omega_Q}{\ln(1-Q)\cdot(1-Q)\cdot\Omega_Q}+\ln(-\ln(1-Q))=\frac{C_K}{B_K}. \quad (7)$$

The parameter of economic loss will be expressed explicitly as:

$$\Omega_Q=\frac{1}{\ln(1-Q)(1-Q)\frac{\left[-A_K y_Q^2+B_K y_Q-C_K\right]}{B_K-2A_K y_Q}-Q}, \quad (8)$$

$$\Omega_Q=\frac{1}{\ln(1-Q)(1-Q)\left[\ln(-\ln(1-Q))-\frac{C_K}{B_K}\right]-Q}. \quad (9)$$

Taking into account the dependence of ratios A_K, B_K and C_K on the ratio of variation of loading V_S (internal force or stress) expressed by a number of dimensionless ratios α_A, α_B and α_C finally, we obtain formulas for the parameter of economic loss when representing the load by a normal law and the double exponential law of Humbel:

$$\Omega_Q=\frac{1}{\ln(1-Q)(1-Q)\left[\frac{-\alpha_A y_Q^2+\alpha_B y_Q+\alpha_C-V_S^{-1}}{\alpha_B-2\alpha_A y_Q}\right]-Q}, \quad (10)$$

$$\Omega_Q=\frac{1}{\ln(1-Q)\cdot(1-Q)\cdot\left[\frac{\alpha_C-V_S^{-1}}{\alpha_B}+\ln(-\ln(1-Q))\right]-Q}. \quad (11)$$

Equations (10) and (11) connect three dimensionless values: Ω_Q parameter of economic damage, the optimal probability of failure Q load and ratio of variation V_S. Figure 3 shows a graphical illustration of the comparison of the dependence $\Omega_Q(Q)$ for two load distribution laws: the normal and dual exponential distribution of the Humbel.

From the analysis of the graphs it is possible to draw the following conclusions:

(1) for the load described by the normal distribution law, the optimal level is always higher than for the load represented by the double exponential law;
(2) at optimal probability (approximately less than 0.01), the dependence of the parameter of economic loss on the probability of failure is not very different from direct proportionality $\lg(\Omega_Q)-\lg(Q)$;
(3) regardless of the load distribution law, with a given ratio of variation there is a limiting value of the economic loss parameter below which the optimal level of reliability does not exist. For example, for the load described by the normal distribution law with the ratio of variation $V_S = 0.9$, the question of the optimal level of reliability is only feasible when the economic loss parameter is more than 3.0;

(4) with an increase in the ratio of variation of external influence, the difference between the two adjacent optimal levels of reliability, which correspond to the accepted parameter of economic damage, decreases;

(5) regardless of the law of the distribution of the load, there always exists a function $\Omega_Q(Q)$, no matter how big the ratio of variation is.

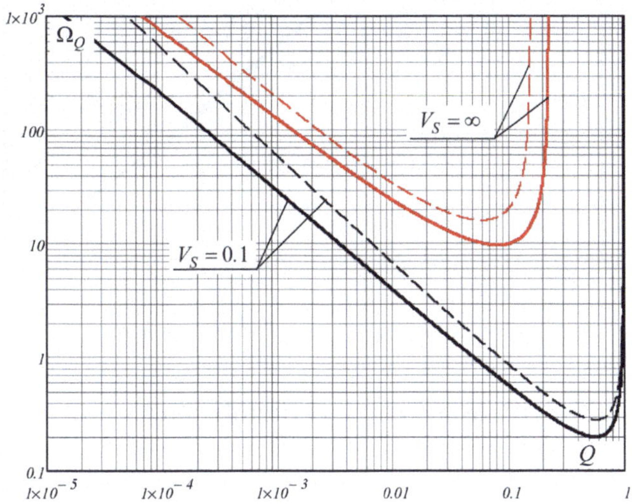

Fig. 3. Comparison of the economic loss parameter as a function of the optimal failure probability for representing a load by a random variable with a normal (solid lines) or double-exponential (dashed lines) distribution

For practical use Eqs. (10) i (11) can be simplified without significant loss of accuracy of the calculations. To do this, we apply the approximate ratio $\ln(1-Q) \approx -Q$ for $Q \to 0$. As a result, we obtain the following formulas:

$$\Omega_Q = \frac{1}{Q} \cdot \frac{\alpha_B - 2\alpha_A \ln[-\ln(1-Q)]}{V_S^{-1} - \alpha_B \ln[-\ln(1-Q)]}, \quad (12)$$

$$\Omega_Q = \frac{1}{Q} \cdot \frac{1}{V_S^{-1} - \ln[\ln(1-Q)]} \quad (13)$$

Comparison of these expressions with reference Eqs. (10) and (11) in double logarithmic coordinates performed on Fig. 4 and testifies to their good consistency.

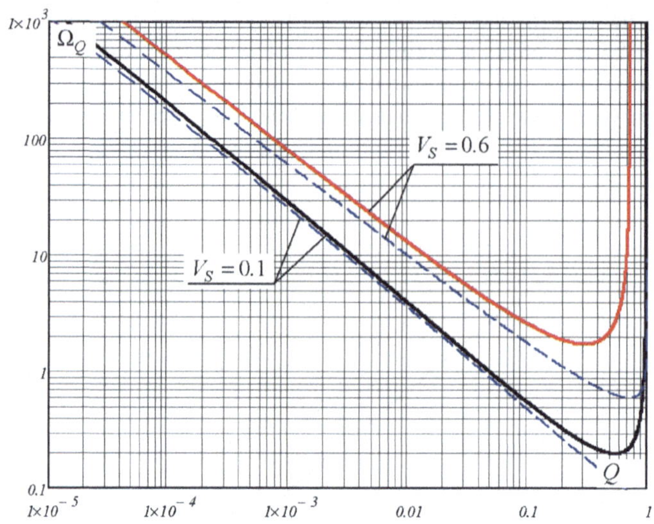

Fig. 4. To support Eq. (11) when applying a normal law

5 Scientific Novelty and Practical Significance

The results of this study provide an engineer with a convenient and a simple algorithm for calculating the minimum level of reliability of silo elements under an uniaxial stress-strain state with regard for economic reasons. At the same time, for the first time, in order to express the reliability function, a formula was used in which there is a series of dimensionless ratios depending on the laws of the distribution of random variables generalizing strength and effort. It can be applied to loads represented by the normal law or the double exponential Humbel's law. Also, the expressions obtained can be used to solve engineering problems in determining the parameters of economic loss.

6 Conclusions

1. The problem of establishing the optimal level of reliability, depending on the construction risk, is formulated. The geometric interpretation of the described approach is provided.
2. In the course of the study, a relationship was found that links the parameter of economic loss, the optimal failure rate, and the ratio of variation of the load. A graphical representation of this dependence for different load distribution laws is given.
3. It was established that the optimum level of reliability depends on the probabilistic reasoning regarding the load.
4. Graphically, it was proved that for any casual load there is a marginal parameter of economic damage that meets the minimum possible level of reliability.

5. For practical use, we have obtained simplified formulas for determining the economic loss parameters for loads represented by the normal law or the double exponential law of Humbel.

References

Bolotin VV (1982) Metody teorii veroiatnostei i teorii nadezhnosti v raschetakh sooruzhenii. Stroiizdat

Brown CJ, Nielsen J (2005) Silos: Fundamentals of Theory, Behaviour, and Design. E - FN Spon

Chen Z, Liab X, Yang Y, Zhao S, Fu Z (2018) Experimental and numerical investigation of the effect of temperature patterns on behavior of large scale silo. Eng Fail Anal 91:543–553. https://doi.org/10.1016/j.engfailanal.2018.04.043

Dooms D, Degrande G, De Roeck G, Reynders E (2006) Finite element modelling of a silo based on experimental modal analysis. Eng Struct 28(4):532–542. https://doi.org/10.1016/j.engstruct.2005.09.008

Gallego E, González-Montellano C, Ramírez A, Ayuga F (2011) A simplified analytical procedure for assessing the worst patch load location on circular steel silos with corrugated walls. Eng Struct 33(6):1940–1954. https://doi.org/10.1016/j.engstruct.2011.02.032

Gallego E, Ruizb A, Aguadob PJ (2015) Simulation of silo filling and discharge using ANSYS and comparison with experimental data. Comput Electron Agric 118:281–289. https://doi.org/10.1016/j.compag.2015.09.014

Goodey RJ, Brown CJ, Rotter JM (2019) Rectangular steel silos: finite element predictions of filling wall pressures. Eng Struct 132:61–69. https://doi.org/10.1016/j.engstruct.2016.11.023

Iwicki P, Sondej M, Tejchman J (2016) Application of linear buckling sensitivity analysis to economic design of cylindrical steel silos composed of corrugated sheets and columns. Eng Fail Anal 70:105–121. https://doi.org/10.1016/j.engfailanal.2016.07.013

Kumamoto H, Henley EJ (2000) Probabilistic risk assessment and management for engineers and scientists. Wiley-IEEE Press, New York

Lapenko O, Makhinko A, Makhinko N (2018) Features of silos calculations at asymetric wind load by using momentless theory. Tehnički glasnik 13(1):12–15. https://doi.org/10.31803/tg-20180522160526

Lemaire M (2009) Structural reliability. ISTE Ltd., London

Makhinko N (2018) Stress-strain state of the storage silos under the action of the asymmetric load. Matec Web Conf Struct Build. Facil 230. https://doi.org/10.1051/matecconf/201823002018

Mehretehran AM, Maleki S (2018) 3D buckling assessment of cylindrical steel silos of uniform thickness under seismic action. Thin-Walled Struct 131:654–667. https://doi.org/10.1016/j.tws.2018.07.040

Kachurenko VV, Bannikov DO (2016) Konstruktivnye resheniia stalnykh emkostei dlia sypuchikh materialov. Novaia ideologiia

Pichugin SF (2010) Nadezhnost stroitelnykh konstruktsii. rabota nauchnoi shkoly prof. Pichugina S.F. ASMI

Pichugin SF, Makhinko N (2018) Features of the silo capacities' calculation at the unsymmetrical wind load. Int J Eng Technol 7(4.8):22–26. https://doi.org/10.14419/ijet.v7i4.8.27208

Rotter MJ (2001) Guide for the economic design of circular metal silos. CRC Press, Boca Raton

Singpurwalla ND (2006) Reliability and risk: a Bayesian perspective. Wiley, Chichester

Venttsel ES (2000) Teoriia sluchainykh protsessov i ee inzhenernye prilozheniia. Vysshaia shkola

Comparative Analysis of Pollution of Atmospheric Air in Cities (an Example of Leipzig and Poltava)

N. Maksiuta(✉) and Yu. Golik

Poltava National Technical Yuri Kondratyuk University,
Persotravnevyi avenue 24, Poltava 36000, Ukraine
mns7000@yahoo.com

Abstract. Clean atmospheric air is one of the most necessary conditions of qualify life of the population and for saving the healthy environment for future generations. In my opinion, the best way to provide a high level of life in every area and field of study is an exchange of experience. So the comparison of atmospheric air quality in Poltava and Leipzig is a great opportunity to find out what is done and what can be improved. With this goal, it is necessary not only compare the state of the air but also discuss which methods are used for monitoring its quality and what can be changed to implement maximum effectiveness in using time, sources and money. Every step should also provide an environment saving tendency. The comparing of a meteorological situation in Poltava and Leipzig is presented. Also, the results of the calculation of atmospheric pollution indexes are provided. There is some difference in limit values of pollutants in atmospheric air in Ukraine and the European Union. Data about stationary observation posts of ambient air is also included. There is a discussion of air pollution level in Leipzig and its connection to the meteorological conditions.

Keywords: Atmospheric air · Indexes · Pollution · Meteorology · Comparative analyze

1 Introduction

Clean air is necessary for the quality of human life. The exchange of experiences is very useful to increase the quality. For this purpose, we need similar data from different cities or parts of the world. For example, to obtain information about the state of the air in Poltava (Ukraine), we can use sources such as "Ecological passport of the Poltava region" or "Regional environmental status report in the Poltava region". Nevertheless, these sources provide only general data on atmospheric air in the region. If you receive such information about the air quality in Leipzig (Germany), we may use databases and reports provided by the source www.umwelt.sachsen.de or by "European Air Quality Report" (general information). For comparative analysis, the use of such different databases is quite difficult. Therefore, in this project the method of calculation of the air pollution index agreed by the Ministry of Ecology and Natural Resources of Ukraine is used. The purpose of this paper carry huge scientific and practical value.

Z. Blikharskyy et al. (Eds.): CEE 2019, LNCE 47, pp. 260–267, 2020.
https://doi.org/10.1007/978-3-030-27011-7_33

This comparative analyze will allow to talk about some similarities or differences of air pollutions in different cities and its connection to some meteorological factor (wind speed, temperature fluctuations, solar radiation).

2 Aim

The primary goal of the project for this moment is to compare the actual state of the atmospheric air pollution level in both Poltava and Leipzig city. Also, it is planned to figure out which methods of assessment of atmospheric air quality are using nowadays in different countries with a purpose for future to find out what the most appropriate methods to providing in cities with different characteristics (land area and population). It is also important to take into account such information as how much time each method takes for assessment and its economic proof.

For this purpose, it is necessary to provide the next steps.

1. Providing information about geographic places of both Poltava and Leipzig cities. Including such data as population and area of the city, basic industrial production e.t.
2. Determination of meteorology state at least for 3–5 years and its comparative analysis.
3. Obtaining information about stationary observation posts by providing important data about it. Namely: geographic characteristics of each post, which contaminants are measured in each post.
4. Providing information about standards in atmospheric quality for both countries.
5. Calculation Indexes of atmospheric air pollution by obtained data from the stationary posts. Comparative analysis of the results of the calculation.

3 Method

The atmospheric pollution index (API) was calculated to define the state of air pollution in Leipzig and Poltava (Ukrainian law 1997). API is calculated as the sum of substances that are normalized for daily average maximum allowable concentration (MAC d/a) and correlated by hazard class of pollutants.

Thus, for several pollutants, this index is calculated by a formula

$$X = \left(\sum_{i=1}^{n} I_i\right) = \sum_{i=1}^{n} \left(\frac{C_i}{MAC_{d/a}}\right)^{K_i} \leq 1$$

At the same time, the MAC d/a is the average daily MAC of the pollutants, mg/m^3. In the case of absence of this indicator, instead of it, take MAC o/t (one-time maximum allowable concentration that means a maximum concentration of the pollutant in the air, which does not cause reflex reactions in the human body and it is determined from samples that are selected for a 20–30 min) or ASLI (approximately safe level of impact).

The coefficient K for the calculation of pollutants, depending on their hazard class, is determined as follows:

1. 1 hazard class: K = 1,7;
2. 2 hazard class: K = 1,3;
3. 3 hazard class: K = 1,0;
4. 4 hazard class: K = 0,9.

The results of this air pollution index's calculation can give us some information about atmospheric air in the city (Table 1).

Table 1. Definition of air pollution index values (Ukrainian law 1997)

Air pollution index	Level of pollution	Effect on inhabitants' health	Ecological state of atmospheric air
<5	Low	Little	Normal
5–8	Elevated	Significant	Risk
8–15	High	Unfavorable	Crisis
>15	Very high	Dangerous	Disaster

4 Results of Comparison

4.1 Geography, Industry, Standards

Leipzig is located in the west of the federal state of Saxony. With a population (about 570,000 people), Leipzig is the largest city in Saxony, and the tenth in Germany, and also the fastest growing large German city.

The main areas of development strategy are the economics of health and biotechnology, the automotive industry and the supply industry, energy, and environmental technology, logistics, as well as the media and creative industries. These economic and scientific areas occupy a central role during the majority of the conference. So Leipzig has established itself as a place suitable for international congresses with clearly defined clusters. The profiles of an economic, scientific and research nature are successful, which makes the city known both at the national and international levels (Infoservice 2018).

Poltava is located in the eastern part of Europe, 301 km east of Kiev. The geographical location is quite advantageous and, from a historical point of view, influenced the development of the city. The city is located on important transport routes and provides links between the major cities of Ukraine - Kiev, Kharkiv and the Dnipro.

According to the State Statistics Committee of January 1, 2010, the population of the city was 300.5 thousand people, that is, it decreased by 14 thousand since 1989. As of August 1, 2018, the people of the city was 280,992 permanent residents and 288,013 of the present population. Poltava is one of the cities in Ukraine with a population density of less than 3,000 people per square kilometer (Statistics 2018).

The industrial complex of the city is formed by 11 sectors of economic activity, among the main ones: mining; production of food, beverages, tobacco products; light industry; pulp and paper production; publishing activities; chemical and petrochemical industry; manufacture of other non-metallic mineral products; engineering; repair and installation of machines and their equipment; metallurgical production and production of finished metal products; production and distribution of electricity, gas and water, and much more (Ecological Passport 2015).

Therefore, we can say that Poltava is a smaller city than Leipzig by area and total population, but it has a more significant density of population. Brief compare is in Table 2.

Table 2. Brief compare of Poltava and Leipzig

	Poltava	Leipzig
Area	103 km^2	297 km^2
Population: total	288013	571088
Population: density	2796/km^2	1922/km^2

For understanding a difference in normative characteristics, there are some limit values of pollutants concentration for European Union (Directive EC 2008) and Ukraine (Ukrainian law 1997) in Table 3.

Table 3. Limit values of pollutants concentration for European Union and Ukraine

	Europe	Ukraine
Sulfur dioxide (SO_2)	350 μg/m^3 (per 1 h)	500 μg/m^3 (One-time)
	125 μg/m^3 (per day)	50 μg/m^3 (Average daily)
Nitrogen dioxide (NO_2)	200 μg/m^3 (per 1 h)	200 μg/m^3 (One-time)
	40 μg/m^3 (per year)	40 μg/m^3 (Average daily)
PM_{10}	50 μg/m^3 (per day)	150 μg/m^3 (One-time)
	40 μg/m^3 (per year)	50 μg/m^3 (Average daily)
Lead (Pb)	0.5 μg/m^3 (per year)	1 μg/m^3 (One-time)
Carbon monoxide (CO)	10 mg/m^3 (Maximum daily 8-hour mean)	5 mg/m^3 (One-time)
Ozone (O_3)	120 μg/m^3 (Target value – Maximum daily 8-hour mean)	160 μg/m^3 (One-time)
Arsenic (As)	6 ng/m^3 (Target value – per year)	3 μg/m^3 (Average daily)
Cadmium (Cd)	5 ng/m^3 (Target value – per year)	0.3 μg/m^3 (Average daily)
Nickel (Ni)	20 ng/m^3 (Target value – per year)	1 μg/m^3 (Average daily)
Polycyclic Aromatic Hydrocarbons	1 ng/m^3 (Target value – per year) (expressed as concentration of Benzo(a) pyrene)	1 ng/m^3 (One-time) (expressed as concentration of Benzo(a)pyrene)

Limit values for pollutants are higher in Ukraine. But, it should be noticed, the European Union limit values was defined in Directive 2008/50/EC on ambient air quality and cleaner air for Europe in 2008, and Ukrainian requirements were an approval in 1997 with some little changing in 2000. Considering this information, we can assume that, perhaps, Ukrainian limit values should be revised based on the latest data about environment state.

4.2 Meteorology

Meteorology data is also very important as well as data of pollutions for comparative analyses of atmospheric air pollution. Because in most cases the level of air pollution depends on meteorological conditions. According to Meteorology Database (2015–2018), it seems that the meteorology situation in both cities is similar. It was noticed that distribution of temperatures by months is similar, but still Poltava has a higher temperature in summer time (approximately on 0,6 °C) and lower one in winter (around on 10 °C) as absolute maximum and minimum temperatures. The precipitation in both situations is similar (average 500–700 mm per year).

5 Main Results

Leipzig. As monitoring of air pollution in Saxony, emission measurement stations are operated. The stations are located in such a way as to ensure comprehensive control of emissions. Currently, 29 automatic stations are running at which the concentrations of gases and solid particles are determined. There are 4 emission measurement stations in Leipzig. The following air pollutants are measured: sulfur dioxide, fine dust <10 or 2.5 μm, ozone, nitrogen oxides, and some volatile organic compounds. Under laboratory conditions, the constituents of the dust (heavy metals, soot, polycyclic aromatic hydrocarbons) are also determined. Meteorological parameters are also measured. Carbon monoxide measurement was discontinued in 2008 due to very low concentrations and the absence of the need for these measurements (Umwelt in Sachsen 2018).

Poltava. Systematic observation of the air pollution level is carried out at four stationary posts "POST-2A" (Poltava Hydrometeorology Centre, 2015–2017). SOP No. 1 is located almost in the center of the city. It‘s surrounded by city relax-park "Peremoga". So the traffic near is poor. SOP No. 3 is located in well-ventilated area on the north part of the city with not very high traffic. SOP No. 6 is located on the south part and it is can be chartered like residential area with private sectors. SOP No. 7 is located in the industrial zone on the west part of the city. Sampling for pollutants contamination was measured according to the "complete program" (the name of the program which means that samples was collected four times a day) almost for every 10 main pollutants: PM_{10}, SO_2, H_2SO_4, CO, NO_2, NO, FH, HCl, NH_3, CH_2O. Determination of heavy metals (SOP No. 6) and benzo(a)pyrene at posts No. 3 and No. 7 was made from a total sample per month.

Pollutants that measured in both cities are PM_{10}, SO_2, NO_2, NO, BaP, Cd, Ni, Pb, Cr. Only in Poltava such pollutants as H_2SO_4, CO, FH, HCl, NH_3, CH_2O are measured. Some more metals are also measured here, namely: Fe, Mn, Cu, Zn. Only in Leipzig such pollutants as $PM_{2,5}$, O_3 and As are measured. Also, some volatile organic compounds are measured here, namely: Benzol, Toluol, and Xylol. Nevertheless, it is a planned in future to figure out the question about this connection and which methods or/and measuring devices are used for air quality assessment.

Based on the obtained data from stationary posts, the database for concentration of pollutants in the atmospheric air was created, and the atmospheric pollution index were calculated for each year (2015–2017) for Leipzig (SO_2 + O_3 + NO_2 + NO + PM_{10}) and for Poltava (CH_2O + CO + NO_2 + NO + PM_{10}).

For Leipzig the concentration of SO_2, NO_2, NO, and PM_{10} is taken from "Mitte" stationary post. The concentration of ozone is taken from "Thekla" stationary post because the ozone is not measured on "Mitte".

Concentrations of pollutants in the atmospheric air of Leipzig the levels of SO_2 and PM_{10} never reached the limited values during the observation period (2015–2017). Other pollutants (O_3, NO, and NO_2) always exceeded limited values. Let's have a look for this three pollutants separated. Ozone exceeded the limited values all years mostly every month. But in cold periods the concentration of it was declined (lower than the limit) and on the contrary, it was the highest in the warm months from May to August. The concentration of NO in 2015 was exceeded the limit values in February, October, November, and December, in 2016 – November and December, in 2017 – only in January. So we can talk about the improvement of the situation of NO concentration in ambient air. The same case of reducing the concentration through the years we have for NO_2. But, unfortunately, the concentration is still high, and in 2017 it exceeded the limit almost half of year. Also, it is necessary to notice that the highest level of NO and NO_2 was in cold periods.

During the collecting necessary data, we faced the problem that in Ukraine there is no good database in open access about daily pollutants concentration in the atmospheric air. Therefore, obtaining the required data for this comparative analyze took a little longer time then it was planned. For Poltava the concentration of measured pollutants is given in multiplicity to MAC, and its already averaged for the city (by using data from all stationary posts) by the Poltava Hydrometeorological Center.

Concentrations of pollutants in the atmospheric air of Poltava the levels of CO and NO never reached the limited values during the observation period (2015–2017). Other pollutants (CH_2O, NO_2, and PM_{10}) exceeded limited values each year. In 2015 and 2016 the concentrations of CH_2O exceeded limit values only in hot months (sometimes in 2 times), but in 2017 it‘s concentration exceeded limit values in winter too and in hot months it exceeded limit in almost in 3 times. Concentration of NO_2 exceeded limit values in 2015 only in July and August and in 2016–2017 – only in August (average in 1.15 times). Concentration of PM_{10} exceeded limit values each year average in 1.5 times almost every month except winter ones. Also in 2016 there is no exceeding by PM_{10} in spring.

Daily and monthly Air Pollution Indexes by formula 1 were calculated for five pollutants (SO_2, O_3, NO, NO_2, PM_{10}) for Leipzig.

The results show that the risk situation (API 5–8) is observed from April to July, mostly in May and April. The ozone and then NO_2 take the biggest part in this index. In general, if we will take into account only year values, we can say that the level of air pollution in the city is normal and it is not dangerous for inhabitants' health. But it still has a high value, and it means that in case of inappropriate meteorological conditions pollution level can increase and become a risk state. We see that through the years the average API was not less than 4,14.

There is also daily API was calculated for Leipzig. It has to be noted that in some days this index acquires to high values. For example:

- in 2015 from 05 June to 06 June the API varied from 8,44 to 8,45, from 02 July to 07 July – 6,1–12,09, from 04 August to 14 August – 5,52–11,62.
- in 2016 on 12 May – 8,4;
- in 2017 on 15 February – 8,49.

As it was mentioned earlier, the API in values 5–8 is for risk pollution level of atmospheric air. And API from 8 to 15 is a crisis level, and it is a dangerous situation for people health.

These results were analyzed according to the meteorological conditions in these days that was obtained from website www.umwelt.sachsen.de. So in 2015 when the maximum values of API were calculated, the high level of concentrations of O_3 and NO_2 was observed. Depending on the meteorological conditions it could be caused by high temperatures (22,0–32,1 °C), high solar radiation (up to 310 W/m^2), and low wind speed (average 1 m/sec). On 12.05.2016 the high API was caused by high concentration of ozone in atmospheric air. The reason for it could be high solar radiation (238 W/m^2), and probably, the sharp change in wind direction (from 330° to 90°), but perhaps the wind is not the reason of high pollution concentration. On 15.02.2017 the slow wind speed (0,7 m/sec) caused an increase of NO, NO_2, and PM_{10} concentration, which in turn contributed to the high value of API in this day. But, perhaps, this is not all reasons that became as conditions for high-level pollution. This question can be more widespread opened in future by using more data.

Daily and monthly Air Pollution Indexes by formula 1 were calculated for five pollutants (CH_2O, CO, NO, NO_2, PM_{10}) for Poltava.

We can see that average API for each year for both cities is quiet similar. But if we take into account months' values, there is definitely higher exceeding in Poltava. In 2015 the API value was higher than normal (>5) from May to September, in 2016 – in summer months, in 2017 – from June to September. In 2015 and 2017 in August API value even reached more than 8, that, according to Table 1, means that the ecological state of ambient air is risk and can cause damage for human health. Unfortunately, the Poltava Hydrometeorological Center did not provide data of pollutants concentration in atmospheric air per every day. Considering this it can be possible to analyze calculation results of API for Poltava as it has done for Leipzig. These month‘s results were analyzed according to the meteorological conditions in these days that was obtained from website www.meteomanz.com. Months with high API for each year were characterized as months with high temperature (27 °C as average and 37.8 °C as absolute maximum). These high temperatures can be caused by high solar radiation in hot periods of year. According to meteorological data from the Poltava Hydrometeorological Center, during these months with high API the meteorological conditions were characterized by low wind speed (average 35% days per months with wind speed lower than 1 m/sec).

6 Conclusions

Main conclusions were presented within this article. As a generalized result of the calculations, we can conclude that the annual API is similar for both cities (4.36 on average). This value describes a normal air quality but is quite high. In adverse weather conditions, the concentration of pollutants can increase significantly, which in turn can lead to a deterioration of the condition in the air and a deterioration in the health of the population. In terms of distribution by months, Poltava tends to a sharper "jump" of the values in the summer months. This can be caused by a sharp increase in the CH_2O concentration relative to the limits in hot periods.

Acknowledgements. The activity presented in the paper is part of the research grant of Deutsche Bundesstiftung Umwelt (Germany) and was carried out in the Leibniz Institute for Tropospheric Research on the Department of Atmospheric Chemistry.

References

Ukrainian state sanitary rules of the protection of atmospheric air of populated areas (from pollution by chemical and biological substances) (1997). Ukrainian low. No. 201 from 09.07.97 (in Ukrainian)

Infoservice (2018) www.leipzig.travel, December 2018 (in German)

The Main Department of Statistics in Poltava (2018) http://pl.ukrstat.gov.ua/, December 2018 (in Ukrainian)

Ecological Passport of Poltava Region (2015) http://old.menr.gov.ua/protection/protection1/poltavska (in Ukrainian)

Directive 2008/50/EC on ambient air quality and cleaner air for Europe

Meteorological data of SYNOPS/BUFR (2018) www.meteomanz.com, December 2018

Environment in Saxony (2018) www.umwelt.sachsen.de, December 2018 (in German)

Internal documentation of Poltava Hydrometeorology Centre (2015–2017) (in Ukrainian)

Correction Factors for the Calculation of Consumable Heat Energy Taking into Consideration Climatic Pattern of the Ukrainian Regions

Viktoria Grankina, Olga Maliavina(✉), Aleksandr Romashko, and Roman Tkachenko

Faculty of Urban Environmental Engineering and Management, O. M. Beketov National University of Urban Economy in Kharkiv, Marshala Bazhanova Street, 17, Kharkiv 61002, Ukraine
olgamalyavina@ukr.net

Abstract. In this work, the correction climatic coefficient for different regions of Ukraine is investigated which is used to determine the share of heat consumed by the consumer when installing heat cost allocators. A numerical analysis of the calculated values of the heat losses of the premises for a residential building according to the regulatory climate data of Ukraine by months of the heating season was carried out. By calculation, for such cities as Kiev, Kharkov, Zaporozhye and Lviv, correction climatic coefficients. The discrepancies between the values of the correction factors and those currently in force in Ukraine are revealed, which confirms the relevance of the conducted studies. According to the data obtained and their comparison with the existing correction factors we can draw a conclusion that for fair calculation of heat cost allocation to heat energy to correct them in accordance with the building regional location taking into consideration climatic zones of Ukraine. The proposed correction factors will ensure the accuracy of accounting for individual consumption and the actual energy consumption at the consumer, which is necessary for the implementation of Article 9 of the European Parliament Directive 2012/27/ EU.

Keywords: Multi-apartment buildings · Energy efficiency · Heat loss · Heat cost allocator

1 Introduction

According to the Law of Ukraine (Law of Ukraine. On Commercial Heat and Water Accounting, 2017) between the consumers co-owners of multi-unit apartment building or another building, where are two or more consumers, the correction factors can be established for the distribution of the heat energy consumed between certain consumers (in corner apartments (premises), apartments (premises) located at the ground and last floors of buildings etc.), (p. 14). In many countries (Robinson and Vogt; 2016) there are coefficients of location of premises for calculation of the consumed heat energy using heat cost allocators. The coefficients are used to correct units of heat consumption

Z. Blikharskyy et al. (Eds.): CEE 2019, LNCE 47, pp. 268–274, 2020.
https://doi.org/10.1007/978-3-030-27011-7_34

in the premises which have worst position in the building in the context of heat loss. These are premises at the ground and last floors, corner premises etc. Reducing premises location coefficients are recommended to calculate on the basis of designed values of heat loss of premises for individual building (for standardized buildings – on the basis ofdesigned data for this definite series). Correction coefficient is used either to some portion of consumption, or to the portion independent from individual consumption. Such scientists in work (Ziemele et al.; 2015) from Riga Technical University's Institute of Energy Systems and Environment in the paper titled « Economy of heat cost allocation in apartment buildings » noted that for fair distribution of heat energy a number of criteria should be taken into consideration. For example, location of apartment in the corners of buildings, as well as at the upper and lower floors. These apartments have more heat loss and require more amount of heat in order to provide such comfortable temperature of the premises as in other apartments. Amount of heat being supplied to a certain apartment, doesn't coincide with the amount of the heat consumed therein, this is because of heat loss between neighbouring apartments and ambient environment. Because of the absence of heat isolation, there is a process of heat transfer between the premises in the neighbouring apartments and rooms with various heat flows (Hrankina and Blahodarna; 2016). This process has been analysed in numerous scientific work (Ziemele, et al.; 2015). In the paper the building (Terés-Zubiaga, et al.; 2018), its operational conditions and conditions of ambient environment were tested by means of TRNSYS building modelling modular program. The building model provides for one actual air set per a zone. Convectional heat balance consists of the components as follows: (1) venting; (2) infiltration; (3) inner heat supply (people, equipment, lightning); (4) convective heat flux from external barriers; (3) heat convective flux from other zones (for example, adjoining apartment; (6) convective flux from heating devices.

Research has shown that (Terés-Zubiaga et al.; 2018) comfortable temperature in apartment with fewer external walls can be almost completely provided with heat of neighbouring apartments. Installation of heat energy allocators is not classified as a method of heat energy reduction, but it allows a consumer to choose optionally and support inner temperature which a consumer is ready to pay for. Researches of the Dutch scientists have shown that (Siggelsten; 2014) the behaviour of consumers influenced greatly the energy consumption (4,2% of energy used for heating), but designed peculiarities of building made the most part of heat consumption. However, this ratio can be different in various types of buildings. Thus, a crucial issue when distributing heat energy costs with the use of heat energy allocators is discovery of key factors, which influence determination by a consumer amount of the energy consumed.

2 The Purpose

In this work we proposed to analyse correction factors which are used to determine amount of heat consumption by heat cost allocators. When defining them, it is proposed to take into consideration influence of climatic conditions of a certain region of Ukraine in accordance with DSTU-N B V.1.1-27 2010 (State Standard) "*Construction Climatology*" and DBN V 2.6.-31-2016 "*Construction Heat Isolation*" (Ukrainian national construction regulation) and places of apartment location in building; to compare and analyse design coefficients with those provided by the Consumed Volume Allocation Methodology between Consumers in Public Utilities Building (Order of Ukraine. On Approval of the Methodology of Distribution among Consumers of Volumes Consumed in the Utilities Building № 315, 2018).

The purpose of this study is to find out correspondence of existing correction factors with regard to the criteria as follows: apartment location in the building corners, at the upper and lower floors; ambient environment.

3 Methodology

Correction factor calculation algorithm is carried out in the following sequence. The calculation is made in the basis of the defined heat loss in the premises for dwelling nine-storied building. When calculating, we took into consideration climatic conditions of a certain region of Ukraine in accordance with DSTU-N B V.1.1-27 2010 "*Construction Climatology*". To implement conditions with regard to the rated value of heat-transfer resistance in order to calculate heat loss of premises we used minimum allowable values of heat-transfer resistance of fencing constructions of dwelling houses in accordance with DBN V 2.6.-31-2016 "*Thermal Insulation of Building*" (Ukrainian national construction regulation).

Design correction factor shall be defined according to the formula:

$$Ki = \frac{Qprom}{Qi}. \tag{1}$$

where: K_i–correction coefficient for premises; Q_{prom}– heat loss of typical adjoining premises on the sunny side, W; Qi –heat loss of premises, W.

The calculation was made for such cities of Ukraine as Kyiv, Lviv, Kharkiv, Zaporizhia. Characteristics of dwelling apartment: total area - 110 m^2; ceiling height - 2,7 m; kitchen area - 14,2 m^2; number of external walls - 7; number of windows - 5, double-glazed.

4 Results and Discussion

Calculation of correction coefficients is carried out in accordance with the methodology mentioned above. Results of calculation of apartment heat loss (ground, mezzanine, last floor) in different cities during the heating season for Kharkiv, Kyiv, Zaporizhia, and Lviv cities are shown in Table 1. We studied 2 variants of apartment location at the last floor: 1 variant – presence of roof premises; 2 variant – absence of roof premises.

According to the results shown in Table 1, we can draw a conclusion that for fair calculation of heat energy metering when using heat cost allocators it is required to take into consideration place of apartment in building and climatic zone of housing development. Therefore, we justified determination of correction climatic factors for each region of Ukraine in accordance with the month of heating season taking into consideration apartment location (ground, mezzanine, last floor). We have calculated correction factors according to the methodology mentioned in the Sect. 2 using the Formula (1). The calculation results are shown in the Table 2.

Table 1. Heat loss (W) of three-room apartment

Month	October	November	December	January	February	March	April
Ground floor	1575*	2636*	3403*	3762*	3631*	2799*	1330*
	1471*	2477*	3191*	3548*	3370*	2623*	1325*
	1156*	2106*	2822*	3196*	3055*	2339*	1078*
	1487	2380*	3142*	3435*	3224*	2558*	1503*
(2nd -8th) floor	1400*	2454*	3217*	3574*	3444*	2617*	1156*
	1302*	2308*	3022*	3379*	3201*	2454*	1156*
	987*	1937*	2653*	3027*	2887*	2170*	909*
	1318*	2211*	2974*	3266*	3055*	2389*	1335*
9th (1 variant) floor	1539*	2593*	3356*	3713*	3583*	2756*	1295*
	1447*	2453*	3167*	3524*	3345*	2599*	1301*
	1146*	2096*	2812*	3186*	3046*	2329*	1068*
	1463*	2355*	3118*	3410*	3199*	2534*	1479*
9th (2 variant) floor	1946*	3001*	3764*	4121*	3991*	3163*	1703*
	1836*	2842*	3556*	3913*	3735*	2988*	1690*
	1561*	2511*	3227*	3600*	3460*	2744*	1483*
	1845*	2737*	3500*	3792*	3581*	2916*	1861*

* Note The first row is for Kharkiv City, the second row is for Kyiv City, the third row is for Zaporizia City, the fourth row is for Lviv City.

Table 2. Correction coefficients

Month	October	November	December	January	February	March	April
Temperature °C	7,5*	1*	−3,7*	−5,9*	−5,1*	0,3*	9,0*
	8,1*	1,9*	−2,5*	−4,7*	−3,6*	1,0*	9,0*
	9,6*	3,5*	−1,1*	−3,5*	−2,6*	2,0*	10,1*
	8,0*	2,5*	−2,2*	−4,0*	−2,7*	1,4*	7,9*
$K_{1\ floor}$	0,91*	0,94*	0,94*	0,95*	0,95*	0,94*	0,9*
	0,9*	0,93*	0,94*	0,95*	0,94*	0,94*	0,91*
	0,88*	0,92*	0,94*	0,94*	0,94*	0,93*	0,88*
	0,9*	0,93*	0,94*	0,94*	0,94*	0,93*	0,9*
$K_{last\ floor}$	0,92*	0,95*	0,95*	0,96*	0,95*	0,94*	0,91*
	0,91*	0,94*	0,95*	0,95*	0,95*	0,94*	0,91*
	0,89*	0,93*	0,95*	0,94*	0,94*	0,93*	0,89*
	0,91*	0,93*	0,94*	0,95*	0,94*	0,94*	0,91*
K_{corner}	0,9*	0,91*	0,91*	0,92*	0,92*	0,91*	0,9*
	0,9*	0,91*	0,92*	0,92*	0,91*	0,91*	0,9*
	0,9*	0,91*	0,92*	0,92*	0,92*	0,91*	0,9*
	0,9*	0,9*	0,91*	0,92*	0,91*	0,9*	0,9*

* Note The first row is for Kharkiv City, the second row is for Kyiv City, the third row is for Zaporizia City, the fourth row is for Lviv City.

Explanations to the Table 2, where: $K_{1\ floor}$ – correction factor for ground floor apartments; $K_{last\ floor}$ – correction factor for last floor apartments; K_{corner}- correction factor for corner premises.

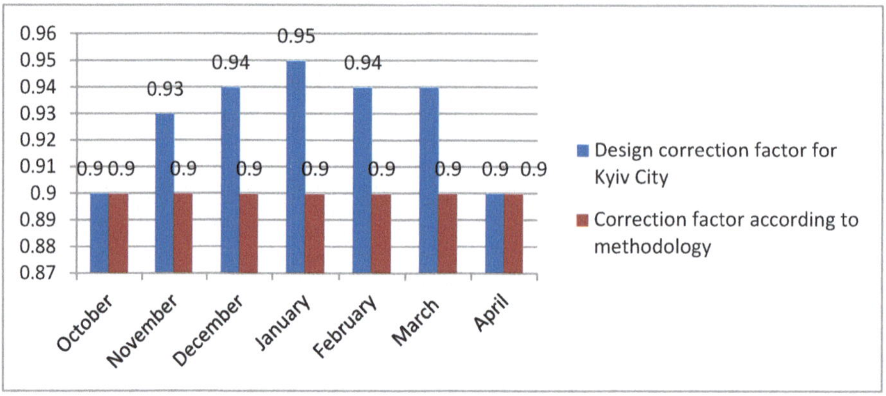

Fig. 1. Correction factors for three-room apartment in Kyiv city when this apartment is at the ground floor.

5 Scientific Novelty

When comparing the data obtained with the correction factors given in the Consumed Volume Allocation Methodology between Consumers in Public Utilities Building approved by the Order of the Ministry of Regional Development, Building and Housing and Communal Services of Ukraine dated 22.10.2018 No. 315 we can see difference (Fig. 1) connected with the fact that the Methodology doesn't cover the building location according to the climatic zones of the Ukrainian regions.

6 Practical Significance

European Directive (European Parliament Directive 2012/27/EU on energy efficiency, 2012) related to the heat energy allocation rules include the clause concerning the accuracy of heat energy cost allocation. In order to avoid unfair allocation of heat cost when using heat cost allocators we propose to take into consideration correction factors, namely, not only apartment place of location in building, but also external climatic conditions of the Ukrainian regions.

7 Conclusions

In this study the numerical analysis has been performed to estimate existing correction factors for the consumed heat energy allocation of heating between individual consumers, specified in the Consumed Volume Allocation Methodology between Consumers in Public Utilities Building approved by the Order of the Ministry of Regional Development, Building and Housing and Communal Services of Ukraine dated 22.10.2018 No. 315. According to the data obtained and their comparison with the existing correction factors we can draw a conclusion that for fair calculation of heat cost allocation to heat energy when using heat cost allocators it is required to correct them in accordance with the building regional location taking into consideration climatic zones of Ukraine.

References

Law of Ukraine. On the commercial area of thermal energy and water supply. № 2119-VIII section 10 (2017)

Enforcement Services. National measurement & regulation office (2014): Heat network (Metering and Billing) regulations. Stanton avenue, Teddington, London, from http://www.gva.co.uk/sustainability/heat-network-regulations/

Robinson, S, Vogt, G (2016): Guidelines on good practice in cost-effective cost allocation and billing of individual consumption of heating, cooling and domestic hot water in multi apartment and multi-purpose buildings, from https://ec.europa.eu/…/MBIC_Guidelines20160530D.pd

Hrankina, VV, Blahodarna, HI (2016): Shchodo pytannia komertsiinoho obliku teplovoi enerhiii vody. Resursozberezhennia ta enerhoefektyvnist inzhenernoi infrastruktury urbanizovanykh terytorii ta promyslovykh pidpryiemstv Ukrayna, pp. 95–97. Retrieved from http://eprints.kname.edu.ua/45096/1/ilovepdf_com-96-98.pdf

Ziemele J, Pakere I, Blumberga D, Zogla G (2015) Economy of heat cost allocation in apartment buildings. Energ Procedia 72:87–94. https://doi.org/10.1016/j.egypro.2015.06.013

Gafsi A, Lefebvre G (2003) Stolen heating or cooling energy evaluation in collective buildings using model inversion techniques. Energ Buildings 35(3):293–303. https://doi.org/10.1016/s0378-7788(02)00093-2

Terés-Zubiaga J, Pérez-Iribarren E, González-Pino I, Sala JM (2018) Effects of individual metering and charging of heating and domestic hot water on energy consumption of buildings in temperate climates. Energ Conversion Manage 171:491–506. https://doi.org/10.1016/j.enconman.2018.06.013

Siggelsten S (2014) Reallocation of heating costs due to heat transfer between adjacent apartments. Energ Buildings 75:256–263. https://doi.org/10.1016/j.enbuild.2014.02.022

Pakanen J, Karjalainen S (2006) Estimating static heat flows in buildings for energy allocation systems. Energ Buildings 38(9):1044–1052. https://doi.org/10.1016/j.enbuild.2005.12.002

Guerra Santin O, Itard L, Visscher H (2009) The effect of occupancy and building characteristics on energy use for space and water heating in Dutch residential stock. Energ Buildings 41(11):1223–1232. https://doi.org/10.1016/j.enbuild.2009.07.002

Order of Ukraine. On approval of the methodology of distribution among consumers of volumes consumed in the utilities building № 315 (2018)

European Parliament Directive 2012/27/EU on energy efficiency. Official Journal of the European Union 2012; 315/1 (2012)

Enforcement services. National measurement & regulation office (2015): Heat network (Metering and Billing) (amendment) regulations. Stanton Avenue, Teddington, London. Retrieved from http://:www.legislation.gov.uk/uksi/2015/…/uksi_20150855_en.pd

The Effect of Porous Pozzolanic Polydisperse Mineral Components on Properties of Concrete

Taras Markiv[1(✉)], Khrystyna Sobol[2], Nadiya Petrovska[2], and Oleksii Hunyak[2]

[1] Department of Building Production, Lviv Polytechnic National University, Bandera Street 12, Lviv 79000, Ukraine
Taras.Y.Markiv@lpnu.ua

[2] Department of Highways and Bridges, Lviv Polytechnic National University, Bandera Street 12, Lviv 79000, Ukraine

Abstract. The role of polydispersity of zeolitic tuff in the realization of its properties as an active mineral addition and optimization of granulometric composition of concrete at the level of the mesostructure have been confirmed. The function of a polydisperse zeolitic tuff and perlite in the creation of water reserve for hydration of cement in conditions of reduced relative humidity and in case of improper curing is substantiated. In dry hardening conditions, concrete containing polydisperse mineral components such as perlite and zeolitic tuff are characterized by a slightly lower percentage of compressive strength reduction in comparison with concrete without additions. Flexural strength of moist cured concrete with polydisperse zeolitic component is 8 and 12% higher than concrete without additives at 28 and 90 days respectively.

Keywords: Zeolitic tuff · Concrete strength · Particle size distribution · Microstructure · Internal curing

1 Introduction

Pozzolanic mineral components of concrete such as fly ash, granulated blast furnace slag, metakaolin, zeolitic tuff (known as mineral additions for Portland cement) have been widely used as a supplementary cementing materials (SCMs) which contribute to the properties of hardened concrete through hydraulic or pozzolanic activity and impart them technical, economical and ecological advantages (Ahmadi and Shekarchi 2010; Bilim 2011; Bostanci et al. 2016; Dvorkin et al. 2012; Zhang et al. 2016). It is well known that mineral components can improve the durability properties, reduce the heat of Portland cement hydration, increase the corrosion resistance and reduce the prime cost of concrete (Najimi et al. 2012; Vejmelková et al. 2015). These pozzolanic mineral components is used in most cases in finely ground state to replace the most energy intensive component such as clinker in Portland cement and latter one in concrete (Sanytsky et al. 2018; Limbachiya et al. 2014). However, the use of fine ground mineral additions in concrete gradually exhausts itself, because their production requires consumption additional energy resources, increases water demand of concrete and as a result the use of higher amount of plasticizer or more effective superplasticizer to obtain

Z. Blikharskyy et al. (Eds.): CEE 2019, LNCE 47, pp. 275–282, 2020.
https://doi.org/10.1007/978-3-030-27011-7_35

concrete with appropriate strength and durability properties. New tendencies of concrete technology development are directed to optimize the particle size distribution. In most cases the lack of particles which diameter lays between particle size distribution of Portland cement and fine aggregate is observed in concrete. This volume can be filled in with mineral additions. In this case the optimization of its micro- and macrostructure takes place and influence on properties of concrete. It allows obtaining targeted strength and durability properties of concrete at lower Portland cement content. It meets the strategy of sustainable development in construction industry.

One of the many types of pozzolans which can be used is zeolitic tuff (Kocak et al. 2013; Sobol et al. 2014; Jo et al. 2012; Markiv et al. 2016). This natural pozzolanic mineral component of concrete has unique properties. Although zeolite which is the main rock forming mineral of zeolitic tuff is crystalline, it shows proper pozzolanic activity (Najimi et al. 2012; Markiv et al. 2016). It also has intracrystalline channel porosity. It becomes the basis for studying the possibility of its use for internal curing of concrete. The possibility of zeolitic tuff use for internal curing of concrete is also being investigated by many researchers. Ghourchian et al. (2013) concluded that the performance of porous aggregates in internal curing depends on their crystalline microstructure and they sometimes need pretreatment before use (Yang et al. 2019).

Thus, the aim of this article is to study the influence of particle size distribution (PSD) of zeolitic tuff coming from Sokyrnytsia (Ukraine) on the properties of concrete and its possible use for internal curing.

2 Materials and Methods

2.1 Materials and Mixture Design

Portland cement CEM I 42.5, which conforms the requirements of EN 197 was used in this study. Two types of mineral additives were used for cement substitution: natural zeolitic tuff (clinoptylolite type) derived from Transcarpathian region of Ukraine and expanded perlite with bulk density of 67 kg/m^3. Retaining on sieve 008 for coarse zeolite (ZC) and coarse perlite (PC) was about 12 and 15% respectively.

Three types of concrete mixtures were prepared. Reference mixture (Ref) contained 400 kg/m^3 of Portland cement. For ZC and PC mixtures 10 wt.% of cement was replaced with polydisperse zeolitic tuff and perlite. Amount of expanded perlite was calculated to occupy the same volume in concrete as zeolitic tuff because of its very low bulk density. The content of fine and coarse aggregates for all mixtures was 600 kg/m^3 and 1250 kg/m^3 respectively. All mixtures also contained commercially available polycarboxylate based superplasticizer. The dosage of superplasticizer was 1.0 wt.% of cement. Consistency class of concrete mixtures was S1 (Slump = 20–40 mm). W/c ratio for concretes were 0.38–0.42.

2.2 Specimen Preparation, Curing and Testing

First, concrete ingredients in the designed proportions have been mixed with water in laboratory pan mixer for 5 min, to let the porous components absorb the mixing water.

After that superplasticizer and additional water were added to receive the target slump (20–40 mm) for all mixtures. Then, prepared fresh concrete were casted into 100 × 100 × 100 mm cube and 100 × 100 × 400 mm prism moulds and compacted by means of laboratory vibrating table.

Concrete specimens were divided into two parts for curing under dry conditions (RH = 68 ± 3%, t = 20 ± 2 °C) and moist curing (RH = 95 ± 5%, t = 20 ± 2 °C). Concrete specimens were exposed to appropriate curing conditions immediately after casting.

Internal relative humidity (IRH) test was carried out using set-up, shown in Fig. 1. After demoulding a hole was drilled in specimens and digital RH probe was installed in each cube. Probes were sealed with sealant to avoid moisture loss. Similar method was proposed by El-Dieb (2007).

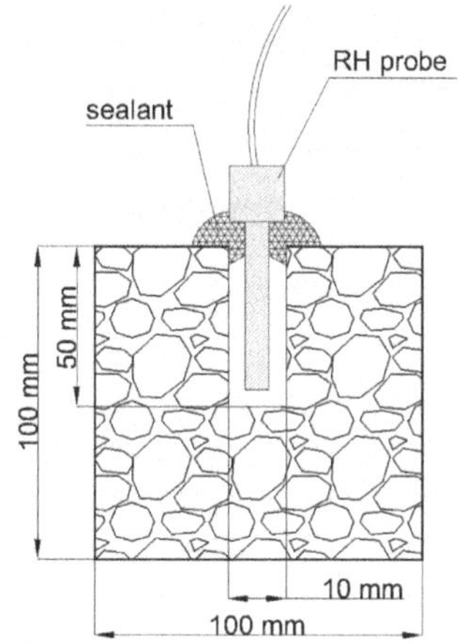

Fig. 1. View of internal relative humidity test

IRH was measured for concrete cured in dry (RH = 68 ± 3%, t = 20 ± 2 °C) and hot (RH = 55 ± 5%, t = 40 ± 2 °C) conditions. Compressive and flexural strengths' tests of concrete were carried out according to EN 12390-3 and EN 12390-5 respectively.

3 Results and Discussion

As it is well-known, high-quality aggregates have low or almost zero content of grains less than 0.16 mm size. Since Portland cement which is produced nowadays is characterized by a significant fineness, there are gaps between the binder and aggregates in the grading curve of concrete ingredients. The diameter of particles of finely dispersed additions ranged from 1 to 60 μm, which corresponds to the PSD of Portland cement. At the same time, diameter of the particles of polydisperse mineral components used in this work are located in the range of 0.4–800 and 6–1250 μ for the polydisperse zeolitic component

(PZC) and polydisperse perlitic component (PPC), respectively. A better continuity of the gradation at the level of concrete mesostructure is provided by introducing a PZC. It is confirmed by the comparison of the distribution curves of the ingredients of studied concrete and the curve of the maximum particle packing density – 0.39 power curve (Fig. 2).

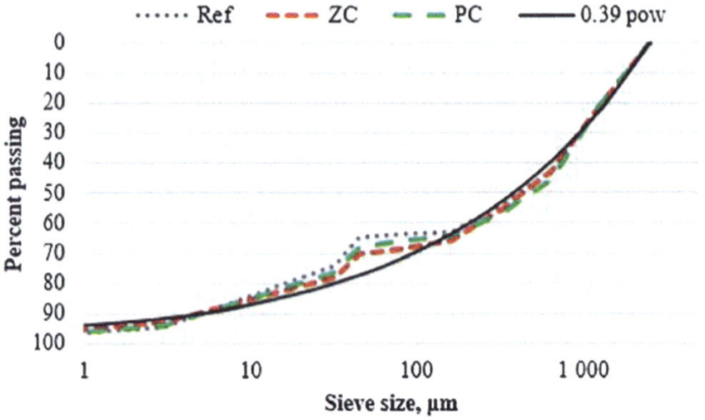

Fig. 2. PSD of concrete ingredients and 0.39 power curve

Generally, incorporation of mineral additions results in decrease of concretes mechanical properties. When additions are used the percentage of concrete strength reduction usually corresponds to the percentage content of additions in concrete. As it can be seen in Table 1, the pattern is retained for concrete with 10% substitution of Portland cement with perlite, it is characterized by 8–10% lower compressive and flexural strength in comparison with Ref concrete. Results of compressive strength test at 28 and 90 days of hardening indicate, that concrete containing zeolitic tuff does not inferior significantly to concrete without additions. At the same time, flexural strength of ZC was 8 and 12% higher than Ref at 28 and 90 days respectively.

Table 1. Flexural and compressive strength of concretes, cured in moist conditions

Mixture identification	Flexural strength, MPa		Compressive strength, MPa		
	28 days	90 days	3 days	28 days	90 days
Ref	6.5	7.6	43.8	69.7	73.2
ZC	7.0	8.5	39.7	66.8	70.9
PC	5.4	6.4	37.8	64.0	66.5

Curing conditions are crucial for gaining an appropriate strength and durability of concrete. Internal water reserve plays very important role in providing a more complete hydration of cement matrix. Figure 3 shows changes of relative humidity for concretes with age. Concrete containing PPC showed the highest internal humidity at all ages and both curing conditions, which corresponds to higher amount of absorbed water than PZC. Retained moisture in porous additions can also contribute to pozzolanic reaction, which takes place at later ages of hardening.

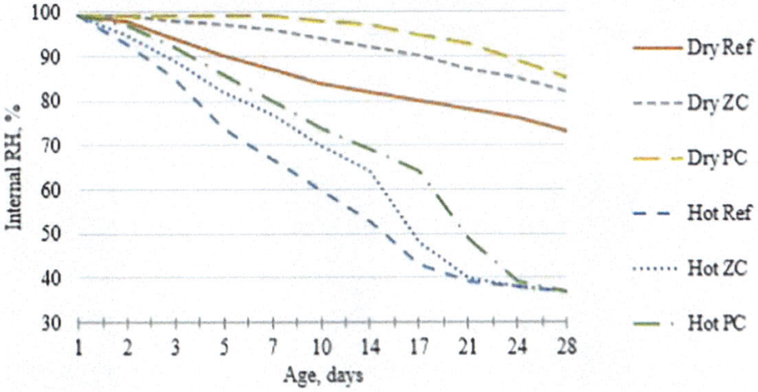

Fig. 3. Internal relative humidity of concretes, cured in dry (RH = 68 ± 3%, t = 20 ± 2 °C) and hot (RH = 55 ± 5%, t = 40 ± 2 °C) conditions

In low RH environment, a self-desiccation process can occur in concrete with low w/c, resulting in incomplete hydration of Portland cement and loss of properties. The influence of porous mineral components on flexural strength tests of concrete, cured in dry conditions, can be seen in Fig. 4. A reduction of flexural strength for Ref concrete at 90 days of hardening corresponds to insufficient moisture content of concrete without additions. The flexural strengths are 32% and 9% higher at 90 days of hardening for ZC and PC concretes respectively in comparison with Ref concrete.

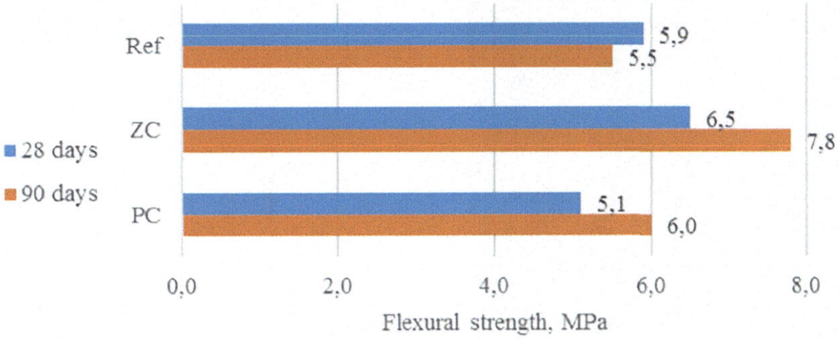

Fig. 4. Flexural strength of concretes, cured in dry conditions

Significantly higher flexural strength of concrete with addition of PZC can be related to both better particle packing and internal curing effect of polydisperse zeolite component as well as its good pozzolanic activity.

Results of compressive strength test are presented in Fig. 5. Ref, ZC and PC concretes are characterized by 6.4, 2.4 and 1% strength reduction in dry curing conditions respectively in comparison with moist cured specimens (Table 1) at 90 days of hardening, which correspond with the tests results of IRH. However, the strength of concrete containing zeolitic component was found to be the highest among concretes under dry curing conditions.

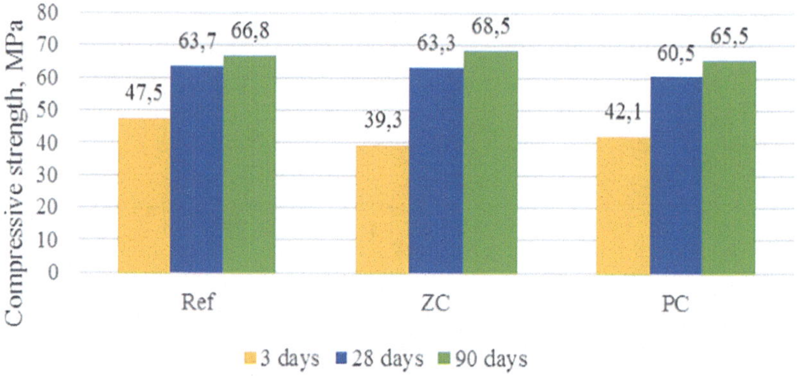

Fig. 5. Compressive strength of concretes, cured in dry conditions

Observation under scanning electron microscope (Fig. 6a) shows, that microstructure of concrete incorporating natural zeolitic tuff has different habitus of crystals of hydraton products. There are no large structural elements of Portlandite $Ca(OH)_2$ inherent to concrete without additions (Fig. 7).

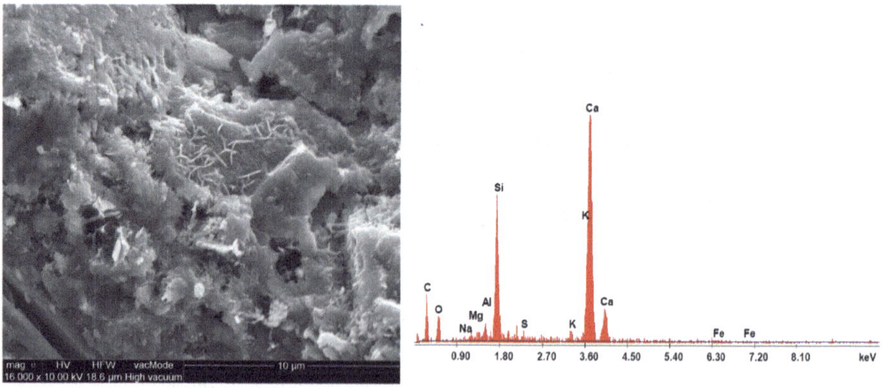

a b

Fig. 6. SEM image (a) and EDX point analysis (b) of concrete containing zeolitic tuff after 28 days of hardening

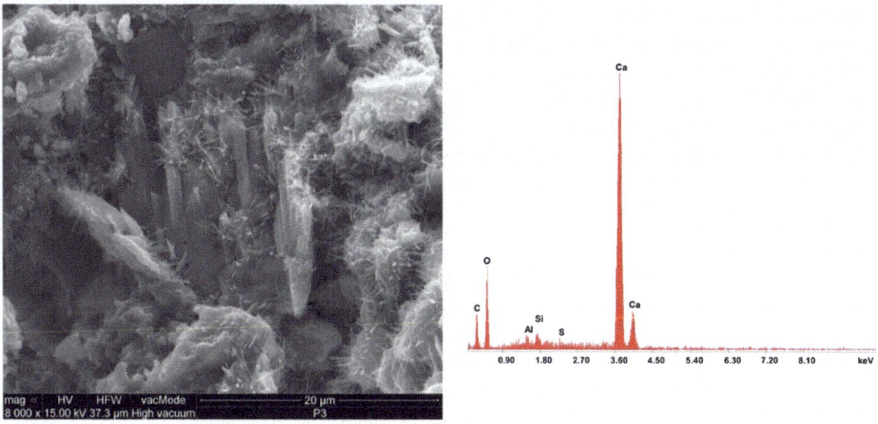

a b

Fig. 7. SEM image (a) and EDX point analysis (b) of Ref concrete after 28 days of hardening

Results of EDX point analysis (Fig. 6b) indicates the presence of low-alkali fiber-like crystals of calcium hydrosilicates CSH(I). They reinforce the contact zones in hardening concrete and provide colmatation of pores, therefore improving concrete's flexural strength.

4 Conclusions

In the present study, the expediency and effectiveness of use of a polydisperse zeolitic component in concrete are proved. The following conclusions can be drawn:

1. The addition of PZC provides better particle packing of concrete ingredients at the mesostructure level and improves strength characteristics of hardened concrete.
2. The presence of low-alkali fiber-like crystals of calcium hydrosilicates CSH(I) in ZC concrete reinforces its contact zones and improve flexural strength.
3. Porous polydisperse components provides internal curing effect by creating additional water reserve for hydration of cement in conditions of reduced relative humidity.

References

Ahmadi B, Shekarchi M (2010) Use of natural zeolite as a supplementary cementitious material. Cem Concr Compos 32:134–141

Bilim C (2011) Properties of cement mortars containing cliniptilolite as a supplementary cementitious material. Constr Build Mater 25:3175–3180

Bostanci SC, Limbachiya M, Kew H (2016) Portland slag and composites cement concretes: engineering and durability properties. J Clean Prod 112(1):542–552

Dvorkin L, Bezusyak A, Lushnikova N, Ribakov Y (2012) Using mathematical modeling for design of self compacting high strength concrete with metakaolin admixture. Constr Build Mater 37:851–864

El-Dieb A (2007) Self-curing concrete: Water retention, hydration and moisture transport. Constr Build Mater 21(6):1282–1287

Ghourchian S, Wyrzykowski M, Lura P, Shekarchi M, Ahmadi B (2013) An investigation on the use of zeolite aggregates for internal curing of concrete. Constr Build Mater 40:135–144

Jo B-W, Choi J-S, Yoon K-W, Park J-H (2012) Material characteristics of zeolite cement mortar. Constr Build Mater 36:1059–1065

Kocak Y, Tasci E, Kaya U (2013) The effect of using natural zeolite on the properties and hydration characteristics of blended cements. Constr Build Mater 47:720–727

Limbachiya M, Bostanci SC, Kew H (2014) Suitability of BS EN 197-1 CEM II and CEM V cement for production of low carbon concrete. Constr Build Mater 71:397–405

Markiv T, Sobol K, Franus M, Franus W (2016) Mechanical and durability properties of concretes incorporating natural zeolite. Arch Civ Mech Eng 16:554–562

Najimi M, Sobhani J, Ahmadi B, Shekarchi M (2012) An experimental study on durability properties of concrete containing zeolite as a highly reactive natural pozzolan. Constr Build Mater 35:1023–1033

Sanytsky M, Kropyvnytska T, Kruts T, Horpynko O, Geviuk I (2018) Design of rapid hardening quaternary zeolite-containing Portland-composite cements. Key Eng Mater 761:193–196

Sobol K, Blikharskyy Z, Petrovska N, Terlyha V (2014) Analysis of structure formation peculiarities during hydration of oil-well cement with zeolitic tuff and metakaolin additives. Chem Chem Technol 8(4):461–465

Vejmelková E, Koňáková D, Kulovaná T, Keppert M, Žumár J, Rovnaníková P, Keršner Z, Sedlmajer M, Černý R (2015) Engineering properties of concrete containing natural zeolite as supplementary cementitious material: strength, toughness, durability, and hygrothermal performance. Cem Concr Compos 55:259–267

Yang L, Guang Y, De Schutter G (2019) Investigation on the potential utilization of zeolite as an internal curing agent for autogenous shrinkage mitigation and the effect of modification. Constr Build Mater 198:669–676

Zhang J, Wang Q, Wang Z (2016) Optimizing design of high strength cement matrix with supplementary cementitious materials. Constr Build Mater 120:123–136

Modernization of Old Steel Tanks Used to Store Liquid Fuels – Selected Tasks Caused by Increasing Environmental Requirements

Mariusz Maslak[1(✉)] and Michal Pazdanowski[2]

[1] Institute of Building Materials and Engineering Structures, Faculty of Civil Engineering, Cracow University of Technology, Warszawska 24, 31-155 Cracow, Poland
mmaslak@pk.edu.pl

[2] Institute for Computational Civil Engineering, Faculty of Civil Engineering, Cracow University of Technology, Warszawska 24, 31-155 Cracow, Poland

Abstract. Selected structural changes which have to be introduced in the existing on the ground tanks used to store liquid petroleum products, should these tanks be further kept in service under the same or modified service regimen, are discussed in this paper. Modernization of these tanks within the scope described by the authors is required due to the increasing environmental requirements. In particular, the improved sealing of the previously used single shell tank bottom structure has been considered. This may be achieved by adding an additional steel or composite second bottom with monitoring space, or alternatively by raising the whole tank and installing underneath it a geomembrane impermeable to petroleum materials and equipped with leak monitoring system. An application of lamination techniques with modern composite materials has to be accompanied with development and installation of a fail-safe system to ground the electrostatic charges generated by friction on the shells modernized in such a manner. The second modernization technology discussed in this paper is related to adaptation of the typical on the ground cylindrical storage tank equipped with floating roof to store the jet fuel. In this case the recommended solution is an addition of a lightweight aluminum dome.

Keywords: Steel tank · Environmental requirements · Modernization works · Lamination techniques · Grounding of electrostatic charge · Modified service regimen

1 Introduction

Approximately 60% of the steel on the ground cylindrical tanks with vertical axis used to store liquid fuels in Poland had been built in the seventies of the twentieth century, and thus their service time has already exceeded forty years. Usually these tanks had been erected according to the typical technical design, and designated to store the petroleum products characterized by high vapour pressure and volumetric weight $\rho \leq 9\ \text{kN/m}^3$. The most often these tanks had been erected as all welded structures with floating roofs and having the capacity of 2000 m^3 or 5000 m^3 or with permanent

Z. Blikharskyy et al. (Eds.): CEE 2019, LNCE 47, pp. 283–290, 2020.
https://doi.org/10.1007/978-3-030-27011-7_36

roofs and having the capacity of 500 m^3 or 1000 m^3. At the moment of entering service these tanks undoubtedly satisfied all the requirements then in force, allowing their formal release for use. But since then important changes have occurred in the companies dealing with the storage and distribution of liquid fuels. Many small, unprofitable storage bases have been closed while only the larger bases specializing in the storage of selected fuel types have been designated for extended service. This resulted in the need to modernize the tanks located in such bases, caused priomarily by the change in the way these facilities are used. The second driving factor generating the need for new structural solutions lies in the tightening of environmental standards, mainly related to prevention or at least effective monitoring of possible leaks of stored fuel (Ziółko 1996). One should note here, that the large number of fuel tanks used for many years does not conform to the current standards. However, every decision on the potential modernization of a tank should be preceded by a comprehensive expertise assessing the possible further application of individual structural components used so far, assuming that the current service regimen is maintained or taking into account designed changes in the service model.

2 Addition of the Second Bottom in the Tank

In the typical fuel storage tanks used in Poland the bottom has been usually made as single layer and its liquidtightness has been controlled via drainage pipes located underneath it. The stored fuel seeping into these drainage pipes signalled a breach in the liquidtightness of the tank bottom. Unfortunately, low volume seepage from under the tank bottom during service time proved to be difficult to discover in real life. The differences in mesurements of the stored fuel level could in such cases remain within the limits of allowable natural losses. In addition, the seeping product could remain under the bottom undiscovered for a long time and show up on the surface of the terrain or in the controll drainage pipes only during the spring thaw or autumn intensive rainfall periods.

The requirements currently in force and listed in the appropriate Technical Conditions (Law Register no 113 dated September 18, 2001; Law Register no 243, dated November 21, 2005) impose on the user of a fuel tank the requirement to equip it with devices or systems signaling the emergence of a leak and preventing the penetration of stored product into the ground, in particular to surface waters. Adaptation to these requirements means in practice the need to modernize the bottom of the previously used tank. Its effective implementation involves selecting one of two possible solution approaches:

- erection of the second bottom with monitoring space within the existing tank structure,
- raising of the tank and spreading a geomembrane impermeable to petroleum products underneath it accompanied by installation of a leak monitoring system.

Should one opt for selection of the first proposed solution, an additional decision must be made on the double bottom installation technology, as at least two competitive technologies are available here depending on whether the second bottom should be made of steel or composite materials. In the case of the steel second bottom installed over the existing steel single bottom of the tank the space created between those two bottoms is subdivided into separate sections to allow for more efficient location of the leaks in case of potential failure. This technology has a significant drawback in that it is very difficult to properly install the second bottom if the first one is warped, and this occurs very frequently in tanks intensively used for a long time. Thus an evaluation of the technical condition of the steel tank is necessary prior the installation of the second bottom followed by required repairs. The second important disadvantage of the modernization of this type lies in the complete lack of visual control over the corrosion progress in the plates of the original steel bottom subsequently permanently covered by the new upper steel bottom. Installation of a new second bottom made of composite materials results in the creation of a layer with monitoring space over the old bottom. Usually this is made of multilayered or hybrid composite. If the multilayered composite is applied, its bottom layers are made of nonconductive epoxy, exhibiting the leakage resistance R_u in the range of 10^{14} Ω. The monitoring space is obtained via the application of spatial laminate, while the electrostatic conductivity is ensured by covering the topmost laminate layer with conducting lining exhibiting the resistance $R_u \leq 10^1$ Ω and having the thickness of 300 ÷ 500 μm.

3 The Problem of New Composite Layer Cooperating with Old Steel Bottom of the Tank

In spite of undoubtful progress made in the lamination technology, and especially improvements in mechanical characteristics of composites used in practice, the structural failures of such structures, manifesting themselves in cracking or local debonding of laminate layers are relatively frequent. During service the steel bottom undergoes deformations due to the changes in the level of liquid stored in the tank. During filling the steel plates of the bottom move downwards, while during emptying these plates gradually return to the initial configuration formed during welding. The changes of this type, at significant difference in deformability of steel and composite material overlying it, relatively quickly result in debonding in the modernized shell. This problem manifests itself predominantly in the relatively thick, and thus quite rigid shells. But even thin laminate layers after sufficient passage of time may debond from the steel substrate especially on the surfaces exhibiting large imperfections (Ziółko and Al-Shaibah 1998). Until relatively recently the laminates equipped with monitoring had been made of mats with unidirectional strands. With such solution substantial stresses in the direction perpendicular to strands could have easily resulted in shell cracking. Currently mats woven with strands running in multiple directions are the standard, and this results in much higher resistance to large longitudinal deformations.

4 The Question of Ensuring Required Liquid Tightness

The most frequently laminate loses the required liquid tightness due to insufficient application of the epoxy resin. Glass strands insufficiently permeated with epoxy are incapable of preventing the stored liquid from seeping into the internal structure of the mat. On the other hand overly thick epoxy layers applied may result in completely filled monitoring space, thus completely preventing the monitoring itself. Thus the laminates with monitoring layer have to be applied in segments, subject to continuous control of the applied layer quality. After hardening of the laminate the monitoring space is pressurized and randomly drilled into, to verify the patency of the system. Should the structure lose its liquid tightness, water vapour may enter the enclosed volume of the tank. This vapour, together with vapours of the stored liquid may condense on the surface of the permeable composite penetrating it throughout. Even groundwater may penetrate the insufficiently liquid tight shell. This groundwater enters the inside of the tank and creates the characteristic rusty colouring on the surface of the composite (Maślak and Siudut 2008). The liquid tightness of the tank equipped with additional composite bottom is ensured when the composite layer is applied to the vertical wall of the tank to the height of approximately 1 m above tank bottom. Such solution, however, exhibits a disadvantage in that it prevents the monitoring of the fillet weld joining the tank bottom with side shell. The alternative solution, where the fillet weld is covered with a layer of transparent epoxy resin constituting in addition a border protecting the edge of the laminated bottom, is devoid of this disadvantage.

5 Monitoring the Technical Condition of the Structure Covered by Laminate

After application of the top layer made of laminate the user is incapable of visually monitoring the technical condition of the steel structural components covered underneath it. The corrosion processes under the laminate do not terminate, but are only retarded to some extent. These processes are especially dangerous in tank bottoms, on the unprotected bottom side directly adjacent to the sand layer. It seems, that the usually applied random checks of the technical condition of steel bottom structure, performed only at preselected locations, may be insufficient to reliably predict the durability of the tank. Thus prior to the decision on laminating the tank bottom precise measurements of the plate thicknesses reduced by the corrosion, checks of the weld liquid tightness as well as all necessary repair works should be undertaken. This is the more important, as currently only the acoustic method (Gołaski et al. 2001), sometimes incapable of sufficiently precise observation, is legally approved to perform nondestructive tests to reliably evaluate the technical condition of steel plates and welds covered by a layer of laminate.

6 Ensuring the Sufficient Electrostatic Conductivity of the Laminate Layer

The electrostatic conductivity of the laminate is achieved by application of top conductive layer as well as carbon bands and/or copper wire creating additional discharge paths to the ground for the electrostatic charges. In addition, the areas of the tank bottom not covered with laminate (for instance the joints between the bottom plates and supports of the floating roof) may be used for this purpose. The conducting layers of composite with substantial addition of the colouring compound are very porous, and this in turn is a factor facilitating the initiation of electrolysis and development of galvanic microcells (Orlikowski et al. 2001). Prevention of adverse chemical reactions between the composite and the fuel filling the tank is an additional task of the conductive layer. In practice it may only slow down the chemical corrosion processes, as the susceptibility of this type is caused not only by the porosity of the layer but also by the diffusion of the stored fuel through the internal structure of the cladding. Therefore the condition of the conducting layer should be monitored periodically. It should be checked against washout or other substantial damage. Often, when the allowed time span between applications of subsequent layers is exceeded, peeling of the applied layer is observed. An increase in the bond between the conductive layer and underlying structural layer may prevent this, and the increased bond may be obtained by sprinkling the top of structural layer with fine grained quartz sand (grain size of 0.2 to 0.5 mm) during the application. It has to be underlined here, that the faulty conductive layer is a source of substantial fire risk (Maślak and Siudut 2005, Maślak and Siudut 2006). According to Polish regulations (PN-E-05204 1994) the recommended time span between subsequent checks of electrostatic protection effectiveness for the zone 0, defined inside the tank, is a quarter of a year, while for the remaining zones is one year. This requirement is difficult to satisfy. The measurements of this type usually may be performed only after emptying the tank, cleaning up the inside and thorough ventilation. In practice measurements of this type are performed only during renovations or modernizations, or when complete emptying the tank is possible.

7 Modernization of the Tank Bottom with Application of the Impermeable Geomembrane

When, in order to ensure the liquid tightness of the tank bottom, a geomembrane impermeable to petroleum products and equipped with appropriate leak monitoring system has to be spread on the subsoil underneath, raising the whole tank is the preferred solution. The tank may be raised using hydraulic jacks. Two approaches are possible here. The smaller tanks, of lower weight, are usually raised as whole, while in the case of larger tanks only the cylindrical shell is raised after separation form the bottom. The geomembrane is usually made of thermoplastic or elastomeric materials. These materials should exhibit high resistance to tear at elongation reaching up to 1000%. A tank bottom leak monitoring system covers the surface of the geomembrane. The geomembrane itself in such a system is designated to prevent the potential leakage of the stored fuel to the environment, while the monitoring system is to automatically generate the alarm signal to facilitate a quick remedial action on part of the operator. Undoubtedly, modernization of this type has the advantage of allowing the repair works on the foundations of the tank. A complete replacement of the corroded tank bottom is also possible, allowing for replacement or evening out the subsoil, thus eliminating the source of uneven settlement.

8 Modernization of a Steel Tank with Floating Roof for Storage of Jet Fuel

Jet fuel is usually stored in tanks with permanent roofs. Permanent roof practically eliminates the risk of contaminating the jet fuel with precipitation water. In tanks equipped with floating roofs, even equipped with the best sealing solutions currently available, complete elimination of the contamination risk by precipitation water is rather impossible. Therefore, should the change of the service regimen be considered in the fuel base to shift to the storage of jet fuel, the only recommended structural solution is to rebuild the floating roof and equip the tank with lightweight aluminium shell cover (Maślak et al., 2015). Thus the simplest solution may be summarized here in the following steps: removal of the steps leading to the top of the floating roof, adaptation of the floating roof to fulfil the role of the internal floating roof, and erection of an additional external lightweight dome roof over the whole tank. Unfortunately, the requirements for tanks configured in such manner are not specified in the legal regulations currently in force in Poland (EN 1993-4-2, 2007; EN 14015, 2010), thus giving the users of fuel depots an unnecessary and avoidable leeway in preparing proprietary, sometimes risky and not proven solutions. Therefore, in search of best solutions one may turn to detailed guidelines contained in the regulations published in other countries, and especially in the standards (API 650, 2013) within the realm of design and use as well as (NFPA 11, 2016) in the domain of fire protection. According to these regulations the whole aluminium structure designed as the cover over a tank should be self supporting and besides climate induced loads (wind, snow) should be able to support a p = 113 kg/m^2 area load corresponding to the localized load generated by a serviceman with toolbox, spread over 1 square meter of the shell sheathing. In addition, the snow cover on the shell may result in the displacement of the sheathing panels up to the angle of 1.98°. In practical terms, for the largest sheet metal panels, this angle results in the displacement in the range of 45 mm with respect to the initial location of the panel, right after the assembly. The snow should be periodically removed, not only from the shell itself but also from the service platforms, especially if the load applied to these platforms exceeds the value of s = 75 kg/m^2, corresponding to the 15 cm thick layer of the wet snow. In addition to all the loads listed above, the shell should be capable of safely supporting the concentrated load applied at its apex, and generated by the lightning rod. The regulations listed above do not stipulate the requirement of permanent access to the top of the internal floating roof. Of course the fuel depot user may add such requirement to the list of requirements on his own. Visual control of the technical condition of the internal floating roof may be performed via hatches in the aluminium shell, but then direct access to the internal roof is possible only in its highest (topmost) location, i.e. when the tank is completely filled with fuel.

9 Concluding Remarks

A change in the service regimen caused by logistical and often just economic considerations as well as increasing requirements for environmental protection force the staff of many domestic fuel depots to plan and carry out comprehensive modernizations of existing steel fuel tank structures. This is correlated with making many decisions regarding eficient technology of rebuilding and renovation works, and seeking such solutions which would ensure sufficient durability of the renovated structure subjected to actions of previously unforeseen type and exposed to the threats specific to these actions. This task may be quite difficult for users of such fuel depots, as so far no explicit and formalized regulations regarding both the design and operation of such tanks exist in Poland. Therefore attempts are being made to adapt the solutions reccomended for use in other countries to domestic requirements. Unfortunately, these solutions often originate in various regulations, sometimes incompatible, yielding somewhat chaotic results, and as a consequence the required safety level or suficient durability of the modernized tank may not always be achieved. Thus it seems, that a certain degree of systematization and evaluation of solutions which may be found in engineering practice should be attempted. The review proposed and discussed in this paper represents an attempt at this approach.

References

API Standard 650 (2013) Welded steel tanks for oil storage, 12th edn. American Petroleum Institute

EN 1993–4–2 (2007) Eurocode 3: Design of steel structures, Part 4–2: Tanks

EN 14015 (2010) Specification for the design and manufacture of site built, vertical, cylindrical, flat-bottomed, above ground, welded, steel tanks for the storage of liquids at ambient temperature and above

Gołaski L, Kozakowski S, Siwicki K (2001) Evaluation of the technical condition of liquid fuel storage tank bottom by the accoustic emmission method. Verification of the research results. In: Proceedings of the 30th National Conference on Nondesttructive Testing, Szczyrk, 22–25.10.2001 (in Polish)

Law Register no 113 dated September 18, 2001, with later changes (2001) Regulation of the Minister of Economy on the technical conditions of technical supervision which should be satisfied by non-pressurized and low-pressure tanks for storage of liquid flammable materials (in Polish)

Law Register no 243 dated November 21, 2005, with later changes (2005) Regulation of the Minister of Economy on the technical conditions which should be satisfied by fuel bases and fuel depots, long range petroleum and petroleum products pipelines and their locations (in Polish)

Maślak M, Pazdanowski M, Siudut J (2015) Element shaping in renovated steel tanks for fuel storage due to changes in service and increased environmental requirements. In: Proceedings of the 9th Scientific – Technical Conference "Structural Shaping. Thin-walled Structures", Rzeszów, 11–12.09.2015 (in Polish)

Maślak M, Siudut J (2005) Fire risk in underground steel fuel storage tanks of horizontal axis. In: Proceedings of the 7th Scientific - Technical Conference "Service Problems of Petroleum Products Storage Depots", Poznań, 10–11.05.2005 (in Polish)

Maślak M, Siudut J (2006) Selected fire protection problems of underground steel tanks for fuel storage. Techn Supervision (Dozór Techniczny) 2(2006):36–42 (in Polish)

Maślak M, Siudut J (2008) Selected problems of ensuring durability of steel tanks for liquid fuel storage modernized by lamination method. Corrosion Protection (Ochrona przed Korozją), 5 s/A/2008, 235–240 (in Polish)

NFPA 11 (2016) Standard for low-, medium-, and high-expansion foam. National Fire Protection Association

Orlikowski J, Krakowiak S, Darowiecki K, Cebulski S (2001) Electrochemical analysis of coatings applied to protect internal surfaces of liquid fuel storage tanks. In: Proceedings of the 10th International Scientific - Technical Conference "Metal Structures", Gdańsk, 6–8.06.2001 (in Polish)

PN-E-05204 (1994) Protection against static electricity – Protection of structures, installations and devices – Requirements (in Polish)

Ziółko J (1996) Protection against soil contamination in modern steel liquid fuel storage tanks. Engineering and Construction (Inżynieria i Budownictwo), 10/1996 (in Polish)

Ziółko J, Al-Shaibah O (1998) Serviceability limit state of steel tank bottoms covered with epoxy laminate. In: Proceedings of Scientific Conference "Limit States Analysis of Metal Structures", Wrocław – Szklarska Poręba, 23-24.10.1998 (in Polish)

An Analysis of Sewage Sludge and Biogas Production at the Zamość WWTP

Adam Masłoń(✉)

Department of Environmental Engineering and Chemistry, Rzeszow University of Technology, 6 Powstańców Warszawy Av, 35-959 Rzeszów, Poland
amaslon@prz.edu.pl

Abstract. The management of sewage sludge at wastewater treatment plants is a very important technical, economic and environmental issue. One of the popular environment-friendly methods of processing sludge is methane fermentation – a complex biochemical process through which organic matter present undergoes biochemical decomposition in anaerobic conditions, with the final products being biogas (a valuable source of energy), as well as a digestate that constitutes an excellent organic fertiliser. Most often, biogas generated provides energy that supplies a wastewater treatment plant's own needs (given that such plants exert a high demand for both electricity and heat). Addressed here as an example is biogas production at the WWTP in the city of Zamość, whose management of sewage sludge is analysed in detail, with account taken of the balance between quantities of sludge and biogas generated. Methane fermentation there is found to represent an effective technology by which energy may be recovered from sewage sludge, with the biogas obtained proving a very valuable fuel that the plant can put to use readily.

Keywords: Wastewater · Wastewater treatment plant · Biogas · Sewage sludge

1 Introduction

A by-product of wastewater treatment processes is sewage sludge, whose quantities and qualitative physicochemical characteristics are matters of great significance to the operators of municipal wastewater treatment plants (WWTPs). Production can be seen to increase as highly-effective and advanced processes of biological and chemical wastewater treatment develop further, but the treatment and disposal of sludge have to be regarded as integral parts of each WWTP's technological processes. Typically, amounts of sewage sludge generated equate to between 2 and 3% by volume of what feeds into the WWTP. This is enough to ensure that the costs of sludge processing may equate to nearly 20% of investment costs at a plant, and as much as 30% of running costs. This leaves the management of sewage sludge as a very important technical, economic and environmental issue, with methane fermentation needing to be seen as one of the popular environment-friendly methods.

The biological stabilisation of sewage sludge under anaerobic conditions, as also called methane fermentation, is a complex biochemical process that sees organic matter

Z. Blikharskyy et al. (Eds.): CEE 2019, LNCE 47, pp. 291–298, 2020.
https://doi.org/10.1007/978-3-030-27011-7_37

present decomposing into sludge. High molecular-weight organic substances present are broken down by bacteria into simple, chemically-stabilised compounds. The final products of the biochemical decomposition are in fact biogas, a valuable potential energy source, as well a digestate also utilisable, given its status as an excellent organic fertiliser (Dzene and Slotina 2013; Felca et al. 2018).

Biogas is actually a mixture of gases whose proportions can vary quite markedly. Typical constituents would be methane and carbon dioxide, along with small amounts of hydrogen sulphide, nitrogen, oxygen and hydrogen (Di Fraia et al. 2016; Dzene and Slotina 2013; Municipal Services Department 2018; Tyagi and Lo 2013). The calorific value of biogas depends on its methane content, but values reported in the literature are in the 16.8–23 MJ/m^3 range. A maximum of 35.7 MJ/m^3 is reached where carbon dioxide and other pollutants are removed. Thus, the energy contained in 1 Nm^3 of such biogas is equivalent to that in 1.25 kg of coal, 1 dm^3 of diesel oil, 0.93 Nm^3 of natural gas or 9.4 kWh of electrical energy (Lewandowski 2010). Biogas energy may be used in electricity generation in spark or turbine engines, in heat production in gas boilers, in the production of electricity and heat in associated units via a cogeneration process, in the production of fuel for vehicles, or as a substrate in technological processes leading to the production of methanol (Silvestre et al. 2015).

Most often, energy from biogas generated at a WWTP is used by that plant to meet its own needs, which are actually quite great where both electricity and heat are concerned (Di Fraia et al. 2016; Makisha and Semenova 2018). Where biogas-powered cogeneration units are used, up to 60% of the demand for electricity at facilities can be met, while heat is usually generated in large surplus. Excess heat can be used effectively in drying sediments, with up to 50% of the demand for energy that process entails being covered in this way (Masłoń 2017). Methane fermentation thus represents an effective technology by which to recover energy from sewage sludge, albeit with some measures needed to increase recovery. One example of the fermentation process being intensified entails co-fermentation of sewage sludge with other organic waste, via the so-called co-digestion process (Masłoń and Tendera 2017; Mata-Alvarez et al. 2014).

This article presents matters relating to the production of biogas at the wastewater treatment plant in Zamość (Poland). Management of sewage sludge at the plant is analysed in detail here, with account taken of the sludge-biogas balance.

2 Characteristics of the Zamość WWTP

The subject of analysis is a mechanical-biological wastewater treatment plant located in Zamość (Poland). This plant, which came into operation in 1995, has a projected 24-h capacity $Q_{d\ av}$ equal to 25,000 m^3/d (in rainless weather), and a maximum daily capacity Q_{dmax} equal to 28,000 m^3/d (in rainy weather). The equivalent number of inhabitants determined for this was 250,000 PE, while the plant's current capacity is at 135,720 PE. Municipal and industrial wastewater is supplied to it, with the share of the latter in treated wastewater at 6–9%. Also transported in to the WWTP, by slurry tanker, are small amounts of wastewater from other sources, accounting for about 1.6% of the total input of raw wastewater (Municipal Services Department 2018).

The technological system used to treat wastewater at the Zamość plant consists of: a screens with clearance of 8.0 mm; an expansion chamber; stepped grating (3 mm); a horizontal sandbox aerated by two chambers; 2 radial pre-settlers 24 m in diameter and with an active volume of 890 m^3 each; 2 activated sludge chambers of active capacity 11,650 m^3 with separate zones that are anoxic (in the predenitrification chamber), anaerobic (the dephosphatation chamber), anoxic (the denitrification chamber) and aerobic (the nitrification chamber); and 2 secondary radial settling tanks of diameter 48 m and an active volume of 6220 m^3 each.

The biological removal of phosphorus from wastewater is further augmented using chemical agents (aluminum or iron salts). Overall, the system applied in the treatment of wastewater ensures integrated removal of carbon, nitrogen and phosphorus compounds from wastewater (Municipal Services Department 2018).

The sedimentary technological line is in turn based on compaction of the initial sludge in a gravity thickener, with excessive sludge thickened in a mechanical compactor. Concentrated sediments are directed to 2 separate fermentation chambers, each of capacity 2030 m^3, and then to an open fermentation basin offering retention capacity of 9500 m^3. Following fermentation, sludge is dewatered on a belt press, hygienised with quicklime and stored on a square prior to being directed for agricultural use.

Biogas generated is in turn desulphurised prior to storage in a 500 m^3 tank. From there it passes for incineration to two gas boilers each of 500 kW capacity, as well as to a 190 kW gas-fired co-generator. All electricity and heat energy generated is used by the WWTP to meet its own needs (in electrical appliances and the heating of premises and the fermentation chambers) (Municipal Services Department 2018).

3 Results and Discussions

Over the analysed period, the plant received very variable loads, with between 9135.5 and 14060.0 m^3/d inflowing in 2015 (mean 11911.4 ± 1250 m^3/d). Concentrations of pollutants in this raw wastewater also varied, e.g. from 227.7–525.2 g O_2/m^3 (mean 398.7 ± 73.6 g O_2/m^3) where BOD_5 was concerned. This translated into loads of 2791.6–6210.1 kg O_2/d (mean 4762.7 ± 1096.3 kg $O_{2/}$d).

Amounts of sewage sludge generated by the plant are presented in detail, along with volumes of biogas from the fermentation process, in Figs. 1 and 2. Most sludge was produced in January (5126 m^3), least in September (3439.7 m^3), with the monthly average figure being of 4290 m^3.

Greater (above-average) generation of sewage sludge characterised the first half of the year as compared with the second. When compressed, sewage sludge directed to fermentation chambers was of stable dry matter content in the 4–5.5% range. Amounts of sludge (primary plus excess from activated-sludge chambers) undergoing fermentation were in the range 114.7–165.4 m^3/d. Across the analysed period, fermentation of the entire amount of sludge allowed between 48789 and 74637 Nm^3/month of biogas to be obtained (mean 2080.9 ± 224.7 Nm^3/d).

Statistical analysis showed the production of both sludge and biogas was correlated with both sewage inflow and BOD_5 loading to the WWTP (Figs. 3 and 4). Near-significant correlations were found between volume of sediments generated and inflow

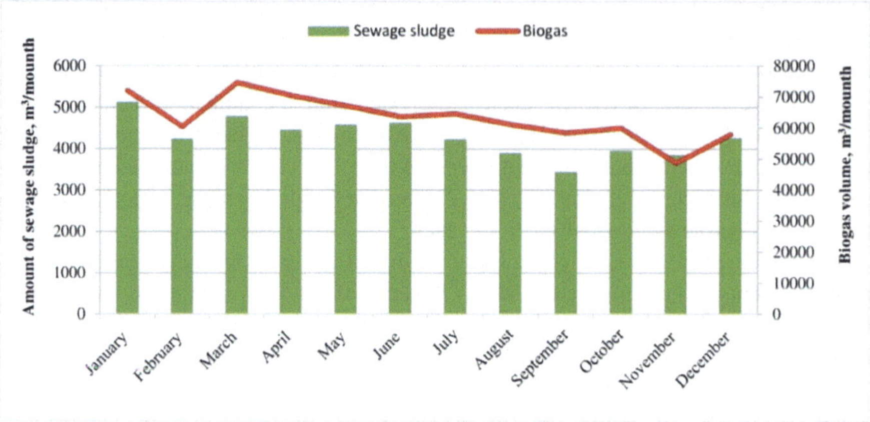

Fig. 1. Monthly generation of sewage sludge and biogas at the WWTP.

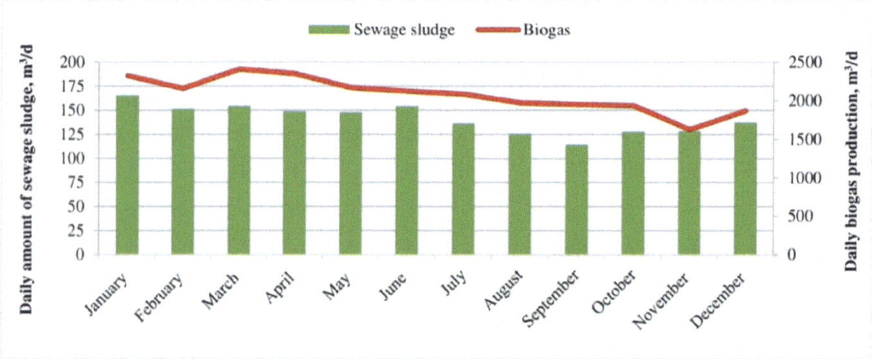

Fig. 2. Mean amounts of sewage sludge and biogas generated daily.

of sewage ($r = 0.46$, $p = 0.1$) and the BOD_5 load removed at the plant ($r = 0.45$, $p = 0.1$). In turn, the relationship between biogas production and sewage inflow was non-significant ($r = 0.36$, $p = 0.25$), while that with BOD_5 load removed at the plant approached significance ($r = 0.46$, $p = 0.1$).

Determined rates of production of sewage sludge in relation to wastewater inflow were of 9.85–13.75 dm^3/m^3 (mean 11.9 ± 1.3 dm^3/m^3). The volume of sewage sludge relative to the inflow of wastewater was lower in the May–July period than in remaining months – a finding explicable in terms of the plant's hydraulic loading. A similar tendency is noticeable in spring, when meltwater reaches WTTPs. For comparison, unit volumes of sewage sludge noted at other Polish plants are 7.41 dm^3/m^3 (at Mielec) (Masłoń et al. 2015) and 24.6 dm^3/m^3 (at Rzeszów) (Masłoń 2017).

Analyses for the Zamość plant point to very high levels of biogas production per unit volume of sludge (i.e. "biogas yield"). For each 1 m^3 of sludge passing into the

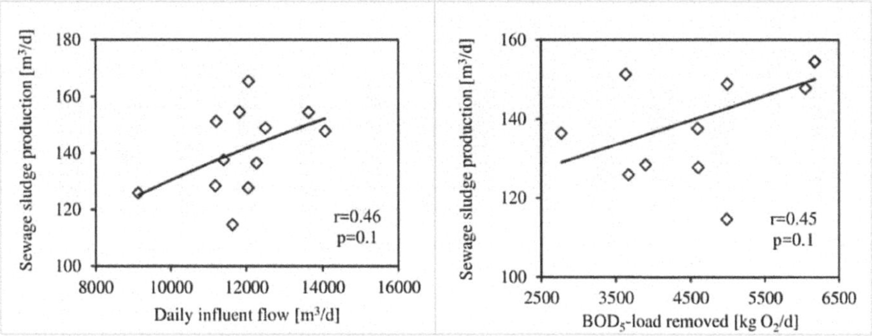

Fig. 3. Amounts of sewage sludge generated in relation to daily inflows of wastewater and BOD_5-loads removed at the WWTP.

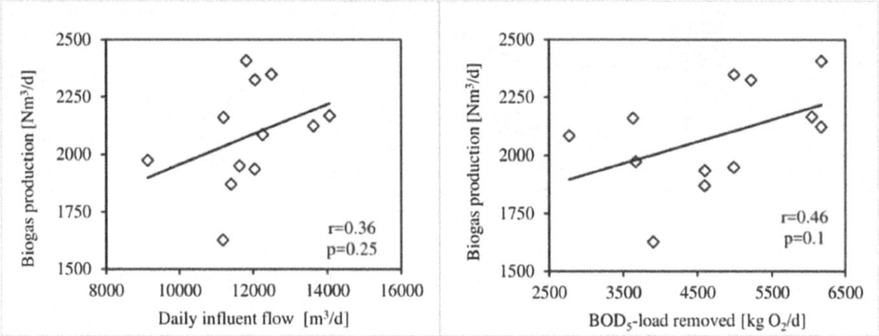

Fig. 4. Volumes of biogas generated in relation to daily inflows of wastewater and BOD_5-loads removed at the WWTP.

separate fermentation chambers, some 12.7–17.0 Nm^3 of biogas were generated (mean 14.8 ± 1.2 Nm^3) (Fig. 5). Similar biogas yields recorded relate to the Rzeszów WWTP (10.5–21.9 Nm^3/m^3 of sludge, mean 14.47) (Masłoń and Tendera 2017) and the one in Mielec (9.6–19.9 Nm^3/m^3 of sludge, mean 14.0) (Masłoń et al. 2015) though both of these installations augment their fermentation processes (engage in co-fermentation) through the addition of waste fats.

Seasonal analysis of biogas production at the Zamość WWTP further reveals highest yields in summer (at 15.9 Nm^3/m^3 of sludge), with figures below 14 Nm^3/m^3 in all other months (Fig. 6). Equally, in absolute terms, a large volume of sludge ensures that most biogas is produced in winter (Fig. 7). Thus overall production of 69073 Nm^3 in January–March inclusive compares with the 55581–67111 Nm^3 generated in other seasons.

Amounts of biogas generated translate readily into electricity and heat output. Unit production of electricity was in the range 1.16–2.21 kWh/Nm^3 biogas (mean 1.9 ± 0.3 kWh/Nm^3) (Fig. 8). At the region's other WWTPs, recorded rates of

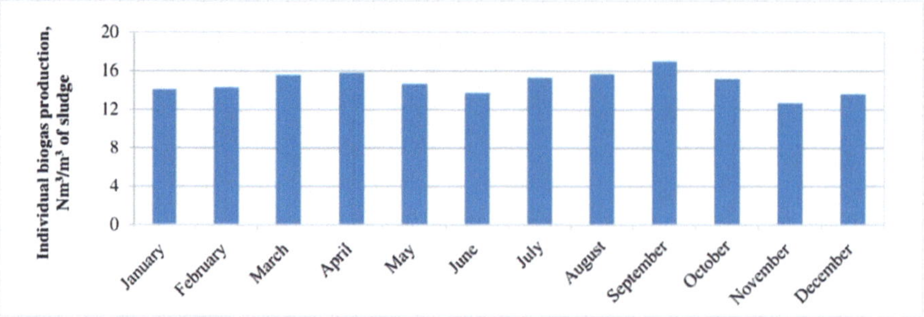

Fig. 5. Unit biogas production per m^3 of sewage sludge.

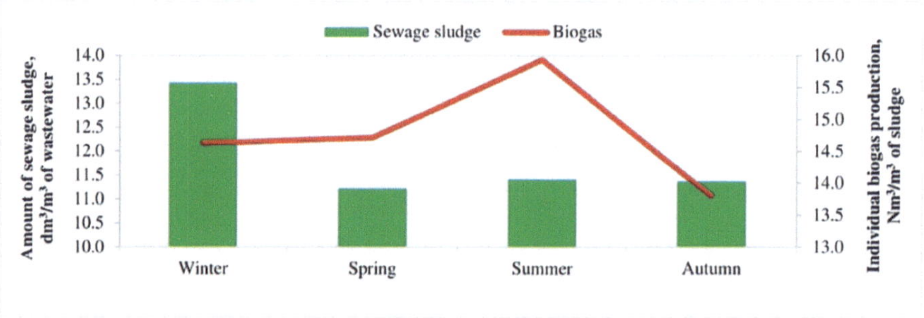

Fig. 6. Average amounts of sewage sludge and biogas generated daily.

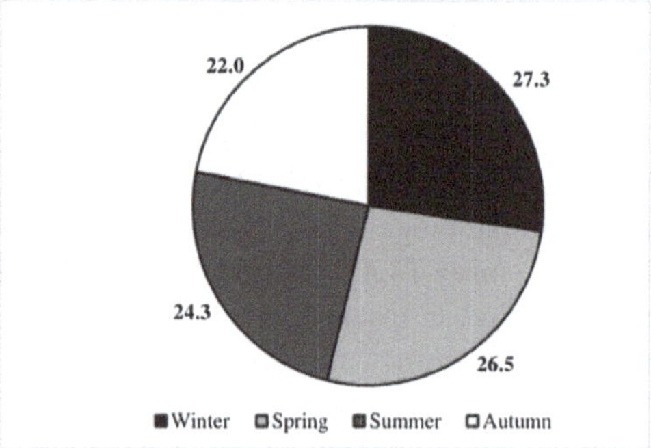

Fig. 7. Percentage shares of biogas production accounted for by different seasons of the year.

generation of electricity per Nm^3 of biogas were of 1.9–4.8 kWh at Mielec (Masłoń et al. 2015), 3.82–4.51 kWh at Krosno (Trojanowicz and Karamus 2016), and 2.02–2.48 kWh (Masłoń 2017). The overall impact of biogas combustion in boilers and in a gas co-boiler was considerable generation of electricity. In 2015, this was in the range 67.1–139.2 MWh/month, with the annual total being 1.45 GWh. The work of the boilers and co-generator in fact allowed 52.2% of demand for electricity at the studied plant to be covered. Nevertheless, by intensifying the sewage-sludge fermentation process through the addition of an organic substrate (e.g. waste fat), it would be possible to increase this biogas yield further, with resultant improved energy efficiency overall at the treatment plant.

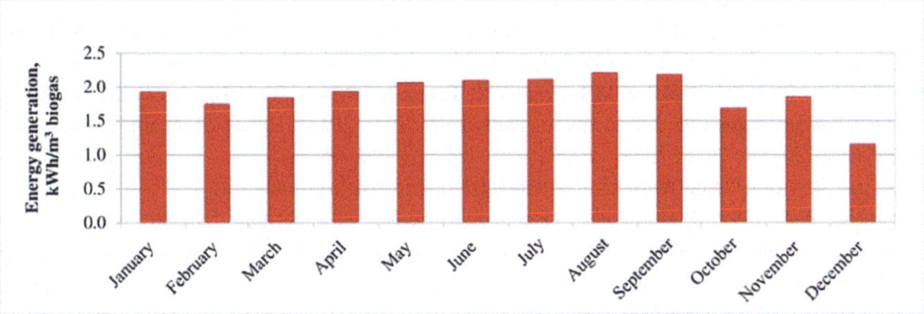

Fig. 8. Electricity generation from biogas.

4 Conclusions

The system by which sewage sludge at the Zamość WWTP is processed operates at high energy efficiency. Analysis indicates very high levels of production of biogas from the sludge (so-called biogas yield). On average, 1 m^3 of sludge introduced into the separate fermentation chambers generates 12.7–17.0 Nm^3 of biogas, with this largely translating into electricity produced (1.16–2.21 kWh/Nm^3 of biogas, average 1.9 ± 0.3 kWh/Nm^3). In 2015, such biogas combustion in boilers and in a gas co-boiler produced some 1.45 GWh of electricity in total, allowing 52.2% of the WWTP's energy needs to be met. Nevertheless, an intensification of the sewage-sludge fermentation process involving the addition of an organic substrate (like waste fat) could raise the biogas yield further, thereby improving the energy efficiency of the whole plant yet further. Methane fermentation thus represents a good technology by which to recover energy from sewage sludge, with the biogas obtained representing a very valuable fuel that can gain use directly at wastewater treatment plants themselves.

References

Di Fraia S, Massarotti N, Vanoli L, Costa M (2016) Thermo-economic analysis of a novel cogeneration system for sewage sludge treatment. Energy 115:1560–1571

Dzene I, Slotina L (2013) Efficient heat use from biogas CHP plants. Case studies from biogas plants in Latvia. Environ Climate Technol 3:45–48

Felca ATA, Barros RM, Filho GLT, dos Santos IFS, Ribeiro EM (2018) Analysis of biogas produced by the anaerobic digestion of sludge generated at wastewater treatment plants in the South of Minas Gerais, Brazil as a potential energy source. Sustain Cities Soc 41:139–153

Lewandowski WM (2010) Pro-ecological renewable energy sources. Wyd. Naukowo-Techniczne, Warszawa (in Polish)
Makisha N, Semenova D (2018) Production of biogas at wastewater treatment plants and its further application. MATEC Web Conf 144:04016
Masłoń A (2017). Analysis of energy consumption at the Rzeszów Wastewater Treatment Plant. E3S Web Conf 22:00115
Masłoń A, Pazdro S, Mroczek W (2015) Sewage sludge management in wastewater treatment plant in Mielec. Forum Eksploatatora 4(79):47–54 (in Polish)
Masłoń A, Tendera K (2017) Sewage sludge management at the Rzeszów WWTP. Forum Eksploatatora 1(88):38–45 (in Polish)
Mata-Alvarez J, Dosta J, Güiza MS, Fonoll X, Peces M, Astals S (2014) A critical review on anaerobic co-digestion achievements between 2010–2013. Renew Sustain Energy Rev 36:412–427
Municipal Services Department in Zamość (2018). Information materials
Silvestre G, Fernández B, Bonmatí A (2015) Significance of anaerobic digestion as a source of clean energy in wastewater treatment plants. Energy Convers Manag 101:255–262
Trojanowicz K, Karamus Ł (2016) Energy utilization of biogas as an element of sewage sludge management in the wastewater treatment plant in Krosno. Forum Eksploatatora 4(85):46–53 (in Polish)
Tyagi VK, Lo SL (2013) Sludge: a waste or renewable source for energy and resources recovery? Renew Sustain Energy Rev 25:708–728

Analysis of the Water Demand in the Slovak Republic and Ukraine

O. Matsiyevska[1(✉)], P. Kapalo[2], and L. Vovk[1]

[1] Department of Hydraulic and Sanitary Engineering, Lviv Polytechnic National University, Bandera str. 12, Lviv 79013, Ukraine
oksanamatsiyevska@gmail.com

[2] Institute of Architectural Engineering, Faculty of Civil Engineering, Technical University of Kosice, Vysokoskolska 4, 042 00 Kosice, Slovak Republic
peter.kapalo@tuke.sk

Abstract. Designing new and reconstructing existing water supply networks are based on water demand. Most of the existing water supply networks in Ukraine are designed for a higher amount of water flow in the pipes, which causes a decrease in the velocity of water, an increase in the duration of stay of water in the network and deterioration of water quality. Ukrainian water demand (average daily per year) of drinking water is higher than Slovakian: up to 35% for residential houses with water supply and sewage without baths; from 11% to 74% for houses with local water heaters; from 59% to 96.5% for houses with centralized hot water supply. For the majority of other consumers, the divergence of normative water consumption is insignificant. A comparison of water consumption in Ukraine and the Slovak Republic has been performed for the selected example in residential houses, industrial plants, irrigation of greenery and street sprinkling of surfaces. For the selected example, the total average water demand in Ukraine is higher than in the Slovak Republic by 44.41%, maximum water demand by 39.88%, minimum water demand by 9.98%.

Keywords: Drinking water · Annual water demand · Water consumption

1 Introduction

Climate change on earth and the way people live on earth reduce the quality of drinking water sources. In some areas on the ground there is a large shortage of drinking water supplies (Chalchisa et al. 2017). According to Roy et al. (2015), in the US there has recently been a greater emphasis on quality supply of drinking water by water companies. Empirical research (Powell and Yurchenko 2019) points to a dynamic picture of the development of private security in urban drinking water.

Designing new and reconstruction existing water supply networks are based on water demand. In particular, the value of demand of drinking water by inhabitants and workers of industrial enterprises (daily water consumption of the residential sector) will influence the determination of optimal diameters of pipelines. Recently, Ukrainians are taking less water from water supply networks due to the increasing tariffs for

Z. Blikharskyy et al. (Eds.): CEE 2019, LNCE 47, pp. 299–306, 2020.
https://doi.org/10.1007/978-3-030-27011-7_38

centralized water supply and sewage. Most of the existing water supply networks in Ukraine are designed for a higher amount of water flow in the pipes, which causes a decrease in the velocity of water, an increase the duration of stay of water in the network and deterioration of water quality. The overheated diameters of the pipes of the distribution network are the reason for the deterioration of the hydraulic performance of its work (Matsiyevska 2015; Goyal and Patel 2015).

Reduction of water consumption and rational use of water is one of the priority tasks of humanity today. A comparison of the current norms of drinking water consumption in Ukraine 100–285 l/day per one inhabitant (SBN V.2.5-64 2012; SBN V.2.5-74 2013) with the previous 125–350 l/day per one inhabitant (SN&R 2.04.02-84*) indicates their decrease by about 25%. In other countries, the average value of specific water consumption in residential buildings fluctuates between 150 and 200 l/day per inhabitant (Novitska and Genich 2014). In particular, l/day per inhabitant: Norway – 175; Austria – 165; Latvija – 170; Estonia – 165; Finland – 151; Lithuania – 140; Spain – 126; Sweden – 194; Belgium – 108 (Maslak et al. 2014).

2 Aim

The aim of the research is to analyze the water demand for different categories of consumers in Ukraine and the Slovak Republic.

3 Method

The research methodology consisted of comparative analysis of the water demand for different categories of consumers in Ukraine and the Slovak Republic, and analysis of calculation results of the total daily water flow rate from the water supply network of the settlement (without taking into account the water demand for fire extinguishing).

4 Results

The values of the estimated daily water demand (average per year) for different consumers of the two countries are given in Tables 1 and 2.

The daily drinking water demand is indicated in Table 1 in a certain range of values because the territory of Ukraine according to architectural and construction climatic conditions is divided into four districts (SSTU-N B V.1.1-27 2010). For further calculations, the northwestern architectural and construction climatic region of Ukraine (district I) has been selected.

In Ukraine the average daily rate of water consumption per inhabitant is 25–60 l/day for residential areas with water consumption from standpipe water supply point (SBN V.2.5-74 2013).

Some standards of water consumption of the Slovak Republic are not taken into account in Ukraine. The water demand can be reduced by 25% if the consumer lives in a mansion house, or water consumption is accounted for separately for each apartment,

Table 1. Drinking water daily demand (average per year), l/day per inhabitant

Residential buildings	Ukraine (SBN V.2.5-64 2012, SBN V.2.5-74 2013)	The Slovak Republic (Decree no. 684 2008)
With plumbing and sewage without baths	100–110	100
The same with gas supply	120–135	
With plumbing, sewage and baths with solid fuel fired water heaters	150–170	135
Same with gas water heaters	210–235	
With centralized hot water and seated baths	230–260	145
The same with baths longer than 1500 mm	250–285	

or a portion of the apartments is not connected to the sewage network. For apartments in a mansion house with over-standard sanitary equipment (for example, a swimming pool), the water demand is increased by 15% (Decree no. 684 2008).

The analysis of Table 1 shows that Ukrainian (taking into account the architecturally-structural and climatic regionalization of the territory of Ukraine) daily water demands (per year) exceed Slovakian by: up to 35% for residential houses with water supply and sewage without baths; from 11% to 74% for houses with local heaters of water; from 59% to 96.5% for houses with hot water supply.

In Ukraine, the demand of drinking water per worker at industrial enterprises is: ordinary (cold) industrial room, 25 l/shift; workshops with a heat output of more than 85 kJ per 1 m^3/h (hot industrial room), 45 l/shift. In addition, for enterprises with a special sanitary regime, taking into account water consumption for one shower q = 500 l/h. It is believed that the employees of the enterprises shower for 45 min after the shift (SBN V.2.5-64 2012).

Some specific water needs for direct use in the Slovak Republic different from Ukrainian: 1 person: for drinking, 5 l/shift; for the kitchen, 25 l/shift. Specific demand in water for indirect use (washing, showering): simultaneously hot and dirty work, 220 l/shift; work with dirty operations and dusty operations or hot and clean operations, 120 l/shift; A company with a purely clean operation, 50 l/shift (Decree no. 684 2008).

Estimated water demands are shown in Table 2 for the main consumers and include all additional consumption (for maintenance staff, showers for maintenance staff, visitors, room cleaning, etc.). There are increasing coefficients with range from 1.1 to 1.2 for the 3rd and 4th climatic regions of Ukraine.

The analysis of Table 2 shows that the divergence of the values of estimated water daily demand (average for year) for the vast majority of consumers is not significant.

Output data and the results of calculating the total daily water consumption from the water supply network of the settlement (without water consumption for fire-fighting) are given in Tables 3, 4, 5 and 6. The calculations were carried out on the basis of normative documents for Ukraine (SBN V.2.5-64 2012) and the Slovak Republic (Decree no. 684 2008).

Table 2. Estimated daily water demand (average for year) l/day per unit of measurement

Consumer	Unit of pressure measurement	Ukraine (SBN V.2.5-64 2012)	The Slovak Republic (Decree no. 684 2008)
Hostels	1 inhabitants	90–140	200
Sanatoriums with baths in all rooms	1 bed	200	250
Sports complexes	1 seats	60–200	60
Special schools with a daytime stay of children	1 child	40–80	60
College	1 student	20	25
University	1 student	20	40
Scientific-research institutes, institutions of management	1 worker	15	60
Food enterprise (canteen, restaurant, etc.)	1 dish	2–10	25
Cinemas, entertainment industry	1 person	8	5
Stadiums and Gyms: – for viewer;	1 seats	3	3
– for athletes (with demand for having bath);	1 person	50/100	60
Laundry	1 kg of dry linen	40–75	60
Sprinkling surfaces of sports facilities	1 m^2	1.5	1.2
Sprinkling of streets	"—"	0.5	1.0
Irrigation of greenery	"—"	3–6	1.0

Two levels of improvement of residential buildings in the settlement are adopted: (1) with water supply, sewage and baths with local water heaters; (2) with centralized hot water supply and baths with a length of more than 1500 mm. Output data and results of calculating the daily water consumption of the inhabitants of the city are shown in Table 3.

Industrial enterprises are within the territory of the city. We accept that the flow of water to the technological needs of industrial enterprises is constant throughout the year and the same for both Ukraine and the Slovak Republic. Therefore, in further calculations and comparisons it is not taken into account.

Output data and results of calculating the daily consumption of drinking water by industrial enterprises are shown in Table 4.

In the Slovak Republic, the average daily water demand is 45% lower than in Ukraine. Also, the maximum daily water requirement is 40% lower.

In selected industrial plants in Ukraine, the annual water demand is only 5.64% less than in the Slovak Republic.

Table 3. Daily water consumption in residential buildings

Level of improvement residential buildings	Number of inhabitants in thousand	Water demand, l/day per 1 inhabitant	Unaccounted consumption		Coefficient of daily non-uniformity consumption		Daily water consumption, m^3/day		
			%	γi	K_{max}	K_{min}	average	maximum	minimum
Ukraine									
1	55	210	10	1.1	1.2	0.8	12705	15246	10164
2	55	250	10	1.1	1.2	0.8	15125	18150	12100
Total	**110**						**27830**	**33396**	**22264**
The Slovak Republic									
1	55	135	—	—	1.3	—	7425	9653	—
2	55	145	—	—	1.3	—	7975	10368	—
Total	**110**						**15400**	**20021**	

Table 4. Daily drinking water consumption in industrial enterprises

Hot industrial room			Cold industrial room			Taking a shower						Total water consumption, m^3/day
Number of workers on shift, person	Water demand, l/shift per worker	Water consumption, m^3/day	Number of workers on shift, person	Water demand, l/shift per worker	Water consumption, m^3/day	Number of worker that taking a shower, person		Number of worker for 1 shower, person		Water demand for 1 shower, l/hour	Water consumption, m^3/day	
						Hot industrial room	Cold industrial room	Hot industrial room	Cold industrial room			
Ukraine												
750	45	33.75	1100	25	27.50	375	330	7	15	500	75.57	**136.82**
The Slovak Republic												
750	120	90	1100	50	55	—	—	—	—	—	—	**145.00**

Output data and calculated results of the daily consumption for irrigation of greenery and sprinkling of street surfaces are shown in Table 5.

Table 5. Daily water consumption on watering surface

Type of surface	Watering area, ha	Water demand, l/m^2	Number of waterings during the day	Daily water consumption on watering surface, m^3/day		
				Average	Maximum	Minimum
Ukraine						
Impervious coverage	5.0	0.5	1	25.0	25.0	0
Greenery surface	5.0	3.0	1	150.0	150.0	0
Total				**175.0**	**175.0**	**0**
The Slovak Republic						
Impervious coverage	5.0	1.0	1	50.0	50.0	0
Greenery surface	5.0	1.0	1	50.0	50.0	0
Total				**100.0**	**100.0**	**0**

For irrigation of greenery and sprinkling of street surfaces in Ukraine there is a yearly need of 42.86% more than in the Slovak Republic.

Total daily water consumption from water supply settlement networks are shown in Table 6.

Table 6. Total daily water consumption from water supply networks

Consumer	Total daily water consumption, m^3		
	Average	Maximum	Minimum
Ukraine			
Residents	27830.00	33396.00	22264.00
Industrial enterprises	136.82	136.82	136.82
Watering	175.00	175.00	0.00
Total	**28141.82**	**33707.82**	**22400.82**
The Slovak Republic			
Residents	15400.00	20021.00	20021.00
Industrial enterprises	145.00	145.00	145.00
Watering	100.00	100.00	0.00
Total	**15645.00**	**20266.00**	**20166.00**

From the results shown in Tables 3, 4, 5 and 6 we can state, that for the selected sample, the total average water demand in Ukraine is higher than in the Slovak Republic by 44.41%. The difference between results for two countries in total maximum water demand is 39.88%, in minimum water demand is 9.98%.

5 Scientific Novelty and Practical Significance

The obtained results outline directions for further research: studying the structure of water consumption for different needs; scientific substantiation of the drinking water demand for residents of houses and workers of industrial enterprises. At the same time, the main attention should be paid to the behavior of the consumer (Romano et al. 2014; Zadeh et al. 2014).

From a practical point of view the reduction of water consumption standards will reduce the diameters of water supply networks and, consequently, substantially reduce the material content of water supply systems and save raw materials and energy resources.

6 Conclusions

Ukrainian water demand (average daily per year) of drinking water is higher than Slovakian: up to 35% for residential houses with water supply and sanitation without baths; from 11% to 74% for houses with local water heaters; from 59% to 96.5% for houses with centralized hot water supply. For the majority of other consumers, the divergence of normative water consumption is insignificant. In the Slovak Republic, the average daily water demand is 45% lower than in Ukraine. Also, the maximum daily water requirement is 40% lower.

In selected industrial plants in Ukraine, the annual water demand is only 5.64% less than in the Slovak Republic.

For irrigation of greenery and sprinkling of street surfaces in Ukraine there is a yearly need of 42.86% more water than in the Slovak Republic.

From the results shown in Tables 3, 4, 5 and 6 we can state, that for the selected sample, the total average water demand in Ukraine is higher than in the Slovak Republic by 44.41%, maximum water demand by 39.88%, minimum water demand by 9.98%.

References

Maslak VN, Nasonkina NN, Gutarova MYu, Yakovenko KA, Chumak AV (2014) Analysis of water consumption standards in Ukraine. MOTROL Comm Mot Energy Agric 16(6):43–52 (in Ukrainian)

Matsiyevska OO (2015) Study of water quality in the distribution network of the centralized water supply system in the city of Lviv. East-Eur J Enterp Technol 6(6):62–70. https://doi.org/10.15587/1729-4061.2015.56225 (in Ukrainian)

Novitskaya OS, Genich OV (2014) Investigation of the structure of water consumption in residential buildings. Bull Natl Univ Water Manag Environ Prot 1:153–159 (in Ukrainian)

Chalchisa D, Megersa M, Beyene A (2017) Assessment of the quality of drinking water in storage tanks and its implication on the safety of urban water supply in developing countries. Environ Syst Res 6(12):1–6. https://doi.org/10.1186/s40068-017-0089-2

Goyal RV, Patel HM (2015) Analysis of residual chlorine in simple drinking water distribution system with intermittent water supply. Appl Water Sci 5:311–319. https://doi.org/10.1007/s13201-014-0193-7

Powell J, Yurchenko Y (2019) The evolution of private provision in urban drinking water: new geographies, institutional ambiguity and the need for political economy. New Polit Econ 1–16. https://doi.org/10.1080/13563467.2018.1562432

Romano G, Salvati N, Guerrini A (2014) Estimating the determinants of residential water demand in Italy. Water 6:2929–2945. https://doi.org/10.3390/w6102929

Roy S, Phetxumphou K, Dietrich AM, Estabrooks AA, You W, Davy BM (2015) An evaluation of the readability of drinking water quality reports: a national assessment. J Water Health 13(3):645–653. https://doi.org/10.2166/wh.2015.194

Decree no. 684/2006 Coll. (2008) Decree of the Ministry of Environment of the Slovak Repulic, which lays down details on technical requirements for the design, project documentation and construction of public water supply and public sewerage systems (in Slovak)

Zadeh SM, Hunt DV, Rogers CD (2014) Socio-technological influences on future water demands. Water 6:1961–1984. https://doi.org/10.3390/w6071961

SBN V.2.5-64 (2012) Internal plumbing and sewerage. Part I. Design. Part-II. Construction (in Ukrainian)

SBN V.2.5-74 (2013) Water supply. Outdoor networks and facilities. Main provisions of the project (in Ukrainian)

SN&R 2.04.02-84 *. Water supply. Outdoor networks and facilities(in Russian)

SSTU-N B V.1.1-27 (2010) Protection from hazardous geological processes, harmful exploitative influences, from fire. Construction Climatology (in Ukrainian)

Sustainability of Buildings and Its Support Through Innovative Technologies

Peter Mésároš(✉), Jana Smetanková, and Katarína Krajníková

Department of Construction Technology and Management,
Faculty of Civil Engineering, Institute of Construction Technology
and Management, Technical University of Košice, Vysokoškolská 4,
042 00 Košice, Slovakia
peter.mesaros@tuke.sk

Abstract. Today, more and more attention is dealt with environmental issues and sustainable construction in society. The construction industry is demanded on the amount of energy consumed and greenhouse gas emissions. Using sustainable construction is a fundamental concern for society to lead to environmentally sustainable future. Sustainability contains social, economic and environmental aspects. In last years, we have seen an increase in innovation in the construction industry sector. Support in buildings construction′s innovation is one of the major ways to realization, manage and operate the building's life cycle. Making these processes more efficient leads to the promotion of sustainable construction. This article is pointing out the different aspects of sustainability and provides an overview of innovative technologies used in the construction sector and focuses on their contribution to promoting sustainable construction. Special attention is dealt with technology of building information modelling, especially its benefits and tools to support building sustainability.

Keywords: Innovation · Construction sustainability · Innovative technologies · Building information modelling · Support the sustainability of buildings · Benefits of BIM

1 Introduction

The concept of sustainability is the endurance of biological systems, so the time in which these systems are diverse and productive. However, the current definition of sustainability is different. The concept of sustainability is understood to support the development of sustainable models that are needed to ensure the survival of the human race and planet Earth (Sustainability Degrees 2019). Sustainability is a development that satisfies the needs of the present without the risk that future generations will not be able to meet their own needs (Brundtland Report - Our Common Future 1987).

Sustainability is the principle that no more can be consumed than can be reproduced, recycled and delivered in the future (Living Sustainably is our future 2019).

The concept of sustainable development is the systematic expansion of human rights by the opening equal opportunities for development for all people on this planet and for future generations (Wiegandt 2018).

Z. Blikharskyy et al. (Eds.): CEE 2019, LNCE 47, pp. 307–314, 2020.
https://doi.org/10.1007/978-3-030-27011-7_39

For this purpose, it is necessary to highlight, specifically:

- environmental protection,
- social responsibility and,
- economic practice (Sustainability Degrees 2019).

At present, the pressure is increased how to promote sustainable business. Companies are facing increasing pressure to develop long-term plans and tactics that will be more respectful for the environment, employee welfare and future generations' perspectives. As it was mentioned, it is also necessary to take into account the possibility of increasing profits, financing innovation and increasing market share (Sustainability Degrees 2019).

The construction industry requires high energy consumption and produces a lot of emissions of greenhouse gas (Soltani 2016).

Applying a sustainable exhibition becomes a fundamental concern that needs to be underpinned in order to move towards an environmentally sustainable future (Ortiz et al. 2009).

2 Innovation and Support of Building Sustainability

Innovations present a key role in both short and long term economic, social and environmental sustainability. Construction sustainability is the application of the principles of sustainable development in the construction industry.

Sustainable construction is a set of processes that help the industry to gain profitability and competitiveness and to:

- improve quality of life and customer satisfaction,
- increased flexibility and potential to meet users' needs,
- promoting environmental and social environments,
- efficient use of resources (Raynsford 2000)

Sustainability includes social, economic and environmental aspects. Some research reported also other sustainability dimensions such as technical sustainability, cultural sustainability, community sustainability and management sustainability (Ofori 1998; Hill and Bowen 1997; CIB 1999; Ashley et al. 2003).

Criteria representing the social dimension of sustainable construction are for example:

- protecting and promoting human health,
- through a healthy and safe working environment,
- communication and promotion of cooperation between participants,
- education and development - training, promoting skills,
- promoting the social inclusion of workers,
- promoting safety and employee satisfaction and the others (Sourani and Sohail 2019).

Criteria representing the economic dimension of sustainable construction are for example:

- financial availability,
- sustainability of economic growth,
- support lifecycle costing,
- creating and maintaining a high stable level of employment,
- increasing investment to support renewable resources,
- investments in organic products and the use of renewable resources,promoting flexibility, profitability, competitiveness, productivity and the other (Sourani and Sohail 2019).

Criteria representing the environmental dimension of sustainable construction are for example:

- saving of energy, water and land conversation,
- preservation of materials, or reuse and recycling,
- use of resources,
- promoting and using renewable energy,
- minimizing water, land and air pollution, including noise elimination,
- maintaining and increasing biodiversity,
- creating a healthy, non-toxic environment,
- waste minimization and the others (Sourani and Sohail 2019).

Promoting innovation of building is one of the most important ways of planning, realization and managing the building's life cycle. Use and implementation of innovation has a positive impact on construction industry (Mandičák and Behúnová 2016). There are a lot of innovative technologies that help for planning, realization build and manage buildings. (Higgins 2019).

Innovative methods include, for example Embedded sensors. Embedded sensors bring a lot of opportunities to acquisition and manage safety, material performance and operational workflow data. Intelligent equipment-sensors and cameras that are fixed on a load-bearing structure are linked to software for building management and provides image of progress in real time. Embedded sensors impact construction process while provided important information needed for operational management. Embedded sensors provide extremely important information to support the environmental sustainability and performance of the construction sector (Higgins 2019).

Another innovative technology is Artificial Intelligence (AL) and Machine Learning. AL systems are machines that simulate intelligent human behavior. The computer system performs tasks that normally require human intelligence, as well as visual perception, choosing etc. (Skymind 2019). The AL system categorizes data faster than the human operator. This will reduce the time which is needed to map issues at the workplace (Higgins 2019).

Machine Learning is a subset of Artificial Intelligence, but not vice versa. Machine Learning is learning knowledge graphs and expert systems that are able to modify more data (dynamic determination) and they do not require human intervention to carry out certain changes (Higgins 2019; Skymind 2019).

Machine Learning creates an opportunity for new technologies, such as for example Smartvid.io. Smartvid.io visualizes information from cameras, which is located into the workplace, and uses artificial intelligence to identify potential risks and security violations (Higgins 2019).

Important innovative technology is virtual reality. Virtual Reality enables real-time comparison of 3D models with real space, supports location data analysis, virtual tutorials, etc. Virtual reality increases adaptation, security, accuracy, and promotes collaboration between building process participants (Higgins 2019).

Building information modeling is an emerging innovative technology. Building information modeling is a digital tool used throughout the building's life cycle. BIM supports the visualization, planning, communication and collaboration of project participants (Eastman et al. 2011; Kymmell 2008)

BIM reproduces physical and functional features and provides the ability to modify and implement changes in project development and realization (Smith 2007).

Building information modelling provides significant technical advantages. A digital representation of the physical and functional features of a device that allows the user transfer design data and specify it between different software applications. Building information modelling tools provide complex information about the building that helps them in the process of designing, realization and using the building, what supporting so-called knowledge management. Another benefit is the delivery of information in standardized formats, what encouraging collaboration between participants of construction. BIM models are created in the Industry Foundation Classes-IFC format. The IFC format is a neutral open format. We can integrate the IFC model into multiple software applications (Hietanen and Final 2006, Lin et al. 2013).

BIM also helps with facility management, warranty and service information monitoring, quality control, monitoring and evaluation, energy management, crisis management and other process (Ghaffarianhoseinia et al. 2017).

3 BIM and Sustainable Construction

Building industry's effort to promote sustainability through building information modelling is growing. BIM makes work easier through an integrated platform that brings technical benefits to project completion, but also improves sustainability throughout the project lifecycle (Gilder and Elmualim 2014).

BIM provides the conditions to promote better design. The aim is to facilitate the needs of society by enabling stakeholders to explore and develop a 3D design and to promote collaboration between construction participants.BIM supports an integrated public procurement system and strengthens social sustainability (Chong et al. 2017; NIBS 2015; Grilo and Goncalves 2011).

In terms of economic sustainability, BIM increases life-cycle cost savings. There are studies that have analyzed the costs and benefits of implementing BIM projects. These researches showed cost savings of 6.92% (Lu et al. 2014)

BIM is an important tool to promote better spatial proposal and related analysis of potential environmental impacts, air circulation, building ecosystems, energy simulation, and so on. (Bonenberg and Xia 2015).

Research with the title BIM adoption towards the sustainability of construction industry in Indonesia (Zhabrinna et al. 2018). Research has focused on exploring the impacts of building information modelling on the sustainability of construction in Indonesia. Research has confirmed that BIM improves the sustainability of the construction industry. The survey focused on 5 areas, specifically reduce construction waste, early detection of conflict in design and planning, reduce final cost, delay avoidance and life asset management improvement. Respondents were addressed and answered how BIM affects those areas. The survey showed the following findings. The implementation of BIM is positively reflected in particular early detection of conflict in design and planning, following areas delay avoidance, life asset management improvement, reduce final cost and reduce construction waste. The summary results of survey are shown in the Table 1.

Table 1. Survey results – BIM adoption toward the sustainability of construction industry in Indonesia (Zhabrinna et al. 2018)

Benefits of BIM	Strongly disagree	Disagree	Neutral	Agree	Strongly agree	Rank
Reduce construction waste	2,5%	0%	40%	38%	20%	5.
Early detection of conflict in design and planning	2,5%	2,5%	23%	38%	35%	1.
Reduce final cost	2,5%	2,5%	25%	48%	22,5%	4.
Delay avoidance	0%	2,5%	23%	50%	25%	2.
Life asset management improvement	0%	2,5%	25%	48%	25%	3.

We can see that 38% of respondents agreed and until 35% strongly agreed that BIM assists in early detection of planning and design conflicts.

Research with the title Building information modelling (BIM) for sustainable building design (Wong and Fan 2013) explored building information modelling on construction sustainability. The main benefits of using BIM are mainly creation constructional design of building with regard to sustainable construction and design optimization. The summary results of survey are shown in the Table 2.

The main benefits of using BIM are creation integrated project and delivery, design optimization, better communication and coordination with more accurate and efficient work. These benefits are reflected in selected parameters such as Energy Reduction, Water conservation, Wastage Lessen, IEQ improvement (Wong and Fan 2013).

Another method which is promoting to increase productivity, efficiency and promote sustainability is creation so-called BIM Object. BIM Object are digital descriptions of products and materials. Individual BIM objects are grouped in so-called BIM Library. BIM Libraries provide differently portals, for example BIM Object Cloud, BIM components, BIM store, Arcat, Revit City and other. The libraries contain

Table 2. Summary of BIM for sustainable design (Wong and Fan 2013)

	Functions	Benefits	Sustainable achievements
BIM Inherency + BIM-based analysis tools	3D Model Visualization Clash detection Compliance with regulations Energy analysis Solar analysis Thermal analysis Lighting design Acoustic analysis Ventilation and air Flow Materials/Resource Management	Integrated project Delivery Design Optimization Better communication and Coordination More accurate and efficient	Energy Reduction Water conservation Wastage lessen IEQ Improvement

individual objects, which, however, contain only graphical characteristics and basic parameters. Components don't provide information about time, cost, sustainability and facility management and other information's.

4 Conclusion

Interest in sustainable construction is progressing. Sustainability is a principle that aims to produce, regenerate and provide future generations with as many resources as possible for their future development. There are a lot of innovative technologies that greatly contribute to building sustainability. Building information modelling is a useful tool in designing, managing and planning a building process. BIM provides the necessary information how to improve the design and realization of buildings. Building information modelling brings significant benefits in the building industry, in economic, social, or environmental aspect. The main aim is still to develop and integrate BIM technology to help protect the environment and natural resources. It is necessary to develop scientific and professional work that directly supports the promotion of sustainable construction in the design, implementation, realization and using of buildings.

Acknowledgements. "This work was supported by the Slovak Research and Development Agency under the contract no. APVV-17-0549". The paper presents the partial research results of project VEGA - 1/0828/17 "Research and application of knowledge-based systems for modelling cost and economic parameters in Building Information Modelling"

References

Ashley R, Blackwood D, Butler D, Davies J, Jowitt P, Smith H (2003) Sustainable decision making for the UK water industry. Proc Inst Civ Eng, Eng Sustain 156(ES1):4149

Bonenberg W, Xia W (2015) Green BIM in sustainable infrastructure. Procedia Manuf 3:1654–1659

Brundtland report - our common future (1987) World commission on the environment and development. https://www.tandfonline.com/doi/abs/10.1080/00139157.1987.9928891?journalCode=venv20. accessed 18 May 2019

Chong HY, LeeC Y, Wang X (2017) A mixed review of the adoption of building information modelling (BIM) for sustainability. J Cleaner Prod 142:4114–4126

CIB (1999) Agenda 21 on Sustainable Construction. CIB Report Publication 237. CIB

Eastman CM, Eastman C, Teicholz P, Sacks R (2011) BIM handbook: a guide to building information modeling for owners, managers, designers, engineers and contractor. Wiley, Hoboken. ISBN 9780470541371

Ghaffarianhoseinia A, Tookeya J, Ghaffarianhoseinib A, Naismitha N, Azhard S, Efimovaa O, Raahemifarb K (2017) Building information modelling (BIM) uptake: clear benefits, understanding its implementation, risks and challenges. Renew Sustain Energ Rev 75: 1046–1053

Gilder A, Elmualim J (2014) BIM: innovation in design management, influence and challenges of implementation. Architect Eng Des Manage 10:183–199

Grilo A, Goncalves RJ (2011) Challenging electronic procurement in the AEC sector: a BIM-based integrated perspective. Auto Constr 20:107–114

Hietanen J, Final S (2006) IFC model view definition formatInt Alliance Inter

Higgins A (2019) Major construction technology trends to watch in 2019. https://connect.bim360.autodesk.com/construction-technology-innovation-2019. accessed 18 May 2019

Hill RC, Bowen PA (1997) Sustainable construction: principles and a framework for attainment. Constr Manage Econ 15(3):223239

Kymmell W (2008) Building information modelling: planning and managing construction projects with 4D CAD and simulations. McGraw Hill Construction, New York. ISBN 9780071494533

Lin Y-H, Liu Y-S, Gao G, Han X-G, Lai C-Y, Gu M (2013) The IFC-based path planning for 3D indoor spaces. Adv Eng Inform 2013(27):189–205

Living sustainably is our future (2019) The new challenge with great opportunities for us all. https://www.sustainability-yes.ch/en/?gclid=EAIaIQobChMIxLfEu8if4gIV1F3Ch3LuwAwEAMYASAAEgJykvD_BwE. accessed 15 May 2019

Lu W et al (2014) Cost-benefit analysis of building information modelling implementation in building projects through demystification of time-effort distribution curves. Build Environ 82:317–327

Mandičák T, Behúnová A (2016) Impact of controlling systems on direct costs of construction projects. Acta Tecnologia 2(2):11–15 ISSN 2453-675X

NIBS (2015) National BIM Standard-United States® (NBIMS-us™) Version 3, Washington

Ofori G (1998) Sustainable construction: principles and a framework for attainment comment. Constr Manage Econ 16(2):141145

Ortiz O, Castells F, Sonnemann G (2009) Sustainability in the construction industry: a review of recent developments based on LCA. Constr Build Mater 23:28–39. https://doi.org/10.1016/j.conbuildmat.2007.11.012

Raynsford N (2000) Sustainable construction: the government's role. Proc Institution Civ Eng, Civ Eng 138(S2):1622

Skymind (2019) Artificial Intelligence (AI) vs. Machine Learning vs. Deep Learning. https://skymind.ai/wiki/ai-vs-machine-learning-vs-deep-learning. accessed 18. May 2019

Smith S (2007) Using BIM for sustainable design. AEC Cafe Weekly. at: http://www10.aeccafe.com/nbc/articles/view_weekly.php?section=Magazine&articleid=386029&printerfriendly=1. accessed 18. May 2019

Soltani S (2016) The contributions of building information modelling to sustainable construction. World J Eng Technol 4(2):193–199

Sourani A, Sohail M (2019) A review of sustainability in construction and its dimensions p. 536–547. http://www.irbnet.de/daten/iconda/CIB6623.pdf. accessed 18 May 2019

Sustainability Degrees (2019) What is Sustainability? https://www.sustainabilitydegrees.com/what-is-sustainability/. accessed 18 May 2019

Wiegandt, K (2018) Uwe Schneidewind. Die grosse transformation - eine einführung in die kunst gesellschaftlichen wandels. Fischer Taschenbuch; Auflage: 2. ISBN-10 3596702593

Wong K, Fan Q (2013) Building information modelling (BIM) for sustainable building design, In: Facilities, vol. 31. No. ¾, pp. 138–157, Emerald Group Publishing Limited 0263-2772, https://doi.org/10.1108/02632771311299412. at: https://www.researchgate.net/publication/263171328_Building_information_modelling_BIM_for_sustainable_building_design. accessed 18 May 2019

Zhabrinna I, Richard J, Davies M, Pratama MA, Yusuf M (2018) BIM adoption towards the sustainability of construction industry in Indonesia. MATEC Web of Conf ICRMCE 195:06003

Optimization of the Sample Preparation Method for the Determination of Biofilm in the Water Supply System

D. Papciak(✉), A. Domoń, A. Wojtuś, and M. Zdeb

Department of Water Purification and Protection,
Rzeszow University of Technology, 12 Powstańców Warszawy Street,
35-959 Rzeszow, Poland
dpapciak@prz.edu.pl

Abstract. In drinking water supply systems, formation of biofilm is associated with technical and hygiene problems which may cause an increased health risk. Techniques of monitoring therefore need to be optimized and adapted to yield valid and informative data. The aim of this research was to develop an effective method of biofilm sampling from the surface of the pipe material in order to perform quantitative analysis and identify the advantages and limitations of the quantitative determination of the number of microorganisms using three different microbiological methods: classical heterotrophic plate counts (HPC), flow cytometry, luminometry. Based on the conducted research, the following was concluded: the best method to detach the biofilm from the surface of the coupons is mechanical separation with a sterile swab. The swab was placed in 30 ml of Ringer's solution and exposed to ultrasounds, the optimal duration of ultrasounds, is 60 s for sonicator power of 28 W. Attention should be given to the volume of solutions, which is selected depending on the operating time and the sonicator power (30 ml is optimal). Due to the simplicity and speed of obtaining results, the luminometric ATP measurement has been established as the best method for the quantification of microorganisms in biofilms.

Keywords: Biofilm · Drinking water distribution system · Quantify biofilm microorganisms

1 Introduction

Water supplied to customers must be of adequate quality, not only on entry into the water supply network but also at the receiver end. Irregular changes in the quality of the supplied water can cause a physicochemical and microbiological destabilization of the pipeline material and the presence of growths. This phenomenon may result in a release of sediments, substances, and microorganisms created over decades, thus posing health and aesthetic problems related to turbidity and discoloration of tap water (Manuel et al. 2010; Gomes et al. 2014). The main cause of changes in the qualitative composition of water during its transport is, apart from the technical condition of the water distribution system, the lack of physical, biological, and chemical stability of the water leaving the

Z. Blikharskyy et al. (Eds.): CEE 2019, LNCE 47, pp. 315–322, 2020.
https://doi.org/10.1007/978-3-030-27011-7_40

waterworks plant (Tsvetanova and Hoekstra 2009; Douterelo et al. 2016; Pietrucha-Urbanik et al. 2017; Papciak et al. 2018; Waller et al. 2018). In order to both obtain and maintain the quality of tap water, materials that are resistant to corrosion phenomena and do not support the growth of microorganisms in the form of biofilms covering the surface of pipelines are sought when building distribution systems (Yu et al. 2010).

In distribution systems for water intended for consumption, over 95% of microorganisms occur on internal surfaces of water pipes in the form of biofilms. The biofilm consists mainly of opportunistic bacteria but its protective and nutritive properties also make it an ideal place for the development of both intestinal viruses, slowly growing potential pathogens, fecal bacteria or Legionella bacteria (Simões et al. 2010). Microorganisms which are relatively pathogenic deposit on biofilms and multiply, posing a threat to the health of tap water consumers (Brember et al. 2001). In addition, microorganisms of biofilms undergo metabolic processes producing substances which can adversely affect water quality (Ying et al. 2010).

Numerous studies aimed at identifying materials with the lowest levels of inhabitance by microorganisms which may affect the health and safety of water consumers are proof of this.

According to available literature, there are several ways to quantify biofilm microorganisms (Zamorska and Papciak 2008). Classical HPC (heterotrophic plate counts) methods, luminometric ATP assay (Douterelo et al. 2016; Waller et al. 2018) as well as direct counting of bacteria using fluorochromes (e.g. acridine orange) are used to determine the number of bacteria in biofilm (Douterelo et al. 2019). The first stage, during which the biofilm is detachment from the material of the installation, is essential but also the most complicated. Physical methods are most commonly used for this purpose. The most popular are: sonication of fragments of material with biofilm (Górka et al. 2008; Waller et al. 2018; Zhang et al. 2018), shaking using glass beads as well as by using a sterile cotton swab, a scraper or a knife (Boe-Hansen et al. 2003; Srinivasan et al. 2008; Jang et al. 2011). However, a doubt remains as to which of them is the most effective at separating biofilm from the porous structure of materials; often, there is a lack of data on the sonicator power and the duration of ultrasounds which makes it impossible to repeat and compare the results of analyses.

The aim of this research was to develop an effective method of biofilm sampling from the surface of the pipe material in order to perform quantitative analysis and identify the advantages and limitations of the quantitative determination of the number of microorganisms using three different microbiological methods.

2 Materials and Methods

Subject of Study: The coupons made of galvanized steel, $1 \times 1 \times 0{,}1$ cm were placed in sterile glass flasks which were filled with 150 ml of untreated surface water. The coupons was stored in a darkness at a temperature of 18–22°C for 40 days to enable the formation of biofilm. Due to the fact that the biofilm formed on both sides of the material, it was assumed that the surface colonized by microorganisms was 2 cm^2 (sides were omitted). The biofilm was separated from the surface of the material by two methods: (1) sonication and (2) a sterile cotton swab + sonication. Different durations of ultrasound and various physiological fluids were used in the study.

Detachment the Biofilm from the Surface of the Material: After 40 days, the coupons were transferred to 30 ml of Ringer's solution (composition: 8.5 g NaCl/l, 0.044 g Na_2HPO_4/l, 0.023 NaH_2PO_4/l) or 30 ml of PBS solution (composition: 8 g NaCl/l, 0.2 g Na_2HPO2/l, 1,44 g NaH_2PO_4/l, 0.2 g KCl/l). Filled beakers were placed in an ice bath to prevent excessive temperature rise during sonication (which has a direct impact on the activity of bacteria) and exposed to ultrasounds for 30 s, 60 s, and 75 s, respectively. A Vibra-cell Labo Plus 28 sonicator was used in the study. The biofilm was detached from the sample surface also using a sterile cotton swab. The swab was placed in 30 ml of Ringer's solution and exposed to ultrasounds (60 s). As in the previous case, during the sonication process, the beaker was placed in an ice bath.

Quantitative Determination of Microorganisms in the Biofilm: The quantitative determination of microorganisms in the biofilm was carried out using different methods:

(1) **ATP measurement** using LuminUltra Photonmaster Luminometer. To determine: (ATPo) - the total number of microorganisms, (ATPz) - the number of dead microorganisms, (ATPo - ATPz) the number of living microorganisms. The result was given in RLU (Relative Light Units). Taking into account the surface of 2 cm^2, the obtained values were converted to RLU/ml/cm^2.
(2) **Flow cytometry** using a Cy Flow Cube 8 cytometer manufactured by Sysmex Partec. In order to determine the total number of microorganisms, a dye called Sybr Green was used, while propidium iodide enabled the number of dead microorganisms to be determined. The result was expressed as the number of particles per 100 μl of the test sample. Taking into account the surface of 2 cm^2, the obtained values were converted to number of particles/1 ml/1 cm^2.
(3) **HPC methods** using A and R2A agar. In order to determine the number of mesophilic bacteria present on the agar, the Petri dishes were placed in an incubator (37°C) for 48 h, while in the case of the psychrophilic bacteria (22°C), samples were incubated for 72 h. The total number of bacteria on the R2A agar was determined with the use of extended incubation over 7 days. The results converted to CFU/ml/cm^2, taking into account the surface of the coupons of 2 cm^2.

3 Research Results and Discussion

3.1 Selection of Optimal Duration of Ultrasounds in Ringer's Solutions

The duration of ultrasounds is an important parameter in detaching the biofilm from the surface of the material, affecting the results of the quantitative analysis of microorganisms. Too long an exposure to ultrasounds causes excessive fragmentation of cell structures, while too short an exposure prevents the complete detachment of the biofilm from the test sample. The highest number of microorganisms, regardless of the counting method, was reported after a sonication time equal to 60 s. Total ATP for the time indicated as optimum was 22263 [RLU/ml/cm^2], while for extracellular ATP (the sample after filtering) it was 8902 [RLU/ml/cm^2]. The difference between total and extracellular ATP, indicating the number of living microorganisms, was 13361 [RLU/ml/cm^2] (Fig. 1). The test results obtained by flow cytometry using SYBR Green fluorescent dye for all tested times (30 s, 60 s, and 75 s) showed minor differences and amounted to, respectively: 9168305, 9067320 and 8873200 [particles/ml/cm^2]. However, the number of living microorganisms was the highest in the case of 60-s sonication, being 8832095 [particles/ml/cm^2] (Fig. 1).

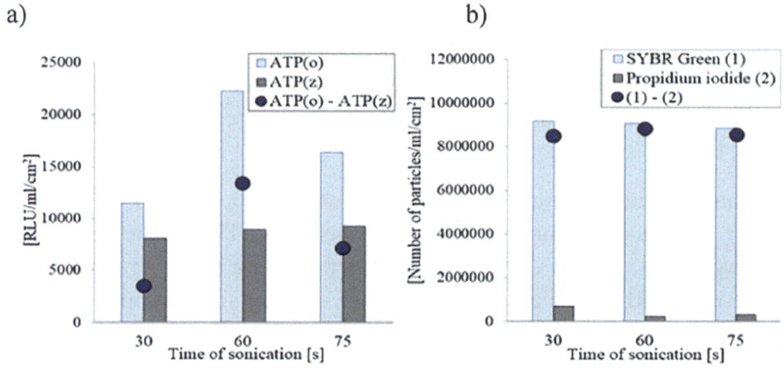

Fig. 1. Measurement of ATP (a) and the number of microorganisms by flow cytometry (b) depending on the time of sample sonication

The number of mesophilic bacteria was highest for the variant including a 60-s sonication and amounted to 45 [CFU/ml/cm^2]. The share of psychrophilic bacteria was higher with their maximum count equal to 490 [CFU/ml/cm^2] (as in the case of mesophilic bacteria) reported after 60 s of ultrasounds (Fig. 2).

R2A agar stimulated the growth of a larger number of microorganisms compared to A agar (incubation time: 2 and 3 days) and the results obtained for both substrates differed by two orders of magnitude. In the case of R2A agar, both the composition of the medium as well as the incubation time (7 days) were different. It should also be considered that sonication applied during separation of the biofilm might have caused damage to some microorganisms remaining on the surface of the steel plate and longer incubation time could have provided better conditions for the growth of damaged and stressed microorganisms. The highest numbers of microorganisms were reported in the sample subjected to the 60-s sonication with the results obtained being 14600 and 12800 [CFU/ml/cm^2] for mesophilic and psychrophilic bacteria (Fig. 2). After analyzing the obtained results, it was found that the optimal time enabling effective separation of the biofilm from the surface of the tested material, while not causing disintegration of bacterial cells, is 60 s.

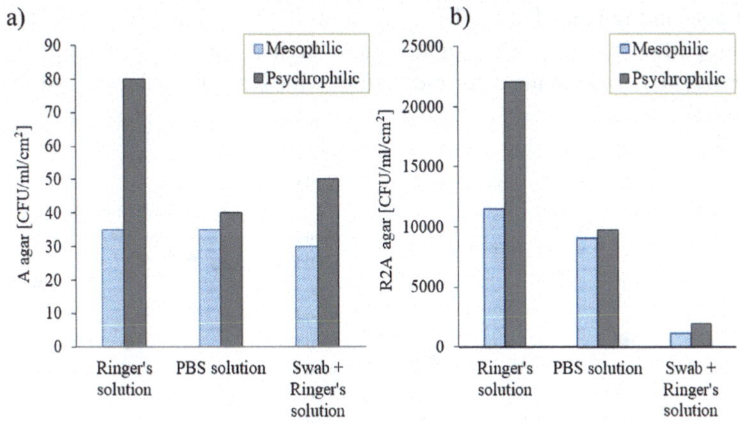

Fig. 2. Number of mesophilic and psychrophilic bacteria on A agar (a) and R2A agar (b) depending on the time of sample sonication

3.2 Comparison of the Effect of Physiological Solutions on the Viability of the Microorganisms Forming the Biofilm

Literature sources report that a high effectiveness of separating the biofilm from the surface of materials was achieved using the sonication method and PBS solution (Waller et al. 2018) as well as the sterile cotton swab method (Boe-Hansen et al. 2003). Due to this, a decision was made to carry out a further stage of the research; testing two different saline solutions using the optimal sonication time of 60 s enabled determining of which is more efficient at maintaining or extending the life of bacterial cells, which is particularly important in the case of HPC methods.

The obtained ATP values showed a clear difference between the applied saline solutions. Far higher values of, respectively, ATP(o) = 22205 [RLU/ml/cm^2] and ATP (z) = 9755 [RLU/ml/cm^2] were achieved for Ringer's solution (Fig. 3). These values were comparable to the results obtained in the previous stage of the research (ATP (o) = 22263 [RLU/ml/cm^2], ATP(z) = 8902 [RLU/ml/cm^2]), which demonstrates the correctness and repeatability of the performed analyses.

The mechanical method of detachment the biofilm from coupon with a sterile cotton swab and 30 ml of Ringer's solution, selected for the study due to greater efficiency, made it possible to obtain good results. The results were similar or even higher than the measurements made for the variant: Ringer's solution + sonication for 60 s. ATP(o) was 22650 [RLU/ml/cm^2], while ATP(z) = 11807 [RLU/ml/cm^2] (Fig. 3). The number of microorganisms determined by flow cytometry, as in the case of ATP analysis, was the highest with the use of Ringer's solution and the variant with a cotton swab (Fig. 3). The PBS solution did not enable obtaining such high values (the total number of living and dead microorganisms was 5115830 [particles/ml/cm^2]), which unambiguously demonstrates the usefulness of Ringer's solution in the quantitative determination of microorganisms in the biofilm. As in the earlier stage of the research, the number of bacteria grown on A agar was significantly lower than in the case of R2A agar. The number of mesophilic bacteria obtained with A agar was similar in all variants and remained at the level of 35–30 [CFU/ml/cm^2] (Fig. 4). The number of psychrophilic bacteria was significantly higher and a maximum value of 80 [CFU/ml/cm^2] was obtained for the variant with Ringer's solution and 60-s sonication (Fig. 4).

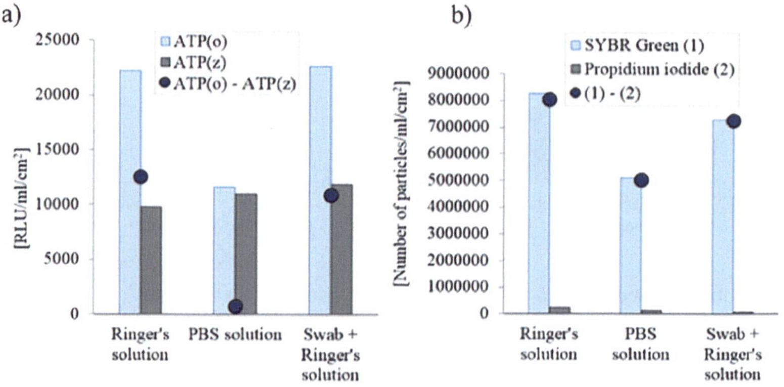

Fig. 3. Measurement of ATP (a) and the number of microorganisms by flow cytometry (b) using three different methods of biofilm detachment

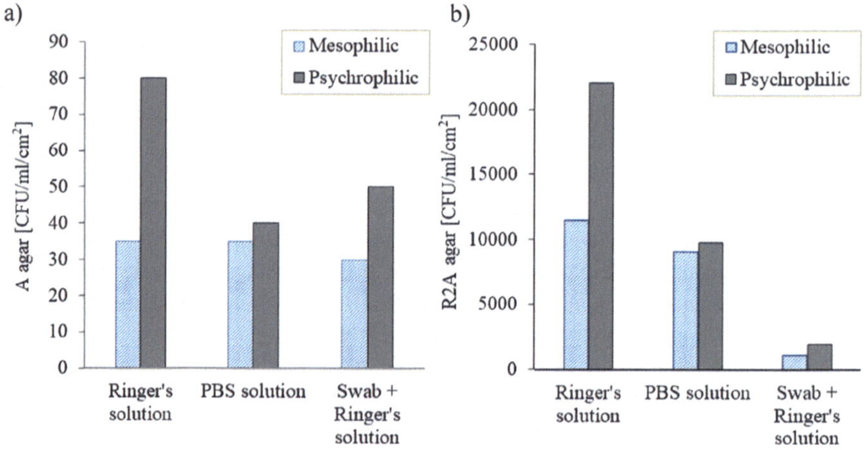

Fig. 4. Number of mesophilic and psychrophilic bacteria on A agar (a) and R2A agar (b) and using three different methods of biofilm detachment

For R2A agar, the obtained values were higher by two orders of magnitude when compared to A agar. The most effective method of separating the biofilm in this case also proved to be the variant with the use of Ringer's solution and 60-s exposure to ultrasounds. The results of the HPC analysis on R2A agar obtained using the swab method, deviated from the other two variants and amounted to, for mesophilic bacteria 1150 [CFU/ml/cm^2] and for psychrophilic bacteria 1950 [CFU/ml/cm^2], respectively (Fig. 4).

4 Conclusions

Quantitative analysis of biofilm can be used to assess the degree of colonization of materials by microorganisms, to determine the speed, type, and amount of biological growth occurring on the internal surfaces of water pipes and to assess the microbiological stability of tap water.

Detaching the biofilm from the surface of the material is the essential but also the most complicated stage of the process. Based on the conducted research, the following was concluded:

- the best method to detach the biofilm from the surface of the coupons is mechanical separation with a sterile swab, which is then subjected to ultrasounds in a solution;
- the optimal duration of ultrasounds, necessary to separate the biofilm from the surface of the material, while at the same time guaranteeing the least destruction of microorganisms, is 60 s for sonicator power of 28 W;
- the environment of normal buffered saline (Ringer's solution) enables extending the life of microorganisms forming the biofilm when compared to PBS solution;
- attention should be given to the volume of solutions, which is selected depending on the operating time and the sonicator power (30 ml is optimal).

The method presented in the research, consisting of the physical separation of the biofilm from the surface of the material using only ultrasounds, may be useful in studies carried out on samples of materials (plates) produced on a laboratory scale. In actual conditions, it may, however, prove impractical and cumbersome. The variant with the use of a sterile swab enables rapid collection of the biofilm from the surface of the ducts forming internal installations or water supply networks. When considering the application of the said method in practice, it would be necessary to create a template from a flexible material which would include rectangular holes with preset dimensions. Flexibility of the element would enable it to fit into the internal surface of the duct and collect the biofilm with a cotton swab.

References

Boe-Hansen R, Martiny AC, Arvin E, Albrechtsen H-J (2003) Monitoring biofilm formation and activity in drinking water distribution networks under oligotrophic conditions. Water Sci Technol 47(5):91–97

Brember PJ, Webster BJ, Wells DB (2001) Biocorrosion of copper in potable water. J Am Water Works Assoc 93(8):82–91

Douterelo I, Jackson M, Solomon C, Boxall J (2016) Microbial analysis of in situ biofilm formation in drinking water distribution systems: implications for monitoring and control of drinking water quality. Appl Microbiol Biotechnol 100:3301–3311

Douterelo I, Sharpe RL, Husband S, Fish KE, Boxall JB (2019) Understanding microbial ecology to improve managent of drinking water distribution systems. WIREs Water. 6:e1325. https://doi.org/10.1002/wat2.1325

Gomes IB, Simões M, Simões LC (2014) An overview on the reactors to study drinking water biofilms. Water Res 62:63–87

Górka A, Papciak D, Zamorska J, Antos D (2008) The influence of biofilm on the effectiveness of ion exchange process. Ind Eng Chem Res 47:7456–7664

Jang H-J, Choi Y-J, Ka J-O (2011) Effects of diverse water pipe materials on bacterial communities and water quality in the annular reactor. J Microbiol Biotechnol 21(2):115–123

Manuel C, Nunes O, Melo L (2010) Unsteady state flow and stagnation in distribution systems affect the biological stability of drinking water. Biofouling 26(2):129–139

Papciak D, Tchórzewska-Cieslak B, Pietrucha-Urbanik K, Pietrzyk A (2018) Analysis of the biological stability of tap water on basis of risk analysis and parameters limiting the secondary growth of microorganisms in water distribution system. Desalination Water Treat 117:1–8

Pietrucha-Urbanik K, Tchórzewska-Cieślak B, Papciak D, Skrzypczak I (2017) Analysis of chemical stability of tap water in terms of required level of technological safety. Arch Environ Prot 43(4):3–12

Simões M, Simões LC, Vieira MJ (2010) A review of current and emergent biofilm control strategies. LWT Food Sci Technol 43(4):573–583

Srinivasan S, Harrington GW, Xagoraraki I, Goel R (2008) Factors affecting bulk to total bacteria ratio in drinking water distribution systems. Water Res 42(13):3393–3404

Tsvetanova ZG, Hoekstra EJ (2009) A study on assessment of biomass production potential of pipe materials in contact with drinking water. Water Sci Technol Water Supply 9(4):423–429

Waller SA, Packman AI, Hausner M (2018) Comparison of biofilm cell quantification methods for drinking water distribution systems. J Microbiol Methods 144:8–21

Ying W, Yang F, Bick A, Oron G, Herzberg M (2010) Extracellular polymeric substances (EPS) in a hybrid growth membrane bioreactor (HG - MBR), viscoelastic and adherence characteristics. Environ Sci Technol 44:8636–8643

Yu J, Kim D, Lee T (2010) Microbial diversity in biofilms on water distribution pipes of different materials. Water Sci Technol 61:163–171

Zamorska J, Papciak D (2008) Activity of nitrifying biofilm in the process of water treatment in diatomite bed. Environ Prot Eng 34:37–52

Zhang J, Li W, Qi W, Wang F, Zhou Y (2018) Impact of biofilm formation and detachment no the transmission of bacterial antibiotic resistance in drinking water distribution systems. Chemosfere 203:368–380

Development of Strip Anchoring for CFRP Strengthening System

Bartosz Piątek(✉) and Tomasz Siwowski

Department of Roads and Bridges, Rzeszow University of Technology, Powstancow Warszawy 12, 35-959 Rzeszow, Poland
piatek@prz.edu.pl

Abstract. Prestressing of reinforced concrete structures with CFRP strips is highly effective and currently developing method of strengthening. Mechanical anchorages are usually used in this technology in order to introduce and permanently maintain prestressing force in the CFRP strips. The paper presents the results of static tests of CFRP strips anchorages used in strengthening system recently developed in Poland. The system consists of two main elements: mechanical steel anchorages mounted on both ends of a single CFRP strip and a relevant tensioning device. The tests included two types of anchorages: N-type with hybrid bonded/riveted joints between steel plates and CFRP strips and S-type with bonded/bolted joints. Both types of anchorages have been tested in axial static tests in five series. The anchorages in subsequent series were modified based on the test results and the conclusions of previous series. The main goal of the tests was to establish the carrying capacity and the efficiency of the anchorages. The experimental results allowed for significant increase of anchorages carrying capacity in comparison with initial design and showed that the S-type anchorages exhibited a higher values of carrying capacity as well as higher level of homogeneity.

Keywords: Anchorages · CFRP strips · Prestressing · Strengthening system

1 Introduction

FRP composites are currently becoming more and more popular in civil engineering applications. One of the fields where they can be used is strengthening of existing concrete, steel, masonry or timber structures (Hollaway and Teng 2008). For this purpose most often CFRP (Carbon Fiber Reinforced Polymer) composites in form of strips, sheets (wraps) or profiles are used. Structural elements can be strengthened in flexure, shear or they can be confined by CFRP products. Flexural strengthening of structural elements subjected to bending loads is performed using sheets or (more often) strips. CFRP strips can be generally fixed to the strengthened element in two ways, as externally bonded (EB) or near surface mounted (NSM) reinforcement (Kim 2014). In the first method CFRP strips are glued to the external (tensioned) surface of the element by epoxy adhesive. The second method uses CFRP strips or profiles glued into grooves, which are previously cut in strengthened elements. In both methods CFRP strips can be applied in untensioned (passive) or tensioned (active) form.

Z. Blikharskyy et al. (Eds.): CEE 2019, LNCE 47, pp. 323–330, 2020.
https://doi.org/10.1007/978-3-030-27011-7_41

Effects of passive strips are limited to improving load-bearing capacity of strengthened elements while active strips can also improve serviceability parameters, such as deflections or crack widths (Bakis et al. 2002; Hollaway 2010). For this reason strengthening systems with tensioned CFRP strips are currently developed and widely tested all over the world (Aslam et al. 2015; Mohee et al. 2016). Most of the systems with tensioned strips uses mechanical anchorages to introduce the prestressing force in the CFRP strips during installation and permanently maintain it during service (Andrä and Maier 2000; Correia et al. 2015). Beside keeping the force, anchorages protect against debonding at the ends of the CFRP strips from the surface of strengthened element, which allows to obtain higher utilisation of CFRP material in the ultimate limit states. However some systems instead of mechanical anchorages use different method to maintain prestressing force in the strips. This method relies on gradual reduction of the force in the strip at its end, which eliminates dangerous stresses in this zone (Kotynia et al. 2011; Correia et al. 2015; Haghani et al. 2015).

The paper presents static tests of the mechanical anchorages used in CFRP strengthening system recently developed in Poland (Piątek and Siwowski 2017; Siwowski and Siwowska 2018). The tests included two types of anchorages: N-type with hybrid bonded/riveted joints between steel plates and CFRP strips and S-type with bonded/bolted joints. The main goal of the tests was to establish the carrying capacity and the efficiency of the anchorages.

2 Strengthening System

2.1 Description of the System

The new system consists of two main elements: special steel anchorages mounted on both ends of a single CFRP strip and a relevant tensioning device (Fig. 1). The system uses high-strength UHS 614 strips with cross-section amounting 1.4 mm × 60 mm with the ultimate tensile strength of 3200 MPa, modulus of elasticity amounting 160 GPa and the strain at failure of about 2%. Besides that, an epoxy adhesive is used to bond a part of CFRP strip between anchorages with the concrete surface. Two kinds of anchorages are used in the system: an active anchorage combined with a tensioning device and a passive anchorage. Single anchorage is made of two steel plates connected together between which CFRP strip end is fixed by mechanical fasteners and epoxy adhesive.

The tensioning device is the second key element of the strengthening system. It consists of three separately installed components: guide rails, carriage (bolted to the active anchorage) and hydraulic jack. The hydraulic jack is driven by a manual pump and it can generate maximum prestressing force of 170 kN. Thanks to division of the device body into three small and light parts (the heaviest element weights 37 kg), its application on-site is very fast and easy.

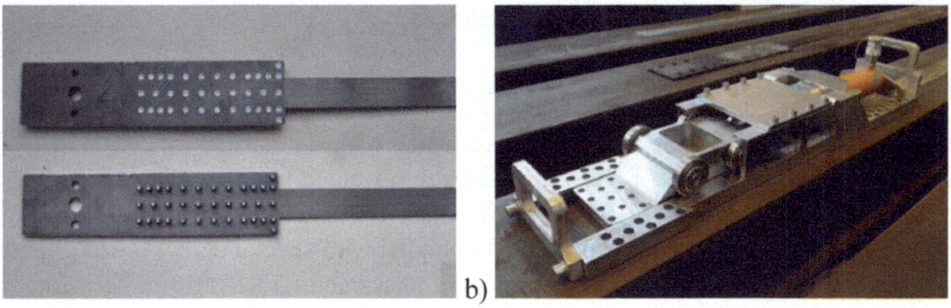

Fig. 1. Main elements of the strengthening system: anchorages (a) and tensioning device (b)

2.2 Strengthening Procedure

The strip with determined length is delivered on site as ready-to-install, i.e. with two prefabricated steel anchorages mounted on both strip ends. It is anchored at the passive side and at the active side guide rails are mounted on the concrete surface. Then, the CFRP strip is attached between the guide rails and the carriage is bolted to the active anchorage. After that, epoxy adhesive is applied along the strip. In the last step, the hydraulic jack with the resisting block are mounted to the guide rails, and the CFRP strip is stretched. Finally, after the relevant strip tensioning, the active anchorage is attached to the concrete beam and tensioning device can be removed. Subsequent actions are showed in Fig. 2.

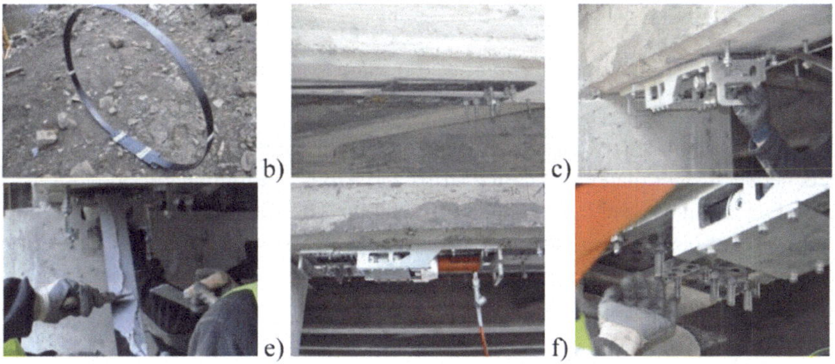

Fig. 2. Strengthening procedure: (a) strips with anchorages delivered in coils, (b) passive anchorage installation, (c) tensioning device installation, (d) epoxy adhesive application, (e) strip tensioning, (f) active anchorage installation

3 Construction of the Anchorages

3.1 General Description

The anchorage is made of two outer steel plates with dimensions of 550 mm × 128 mm and inner steel spacer of 249 mm × 126 mm welded together along their edges to create a pocket, in which CFRP strip end is fixed. The end of the strip (300 mm length) is placed in the steel anchorage pocket and bonded with special epoxy-based glue. It is followed by gripping the both materials (steel plates and CFRP strip in-between) with mechanical fasteners (rivets or bolts). The anchorages transfer the tension force from tensioning device to the strip by internal bonding, gripping and friction simultaneously. The anchorage has two functional areas: external and internal (Fig. 3). The strip is clamped, bonded and gripped within the internal area. The external area comprises holes for attaching the plate to the concrete surface by anchors and threaded holes for mounting the tensioning device (in the active anchorage only).

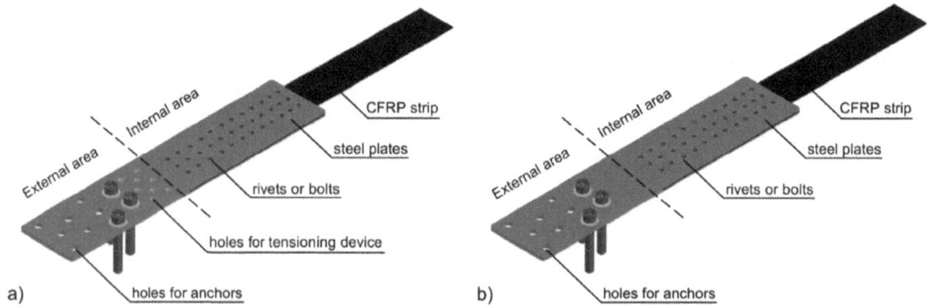

Fig. 3. Scheme of active (a) and passive (b) anchorage

The tests included two types of anchorages: N-type with hybrid bonded/riveted joints between steel plates and CFRP strips and S-type with bonded/bolted joints. Both types of anchorages have been tested in five series. The anchorages in subsequent series were modified based on the test results and the conclusions of previous series in order to increase the carrying capacity of the anchorages.

3.2 N-Type Anchorages

N-type anchorages uses epoxy adhesive and rivets (mild steel grade S215, diameter of 4 mm) for joining steel plates and CFRP strip. In the two first test series (S1N, S2N), the anchorages used in the previous version of the prestressing system (NPS) were examined as a reference (Fig. 4a). In the next three series (S3N-S5N), novel bonded/riveted (N-type) anchorages were tested. In comparison with the old-type anchorages (NPS), the thickness of the steel plates was enlarged from 2 mm to 3 mm, the adhesive was improved and the number as well as layout of the rivets were changed (Fig. 4b).

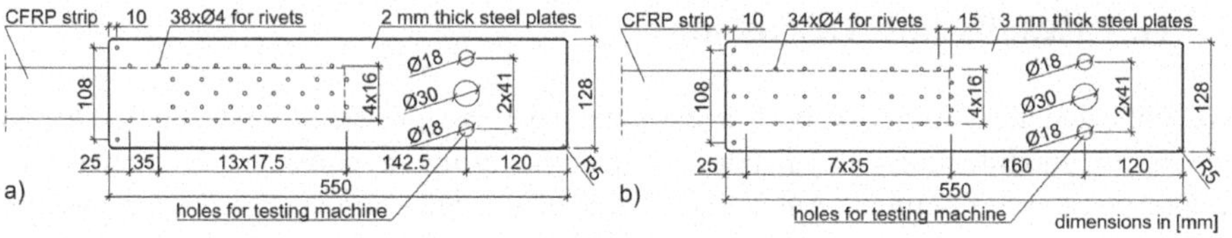

Fig. 4. Layout of mechanical fasteners in the anchorages in series: (a) S1N-S2N, (b) S3N-S5N

3.3 S-Type Anchorages

Series S1S-S5S included bonded/bolted (S-type) anchorages which joins steel plates and CFRP strip using epoxy resin and small high-strength (steel grade 10.9) bolts with diameter of 6 mm. Modifications in subsequent series comprised several attempts to enhance the shear capacity of the bonded steel and composite surfaces by modifying epoxy adhesive curing condition, and trying to strengthen the composite layer with an additional 2D CFRP sheet (series S1S – one layer, series S2S – two layers, series S3S-S5S – without CFRP sheet) as well as to check the different arrangement of the bolts in series S5S (Fig. 5a, b).

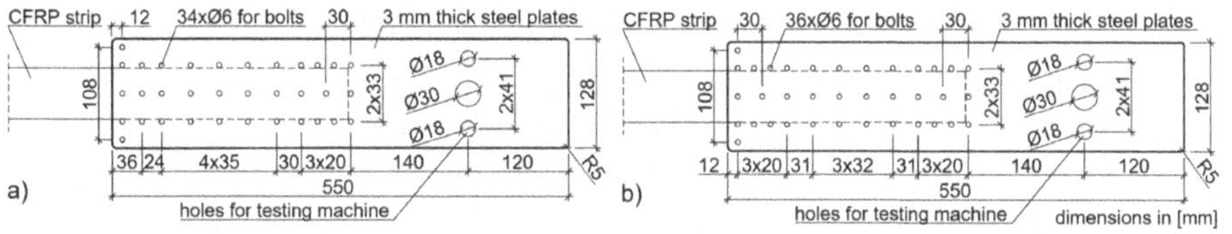

Fig. 5. Layout of mechanical fasteners in the anchorages in series: (a) S1S-S4S, (b) S5S

4 Test Setup and Instrumentation

All specimens were made as a one-sided anchorage (Fig. 6). The free ends of the strip were protected by aluminium plates and placed between the jaws of the testing machine. The anchorage specimens were installed in the testing machine Instron J1D 1200 kN using special jaws to ensure uniaxial loading. The tests were conducted with displacement controlled rate of 2 mm/min. During the tests, increment of force and displacement were measured by the set of sensors built in the testing machine. Additionally, tests were recorded by the Vision Research Phantom v640 camera, which can save more than 2000 frames per second. This enabled a detailed identification of failure modes.

a) b)

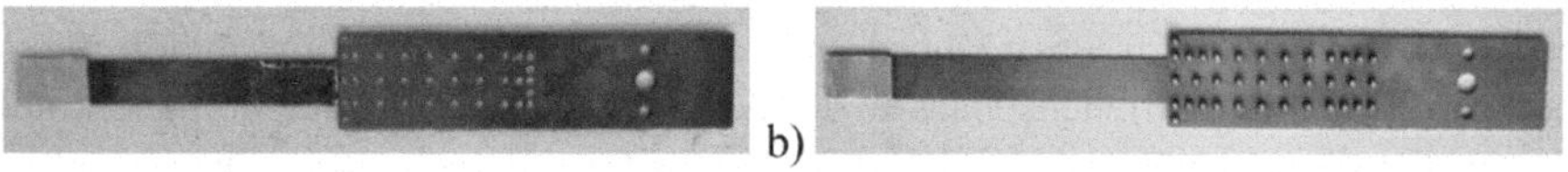

Fig. 6. Specimen of N-type (a) and S-type (b) anchorage

5 Test Results and Discussion

The collected test results of both anchorage types are presented in Table 1. Due to successive modifications in anchorages manufacturing, a considerable increase of failure force was obtained. Anchorages used in the previous version of the system, tested in the two first series (S1N-S2N), exhibited a mean failure load of approximately 110 kN. In the last test series of N-type anchorages a mean failure load was 185 kN (increase of 68%). N-type anchorages showed relatively large dispersion of results. S-type anchorages exhibited greater values of failure load, which reached almost 200 kN (increase of 81%) in the last series (S5S). Furthermore, during S-type anchorages, a smaller dispersion of results was observed. The coefficients of variation equalled 1.7% and 0.4% in the last two series, which indicates that the tested anchorage specimens had a high level of homogeneity. In the last series a repeatable anchorage efficiency (calculated as a ratio of maximum failure force of the anchorage to the mean CFRP tensile strength (force), which is 276 kN for the UHS 614 CFRP strips) higher than 70% was obtained. It is a sufficient value for post-tensioning system for strengthening concrete structures because optimal strengthening effects are obtained with strip pre-stressing level amounting about 60% of CFRP tensile strength (Meier 1995).

Table 1. Collective testing results of anchorages

Series No.	Failure load of specimen:			Mean failure load [kN]	Standard deviation [kN]	Variability coefficient [%]	CFRP efficiency [%]
	P1 [kN]	P2 [kN]	P3 [kN]				
S1 N	91	111	129	110	19.4	17.5	47
S2N	103	120	103	109	9.6	8.8	39
S3N	126	158	151	145	16.7	11.5	52
S4N	166	144	156	155	11.4	7.3	56
S5N	177	160	217	185	29.2	15.8	67
S1S	175	174	166	171	5.2	3.0	62
S2S	147	161	208	-	-	-	53/58/75
S3S	177	221	-	199	31.3	15.7	72
S4S	189	194	188	190	3.2	1.7	69
S5S	199	198	-	199	0.8	0.4	72

Figure 7 shows the load-displacement plots for specimens of each series with the maximum value of the failure load. Characteristic abrupt decreases on the plots are connected with loss of adhesive between CFRP strip and steel plates or longitudinal splitting of the strip. In modified versions of anchorages number of mechanical fasteners crossing the strip was reduced and located in one row in the middle of strip (compare Fig. 4a and b). This leads to a reduction in stress concentration and number of potential shear planes in the CFRP strip. In new type of anchorages, there is only one shear plane in the middle.

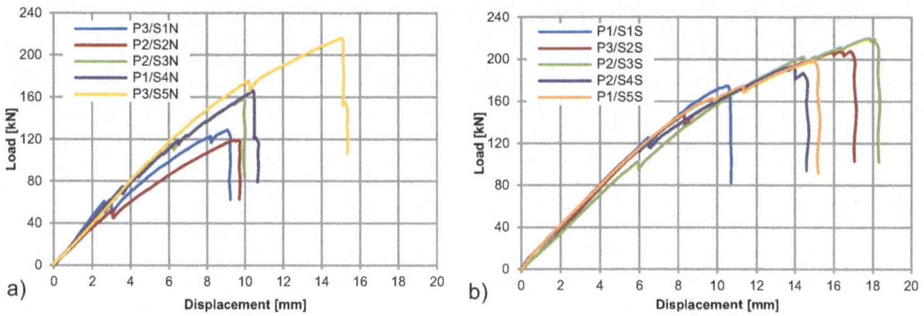

Fig. 7. Maximum tensile load-displacement plots for series: (a) S1N-S5N, (b) S1S-S5S

Introduction of subsequent modifications was possible, among others, thanks to detailed identification of failure modes using high-speed camera. Typical sequence of failure mode is shown in Fig. 8. Failure of the anchorage occurs due to longitudinal splitting the strip along the middle row of mechanical fasteners followed by a sudden pull out mechanism of the bisected strip from the external steel plates.

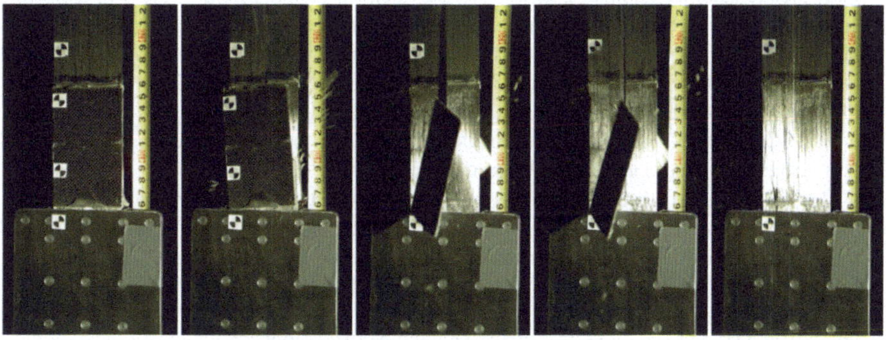

Fig. 8. Sequence of anchorage failure

6 Conclusions

The anchoring method combines adhesive bonding with gripping and friction action and thus gathers the advantages of these three joining methods. Continuous development of anchoring technology for newly developed strengthening system (NPS II) allowed to obtain a significant increase of failure force 68% for N-type anchorages and 81% for S-type anchorages in comparison with anchorages used in previous version of the system (NPS). Thanks to subsequent modifications higher level of homogeneity for S-type anchorages was obtained. After described research strengthening system with tested anchorages was examined on series of reinforced concrete and steel beams in the laboratory as well as two bridges on-site and now it is used in practical applications.

Acknowledgements. The activity presented in the paper is part of the research grant "Innovative system for strengthening building structures using tensioned CFRP strips" (POIG.01.03.01-18-010/12) co-financed by the National Centre for Research and Development (NCBiR).

References

Andrä HP, Maier M (2000) Post-strengthening with externally bonded prestressed CFRP strips. IABSE Congr Rep 16(7):1507–1514

Aslam M, Shafigh P, Jumaat MZ, Shah SNR (2015) Strengthening of RC beams using prestressed fiber reinforced polymers – a review. Constr Build Mater 82:235–256. https://doi.org/10.1016/j.conbuildmat.2015.02.051

Bakis CE, Bank LC, Brown VL, Cosenza E, Davalos JF, Lesko JJ, Machida A, Rizkalla SH, Triantafillou TC (2002) Fiber-reinforced polymer composites for construction – state-of-the-art review. J Compos Constr 6(2):73–87. https://doi.org/10.1061/(ASCE)1090-0268(2002)6:2(73)

Correia L, Teixeira T, Michels J, Almeida JA, Sena-Cruz J (2015) Flexural behaviour of RC slabs strengthened with prestressed CFRP strips using different anchorage systems. Compos B Eng 81:158–170. https://doi.org/10.1016/j.compositesb.2015.07.011

Haghani R, Al-Emrani M, Kliger R (2015). A new method for strengtheng concrete structures using prestressed FRP laminates. In: Saha S, Lloyd N, Yazdani S, Singh A (eds) Implementing Innovative Ideas in Structural Engineering and Project Management. ISEC Press

Hollaway LC (2010) A review of the present and future utilisation of FRP composites in the civil infrastructure with reference to their important in-service properties. Constr Build Mater 24(12):2419–2445. https://doi.org/10.1016/j.conbuildmat.2010.04.062

Hollaway LC, Teng JG (eds) (2008) Strengthening and rehabilitation of civil infrastructures using fibre-reinforced polymer (FRP) composites. Elsevier

Kim YJ (ed) (2014) Advanced composites in bridge construction and repair (No 50). Elsevier

Kotynia R, Walendziak R, Stoecklin I, Meier U (2011) RC slabs strengthened with prestressed and gradually anchored CFRP strips under monotonic and cyclic loading. J Compos Constr 15(2):168–180. https://doi.org/10.1061/(ASCE)CC.1943-5614.0000081

Meier U (1995) Strengthening of structures using carbon fibre/epoxy composites. Constr Build Mater 9(6):341–351. https://doi.org/10.1016/0950-0618(95)00071-2

Mohee FM, Al-Mayah A, Plumtree A (2016) Anchors for CFRP plates: state-of-the-art review and future potential. Compos B Eng 90:432–442. https://doi.org/10.1016/j.compositesb.2016.01.011

Piątek B, Siwowski T (2017) Research on the new CFRP prestressing system for strengthening of RC structures. Architect Civ Eng Environ 10(3):81–87

Siwowski T, Siwowska P (2018) Experimental study on CFRP-strengthened steel beams. Compos B Eng 149(15):12–21. https://doi.org/10.1016/j.compositesb.2018.04.060

Experimental Tests of Joints in Scaffolding System

Z. Pisarek(✉) and P. Sudoł

Department of Building Structures, Rzeszow University of Technology,
Poznanska 2 Street, 35-084 Rzeszow, Poland
pisarzdz@prz.edu.pl

Abstract. Scaffolding is a special structural system used in buildings. Despite their simple construction, they require proper design. In the analysis of scaffolds, not only the strength of the elements themselves plays an important role, but also the knowledge of the characteristics of system nodes. Currently used programs for designing and calculating structures are able to take into account the influence of joint characteristics. That is why some manufacturers give flexibility properties on the details of their solutions. The article presents the results of experimental tests of modular scaffold nodes. A series of four types of tests was carried out; tensile tests of the joint when the force is applied to the ledgers, tensile tests of the joint when the force is applied to the diagonal of the brace members, tests the joint for bending in the plane of the system standards and ledgers, and bending tests in a plane perpendicular to the plane of the system. The test results were compared with the values given by the General Building Authority Approval.

Keywords: Modular scaffolding · Laboratory test · Joints flexibility · Joint strength

1 Introduction

During erection or renovation of many building objects it is necessary to use scaffolding. Scaffolds are designed so that it can be create a model of a spatial structure of any shape. It allows constructing scaffolding around objects with complicated shapes, such as industrial facilities or historical monuments. Modular scaffolding perfectly fulfills this function. It is possible to quickly adapt them to the shape of the structure, modification or extension to the superstructure for the needs of employees.

The modular scaffolding system is composed of standards, ledgers, diagonal braces and decks. They are the fundamental components of the system. Additionally, in the modular scaffolding system there are: side protection, access and supplementary components. The standards, ledgers and diagonal braces are joined together using special connectors. The scaffolding connectors consist of a rosette welded to a standard, and of connecting heads welded to ledgers or swivel-mounted to vertical diagonal braces. The connecting heads surround the rosette, and are wedged to the rosette by hammering in a permanently attached wedge in such a way that the connecting head is

Z. Blikharskyy et al. (Eds.): CEE 2019, LNCE 47, pp. 331–339, 2020.
https://doi.org/10.1007/978-3-030-27011-7_42

forced against the standard. The horizontal diagonal braces are connected to these by inserting a bolt into the hole of the rosette.

Manufacturers of scaffolding systems provide design requirements in the technical documentation, but when the structure of the scaffolding deviates from a typical set-up, it requires adequate calculations of the strength and stability of the scaffolding. A well-designed construction requires modeling of the connection, which illustrates cooperation between individual elements. The scaffolding connections, due to their properties, cannot be classified as rigid or pinned, but should be taken as semirigid. The strength and stiffness parameters of the joints should be given by the manufacturer, but they are not always specified. That causes problems during the design. In order to correctly determine the joint characteristics, it is necessary to analyze the form of damage under the influence of various loads.

2 Issues of the Design and Calculation of the Scaffolding

According to the (Regulation 2003), the scaffolding should be made in accordance with the design or manufacturer's assembly instructions. The type of documentation that the scaffolding has depends mainly on the shapes of the structure. The instruction given by the manufacturer is sufficient to use for assembly when:

- all elements of the system scaffolding have been provided by the manufacturer for tests in accordance with material and construction requirements in accordance with current standards that define products in terms of safety
- the system scaffolding structure will be typical, that is, in accordance with the manufacturer's instructions contained in the Scaffold Installation Manual.

If any of these conditions are not met, an individual project should be carried out.

At present, due to the current European Standard (EN 1993-1-1 2006), in the structure analysis the effect of the joint behavior it is necessary to take into account. The stiffness of the joint influence on the distribution of internal forces and moments within the structure, and on the overall deformation of the structure. This fact can be found in research and publications (Kozłowski 1996), (Błazik-Borowa et al. 2017), (Chandrangsu and Rasmussen 2011). Scaffold manufacturers also provide analytical models for determining the stiffness characteristics of the joints of their products (Layher 2007).

Some assumptions regarding the stiffness of connections in the guidelines included in the subject standards. The (EN 12810-2 2010) standard recommends that:

- connections between pipe elements in which the length of the connecting rod is at least 150 mm without latch and 100 mm with latch and when the clearance between the pipe and the stem is less than or equal to 4 mm to be treated as rigid,
- swivel joints as nominally pinned,
- the cross connectors according to the characteristics specified by the manufacturer,
- the wedge connector as non-torsional flexible connections.

The aim of the work was to experimentally determine the dependence of force - displacement or moment - rotation of some nodes of a selected type of modular scaffolding. The results of experimental tests were compared with the analytical model given by the scaffold system manufacturer in the Technical Approval (Layher 2007).

3 Experimental Tests

3.1 Scope and Methodology of the Test

Due to the limited number of specimens, the tests were limited to four of the six forces present in the node. The following tests were accepted:

- tensile strength of the joint in the horizontal plane,
- bending strength of the joint for moment in the plane of the stand and ledger,
- bending strength of the joint for moment in a plane perpendicular to the plane of the stand and ledger,
- tensile strength of the joint when the force is applied to the diagonal bracing.

Each type of load was performed on three specimens. The tests were carried out in the testing machine Instron 1200kNJ1D. The load was carried out assuming a constant displacement speed. Due to the limited size as well as the shape of the nodes, special holders had to be made to carry out the laboratory tests, enabling load application. Each type of examination required the development of a different type of handle. All elements were exploited, as evidenced by the worn paint that was used as the marking for the company making the samples available. The proposed solutions were own author`s solutions.

3.2 Tests Setup and Types of Failure

The first of the considered load systems is the tension of the ledgers. The ledger elements were placed directly in the jaws of the testing machine (Fig. 1). The problem was to secure the hollow section of the ledger against crushing. This has been solved by filling the profile with special pins.

Already during the assembly, one could notice a tendency to permanent deformations consisting in the plasticization of the steel at the interface between the rosette and the wedge. The ultimate type of destruction was the shear rosette by the wedge (Fig. 2). The wedge was also permanently plasticized.

In the tests of the joint between diagonal brace and a standard the biggest problem was the design of the handle so as to enable axial load of the connection. The brace element is not in the axis of the node, but on some eccentricity (Fig. 3). In addition, the handle had to be prevented with a joint before turning the node around its own axis. The angle between the brace and the standard was chosen so that to reflect the bracing of a field of scaffolding with a width of 2.57 m and a height of 2 m.

The type of failure in this case was the breaking of the oblate tube of the brace weakened by the hole on the pin (Fig. 4). The wedge also deformed in a slightly small deformation while the rosette was undergoing.

Fig. 1. Ledgers in tension.

Fig. 2. Rosette sheared by the wedge.

Fig. 3. Joint between brace and a standard.

Fig. 4. Failure of the joint elements.

Tests of the joints for bending in the system plane of the ledger - standard was aimed at determining the influence of the bending moment in the vertical plane on the deformation of the joint. The testing rig has been designed to eliminate bending of the ledgers. This was done by cutting the swivel joint in half and in the place where the rivet was placed, a mounting bolt for the sheet of the test rig, allowing rotation in the test plane. The load was applied to the fragment of the standard (Fig. 5).

Fig. 5. Specimen of the joint between ledgers and standard on the test rig.

Fig. 6. Deformations of the whole joint and its elements.

The Failure in this test was in the form of deformations hollow section of the standard caused by the clamping of the connecting heads of ledgers pipe wall (Fig. 6). It can see the characteristic square shape of the contact point of these elements. The rosette was deformed under the forces due the deformation of whole joint. joint. The wedge which was bent has also been deformed.

Testing the joints for bending in a horizontal plane was carried out on a test rig prepared for testing the joints for bending in the vertical plane. The difference was to rotate the specimen 90° so that the standard was horizontal (Fig. 7). The load was applied by an additional ledger set perpendicular to the ledgers being tested.

Similarly, to bending in the vertical plane, the elements were not separated (Fig. 8). The wedges were twisted around their axis and deformed by the pressure on the rosette. The deformation of the rosette can also be seen in the places where the ledgers are attached. The final failure was in the form of a shearing rosette through the wedge, as well as in the tension tests of the ledger joints.

3.3 Tests Results

In the ledger tensile test, the value of applied force and displacement in the joint was measured. Figure 9 shows charts for 4 test specimens. In two samples, the bolts were located in the main sockets. In the remaining samples, in places designated for bracing. In practice, these sockets are used in cases of irregular grid scaffolding. Differences in results may result from the difference in strength with which wedges have been driven in and the deformations that the elements possessed.

The relationship of force - displacement in the experimental test of the joints between the standard and the bracing is shown in Fig. 10. The tests showed a clearance in the connection of 2.4 mm. Then the graph was almost linear until the yield point of the most stressed section was reached.

Fig. 7. Specimen on the test rig for bending of the joint in horizontal plane.

Fig. 8. Deformations of the whole joint and wedge.

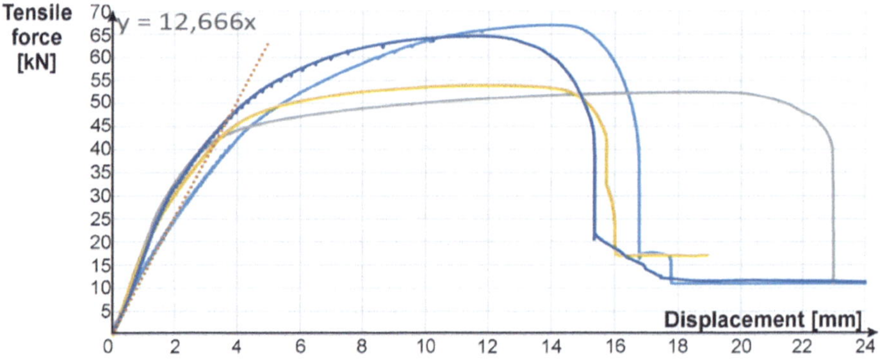

Fig. 9. Force – displacement characteristics for ledger tensile tests.

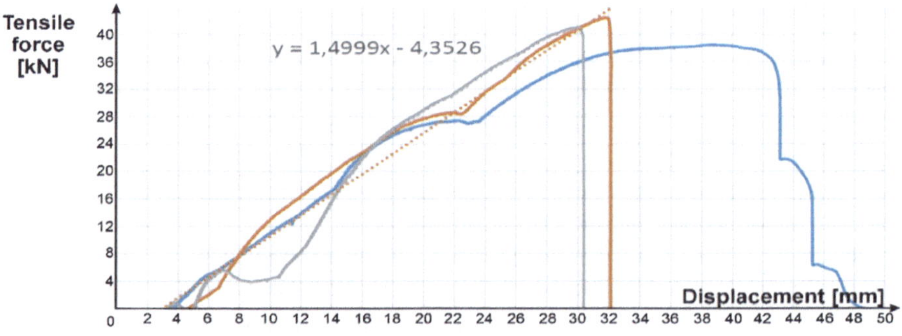

Fig. 10. Force – displacement characteristics for joints between the standard and the bracing.

The tests of the joint between the ledger and the standard was to show the influence of the bending moment on the rotation of the joint. The dependence moment - rotation is shown in Fig. 11. During the test, the elements were not separated. Obtained maximum angle of rotation of the joint 0.4 rad it has no physical meaning. The initial range that presents typical characteristics for semi-rigid joints is interesting. The size of welded head of the ledger determined the rotational strength of the joint.

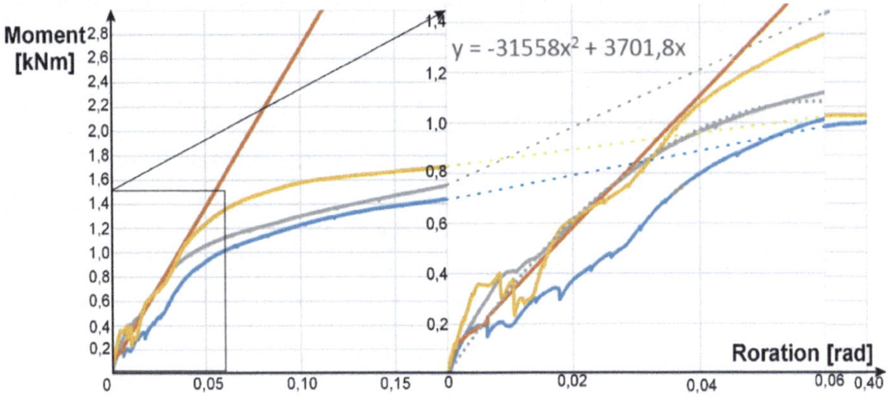

Fig. 11. Moment – rotation characteristics for joints between the standard and the ledger.

Bending test of the joint's ledger with the standard in a horizontal plane. Analyzing the test results, it can be concluded that the graph of the rotation angle from the moment loading the joint (Fig. 12) to the value of 0.23 rad it's almost linear. Exceeding this value would cause, in real construction, working platforms to be removed from the handles. This value also properly exhausts the moment capacity of the connection.

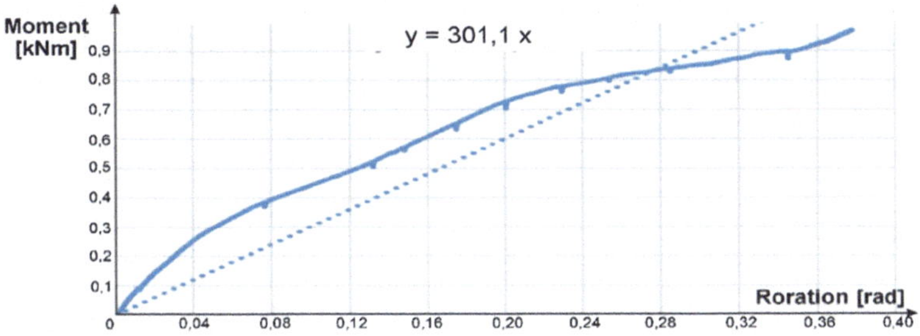

Fig. 12. Moment – rotation characteristics for joints between the standard and the ledger.

4 Comparison of Experimental Results with Manufacturer's Data

The results of experimental tests were compared with the design recommendations given by the manufacturer (Layher 2007). The results of the values of resistance and linear or rotational stiffness are presented in the Table 1. Also, in Figs. 9, 10, 11 and 12 the dot pattern was give characteristics determined by the producer recommended by the manufacturer. In general, it can be concluded that the tests confirmed the correctness of the guidelines given in the catalogs.

Table 1. Comparison of experimental results with manufacturer's data

Type of test	Parameter	Test	Guidelines
Ledgers in tension	N_{Rd} [kN]	56,4	31,0
	$S_{j.ini}$ [kN/mm]	18,56	12,67
Bracing in tension	N_{Rd} [kN]	27,6	17,9
	$S_{j.ini}$ [kN/mm]	14,97	45,00
Bending in vertical plane	M_{Rd} [kNm]	1,30	1,01
	$S_{j.ini}$ [kNm/rad]	56,1	91,4
Bending in horizontal plane	M_{Rd} [kNm]	0,79	0,37
	$S_{j.ini}$ [kNm/rad]	9,2	5,0

5 Conclusions

Based on the results of the conducted research, it can be concluded that the design of scaffoldings according to the recommendations given in the manufacturers' standards and guidelines ensures an adequate level of safety. However, it should be remembered about the clearances in the joints and the significant influence of the force of driving the wedges in the joints. Joints working on bending, despite considerable deformations, protect the structure against the separation of elements. Only the connections between the standard and the bracing rod were destroyed, but under a load much higher than allowed.

References

Regulation of the Minister of Infrastructure of 6.02.2003 on health and safety during construction work. J Law 2003 No 47, item 401 (in Polish)

EN 12810-2 (2010) Façade scaffolds made of prefabricated components. Part 2: Particular methods of structural design. Brussels, CEN

EN 1993-1-1 (2006) Eurocode 3. Design of steel structures. Part 1-1: General rules and rules for buildings. Brussels, CEN

Kozłowski A (1996) A review of models of semi-rigid steel column-beam connections

Błazik-Borowa E, Pieńko M, Robak A, Borowa A, Jamińska-Gadomska P (2017) Analysis of ledger-stand joints in the aluminum modular scaffold. Arch Civil Eng LXIII(1):17–31

Chandrangsu T, Rasmussen KJR (2011) Investigation of geometric imperfections and joint stiffness of support scaffold systems. J Constr Steel Res 67:576–584

Layher Allround Scaffolding Modular System Allround Steel (2007) General Building Authority. Approval Z-8.22–64. Deutsches Institut für Bautechnik

Comments on Eurocode 2 Crack Control Reinforcement for T-Beams Under Flexure

Zbigniew Plewako(✉), Grzegorz Bajorek, and Zakarya Kamel

Department of Building Structures, Rzeszow University of Technology, Al. PowstańcówWarszawy 12, Rzeszów, Poland
plewakoz@prz.edu.pl

Abstract. The paper presents background to method for calculating the minimum reinforcement controlling the crack width and points out some important factors, which should be considered when applying the simplified method for controlling crack width in beams subjected for bending resulting from direct loading. Special attention was put to proper use and limits of Table 7.2N in Eurocode 2.

Keywords: Concrete structures · Cracking of concrete

1 Introduction

Besides the requirements for minimum tension reinforcement preventing brittle fracture covered by Section 9.2.1.1 of Eurocode 2, in cases where crack width shall be limited, additional conditions for minimum tension reinforcement are included in Section 7.3.2. and expressed by Formulae (7.1) of this code cited below:

$$A_{s,min}\sigma_s = k_c k f_{ct,eff} A_{ct} \tag{1}$$

Expression on the right represents the tensile force arose in cross-section when tension in concrete reach its tensile strength $f_{ct,eff}$. To control the width of cracks arising after cracking of concrete (subjected to mentioned tension force), the stress σ_s in reinforcement shall be often below yield strength. For applied σ_s, the area of steel $A_{s,min}$ could be found to substitute the missing tension capacity of concrete.

To avoid time consuming direct crack width checking calculations for $A_{s,min}$ (resulting in steel stress σ_s), provisions of Section 7.3.2 refers to Table 7.2N where to control the limited crack width for assumed stress in tensioned steel allowable maximum diameter of bars could be find. The values in Table 7.2N were derived by procedure clarified in Commentary to Eurocode 2. However, there can be formulated some important remarks, that shall be observed calculating the minimum reinforcement for T-beams.

Besides precast beams, T-sections could be considered for typical in-situ floors, where solid concrete slab is supported on beams, so this topic seems to be important especially for designers with limited theoretical background.

Z. Blikharskyy et al. (Eds.): CEE 2019, LNCE 47, pp. 340–345, 2020.
https://doi.org/10.1007/978-3-030-27011-7_43

Eurocode 2 Table 7.2N (together with Table 7.3N which define maximum spacing of bars) could be used as simplified calculation alternative to control crack for beams or slabs subjected to bending moment resulting from loading. But often popular in the designing practice uncritical use of the values given in this table can sometimes lead to excessive amount of reinforcement. To avoid this situation, the conditions that limit the suitability of using Table 7.2N to determine the reinforcement ensuring SLS of cracking are presented.

2 Eurocode 2 Table 7.2N Used to Calculate the Minimum Reinforcement for T-Beams

2.1 Basics of Developing the Table 7.2

Maximum bar sizes to control crack width presented in Table 7.2N were found based on general formulae for crack width assuming, that maximum crack spacing $s_{r,max}$ is calculated for reinforcement spacing not exceeding $5(c + \phi/2)$ and the effective area of concrete in tension surrounding the reinforcement: $A_{c,eff} = 2,5(h-d)b$.

$$w_k = s_{r,max}(\varepsilon_{sm} - \varepsilon_{cm}) = \left(k_3 c + k_1 k_2 k_4 \frac{\phi}{\rho_{p,eff}}\right) \frac{\sigma_s - k_t \sigma_{sr}}{E_s} \tag{2}$$

As the Table 7.2N is intended to facilitate the determination of the minimum reinforcement in case of pure bending, the considered load is equal M_{cr}. Consequently, the steel stress in cracked section is σ_{sr} and the tension steel area is equal to $A_{s,min}$.

Lets express σ_{sr} using the condition stated in Eq. 1:

$$\sigma_{sr} = \frac{f_{ct,eff} k_c k A_{ct}}{A_{s,min}} = \frac{f_{ct,eff} k_c k}{\rho_{p,eff}} \frac{A_{ct}}{A_{c,eff}} = \frac{f_{ct,eff} k_c k}{\rho_{p,eff}} \frac{b h_{cr}}{2,5(h-d)b} \tag{3}$$

Solving Eq. 3 for $\rho_{p,eff}$:

$$\rho_{p,eff} = \frac{f_{ct,eff} k_c k}{\sigma_{sr}} \frac{h_{cr}}{2,5(h-d)} \tag{4}$$

Substituting $\rho_{p,eff}$, replacing $\sigma_s = \sigma_{sr}$ and solving Eq. 2 for ϕ:

$$\phi = \left[w_k \frac{E_s}{(1-k_t)\sigma_{sr}} - k_3 c\right] \frac{f_{ct,eff} k_c k}{\sigma_{sr} k_1 k_2 k_4} \frac{h_{cr}}{2,5(h-d)} \tag{5}$$

Taking basic factors values recommended by Eurocode 2: $k_3 = 3,4$; $k_1 k_2 k_4 = 0,17$; we obtain the general form of dependence of maximum bar size ϕ_s^* on crack width w_k and variable actual parameters of rectangular cross-section subjected to cracking moment M_{cr}:

$$\varphi_s^* = \left[w_k \frac{E_s}{(1-k_t)\sigma_{sr}} - 3,4c \right] \frac{f_{ct,eff} k_c k}{0,17\sigma_{sr}} \frac{h_{cr}}{2,5(h-d)} \tag{6}$$

Table 7.2N in Eurocode 2 was derived based on following values of cross-section parameters (presented in Table 7.2N notes):

$c = 25\,\text{mm};\ f_{ct,eff} = 2,9\,\text{MPa};\ h_{cr} = 0,5\,\text{h};\ (h-d) = 0,1\,\text{h};\ k_c = 0,4;$
$k = 1,0;\ k_t = 0,4$

Exact functional relation between maximum bar size ϕ_s^* and steel stress $\sigma_s = \sigma_{sr}$ for defined crack width based on presented assumptions is plotted on Fig. 1.

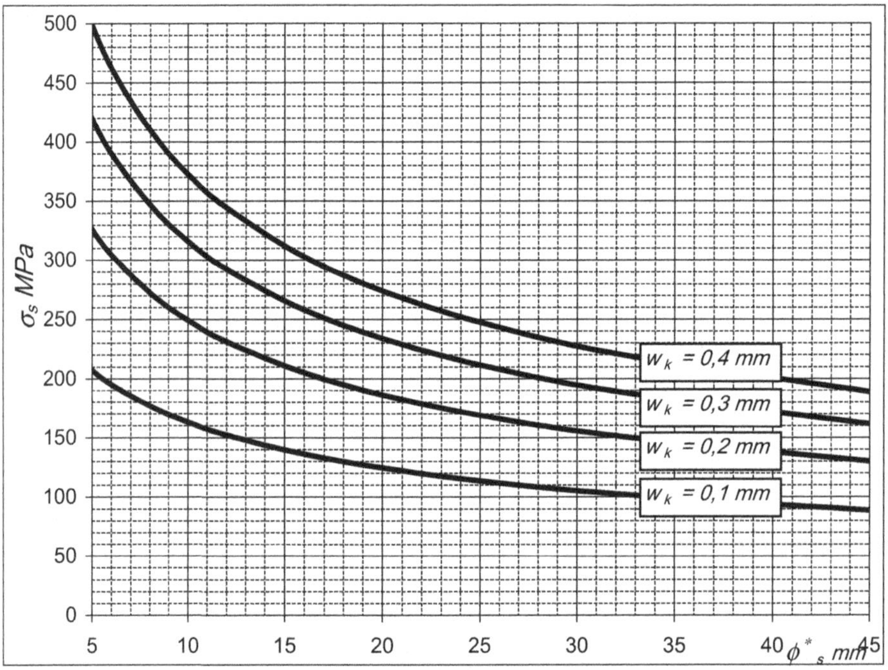

Fig. 1. Maximum bar diameters ϕ_s^* for crack control

In case of load cracks, maximum bar diameters taken from Table 7.2N on presented plot can be adjusted to actual section parameters by Eq. 7:

$$\phi_s = \phi_s^* \frac{f_{ct,eff}}{2,9} \frac{k_c h_{cr}}{2(h-d)} \tag{7}$$

2.2 Cross-Section Parameters K_c and H_{cr} in T-Beams Subjected to Pure Bending Due to Loads

In T-beams, special attention shall be put for calculating height of tension zone prior to first crack, h_{cr}, and coefficient k_c taking account stress distribution within the section immediately prior to cracking and the change of the lever arm due to cracking. Even without defined axial force N_{Ed}, the mean concrete stress in the T-beam web prior to cracking $\sigma_{c,web}$, will differ from 0. So, according to relations in Eurocode 2: $k_c \neq 0, 4$.

To find the mean stress we need to calculate the depth of the tensile zone immediately prior to cracking h_{cr}. For T-beams with flange in compression, the simplified solution (neglecting reinforcement) is given by Eqs. (8 and 9) (Fig. 2).

$$x_{cr} = \frac{1 + (\beta_b - 1)\beta_h^2}{1 + (\beta_b - 1)\beta_h} \frac{h}{2} \tag{8}$$

$$h_{cr} = h - x_{cr} \tag{9}$$

where:

$\beta_b = (b_{eff} - b_w)/b_w$,
$\beta_h = h_f/h$

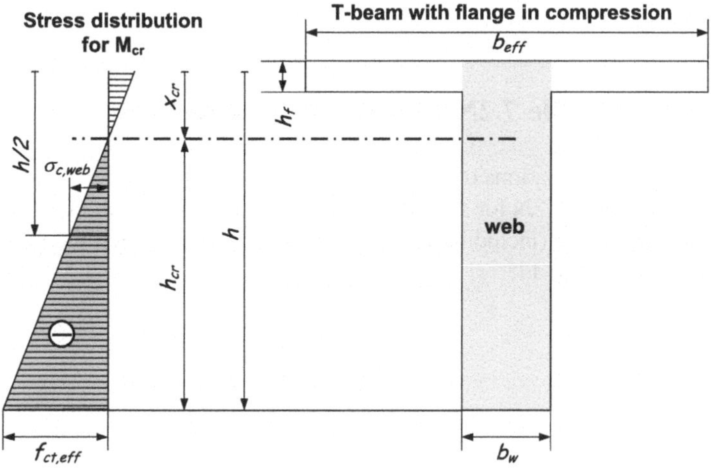

Fig. 2. Stress distribution in T-section subjected to cracking moment

And for maximum tension stress (negative) equal to $f_{c,eff}$:

$$\sigma_{c,web} = -f_{ct,eff} \frac{2h_{cr} - h}{2h_{cr}} \tag{10}$$

Now k_c could be plotted for all values of $\sigma_{c,web}$, according to formulae (7.2) in Eurocode 2, as shown on Fig. 3.

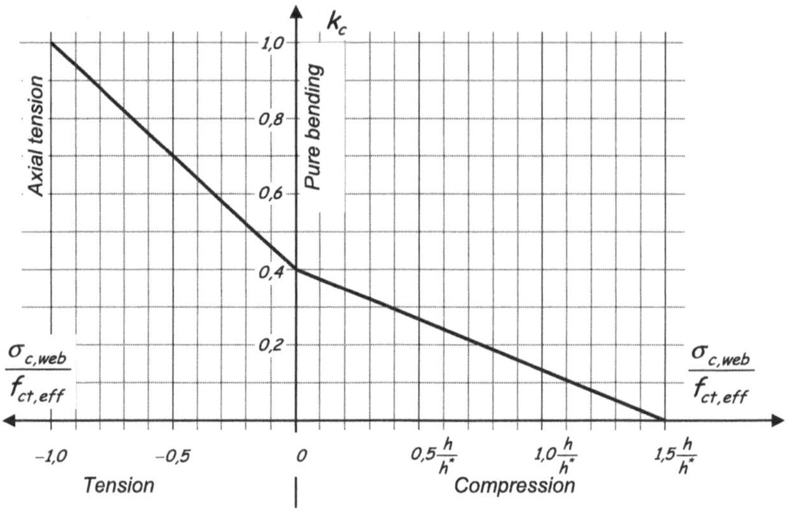

Fig. 3. Value of k_c in relation to average stress in the web of T-beam $\sigma_{c,web}$

It also should be reminded, that value of coefficient $k < 1$, could be only applied if effect of non-uniform self-equilibrating stresses, which lead to a reduction of restraint forces, is considered. In all other cases, e.g. external restrains or loads, $k = 1$ for all section dimensions.

3 Eurocode 2 Table 7.2N Used for Load Cracks

To avoid laborious calculations of crack caused by loading for $M_{Ed} > M_{cr}$, Eurocode 2 allows to use the Table 7.2N for adjusting applied bar diameters with Eq. 7 to assumed cross-section parameters (including tension steel area A_s), for which the stress in bars can be calculated by Eq. 11:

$$\sigma_s = \frac{M_{Ed}}{A_s\left(d - \frac{x}{3}\right)} \tag{11}$$

Where

$$\begin{gathered} x = \left[\sqrt{\left(\alpha_e\rho_f\right)^2 + 2\alpha_e\rho_f} - \alpha_e\rho_f\right] d \text{ for } x \leq h_f \\ x = \left[\sqrt{\left[\alpha_e\rho + \beta_b\beta_d\right]^2 + 2\alpha_e\rho + \beta_b\beta_d^2} - \left(\alpha_e\rho + \beta_b\beta_d\right)\right] d \text{ for } x > h_f \end{gathered} \tag{12}$$

In Eq. 12: $\rho = A_s/(bd); \rho_f = A_s/(b_{eff}d); \beta_d = h_f/d; \beta_b = b_{eff}/b$
where:

d, b – effective depth and width of web, respectively,
h_f, b_{eff} – height and width of top (compression) flange, respectively

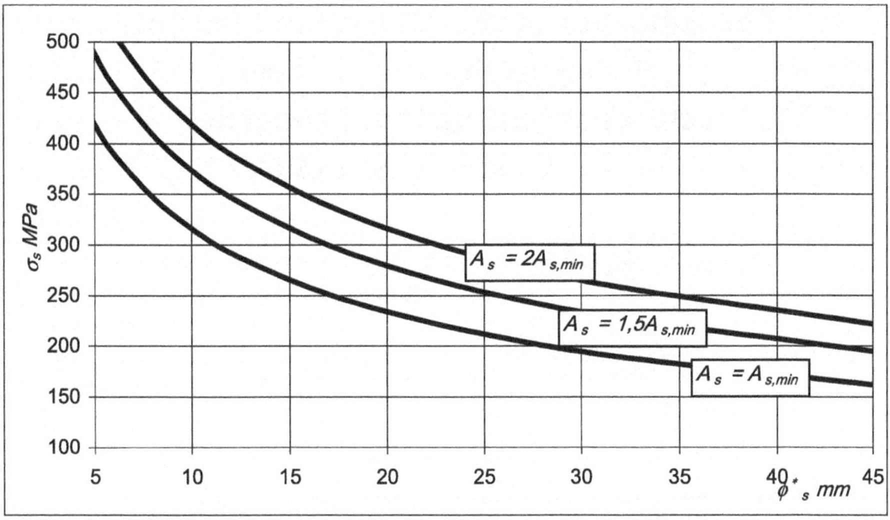

Fig. 4. Maximum bar diameters ϕ_s^* for 0, 3 mm crack control in relation to ratio $A_s/A_{s,min}$

But cross-section parameters adjustment according to Eq. 7 does not fully consider the fact, that assumed tension steel area A_s in many cases is greater that $A_{s,min}$ applied for developing of the Table 7.2N, as $M_{Ed} > M_{cr}$.

If proportion ($A_s/A_{s,min}$) to keep and the steel stress $\sigma_s \approx \sigma_{sr}$ (compatible to (M_{Ed}/M_{cr})) maximum bar diameter ϕ_s (adjusted by Eq. 7) seems to be correct as $(\varepsilon_{sm} - \varepsilon_{cm})$ in Eq. 2 do not change.

However, the steel area A_s bigger than $A_{s,min}$, has a significant impact on maximum crack spacing $s_{r,max}$ for close reinforcement distribution. The reinforcement ratio for effective area of tensioned concrete $\rho_{p,eff}$ calculated for A_s is greater than for $A_{s,min}$, so, according to definition in Eq. 2, spacing $s_{r,max}$ for A_s is shorter than for $A_{s,min}$. This means, that for A_s the crack width will be narrower than for $A_{s,min}$, even for the same steel stress.

It was confirmed by numerical calculations, where for $A_s > A_{s,min}$, using the procedure described in Sect. 2.1 with the same other assumptions, higher values of σ_s, to control the cracking were obtained (see Fig. 4 for $w_k = 0, 3$ mm).

The increase in allowable steel stress of Table 7.2N to control the load cracks is significant, varying within range 0,08 to 0,20 of ratio $A_s/A_{s,min}$. For this reason, adjusting the amount of tension reinforcement to control the cracks directly based on size-stress relations presented in this table could lead to uneconomical increase of tension reinforcement amount in elements subjected to bending.

References

EN 1992-1-1 Eurocode 2 (2004) Design of concrete structures - Part 1-1: general rules and rules for buildings. European Committee for Standardization

Eurocode 2 Commentary (2008) European Concrete Platform ASBL

The Influence of the Molecular Structure of Polyurethane on Vibro– and Electroinsulation Properties of the Tramway Structures

Andrii A. Plugin(✉), Dmytro A. Plugin, Oleksii A. Pluhin, and Olga S. Borziak

Ukrainian State University of Railway Transport, Feuerbach Sq. 7, Kharkiv 61050, Ukraine
aaplugin@gmail.com

Abstract. Theoretical and experimental investigations of the influence of the molecular structure of polyurethane on its vibro- and electroinsulation properties have been carried out. It was established that the rheological model of polyurethane and rubber corresponds to the Kelvin resilient-&-elastic solid. However, due to the lateral electrostatic repulsion between the $C = O$ groups polyurethane has a higher resilient component under compression and subsequent reverse deformation and a higher elastic constituent during the reset after the tension. It conditions the enhanced ability of polyurethane to absorb vibrations in comparison with rubber. Due to the energonegativity, the $C = O$ groups provide a high electrical resistance for the polyurethanes. The experimental data show that polyurethane compositions provide a decrease in the vibration velocity of the rail seat plate in comparison with the track on rubber rail pads and an increase in the electrical resistance between the rail and the plate. The availability of the $C = O$ groups in polyurethane compositions is confirmed by the analysis of the IR-spectra of absorption. When polyurethane is operated under insulation conditions the intensity of $C = O$ lines is decreased with time and it is indicative of the rupture of double bonds, loss of those groups and the degradation of rheological and electric properties.

Keywords: Polyurethane · Rubber · Rheological properties · Railway · Vibration isolation · And electrical insulation

1 Introduction

The motor transport, railway transport, subway and the trams are the sources of intense noise, vibrations, aggressive leakage currents and excess electric charges. The noise and vibrations cause adverse health effect and the vibrations have a destructive effect on building structures and the buildings. Therefore, starting from the latter half of the twentieth century the Europe began to pay much attention to a decrease of noise and vibrations induced by the transport. As for the highways, this problem is solved to a large extent through the arrangement of noise barriers. However, these noise barriers

Z. Blikharskyy et al. (Eds.): CEE 2019, LNCE 47, pp. 346–353, 2020.
https://doi.org/10.1007/978-3-030-27011-7_44

are efficient to a lesser extent with regard to the rail transport. According to Vuychak (2009) a more efficient approach is the insulation of the rails and rail seats from the soil basement using vibration absorption materials. Skripinets et al. (2015) have established empirically that the polyurethanes and polyurethane-based compositions are the most effective vibration absorption materials. The structure of polyurethane is studied in detail by Lipatov (1982), Wang (2013). Palant et al. (2018) and Plugin et al. (2018) have shown that such polyurethane-based structures provide an efficient electric insulation that reliably protects from the action of leakage currents and excess electric charges, and it also prevents the electrocorrosion, loss of traction currents and the failure of automatic shut-off systems, provides high durability of tramway structures in the face of these influences. However, the nature of a high efficiency of polyurethane as a vibro-and electroinsulation material is still underinvestigated. Therefore, the investigation of the influence of the molecular structure of polyurethane on its vibration and electric action protective properties remains to be a vital task. This scientific paper is the continuation of the research done in works Palant et al. (2018) and Plugin et al. (2018). The authors express their gratitude to Olena Palant for the active participation in experimental investigations.

2 Materials and Research Methods

Polyurethane-based two-component (with the hardening agent) compositions designed for the rail insulation EdilonCorkelast® (Edilon Sedra B.V., Netherlands), Icosit C340/45F (Sika AG, Switzerland), and the "X" composition (Kharkiv, Ukraine) have been investigated.

The vibro- and electroinsulation properties of polyurethane compositions were studied using the specimens that simulate the insulation of the rail fixed by the composition in the rail duct from the reinforced concrete plate in which this duct is arranged (Fig. 1a). The vibration parameters, in particular the vibratory displacement and the vibration rate of the reinforced concrete plate exposed to the action of the running rolling stock or artificial vibration generator were defined according to the scheme given in Fig. 1a using the vibration analyzer Vibran-2.0. The technical performances of the Vibran-2.0 are as follows: the vibration velocity measurement range is 0.1 to 300 mm/s; the vibratory displacement measurement range is 0.01 to 10 mm; and the working frequency range is 0.5 to 1000 Hz.

The electric resistance was measured according to the scheme given in Fig. 1b using the digital multimeter Sanwa PC510 with the input resistance of 10 MOhm and a special electrode (Fig. 1c) in which the contact with the concrete surface was provided by the porous sponge impregnated with the saturated solution of copper sulfate $CuSO_4 \times 5H_2O$.

The availability of functional groups with double bonds in polyurethane compositions and the rubbers was defined by analyzing infra-red absorption spectra that were obtained by the reflection survey using the IR-Fourier Spectrometer Bruker Alpha. The IR-spectra were taken in the domain of wave numbers 480 to 4000 cm^{-1}.

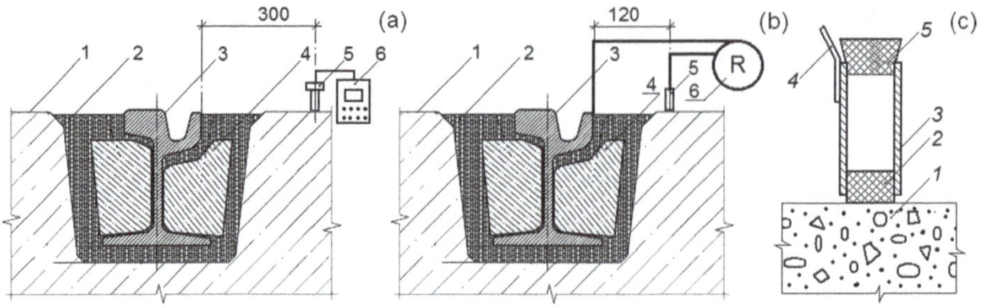

Fig. 1. Investigation of the vibro- and electroinsulation properties of polyurethanes: the measurement circuits of the vibrations (a) and the electrical resistance (b) in the rail-concrete loop: 1 – the reinforced concrete rail seat; 2 – the polyurethane paste; 3 – the rail; 4 – the rail liner; 5 – the vibration sensor (a) or special electrode for the concrete contact (b); 6 – the vibration analyzer (a) or digital multimeter (b); special electrode (c): 1 – the reinforced concrete rail seat; 2 – the sponge impregnated with the saturated copper sulfate solution; 3 – the casing made of stainless steel; 4 – the wire; 5 – the rubber plug

3 Analytical Review of Literature Sources and Theoretical Research

Lipatov et al. (1982) have defined the polyurethane structure as that of the domain type (Fig. 2a) that consists of the output rigid-chain domains 1 and flexible-chain matrix 2. Such a structure corresponds to the structure of disperse system with the solid disperse phase and disperse environment created by flexible fibers. Such a pattern of the structure is confirmed by the polyurethane images that were obtained with the scanning electron microscopy (Lian, Huimin and Zhaofei, 2014) and atomic force microscopy (Anandhan and Lee, 2014), Fig. 2b, c.

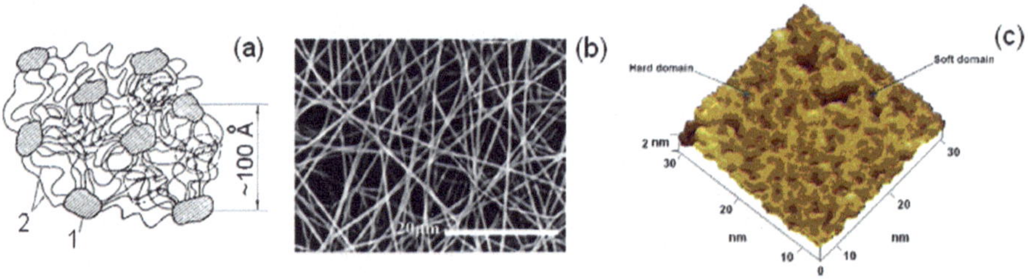

Fig. 2. The polyurethane structure: (a) – the model of the microdomain structure of polyurethane according to Lipatov et al. (1982); (b) – the soft fine-fibrous phase, scanning electron microscope (Lian, Huimin and Zhaofei, 2014); (c) – the 3D image of thermopolyurethane, atomic force microscope (Anandhan and Lee, 2014)

The rubber is characterized by the analogous structure. Such a structure conditions the correspondence of the rheological properties of polyurethane and rubber for the Kelvin resilient solid model (Fig. 3) that is characterized by elastic aftereffect strains and the vibration absorption capacity (Plugin et al. 2012). Plugin et al. (2018) assume that the rheological properties of polyurethanes are subjected to the influence of the functional groups > $C{=}O$ with double bonds that are available in the polyurethanes (Fig. 4) and unavailable in the rubbers. These groups are characterized by an essential dipole moment of 2.5 D and their negative pole is turned outside of the macromolecule inducing an essential negative surface charge on it. The availability of $C = O$ groups on the surfaces of structural elements (globules and fibers) defines the difference in the rheological properties of polyurethane and rubber. However, Plugin et al. (2018) have not suggested the rheological model that would take into account the effect of these groups.

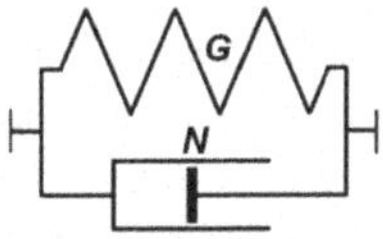

Fig. 3. The rheological model of polyurethane and rubber, i.e. Kelvin resilient solid model that consists of the ordinary rheological G – Hookean and N – Newton solids

We believe that the rheological model of polyurethane is more complicated in comparison with the Kelvin model, as cited in Plugin et al. (2012), and particularly this specific feature conditions its much higher vibration absorption capacity.

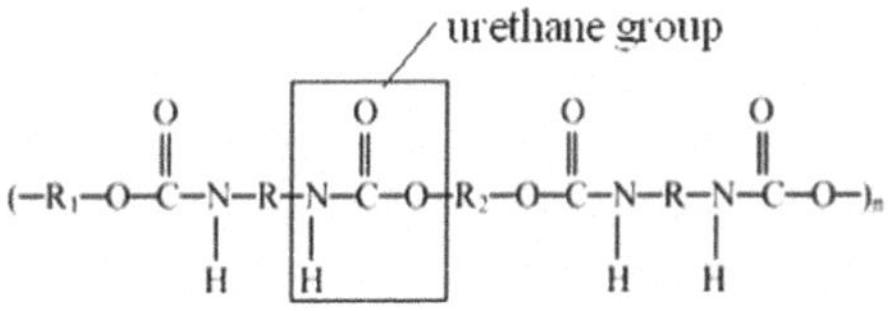

Fig. 4. Polyurethane molecular structure

Let's consider the operation of the rheological model of polyurethane taking into consideration the lateral electrostatic (dipole-dipole) repulsion between the $C = O$ groups. During the compression of the structural element of polyurethane its surface parallel-oriented $C = O$ dipoles enhance the elastic component due to the lateral electrostatic repulsion. It can be simulated using additional rheological Hookean solids (Fig. 5a).

The removal of load results in the reverse deformation of elastic after-action during which electrostatic repulsion forces try to reset the structural element. With the tension of structural element (Fig. 5b) the repulsion between its $C = O$ dipoles mitigates. Therefore, during the tension of the structural element the deformation occurs at a lower resistance in comparison to that of the compression. After the removal of load during the reverse deformation caused by elastic after-action the structural element tries to reset under the action of cohesive forces. However, due to this, the repulsion

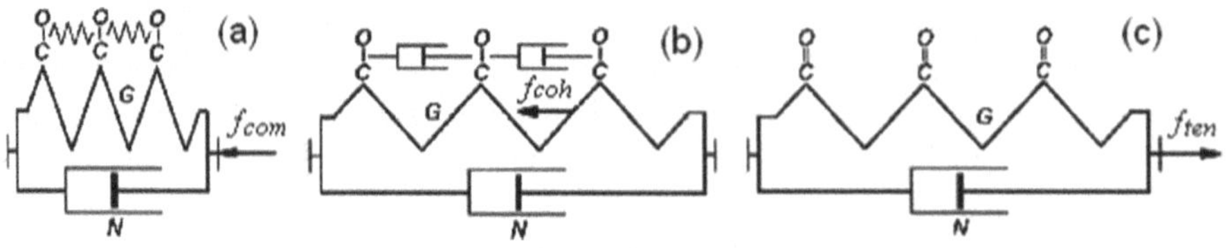

Fig. 5. Rheological models of the deformation of the structural polyurethane element (globules, fibers): (a) – compression and the reverse deformation after the removal of compressing stress; (b) – tension; (c) – reverse deformation after the removal of tensile loading. *G* – Hoogean rheological solid; *N* – Newton rheological solid; f_{com} – compressive force; f_{ten} – tensile force; f_{coh} – cohesion forces

between the $C = O$ dipoles is intensified and the reset is retarded. It can be simulated using additional rheological Newton solids (Fig. 5a).

In aggregate, it corresponds to the resilient-elastic deformation of the structural element with an enhanced resilient component during the compression and the reverse deformation after it and enhanced Newton deformation component during the reset after the tension and it conditions high vibration absorption of polyurethane in comparison with rubber.

We also believe that polyurethane provides a higher electrical resistance for isolating elements and the parts in comparison with rubber and it is also able to prevent the accumulation of excess negative charge (macropolarization) in isolated structures due to the availability of $C = O$ groups with a high electric negativity (Fig. 4). These groups provide a negative charge for porous space surfaces that prevents the transfer of the carriers of electric charge though it, i.e. the anions (due to the repulsion) and the cations (due to the absorption). Hence, polyurethane is characterized by a maximum ability to absorb vibrations due to the lateral electrostatic repulsion between the $C = O$ groups with double bonds and a high electrical resistance and the ability to prevent the accumulation of excess electric charges due to the electric negativity of those groups.

4 Experimental Investigation Data

To confirm the conclusions made about the influence of $C = O$ group on the rheological properties of polyurethanes we analyzed the IR-spectra of polyurethanes and rubbers and the most typical of them are given in Fig. 6. Wang et al. (2013) have shown in that the valence vibrations of $C = O$ group that characteristic of non-bonded urethane stretching are defined by the line that is situated in the range of wave numbers from 1650 to 1750 cm^{-1}. The intensive line is observed in this range in the IR spectra of all the tested polyurethanes (Fig. 6a, b, Table 1), while it is unavailable for rubber spectra (Fig. 6c, Table 1). The spectra lines confirm the availability of the $C = O$ group in the polyurethanes and as a consequence its influence on rheological and electric properties that differ from those of the rubbers. A decrease in the intensity of the lines of $C = O$ group in the polyurethanes is also noticed with time (Fig. 6b, Table 1) in case of their operation under insolation conditions and it is indicative of the rupture of double bonds, the loss of those groups and as a consequence the degradation of rheological and electric properties.

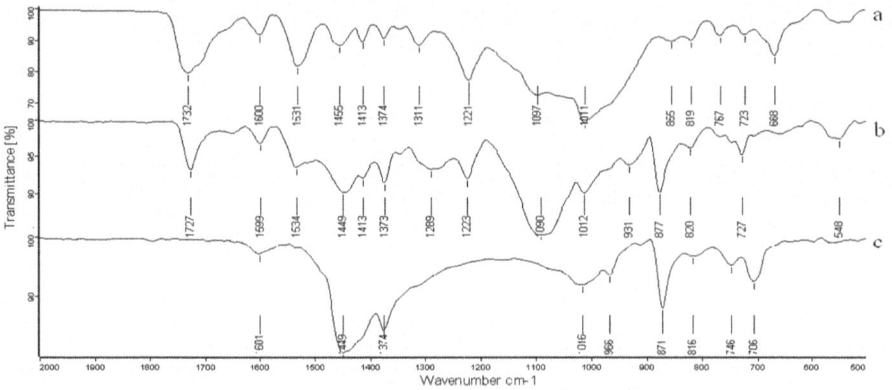

Fig. 6. IR-spectra: (a) – a new polyurethane composition EdilonCorkelast; (b) – after the six years of operation; (c) – the rail lining rubber for the traditional rail track structure

Table 1. Characteristics of the lines of $C = O$ group in the IR-spectra obtained for the polyurethanes and rubbers

Specimen	The wave number for the lines of $C = O$ group	Absorption intensity, %
Edilon Corkelast	1732	23
Edilon Corkelast after the 6 years of operation	1727	8
Icosit C340/45F	1728	20
«X» polyurethane composition	1726	55
Rail lining rubber	Unavailable	
Raw rubber	Unavailable	

To confirm theoretical research data we carried out the experimental investigations of the vibration parameters and the electrical resistance of polyurethane compositions and the rubbers. Table 2 and Fig. 7 give the research data on vibration parameters and the electrical resistance.

Table 2 and Fig. 7 show that polyurethane compositions provide the 27% decrease in the oscillation rate of the vibrations of the plate of rail seat in comparison with the rail-track on rail lining rubber and the electric resistance between the rail and the plate is increased by a factor of 3.9. This fact confirms the conclusion that particularly the availability of the $C = O$ group with double bonds in polyurethane provides for it a high vibration damping capacity and high electric resistance.

Table 2. Full-scale test data obtained for tram track vibration parameters and the electrical resistance of it

Parameter description	Unit of measurement	A value for the rail		Parameter improvement, %
		With Edilon Corkelast isolated rail	Traditional with rubber liners	
Oscillation rate	mm/s	0,62	0,85	27
Vibratory displacement	mm	0,22	0,27	19
Electric resistance in the loops:				
– rail-plate	MOhm	3,9	1	290
– rail-ground	«	–	0,016	–

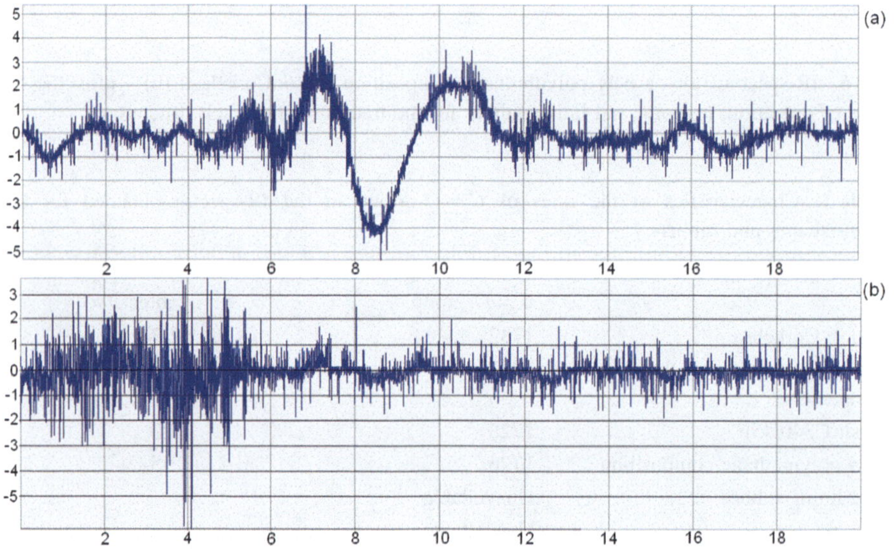

Fig. 7. The vibrogram of the vibrations of reinforced concrete plate during the tram traffic: (a) – with the Edilon Corkelast isolated rail; (b) – traditional with rubber liners

5 Conclusions

The rheological model of polyurethane and rubber corresponds to the Kelvin resilient-elastic solid model. However, due to the lateral electrostatic repulsion (dipole-dipole) between the $C = O$ groups with double bonds polyurethane shows an enhanced resilient component during the compression and the reverse deformation after it and enhanced Newton component during the reset after the tension. It conditions a higher vibration absorption capacity of polyurethane in comparison to rubber. Due to their electro-negativity the $C = O$ groups provide a high electrical resistance for the polyurethanes.

The availability of $C = O$ groups in polyurethane compositions is confirmed by the availability of the intensive line in IR-spectra in the range of wave numbers 1650 to 1750 cm^{-1} that is unavailable in rubber spectra. Polyurethane compositions provide the 27% decrease in the oscillation rate of the vibrations of the rail tracked on the rail seat plate in comparison with the rail tracked on rubber rail liners and the electrical resistance between the rail and the plate is increased by a factor of 3.9. When polyurethane is held (operated) under insolation conditions the intensity of this line is decreased and it is indicative of the rupture of double bonds, the loss of those groups and as a consequence the degradation of rheological and electric properties.

Acknowledgements. The activity presented in the paper is part of the research grant «Theoretical and experimental bases of determining, predicting and ensuring the carrying capacity and durability of transport facilities under aggressive conditions».

References

Anandhan S, Lee HS (2014) Influence of organically modified clay mineral on domain structure and properties of segmented thermoplastic polyurethane elastomer. J Elastomers Plast 46 (3):217–232

Lian Z, Huimin L, Zhaofei O (2014) In situ crystal growth of zeolitic imidazolate frameworks (ZIF) on electrospun polyurethane nanofibers. Dalton Trans 43:6684–6688

Lipatov VS, Shilov VV, Gomza YuP, Kruglyak NE (1982) X-ray methods for studying polymer systems. Kyiv, Naukova dumka (in Russian)

Palant O, Plugin D, Plugin A, Lobiak A, Pluhin O (2018) Use of liners made of concrete on quartz aggregates and polyurethane composition to reduce vibration, noise and improve the stability of tramway track. Matec Web of Conf 230:03014

Plugin AN, Palant OV, Plugin DA, Pluhin OA, Borziak OS (2018) Mechanism of protective properties of polyurethane and compositions on the basis of liquid glass against eletric and vibrational effects. Bulletin National Tech Univ "KhPI" Series: Chem Chem Technol Ecol 35 (1311):25–28

Plugin AN, Plugin AA, Trikoz LV (2012) Fundamentals of the theory of hardening, strength, destruction and durability of Portland cement, concrete and structures of them. Naukova dumka, Kyiv (in Russian)

Skripinets A, Dachenko Y, Kabus A (2015) A research on technological and physicochemical laws of manufacturing vibration-absorbing products based on epoxy-urethane polymer compositions. Eastern-Eur J Enterp Technol 3(11):4–8

Vuychak I (2009) Modern constructions of a ballast-free canvas as a solution to the operational problems of rail roads. Collected Sci Works Ukrainian State Univ Railway Transp 109:93–113 (in Russian)

Wang H, Liu Y, Tian J, Sun B, Huang S (2013) FT-IR analysis of molecular structure evolvement of poly (Ether urethanes) in Ozone atmosphere. Adv Mater Res 742:215–219

Porous Thermal Insulation Materials on Organic and Mineral Fillers

Petro V. Novosad[1], Oksana R. Pozniak[1(✉)], Volodymyr M. Melnyk[2], and Serhii P. Braichenko[1]

[1] Department of Building Production (DBP), Lviv Polytechnic National University, Bandera Street, 12, Lviv 79013, Ukraine
pozniak@ukr.net

[2] Department of Business Economics and Investment (DBEI), Lviv Polytechnic National University, Bandera Street, 12, Lviv 79013, Ukraine

Abstract. The article presents the results of research aerated concrete properties, in which a part of the mineral filler is replaced by a light, environmentally friendly organic filler – flax straw. The flax straw is a renewable and 100% natural waste of the agricultural industry, which can be widely used to obtain thermal insulation materials. It has been shown that the partial replacement of the mineral filler in aerated concrete by the flax straw can lower its average density by 44 and 51%, respectively. The influence of water-solids ratio on strength and average density of flax aerated concrete is investigated. It is established that the optimum value of water-solids ratio is 0.8–1, at which the average density of aerated concrete is 300–400 kg/m^3 and the compressive strength is 0.62–1.1 MPa. It is shown that in order to decrease the average density of aerated concrete on the basis of vegetative fillers, the temperature of the aerated concrete mixture should be 35–45 °C. Thus, the developed thermal insulation material – aerated concrete on the basis of organic and mineral fillers can be used for the construction of low-energy buildings, which is consistent with the principles of sustainable development in construction.

Keywords: Aerated concrete · Flax straw · Sustainable development

1 Introduction

Minimizing cost of energy and material resources in the process of construction and operation of building structures with a reduction of negative environmental impact in accordance with the principles of sustainable development in construction (International Conference on Environment and Development 1992), as well as the provisions of the International Climate Agreement (United Nations Climate Change Conference, COP 21 2015), aimed at reducing and, in the long term, eliminating greenhouse gas emissions, will become a priority in the new construction, the reconstruction and modernization of the existing housing stock (Sanytsky et al. 2013; Brzyski et al. 2017).

The need to restrict the consumption of traditional energy resources globally, reducing the negative impact on the environment, as well as increasing the energy and operating cost of a building, result in the domination of providing the comfort of

Z. Blikharskyy et al. (Eds.): CEE 2019, LNCE 47, pp. 354–360, 2020.
https://doi.org/10.1007/978-3-030-27011-7_45

costumers with the least energy consumption in modern construction (Kaniszewska 2013; Savytskyi et al. 2015; Marushchak et al. 2018).

Ensuring energy saving in the housing and communal sector of Ukraine is achieved by the construction of energy-efficient buildings at the expense of materials with improved thermal insulation properties, in particular aerated concrete (Pozniak and Melnyk 2014; Pozniak et al. 2018). In order with that aim, it is necessary to apply environmentally friendly solutions, such as the materials derived from plants that absorbed CO_2 during the entire period of growth (Arrigoni et al. 2017). These materials are capable to mitigate the impact on environment of other components, such as Portland cement, the manufacturing process of which involves substantial CO_2 emissions. The using of thermal insulation materials reduces the heating demand, thus decreasing the ecological footprint of the building during its operation. Thermal insulation materials can be manufactured using materials derived from such plants as flax (Babenko et al. 2018), hemp (Arrigoni et al. 2017; Schwarzova et al. 2017a, b; Preikss et al. 2013), sunflower (Binici et al. 2014). The use of flax waste in the technology of aerated concrete allows reduce its average density, with creating a cellular and fibrous structure of aerated concrete (Bernatsky and Smirnova 2007).

The aim of this work is to study the influence of flax straw on the rheological properties of aerated concrete mixes and the properties of aerated concrete.

The scientific novelty of the presented research is the possibility of obtaining thermal insulation materials by partial or complete replacement of a mineral filler in obtaining aerated concrete with an ecologically pure agricultural waste – a flax straw. Flax fibre reinforced aerated concrete can be used for construction of low-energy buildings, which corresponds to the principles of sustainable development in construction.

2 Materials and Methods

To conduct the study, the Portland cement CEM I 42,5 R produced at PSH “Ivano-Frankivsk cement” (Yamnitsa, Ivano-Frankivsk region, Ukraine) with specific surface of 342 m^2/kg, the retaining on sieve № 008 is 0.4%, the initial setting time is 1 h 10 min, the final setting time is 5 h 30 min was used. Natural sand (fineness modulus of 1.4–1.5) and fly ash (FA) from the Burshtyn TPP were used for aerated concrete production as a finely dispersed filler. Fly ash from Burshtynska thermal power plant had the following properties: true density - 2.21 g/cm^3; bulk density - 870 kg/m^3; retaining on sieve № 008-8.7 mass.%; chemical composition, mass.%: $SiO_{2\text{-}5}4$; Al_2O_3 – 23.75; Fe_2O_3 + FeO – 13.8; MgO – 1.91; CaO – 4.98; SO_3 – 0.53; K_2O + Na_2O – 0.25. Flax straw was used as an organic filler (Fig. 1). The organic filler exhibits low density (roughly 100 kg/m^3) and high porosity (approximately 75%). To obtain an aerated structure, as a gas forming agent, the aluminium powder (content of active aluminium is 82%, fineness of grinding is 5000…6000 cm^2/g) was used.

In this work, the compositions containing flax straw as a filler, as well as compositions, where the part of the flax straw was replaced by the fly ash and quartz sand in obtaining a non-autoclaved aerated concrete were investigated.

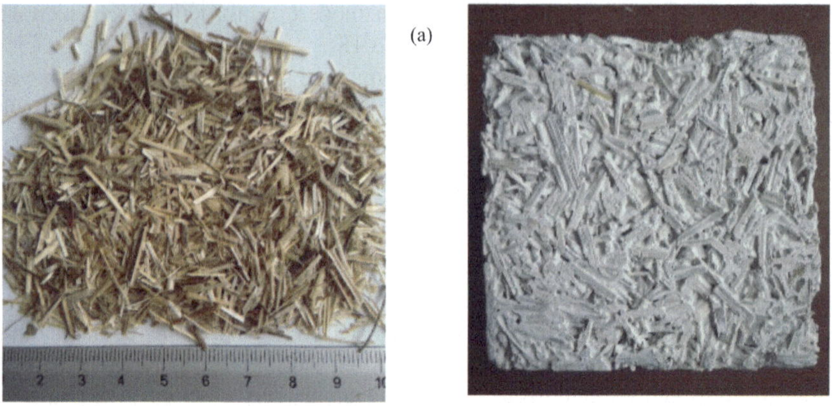

Fig. 1. Flax straw (a) and flax straw porous thermal insulation materials (b)

For this purpose samples of size 100 × 100 × 100 mm (Fig. 1) were moulded. Value of filler-binder ratio varied within 1:1–1:1.5 (wt. %). In order to provide the alkaline environment and the process of gas emission, lime was added to the mixture. The properties of specimens were determined after 28 days of curing. The density and compressive strength were carried out according to national standards.

3 Results and Discussion

The solution of problem of reducing the average density of aerated concrete is possible by using components with less density. As a component of widespread use plant-based fillers have been introduced (Pacheco-Torgal and Jalali 2011). In this work the properties of aerated concrete in which the siliceous component (quartz sand, fly ash) was partially or completely replaced by flax straw were investigated. As can be seen from Fig. 2, the aerated concrete obtained on the basis of composition with filler-binder ratio 1:1, in which the filler is the flax straw or flax straw with fly ash is characterized by the lowest average density.

Thus, the average density of aerated concrete with a filler-binder ratio 1:1 on the basis of fly ash is 535 kg/m^3. Replacing the part of fly ash with flax straw allows to obtain aerated concrete with density of 300 kg/m^3, while the density of aerated concrete on the flax straw filler is 290 kg/m^3. The replacement of fly ash for sand in the composition where as a filler flax straw and fly ash were used results in increasing in the average density of aerated concrete to 355 kg/m^3. The average density of aerated concrete with filler-binder ratio 1:1 on the basis of sand is 720 kg/m^3. It should be noted that composition with filler-binder ratio 1:1,5 is characterized by the highest compressive strength (R = 2.8 MPa) after 28 days of hardening in normal conditions (Fig. 3). The compressive strength of aerated concrete composition with a filler-binder ratio 1:1, in which as a filler fly ash is used reaches 1.49 MPa, sand – 1.92 MPa, flax straw – 0.75 MPa, flax straw and fly ash – 0.7 MPa, flax straw and sand – 1.8 MPa.

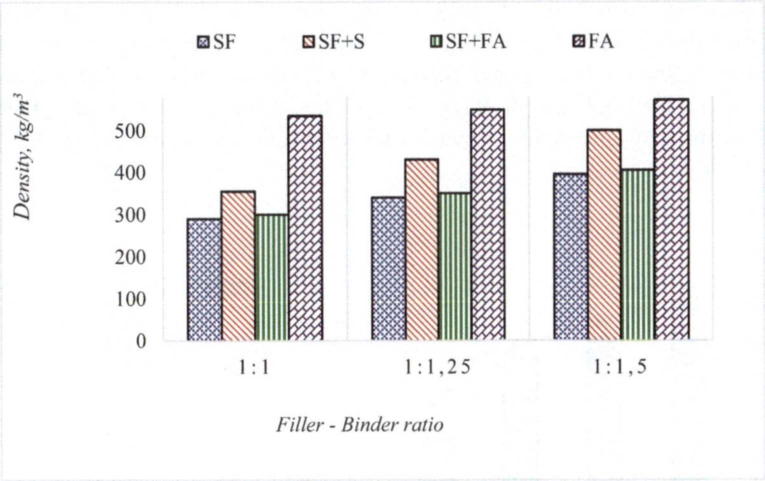

Fig. 2. Dependence of the density of the aerated concrete on the type of filler (SF – flax straw; S – sand; FA – fly ash)

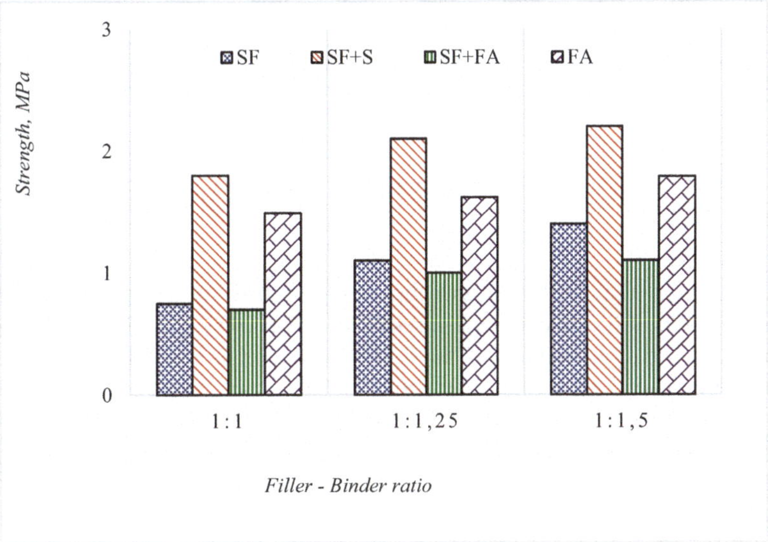

Fig. 3. Dependence of the compressive strength of the aerated concrete on the type of filler (SF – flax straw; S – sand; FA – fly ash)

Important parameters that provide the rheological properties of the concrete mixture and the properties of the aerated concrete are water-solids ratio (W/S) and the temperature of the aerated concrete mixture. Studies of the change in the flowability of aerated concrete mixtures on an organic filler (flax straw), which is determined by the diameter of aerated concrete mixture spill using Suttard cylinder, found that with a water-solids ratio of 0.8; 0.9; 1; 1.05, the diameter of aerated concrete mixture spill incorporating vegetable filler was 14.5; 19; 23 and 27 cm. The recommended values of the diameter of the aerated concrete mixture spill for the production of aerated concrete are 18–24 cm (Pozniak and Melnyk 2014). The average density and compressive strength of the aerated concrete versus water-solid ratio are shown in Fig. 4.

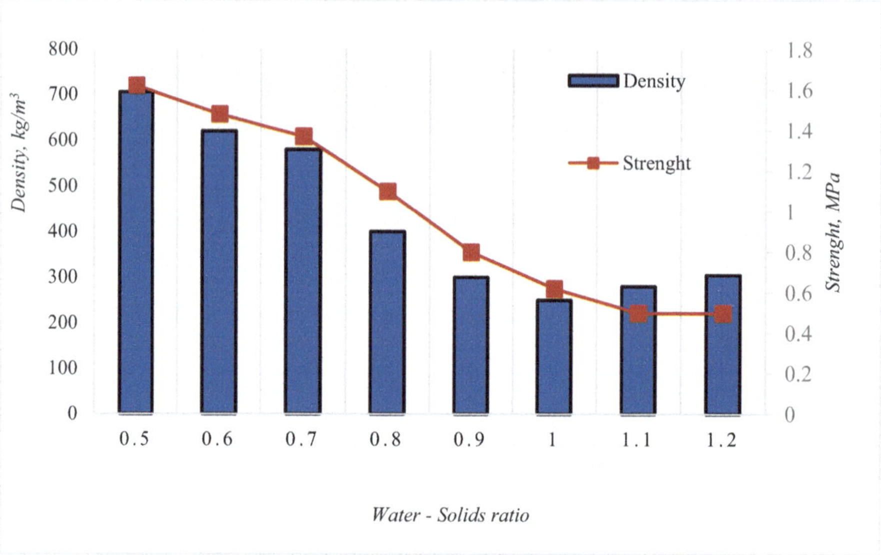

Fig. 4. Dependence of the density and compressive strength of the aerated concrete on the water-solids ratio

Data analysis (Fig. 4) showed that the average density of aerated concrete with a water-solids ratio of less than 0.8 increases from 400 to 705 kg/m^3. It is explained by a smaller multiplicity of swelling of the aerated concrete as a result of increasing its viscosity. Increasing water-solids ratio above 1.05 does not lead to a noticeable change in the average density of aerated concrete, but compressive strength significantly decreases. Thus, the compressive strength of aerated concrete on an organic filler with water-solids ratio 0.8 is 1.1 MPa and an average density of aerated concrete is 400 kg/m^3, with W/S = 1.1 compressive strength is 0.5 MPa and an average density is 280 kg/m^3.

The results of studies of the effect of the aerated concrete mixture temperature on both the average density and the strength of the aerated concrete incorporating flax straw are presented in Fig. 5.

It is established that the optimum for swelling of an aerated concrete mixture is the temperature of 35–42 °C. With a further increasing of temperature, the time of setting of the mixture decreases, what prevents it from swelling. Reducing the temperature of the mixture leads to a slowing down of the gas emission reaction and setting of the mixture, which results in shrinking swelling mass.

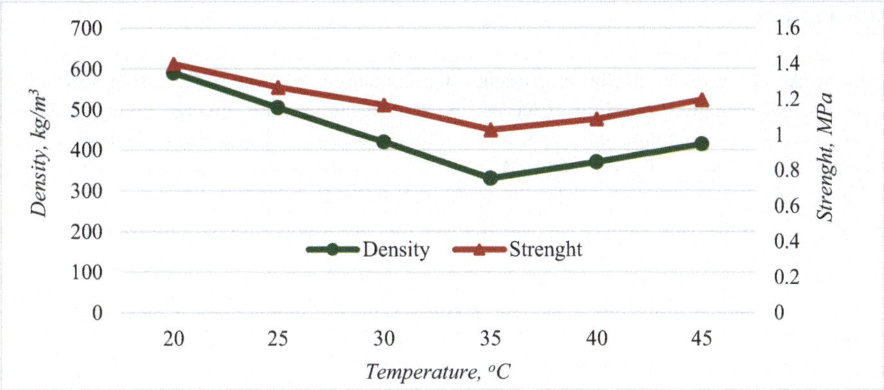

Fig. 5. The influence of the temperature of the aerated concrete mixture on the average density and strength of aerated concrete on organic filler

Thus, designed flax straw reinforced aerated concrete can be used to build low-energy buildings, which provides both the reduction of the natural raw materials using and utilisation of agricultural wastes and as a result the reduction carbon dioxide emission into the environment.

4 Conclusions

The presented researches determine the possibility of agricultural waste using, in particular flax straw, for the production of porous, environmentally friendly insulation materials. The conducted researches have been established that partial replacement in the composition of aerated concrete fly ash or quartz sand by the flax straw can reduce its average density by 44 and 51% respectively, but the compressive strength time decreases at the same. Thus, both the fly ash and quartz sand as fillers in aerated concrete mixtures can be partially or completely replaced by the flax straw for the production of aerated concrete. It is shown that the optimum value of water-solids ratio is 0.8–1, at which the samples of flax fibre reinforced aerated concrete have an average density of 300–400 kg/m^3 and compressive strength of 0.62–1.1 MPa. The diameter of the aerated concrete mixture spill at these values of water-solids ratio is equal 14.5–23 cm. To reduce the average density of the aerated concrete containing vegetable filler, it is necessary to increase the temperature of the mixture up to 35–45 °C. The developed aerated concrete can be used in the construction of low-energy buildings.

References

Arrigoni A, Pelosato R, Melia P, Ruggieri G, Sabbadini S, Dotelli G (2017) Life cycle assessment of natural building materials: the role of carbonation, mixture components and transport in the environmental impacts of hempcrete blocks. J Clean Prod 149:1051–1061

Babenko M, Estokova A, Savytskyi M, Unčнk S (2018) Study of thermal properties of lightweight Insulation made of flax straw. Slovak J Civ Eng 26(2):9–14

Bernatsky A, Smirnova O (2007) The use of waste flax processing for the production of thermal insulation products. News Univ Constr 3:42–46 [in Russian]

Binici H, Eken M, Dolaz M, Aksogan O, Kara M (2014) An environmentally friendly thermal insulation material from sunflower stalk, textile waste and stubble fibres. Constr Build Mater 51:24–33

Brzyski P, Barnat-Hunek D, Suchorab Z, Łagyd G (2017) Composite materials based on hemp and flax for low-energy buildings. Materials 10:510

Kaniszewska A (2013) Introduction to energy-saving and passive construction. http://www.chronmyklimat.pl/projekty/energooszczedne [in Polish]

Marushchak U, Sydor N, Sanytsky M, Braychenko S (2018) Research of impact resistance of nanomodified fiber reinforced concrete. MATEC Web of Conf 230:03012

Pacheco-Torgal F, Jalali S (2011) Cementitious building materials reinforced with vegetable fibres: a review. Constr Build Mater 25:575–581

Pozniak O, Melnyk A (2014) Non-autoclave aerated concrete from modified binders composition containing supplementary cementitious materials. Civ Eng Arch Lub Univ Technol 13(2):127–134

Pozniak O, Sanytsky M, Zavadsky I, Braichenko S, Melnyk A (2018) Research into structure formation and properties of the fiber-reinforced aerated concrete obtained by the non-autoclaved hardening. East Eur J Enterp Technol 3/6(93):39–46

Preikss I, Skujans J, Adamovics A, Iljins U (2013) Evaluation of hemp (Cannabis Sativa L.) quality parameters for building materials from foam gypsum products. Chem Eng Trans 32:1639–1644

Sanytsky M, Poznyak O, Marushchak U (2013) Energy-saving technologies in construction. Lviv Polytechnic Publishing House, Lviv [in Ukrainian]

Savytskyi M, Ozhyshchenko O, Babenko M, Koval A (2015) Integrated use of agricultural waste for the sustainable development of agricultural socio-eco-complexes as closed ecosystem. Constr Mater Sci Mech Eng 82:179–186

Schwarzova I, Stevulova N, Melichar T (2017a) Lightweight composites based on technical hemp hurds in construction industry. Chem Eng Trans 57:1369–1374

Schwarzova I, Stevulova N, Junak J, Hospodarova V (2017b) Sustainable hemp-based composites for the building industry application. AIP Conf Proc 1866(1):040036

The Application of the Superpave Method of Climatic Zones Analysis in Poland with Regard to Bitumen Performance Grading

Marek Pszczola(✉), Dawid Rys, and Piotr Jaskula

Department of Highway and Transportation Engineering,
Gdansk University of Technology, Narutowicza 11, 80-233 Gdansk, Poland
marek.pszczola@pg.edu.pl

Abstract. Currently in Poland, similarly as in the other EU countries, road bitumens are tested and classified with regard to mainly the penetration value determined at the temperature of 25 °C. It should be noted that this classification is not correlated with the climatic conditions in which the bitumens are to serve in the road pavement. Towards the end of the last century a new system of bitumen grading was developed and implemented as part of the American Strategic Highway Research Program (SHRP). This paper presents the results of analyses concerning the determination of the temperatures at which bituminous binders serve in the Polish climatic conditions. A division of the area of Poland into climatic zones depending on the required performance grades (PGs) determined for bitumens on the basis of climatic data from meteorological stations for a period of minimum 20 years is presented. It is proposed to select bitumens depending on the climatic zone in Poland, taking into account the proper probability level which follows from the road class.

Keywords: Bitumen grade · Asphalt pavement · Performance grade (PG) · Climatic zones · Low and high temperatures

1 Introduction

In Poland, similarly as in the other EU countries, bitumens intended for road purposes are tested and classified with regard to their penetration value at 25 °C, irrespective of the climatic conditions. According to current standard PN-EN 12591: 2010, road bitumens are classified not only depending on the penetration test result, but also on the basis of the results of: the Ring and Ball Softening Point test, the Fraass Breaking Point test, the Flash Point test and the Solubility test. In the case of bitumens modified with the SBS elastomer in accordance with standard PN-EN 14023: 2011/Ap1:2014-04, besides the tests mentioned above also elastic recovery at 25 °C and storage stability are tested. In the 1900s in the USA a new system of classifying bitumens, based on asphalt binder performance tests, was developed. By now the system has been introduced in most of the US states (Floyd et al. 2013) and also in Canada (Mills et al. 2009). Relevant research has also been conducted in other countries around the world, e.g. in Pakistan (Waseem Mirza et al. 2011) and Sri Lanka (Mampearachchi et al. 2012). Also in the East-European countries: Latvia, Estonia, Republic of Belarus (Leonovich and Melnikova 2012) and Ukraine

Z. Blikharskyy et al. (Eds.): CEE 2019, LNCE 47, pp. 361–368, 2020.
https://doi.org/10.1007/978-3-030-27011-7_46

(Zolotareyov and Pyrig 2018) the studies have been started to support their national classification by PG system according to climatic conditions that are typical for each country. The example of research conducted for countries: (a) Republic of Belarusand (b) Ukraine was presented in Fig. 1.

(a) (b)

Fig. 1. Climatic zones according to Superpave method established for: (a) Republic of Belarus (b) Ukraine

In Poland the first analysis was conducted in 2017 (Pszczola et al. 2017). The new system of classifying bitumens is called the Superpave Performance Grading (PG) system. The idea behind the new system of grading the performance of bitumens was to better adjust the types of tests and the requirements to the actual conditions in which bituminous binders serve in road pavements. The performance grading of bitumens is the result of the Strategic Highway Research Program (SHRP), approved by the US Congress in 1987, aimed at improving the quality and durability of roads. One of the principal tasks of SHRP was to develop new asphalt specifications which would more closely correlate laboratory tests with the actual performance of bitumen in the asphalt pavement. The research resulted in the Superpave (Superior Performing Asphalt Pavements) mix design method. As part of this method a methodology for testing asphalt binders, comprising such tests as the Dynamic Shear Rheometer (DSR) Test, the Direct Tension Test (DTT) and the Bending Beam Rheometer (BBR) Test was developed. The laboratory test methods were needed in order to determine the performance of bitumens depending on the pavement service temperature. The bitumen specifications were developed by determining the asphalt pavement service temperatures in the particular climatic zones, under which the bitumen must meet the specific performance requirements. This paper presents the determination of the service temperatures of asphalt binders in the bituminous layers of road pavements in Poland in accordance with the original Superpave methodology. An analysis of the air and pavement temperatures in Poland is carried out and maps of temperature zones according to PG for the particular bituminous layers of the pavement structure are plotted.

2 Methodology for Determining Performance Grade Temperatures

The minimum and maximum pavement temperatures are the basis for establish of bitumen performance grade PG X-Y (Solaimanian 1994 and Superpave Performance Graded Asphalt Binder Specification and Testing 1995), where:

- X corresponds to the maximum asphalt pavement service temperature and is calculated as the highest averaged daily maximum pavement temperatures occurring for 7 consecutive days in a year,
- Y is the lowest pavement temperature.

The minimum period to be analysed in order to determine pavement temperatures X and Y is 20 years. Bitumen performance is graded at every 6 °C. For example, PG 58-16 means that the asphalt binder placed in a given bituminous pavement layer will perform properly in a temperature range from minimum –16 °C to maximum +58 °C at the specified level of probability. In Poland the methodology ultimately adopted for determining the PG of bitumens on the basis of climatic data in accordance with the original Superpave method and its later modified version LTPP-SMP (Mohseni 1998), covered:

(1) The adoption of a data analysis period of max 30 years (1986–2015) and the acquisition of temperature data for this period from meteorological stations located within Poland's borders. The available data measurement period could not be shorter than 20 years.
(2) The qualitative analysis of the air temperature data obtained from the particular meteorological stations for which the data were found to be incomplete or whose location was useless for the purpose (e.g. meteorological stations located on mountain peaks).
(3) The determination of the (lowest recorded) minimum air temperatures for a considered year.
(4) The determination of the maximum values of the 7-day average of the highest daily air temperatures occurring for 7 consecutive days in the course of each analysed year.
(5) The adoption of different levels of probability: 50%, 80%, 85%, 90%, 95% and 98%.
(6) The determination of the occurrence of the maximum averages and standard deviations for the maximum annual 7-day highest air temperatures and the minimum annual air temperatures in the whole data analysis period (minimum 20 years).
(7) Checking whether the air temperature distributions are normal distributions.
(8) Specifying at what depth in the pavement structure the bituminous layer is to be analysed. The depth at which the top surface of a given bituminous layer is situated was adopted for determining the minimum pavement temperature, whereas the depth of 20 mm below the top surface of the given bituminous layer was used to calculate the maximum pavement temperature.

The calculation of the PG X-Y values for the maximum (X) and minimum pavement temperatures (Y) in a given pavement course from Eqs. (1) and (2).

$$T^d_{max} = 54,32 + 0,78 \cdot T_{air} - 0,0025 \cdot \phi^2 - 15,14 \cdot \log_{10}(d+25) + z\left(9 + 0,61(\sigma_{air})^2\right)^{0,5} \quad (1)$$

where:

T^d_{max} - the maximum pavement temperature at depth d in the pavement, the upper PG value [°C], T_{air} - the mean value calculated from the maximum 7-day averaged daily the highest air temperatures for the particular years [°C],
ϕ - the latitude of the meteorological station [°],
d - the design depth of the layer temperature [mm],
σ_{air} - the standard deviation calculated from the maximum 7-day averaged daily the highest air temperatures for the particular years [°C],
z - statistical quantity resulting from the normal distribution of the temperature values, e.g. z = 0 for P = 50%, z = 0.84 for P = 80% and z = 2.05 for P = 98%.

$$T^d_{min} = -1,56 + 0,72 \cdot T_{air} - 0,004 \cdot \phi^2 + 6,26 \cdot \log_{10}(d+25) - z\left(4,4 + 0,52 \cdot (\sigma_{air})^2\right)^{0,5} \quad (2)$$

where:

T^d_{min} - the minimum pavement temperature at depth d [°C],
T_{air} - the mean calculated from the minimum annual air temperatures [°C],
ϕ - the latitude of the meteorological station [°],
d - the design depth of the layer temperature, according to Fig. 1 [mm],
σ_{air} - the standard deviation calculated from the minimum annual air temperatures in the particular years [°C],
z - statistical quantity resulting from the normal distribution, the values the same as for the maximum temperature calculations.

The air temperature data required to calculate the PGs of bitumens were obtained from the Polish Institute of Meteorology and Water Management (IMGW). The data covered all the accessible meteorological stations located within Poland's borders and the climatic data measurement period from 1 January 1986 to 31 December 2015 (Pszczola et al. 2016). Originally, the air temperature data came from over 70 meteorological stations and were received from IMGW in the form of over 140 text files. As a result of original data verification the meteorological stations in whose data there were too large gaps in the analysed period of 1986–2015 or in the case of which the measurement period was shorter than the required minimum 20 years were excluded from further investigations. The meteorological stations located on the mountain peaks were also excluded. Ultimately, 61 meteorological stations uniformly located within Poland were selected for further analyses. The maximum air temperatures were determined as a moving average of the calculated highest air temperatures during 7 consecutive days. As the example, the average maximum temperature analysed for meteorological station in Bialystok (east of Poland) in the years 1986–2015 was T_{air} = + 28 °C and the standard deviation for the 30-year period amounted to $\sigma_{air} = 2.2$ °C. At the probability level of 50% the maximum air temperature used to calculate PG-X is +28 °C, whereas at the probability level of 98% it amounts to +32.4 °C. A map of the maximum air temperatures calculated for the particular meteorological stations in Poland is shown in Fig. 2. The maximum temperatures range from +24 °C in the coastal belt to +30 °C over a considerable area of the central and southern part of Poland.

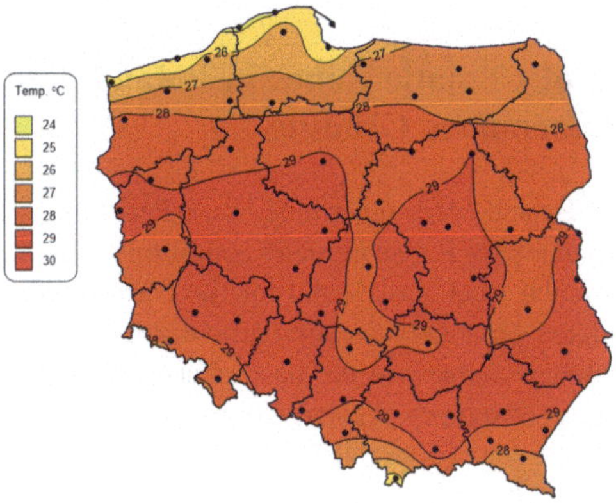

Fig. 2. Climatic zones in Poland determined on basis of max 7-day averages of the highest air temperatures

The minimum air temperature value was determined as the lowest recorded air temperature in each of the analysed year. An average of the minimum air temperatures for the period of 30 years is substituted into Eq. (2). Figure 3 shows a map of the averages calculated from the minimum air temperatures for the particular meteorological stations in Poland. The minimum temperatures range from –11 °C in the coastal belt to –22 °C in the Podlasie region and in the Lublin region.

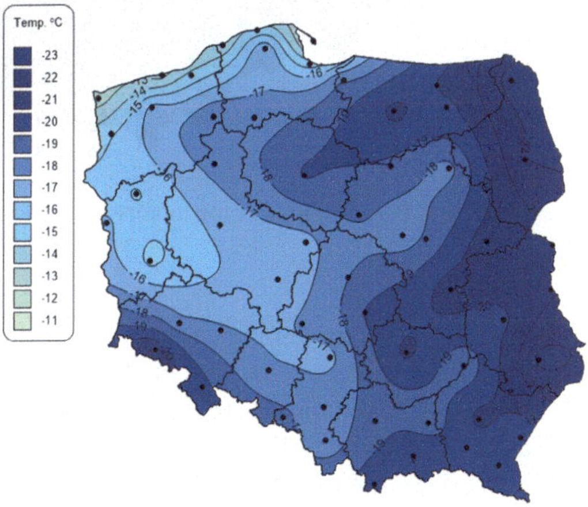

Fig. 3. Climatic zones in Poland determined on basis of averages of annual minimum air temperatures

In order to select a probability level proper for a given technical road class one should take into account not only the risk that more adverse temperatures, and so adverse consequences for pavement condition, may occur, but also the economic aspect and the capability to produce bitumens meeting the requirements specific for the given PG values. The adoption of a higher probability level necessitates the production of more expensive bitumens (e.g. modified bitumens), but low-temperature cracking and rutting are less likely to occur.

3 Determination of Climatic Zones in Poland with Regard to Asphalt Performance Grading

The climatic zones in Poland were determined and Performance Grade values were assigned as the requirements for Polish climatic conditions. The zones were determined separately for the particular road classes and for the corresponding probability levels. They were also calculated separately for the particular layer of asphalt pavement: the wearing course, the binder course and the base course. As the example, the climatic zones for wearing course of motorways and expressways are presented in Fig. 4.

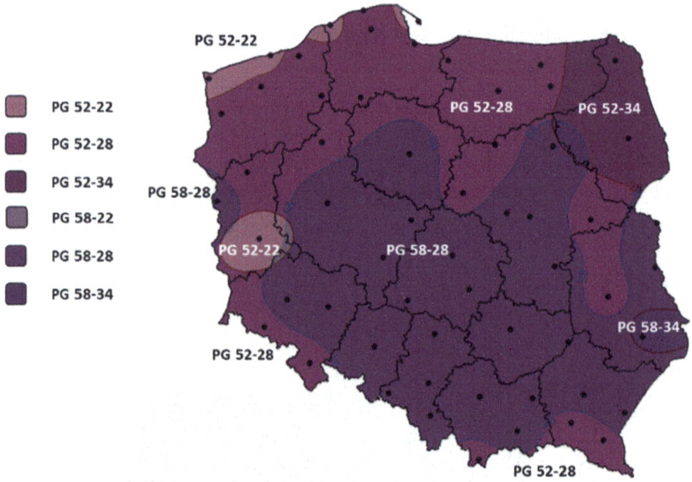

Fig. 4. Climatic zones for selecting PG of bitumens for motorways and expressways, wearing course (P = 98%)

The Superpave method takes into account increased traffic loading by raising upper temperature PG-X since the latter increases the risk of the permanent deformation of the bituminous layers. Depending on the forecasted design traffic and the average speed of vehicles, the upper value of temperature PG-X is increased by 1 or 2 levels, i.e. by 6 °C or 12 °C. It is the Road Authority which decides whether the value of PG-X should be increased. In 2014 as part of the continued work on the grading of bitumens an extended notation of PG, incorporating the results of the Multiple Stress Creep Recovery (MSCR) Test, was proposed in AASHTO M 332. In this notation a letter symbol representing traffic density is added to the PG X-Y designation:

- PG X-Y S – stands for standard traffic below 10 million ESAL and standard traffic loading,
- PG X-Y H – stands for heavy traffic of 10-30 million ESAL or slow traffic,
- PG X-Y V – stands for very heavy traffic above 30 million or traffic standstill,
- PG X-Y E – stands for extremely heavy traffic above 30 million ESAL or traffic standstill.

4 Conclusions

Based on the results of conducted analysis, the following conclusions can be drawn:

1. Taking into account the distribution of temperature along the depth of the asphalt pavement, the performance grades of bitumens for the wearing course, the binder course and the base course layers were determined. The proposed division of the area of Poland into climatic zones according to the bitumen PG values determined on the basis of the climatic data can be useful in the design of road pavements depending on their location.
2. Through the use of the different probability levels proposed by the Superpave method in the analyses, the risk associated with the possible occurrence of extreme pavement temperatures is taken into account. This is particularly important for the selection of a proper kind of bitumen in order to prevent permanent deformations or low-temperature cracking of the pavement. The probability levels can be accepted by the Road Authority or altered depending on the accepted level of risk associated with the probability of occurrence of permanent deformations or low-temperature cracking. The adoption of a higher probability level means a lower risk of such damage to the pavement, but it increases the investment costs.
3. When selecting the PG value of bitumen the pavement's traffic loading should be also taken into account. The decision about raising the upper value of temperature PG-X ultimately rests with the Road Authority.

Acknowledgements. The activity presented in the paper is part of the research grantof the Road Innovations Development (RID-IB) entitled "Paving-grade and modified bitumens in Polish climate conditions", (2016–2018) established by the National Centre of Research and Development and the General Directorate for National Roads and Motorways in Poland.

References

AASHTO MP 332: Standard Specification for Performance-Graded Asphalt Binder using Multiple Stress Creep Recovery (MSCR) Test

Floyd J, Golden K, McMurry R (2013) Standard Specifications Construction of Transportation Systems, Georgia Department of Transportation

Leonovich I, Melnikova I (2012) Influence of temperature on the formation of damages in asphalt concrete pavements under climatic conditions of the Republic of Belarus. Baltic J Road Bridge Eng 7(1):42–47. https://doi.org/10.3846/bjrbe.2012.06

Mampearachchi WK, Mihirani GS, Binduhewa BWP, Lalithya GDD (2012) Review of asphalt binder grading systems for hot mix asphalt pavements in Sri Lanka, University of Moratuwa

Mills BN, Tighe SL, Andrey J, Smith JT, Huen K (2009) Climate change implications for flexible pavement design and performance in Southern Canada. J Transp Eng, ASCE

Mohseni A (1998) LTPP Seasonal Asphalt Concrete (AC) Pavement Temperature Models, Report FHWA-RD-97-103

Nam K, Bahia HU (2004) Development of guidelines for PG binder selection for wisconsin, WisDOT Highway Research Study 0092-01-01, University of Wisconsin – Madison, Wisconsin Department of Transportation

Pszczola M, Judycki J, Rys D (2016) Evaluation of pavement temperatures in Poland during winter conditions. Transp Res Procedia 14(2016):738–747. https://doi.org/10.1016/j.trpro.2016.05.342

Pszczola M, Rys D, Jaskula P (2017) Analysis of climatic zones in Poland with regard to asphalt performace grading. Roads Bridges Drogi i Mosty 16(4):245–269. https://doi.org/10.7409/rabdim.017.016

Solaimanian M (1994) Development of SHRP Asphalt Research Program Climatic Databases, Report SHRP-A-685, Strategic Highway Research Program, National Research Council, Washington, D.C

Solaimanian M, Bolzan P (1993) Analysis of the Integrated Model of Climatic Effects on Pavements, Report No. SHRP-A-637, Strategic Highway Research Program. National Research Council. Washington, D.C

Superpave Performance Graded Asphalt Binder Specification and Testing (1995) Asphalt Institute, Superpave Series No. 1 (SP-1)

The Asphalt Binder Handbook (2011) Asphalt Institute. Manual Series No. 26 (MS-26)

Waseem Mirza M, Zahid Abbas, Ali Rizvi M (2011) Temperature zoning of Pakistan for asphalt mix design. Pak J Eng Appl Sci 8:49–60

Zolotareyov V, Pyrig Y (2018) The application of the method of selecting a brand of bituminous binder, in accordance with the Superpave system, in the conditions of Ukraine. ДорожныеМатериалы, УДК 665:775. https://doi.org/10.30977/BUL.2219-5548.2018.82.0.119

Selected Aspects of the Water Supply System Safety

J. R. Rak and J. Żywiec(✉)

Department of Water Supply and Sewage Systems, Faculty of Civil and Environmental Engineering and Architecture, Rzeszow University of Technology, al. Powstańców Warszawy 12, 35-959 Rzeszów, Poland
j.zywiec@prz.edu.pl

Abstract. The work has the cognitive goal, expressed by the desire to expand knowledge and re-evaluate some of current views on the safety of collective water supply systems (CWSS). These systems belong to the critical infrastructure of urban agglomerations functioning. The work presents the systemic approach to the problems of threats. The interpretation of the Pareto principle in the safety assessment was presented. The system resilience indicator was defined on the basis of direct and indirect risk. Global unreliability indicators referring to the supply of water was presented. The mentioned issue of preliminary analysis of the water supply services level is aimed at developing standards in this area.

Keywords: Water supply · Threat · Safety

1 Introduction

A collective water supply system (CWSS) is an item representing infrastructure critical to the functioning of an urban agglomerations, given that the final product it has to offer is tap water, and given the need to heed the concept of quality (Sienkiewicz and Górny 1999; Szpak and Tchórzewska-Cieślak 2015). Broad analysis of how safety is defined in various scientific disciplines suggests that definitions are assigned to specific research areas (Kowalski et al. 2015). From the philosophical point of view, we can distinguish safety as either permanent or variable (Rak 2009). Safety can likewise be considered from the angle of criteria of a technical, economic, useful, ethical, or whatever, nature (Tchórzewska-Cieślak et al. 2018).

The main purpose of the work is to present the multifaceted nature of safety as it applies to water-supply systems. Causes of failure are presented, along with estimators of water-supply failure rates and determinations of levels of water-supply services.

2 Threats According to the Systemic Approach

After Tchórzewska-Cieślak and Boryczko (2010), a threat within the framework of the systemic approach is defined as:

Z. Blikharskyy et al. (Eds.): CEE 2019, LNCE 47, pp. 369–376, 2020.
https://doi.org/10.1007/978-3-030-27011-7_47

- an accumulation of phenomena inside the system and its environment that increase the likelihood of a crisis situation arising,
- the possibility of a situation occurring in which the system loses its ability to function autonomously,
- the impact of a source of danger on an object or system,
- A threats can generate a so-called crisis situation – characterised by the occurrence of events (phenomena, processes) that are undesirable (destructive); as well as the possibility of such events accumulating, to the point at which the system is unable to discharge its basic functions. After Rak et al. (2019), it is possible to identify anti-crisis strategies that are:
- active (anticipatory or preventive),
- reactive (repulsive or liquidating).

Each system can be assigned a negative potential (relating to destruction) – U_N; as well as a positive potential (resilience) – U_P (Shuang et al. 2017). Examples of destructive and resilience potencies of a CWSS are as presented in Table 1.

Table 1. Destructive and resilience potentials of CWSS

Destruction potential - U_N	Resilience potential - U_P
Incidental water pollution in the water intake cross-section	Alternative water treatment technology
Failure of 2° pumping subsystem	Supplying the water distribution subsystem with the water accumulated in the network tanks (emergency volume)
Failure of the water supply pipe	The ring system of the water-pipe network allowing for supplying from two directions
Failure of the water supply subsystem	Water supply from at least two sources

A measure of CWSS safety may be offered by the critical difference between these two potentials, as Eq. 1 shows:

$$U_N - U_P \leq \Delta U_{CR} \tag{1}$$

Many concepts relating to failures speak of suddenness and randomness of occurrence, with this offering support for the old thesis that failures are caused by the action of "force majeure". In the deterministic approach, the relationship between a failure and its cause comes down to analysis of the chain of undesirable events. The domino theory assumes that a failure is an end event, with previous events forming a chain of causes (Rak 2008).

Causality is somehow "inherited", given that the result of one situation is also the cause of the next one, leading to the failure. This is thus a so-called extension events chain. In line with the current state of knowledge, failures can be classified (Szpak and Tchórzewska-Cieślak 2014) as:

- direct (relating to a sudden undesirable event),
- chain (involving a domino effect of undesirable events),
- crossover (where several undesirable events occur in one place at one time),
- telescopic (with failure occurring in just a short time, following a sequence of disorder events).

An interesting interpretation option where safety assessment is concerned involves the use of the Pareto principle – a method developed by Italian economist Vilfredo Pareto (1848–1923) on the basis of research into the income of the populations of various countries. As a result of further research, it was found that many other statistical values have a distribution analogous to the Pareto distribution (Sienkiewicz and Górny 1999). Basically, a small part of any population may be said to group a significant part of the value of one of its characteristics. The Pareto law is applied to the description of many practical problems with operating technical systems or engaging in everyday life, e.g.:

- 80% of technical failures are caused by 20% of causes,
- 80% of losses are generated by 20% of emergency scenarios,
- 80% of tap-water deficits are caused by 20% of causes,
- 80% of those injured in a catastrophe have injuries caused by just 20% of the destructive factors.

In the literature, Pareto's law is presented graphically using the Pareto-Lorenzo Diagram, which provides for the ranking of factors affecting a studied phenomenon. It thus shows in a graphic way the relative and absolute distributions of types of defect (failure), and their causes. The Pareto concept assumes the division of the field under the Pareto-Lorenzo diagram into:

- area A - 20% of causes groups 80% of cumulative effects,
- area B - the next 30% of causes groups the next 10% of cumulative effects,
- area C - the remaining 50% of causes group another 10% of cumulative effects.

Figure 1 is a graphical presentation of the Pareto concept.

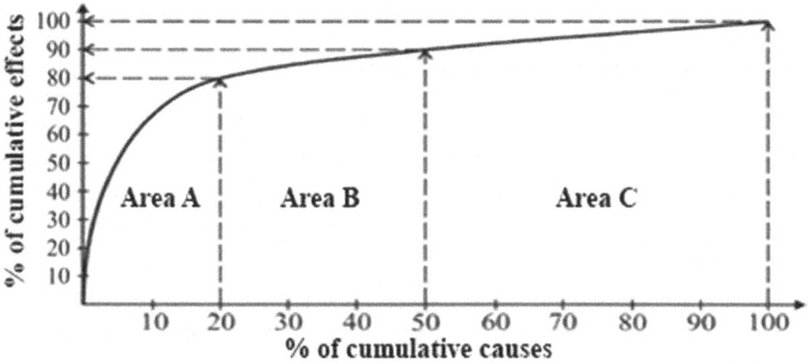

Fig. 1. Diagram Pareto - Lorenzo.

In practice, the Pareto method is used to classify failures occurring in production processes, with a view to the risk of losing security being analysed and assessed. Failures can be divided into:

- group A – of strategic importance (representative), having a significant impact on the functioning of the CWSS, and therefore associated with the concept of protective measures being taken first, in combination with prevention or significant reduction,
- group B – of lower-level significance, to the extent that occurrences should only be limited in second place,
- group C – of low-level significance, to the extent that reparative actions are unjustified for economic reasons; and also random in nature, given that their occurrence is an inherent aspect of system functioning, while the impact of any negative effects is limited.

3 Global Indicators of the Unreliability of Water Supply

3.1 The Probability of Insufficient Water Production to Meet Demand

This is the oldest and most classic reliability indicator resulting from definition. It is defined as the probability that demand for water will exceed the capacity to produce characterising a particular system. The disadvantage of this reliability indicator is the way it defines the credibility of a deficit only, and not its scale.

3.2 Expected Duration of Deficits in Water Production Over a Given Period of Time

This index is applied commonly in analyses relating to the development of a CWSS. It is defined as the expected (average) number of days (hours) during which production does not cover the demand for water. This indicator has a physical interpretation, because it concerns the deficit in water production per unit of time (e.g. day/year, hour/year).

3.3 The Expected Value of the Water Deficit

This is the expected value for the amount of water going undelivered to recipients over a considered period of time. This indicator offers a classic reflection of risk, because it shows probability of deficits of given sizes occurring. As a physical quantity, it is given in m^3/year or m^3/day. This indicator is commonly used to assess effectiveness of CWSS functioning.

3.4 The Indicator of Frequency and Duration Relating to a Shortfall in the Supply of Water

This complements the expected value of the indicator of water deficit (Sect. 3.3). This indicator explains two additional physical characteristics of any deficit as regards the supply of water. It determines the expected frequency of occurrence of deficits, as well as the expected duration of any single deficit. For example, if the expected duration of water production deficits (Sect. 3.2) is 20 h/year, then:

- the frequency can be F = 10 1/year where average duration of a water-supply deficit is 2 h,
- the frequency can be F = 5 1/year with average duration of a water-supply deficit is 4 h.

3.5 The Water-Supply Delivery Indicator

Defined as the ratio of expected water supply to demand for water over a considered period of time. An alternative is the water-supply deficit indicator defined as the ratio of expected deficit to demand for water demand in a considered period of time. The sum of both indicators gives 1. Given that they present relative values, these indicators allow for comparison between the reliability of water production in a small or large CWSS, as well as for observation of changes in the reliability of any single CWSS.

3.6 Indicator of the Single Water Deficit Expected Value

This is equal to the ratio between the expected value for the water deficit (Sect. 3.3) and the probability of occurrence of deficits during the year (Sect. 3.1), over the considered period of time. For example, if the expected value for the water deficit (Sect. 3.3) equals 100 m^3/d, and the probability of occurrence of deficits during the year (Sect. 3.1) is 0.01, then the single water-deficit expected value indicator (Sect. 3.6) is given by 100/0.01 = 10000 m^3/d.

4 Indicator of System Resilience

Technical systems are exposed to two basic kinds of risk: direct or indirect. Direct risk (r_d) is related to threats generated within the system. Indirect risk (r_i) is related to threats generated in the system's surroundings, and having a negative impact on its functioning (Szpak and Tchórzewska-Cieślak 2016). The measure of system resilience to these two types of risk is resilience indicator I_r, as defined as follows under Eq. 2:

$$I_r = \frac{r_d}{r_d + r_i} \tag{2}$$

The r_d risk is identified with the so-called cognitive risk, while r_i risk is identified with random risk. We can interpret the extreme values of I_r as follows:

- $I_r = 0$, then $r_d = 0$, the system is exposed only to external (indirect) risk, which is not very knowable and often random. The system is considered to be vulnerable to threats.
- $I_r = 1$, then $r_i = 0$, i.e. the system is only exposed to internal (direct) risk, which is knowable and which can be reduced. The system is considered immune to threats.

The closer the value of I_r to 1, the more immune the CWSS is to threats.

5 The Level of Water-Supply Services

The standard is determined for a given criterion. For example, if the criterion is the concentration of a given substance in water, then the standard is the acceptable concentration for such an index of water quality. After Boryczko (2010); as well as Pietrucha-Urbanik and Studziński (2012) criteria and standards relating to customer service can be presented in relation to:

- address and opening hours of the customer service office - on the website,
- forms of contact with the service office - direct, postal, online, by phone,
- service for disabled and elderly customers,
- reply to a written inquiry or complaint regarding bills - 14 business days,
- correction of incorrect invoices - 14 business days,
- time for the consideration of other written inquiries and complaints - 14 business days,
- time to have a new water connection installed - from 6 months to 1 year,
- time taken to install a new water meter - 14 working days from the application,
- cessation of supply due to unregulated bills - 14 business days on from the customer being informed,
- the occurrence of breaks in water supply may occur in the case of planned technical maintenance or failure,
- planned technical breaks in water supply:
- – 1–4 h - advance notification, with a minimum of 24 h, in written form or by telephone,
- – above 4 h - advance notification, with a minimum of 48 h, in written form, by phone or via media, and with a maximum notification period set on 72 h,
- in the case of a break in the water supply of more than 6 h, the water-supply company provides a substitute source and informs recipients as to its location. In multi-family buildings, information about a break in water supply is first given to owners and property managers,
- deviations in water quality - notification 48–72 h in advance, and information on remedies,
- fines for not maintaining water-quality standards,
- total permitted duration of a break in water supply within 1 year - no more than 24 to 72 h,
- repair time:
- – $\phi < 300$ mm – up to 4 h,
- – $\phi = 300 \div 500$ mm – up to 6 h,
- – $\phi > 500$ mm – up to 24 h.
- water volume accumulated in network tanks to a minimum of 50% of the average daily demand for water,
- a minimum value for pressure at the water connection (water meter) equal to 0.2 MPa,
- allowable breaks in water supply - B = 0.1%:

$$B = \frac{ND \cdot HD}{NU \cdot 24 \cdot 365} \cdot 100 \quad (3)$$

where:

- ND is the number of users affected by the water supply deficit,
- HD is the number of hours of duration of the water supply deficit,
- NU is the number of users of a water network.

- maintenance of failure rate at an appropriate level thanks to a strategy for repair along the water supply network:
  - main network ≤ 0.1 failure/km·year,
  - distribution network ≤ 0.5 failure/km·year,
  - water connections ≤ 0.7 failure/km·year.
- maintenance at a maximum level of 15% of the rate of loss of water produced and pumped into the system,
- publication by a water company of quarterly water-quality indexes, as recorded at selected points along the water network, and following consultation with recipients.

6 Summary

- The current standard achieved by operations of water-supply companies is helps maintain the level of safety of a CWSS. This involves the supply of water of the required stable quality parameters, continuously, and in amounts required by customers; and with unit cost per m^3 of water kept at an acceptable level.
- A reliable and safe CWSS is the basis for the health, hygiene and wellbeing of consumers. Water-quality monitoring should be performed at every step of the water-supply process, i.e. at the time of obtainment and treatment of raw water, and during the transport, accumulation and distribution of water, until the final product being supplied to the consumer. There should be an obligation in law that consumers should be informed periodically about the quality of water in the tap.
- The global indicators for the unreliability of a CWSS provide for the apportioning of reasons for breaks in water supply, in relation to individual indicators on whose basis it is possible to estimate the costs attendant upon breaks in the supply of water.
- Advanced criteria and standards characterising water-supply services were presented, with particular emphasis placed on the reliability of water supply.

References

Boryczko K (2010) Analysis of risk of interruption in water supply to consumers. J Konbin 13(1):79–90

Kowalski D, Kowalska B, Kwietniewski M (2015) Monitoring of water distribution system effectiveness using fractal geometry. Bull Pol Acad Sci 63:155–161

Pietrucha-Urbanik K, Studziński A (2012) Standards of water services quality levels with regard to the reliability of water supply to the recipients. J Konbin 24(1):71–78

Rak JR (2008) Methods of reliability index determination concerning municipal water quality. J Konbin 5(2):157–173

Rak JR (2009) Selected problems of water supply safety. Environ Prot Eng 35(2):23–28

Rak JR, Tchórzewska-Cieślak B, Pietrucha-Urbanik K (2019) A hazard assessment method for waterworks systems operating in self-government units. Int J Environ Res Public Health 16(5):767

Shuang Q, Liu YS, Tang YZ, Liu J, Shuang K (2017) System reliability evaluation in water distribution networks with the impact of valves experiencing cascading failures. Water 9(6):413

Sienkiewicz P, Górny P (1999) An expert system of decision support in emergency situations. Mat. konf. "KONBIN ~ 99". Publ. ITWL Warszawa, Zakopane – Koscielisko, pp 132–144 (in Polish)

Szpak D, Tchórzewska-Cieslak B (2014) Assessment of the failure rate of water supply system in terms of safety of critical infrastructure. Chemik 2014(6):862–867

Szpak D, Tchórzewska-Cieślak B (2015) Analysis and assessment of the security method against incidental contamination in the collective water supply system. J Konbin 34(1):49–58

Szpak D, Tchórzewska-Cieślak B (2016) Water producers risk analysis connected with collective water supply system functioning. Adv Intell Syst Comput 470:479–489

Tchorzewska-Cieslak B, Boryczko K (2010) Analysis of undesirable events scenarios in water supply system by means of fault tree method. J Konbin 14:309–320

Tchorzewska-Cieslak B, Szpak D, Piegdon I, Szlachta A (2018) Quality analysis of water network failure. J Konbin 47(1):67–85

Experimental Test Method for Structures Under the Impact of Temperature Actions

Valeriy Shmukler[1], Feirusha Salih Hamad[2], and Petro Reznik[1(✉)]

[1] Department of Building Structures, O.M. Beketov National University of Municipal Economy in Kharkiv, 17 Marshal Bazhanov str., Kharkiv, Ukraine
Petro.Reznik@kname.edu.ua

[2] Department of Civil Engineering, College of Engineering, Salahaddin University, Erbil Kurdistan, Iraq

Abstract. A method has been developed that, with the availability of appropriate technical types of supplies, is an effective technology for conducting full-scale studies of structures exposed to temperature fields without the need to create temperature effects. Experimental effective test method, with the availability of appropriate technical types of supplies, for conducting full-scale studies of structures under the impact of temperature actions without the need to create temperature effects was developed. Its distinctive features are the simplicity of experimental part, the low cost of test itself, the physical clarity and logical foundation of results being obtained. The theoretical foundations of the method are based on the Betty-Maxwell energy reciprocity principle and consists in the replacement of a heated or cooled object with an unheated one. New experimental results shows that thermal displacement of the element (structure) in a predetermined direction and at a given time depends on the modulus of deformation of the 1st kind E, coefficient of transverse deformations ν, the thermal expansion coefficient α and the 1st invariant of deformation tensor defined in a predetermined number of points of the heated (cooled) element loaded by a single force. A program of experimental testing was suggested.

Keywords: Temperature effects · Building structures · Numerical methods · Photogrammetry · Frame systems

1 Introduction

The issue of fire resistance of building structures is extremely important (Kalmykov et al. 2017; Bailey 2002). European building codes regulate the design of reinforced concrete, steel and timber structures to meet the requirements of the appropriate level of fire resistance. As well, many national standards (Eurocode 2 2005; Eurocode 3 2005; Eurocode 5 2005) dictate to conduct full-scale fire tests. As is known, laboratory, and in particular, full-scale tests of structures exposed to high temperatures, are complex and high budget procedures. Nowadays, photogrammetric methods are used to test building structures (Luhmann et al. 2013; Schreier et al. 2009; Besnard et al. 2006). For example, in (Al-Kamaki et al. 2015) almost full-sized columns were considered under the influence of temperature and a photogrammetric method using the digital image

Z. Blikharskyy et al. (Eds.): CEE 2019, LNCE 47, pp. 377–383, 2020.
https://doi.org/10.1007/978-3-030-27011-7_48

correlation technique was used to measure the strain variation. Nevertheless, the development of an experimental test method for structures under the impact of temperature actions is an important task. We propose experimental test method for structures under the impacts of positive (negative) temperatures. (Shmukler 2016) Its specific feature is the fact that the experiments are carried out on "cold" samples. The goal of the method is experimental determination of components of displacement vector of structures under the action of positive or negative temperatures. At the same time, there are no restrictions with respect to geometric dimensions of the tested elements or the structural system in general.

2 Hypotheses and Assumptions

The procedure of experiments within the presented approach is predetermined below. Hypotheses and assumptions include as follows:

- full equivalence, in terms of geometry (shape, size) and materials, heated and unheated bodies;
- complete equivalence of the support conditions (boundary conditions) of the heated and non-heated bodies;
- representation of the structure as a non-simply-connected heterogeneous isotropic body, which characteristics are functions of coordinates and time E = E (x, y, z, t); ν = ν (x, y, z, t), where E is the strain modulus of the first kind; ν is the coefficient of transverse deformations; x, y, z are coordinates of the Cartesian system; t is time;
- the ability to establish temperature distribution T in the form of stationary or non-stationary fields. i.e. T (x, y, z, t);
- the possibility of approximate identification of the relationship between strain intensity and stress intensity with the relationship between stress and strain, obtained experimentally for uniaxial loaded standard sample (similarity to the hypothesis of the theory of small elastic-plastic strains).

The formula obtained in (Shmukler et al. 2008) is used as the theoretical basis of the method. The above formula is based on the above hypotheses and assumptions, as well as the concept of Betty-Maxwell (Meisel 1961). The formula allows you to determine the movement at any point of the studied heated (cooled) body. For this purpose we introduce the same but unheated construct loaded, instead of a body (structure) under the impact of given temperature, with a unit force applied at an arbitrary point C and in the direction of the desired temperature displacement. The formula in Cartesian coordinates has the following view:

$$Q_c = \iiint\limits_{V_B} \frac{\alpha E}{1 - 2\upsilon} \cdot T \cdot \theta^* dV_B. \quad (1)$$

where Q_c is structure displacement in C point; α – thermal-expansion coefficient α(x, y, z, t); T - established field T(x, y, z, t); θ^* - single volume strain under impact of single forced applied in C point; V_B – volume occupied by the structure.

3 Mathematic Model

The Eq. (1) is a ground for creation of proposed method. Let's introduce the designation:

$$f(x,y,z,t) = \frac{\alpha(x,y,z,t) \cdot T(x,y,z,t)}{1 - \upsilon(x,y,z,t)}. \quad (2)$$

Then Eq. (1) will have the following:

$$Q_c = \iiint\limits_{V_B} f(x,y,z,t) \cdot divR \cdot dV_B. \quad (3)$$

where $\{R\}^T = \{u, v, w\}$; u = u(x, y, z, t); v = v(x, y, z, t); w = w(x, y, z, t); divR = I_1; I_1 is the first invariant of strain tensor; u, v, w are projections of vector of displacement along X, Y, Z axes; f, u, v, w $\in C^1$ respectively.

In the case of a given stationary temperature field T = const. If the body is isotropic and homogeneous, then E = const, ν = const, α = const, and, as a result, f = const, then Eq. (3) takes the form:

$$Q_c = f \cdot \iiint\limits_{V_B} divR \cdot dV_B. \quad (4)$$

Let's use Gauss-Ostrogradsky formula:

$$Q_c = f \cdot \iiint\limits_{V_B} divR \cdot dV_B = \iint\limits_S (Ul + Vm + Wn)dS. \quad (5)$$

where S is surface area of considered body; l, m, n is direction cosines of the outer normal and the integral on the right-hand side is surface integral of the first kind.

Thus, Eq. (5) allows the determination of set point displacement of heated (cooled) body in given direction and moment only by movements that occur on the surface of the heated (cooled) body loaded with single force. For the general case, when a non-stationary field is specified, the formula (3) after integration in parts will have the following view:

$$\begin{aligned} Q_c &= \iiint\limits_{V_B} f(x,y,z,t) divR dV_B \\ &= \iint\limits_S (Ul + Vm + Wn)dS + \iiint\limits_V \left[U\frac{\partial f}{\partial x} + V\frac{\partial f}{\partial y} + W\frac{\partial f}{\partial z} \right] dV_B \end{aligned} \quad (6)$$

The presence of the second term (volume integral) in the right side of Eq. (6) complicates the task compared to the procedure given by Eq. (5). That is, if temperature is constant and material is homogenous, it is sufficient to measure the displacements of body points located on the surface, and then determine Eq. (5) the temperature displacement.

As for the general case (non-stationary temperature field), the following algorithm is proposed for the implementation of the sequence of actions for Eq. (6):

- discretization of the calculated model of the body by the finite element method (FEM);
- the definition of the function f(x, y, z, t) based on temperature and physical-mechanical parameters;
- analytical (in the case of approximation) or numerical calculation of f derivatives on coordinates x, y, z;
- determination of FEM (if necessary, considering physical and geometric non-linearities)for displacements in all nodes of finite element grid (FE) under impact of single force applied at a given point;
- experimental measurement of displacements in nodes of FE grid located only on body surface;
- calculation of difference in displacements in nodes of FE grid located on the surface of the body, measured experimentally and calculated theoretically;
- calculation of displacements in all nodes of the FE grid based on found difference of displacements applied to the body surface;
- calculation of the total displacements at all nodes of the FE mesh due to the action of the unit force and the surface displacements difference;
- determination of the desired temperature displacement according to Eq. (6).

Adopted experimental-theoretical approach makes it possible to correct the displacement field determined by calculation in such a way that the displacements on the body surface (structure) are experimentally measured.

This algorithm substantiates the proposed experimental-theoretical method in the general case. In some private, but practically important cases, the procedure of the method can be simplified. They, first of all, include constructs in the form of thin plates and shells. In this case, special feature is that the first invariant of the strain tensor can be expressed through the vertical displacements (deflections) of the middle surface of the unheated plate under impact of a single vertical force.

Placing a grid of points on the bottom surface of the plate (the Bernoulli-Kirchhoff-Love hypotheses are considered fair) and replacing the harmonic operator with its finite difference representation, instead of Eq. (3), we will obtain:

$$Q_c = -\sum_{i=1}^{N}\sum_{j=1}^{M}\frac{\alpha_{ij}\cdot E_{ij}}{1-\nu_{ij}}\cdot\nabla_{ij}W\left[\int\limits_{-h/2}^{h/2} T(x,y,z,t)zdz\right]. \tag{7}$$

where N, M are quantity of points on the surface of plate along entire X and Y axis, respectively; $\nabla_{ij}W$ is discrete (finite difference) analogue of Laplace harmonic operator; $i \in [1, N]$ and $j \in [1, M]$ are current indexes.

A similar technique can be used for shells, and, considering some additional prerequisites, thick plates.

4 Experimental Component of the Method

Let's denote its attributes. *Object of study* is the structure (body), which is under temperature effect (field). *Subject of study* is the temperature displacement of the body under study. *Purpose of method* is to determine the temperature displacements of a structure by experimentally and theoretically studying its unheated counterpart. This method assumes the presence of special technical and software (Janowski et al. 2016; Nonis et al. 2013). Technical support includes a loading system, a measurement system and a computer system (PC).

Structural diagram is shown on Fig. 1.

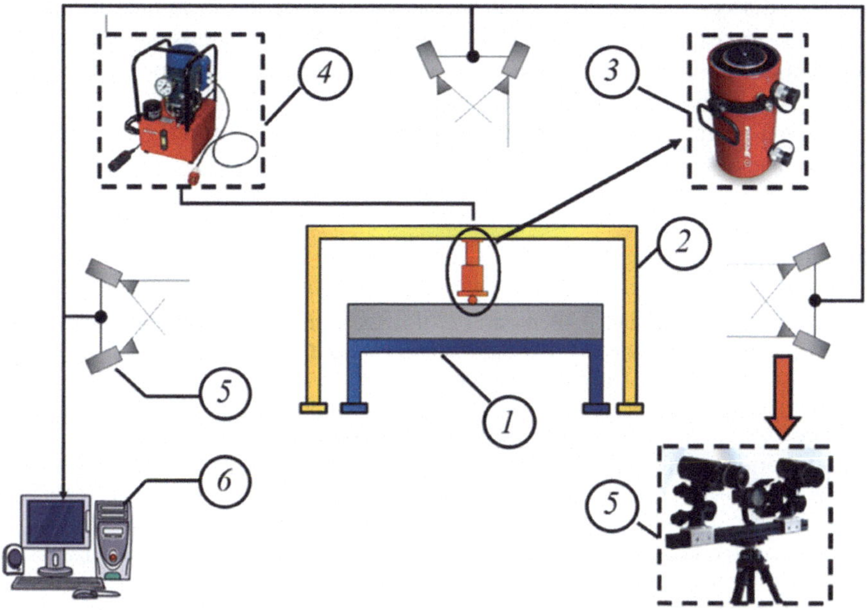

Fig. 1. Block diagram of the method: 1 - object of study; 2 - support frame; 3 - hydraulic jack; 4 - oil station; 5 - photogrammetric or laser installation, type VIC-3D; 6 is personal computer equipped with FEM software such as Lira-Sapr and Ansys.

5 Scientific Novelty and Practical Significance

Thus, the scientific novelty of the method is represented by extension of Maxwell-Betti reciprocal work theorem in case of equalization of work of statically applied force under temperature displacement at a given point, as well as work of temperature at point of application of indicated static force.

Practical significance of the method is determined by a logically composed and justified procedure for conducting technological operations, which, together with the universal technical and created information and software, predetermine a new low-budget method of full-scale and/or laboratory testing of structures for temperature impacts.

6 Conclusions

The developed method is a progressive technology for conducting studies of structures exposed to virtual temperature fields. Its distinguishing features are the simplicity of the experimental part, the low cost of experiment, the physical transparency and the logical validity of obtained results. This is determined by the performance of the necessary procedures on unheated body, loaded in the simplest way, i.e. single force. The capabilities of the method can be significantly expanded by expanding the technical base (i.e. devices that create a simultaneous effect from forces applied in different directions, for example, experimental modeling of the 'compression center'). The important point here is also the possibility of testing structures of various types: arrays, shells, plates, rods, their combinations, etc.

There is no restriction on the qualitative and quantitative parameters of the temperature field.

References

Al-Kamaki YSS, Al-Mahaidi R, Bennetts I (2015) Experimental and numerical study of the behaviour of heat-damaged RC circular columns confined with CFRP fabric. Compos Struct 133:679–690. https://doi.org/10.1016/j.compstruct.2015.07.116

Bailey C (2002) Holistic behavior of concrete buildings in fire. Proc Inst Civ Eng Struct Build 152(3):199–212. https://doi.org/10.1680/stbu.152.3.199.38988

Besnard G, Hild F, Roux S (2006) "Finite-element" displacement fields analysis from digital images: application to Portevin-Le Châtelier bands. Exp Mech 46(6):789–803. https://doi.org/10.1007/s11340-006-9824-8

Eurocode 2 (2005) Design of concrete structures - Part 1-2: general rule - structural fire design. EN 1992-1-2. CEN, Brussels

Eurocode 3 (2005): Design of steel structures - Part 1-2: general rules - structural fire design. EN 1993-1-2, CEN, Brussels

Eurocode 5 (2005) Design of timber structures - Part 1-2: general - structural fire design. EN 1995-1-2, CEN, Brussels

Janowski A, Nagrodzka-Godycka K, Szulwic J, Ziółkowski P (2016) Remote sensing and photogrammetry techniques in diagnostics of concrete structures. Comput Concr 18(3):405–420. https://doi.org/10.12989/cac.2016.18.3.405

Kalmykov O, Gaponova L, Reznik, P, Grebenchuk S (2017) Study of fire-resistance of reinforced concrete slab of a new type. MATEC Web Conf, 116. https://doi.org/10.1051/matecconf/201711602018

Luhmann T, Robson S, Kyle S, Boehm J (2013) Close-range photogrammetry and 3d imaging. De Gruyter, Berlin

Meisel VM (1961) Temperature problem in the theory of elasticity. Publishing House of Academy of Sciences of the Ukrainian SSR, Kyiv

Nonis C, Niezrecki C, Yu T-Y, Ahmed S, Su C-F, Schmidt T (2013) Structural health monitoring of bridges using digital image correlation. In: Proceedings of SPIE - the international society for optical engineering, vol 8695. https://doi.org/10.1117/12.2009647

Schreier H, Orteu J-J, Sutton MA (2009) Image correlation for shape, motion and deformation measurements: basic concepts, theory and applications. Springer, New York

Shmukler V, Klimov Yu, Burak N (2008) Karkasnyye sistemy oblegchennogo tipa. Zolotyie stranitsy, Kharkiv

Shmukler V (2016) Patent of Ukraine 112223. State Patent Office of Ukraine, Kyiv

Current Tendencies of Water Resources Using for Providing the Energy Needs of Society

O. A. Riabenko(✉), O. O. Klyuha, V. S. Tymoshchuk, and O. O. Halych

Department of Hydro-, Heat Power Engineering and Hydraulic Machines, National University of Water and Environmental Engineering, Soborna St. 11, Rivne 33928, Ukraine
o.a.riabenko@nuwm.edu.ua

Abstract. The article describes the studying of objective tendencies of hydropower engineering development for improving the level of providing the increasing energy needs of society. The research method is based on detail analysis of existed information on water resources using for electricity production and discovery of promising avenue of hydropower engineering development. It is highlighted the role of hydraulic energy as one of main types of renewable energy. The article elaborates on the four main tendencies for hydropower development, i.e. intense increasing of hydropower capacity, expansion of the HPP and PSPP functions, construction of the high head HPP and PSPP, enhancing role of small-scale hydropower plants. These tendencies are relevant for global and Ukrainian hydropower engineering. The authors' view on perspective trends of hydropower development are shown. Among these trends the four main tendencies are emphasized. The received results are useful during considering the specific tasks for establishing hydropower facilities and attracting appropriate investors for their implementation.

Keywords: Electrical energy · Hydropower engineering · Hydropower plant · Pumped storage power plant · Water resources

1 Introduction

Hydraulic energy is one of the main types of renewable energy that is used for providing regularly increased industry and household needs of society. Nowadays hydraulic energy produces about 76% from total global renewable energy.

Water resources are collected unevenly across the Earth's surface. To countries which are the most independent in water resources are Brazil (8233 km^3), Russia (4508 km^3), the USA (3051 km^3), Canada (2902 km^3), Indonesia (2838 km^3), China (2830 km^3), Colombia (2212 km^3), Peru (1913 km^3), India (1880 km^3), Congo (1283 km^3), Venezuela (1233 km^3), Bangladesh (1211 km^3), Burma (1046 km^3). At the same time, there are large areas (North Africa, Middle East, some regions of China and India) with moderate or severe water scarcity. About 40% of world population live on these territories (2.5 billion people). It is assumed, up to 2025 this number will increase to 5.5 billion people and equal two thirds of world population (Sorokina and

Z. Blikharskyy et al. (Eds.): CEE 2019, LNCE 47, pp. 384–391, 2020.
https://doi.org/10.1007/978-3-030-27011-7_49

Kucheriavaia 2016). Ukraine has relatively small amount of water resources, estimated in 209.8 km^3.

Existed water availability is used for electricity production, drinking and technical water supply, providing needs of industry, agriculture (including irrigation), and household consumption.

The aim of the article is to study the objective tendencies of hydropower engineering development for improvement of providing level of energy needs of society. The research method is based on detail analysis of existed information on water resources using for electricity production and discovery of promising avenue of hydropower engineering development.

2 The Usage of Water Resources for Electricity Production

Nowadays about 15% of world used water resources is applied for electricity production. It is necessary to emphasize, hydropower engineering belongs to the category of water-users that use water for electricity production and return water completely into the river. This fact makes hydropower engineering differ from other branches of the national economy that belong to category of water consumers and use water for their needs without returning it into the river.

Hydropower engineering provides about 20% of world electricity production. At the end of 2014 the capacity of HPP in the world was 1036 GW, including PSPP – 142.1 GW (13.7%). The leaders of electricity production by water are Chine – 279.4 GW (27.0%), Brazil – 89.3 GW (8.6%), the USA – 79.3 GW (7.7%), Canada – 77.8 GW (7,5%) (World Energy Council 2015). Moreover, countries such as Norway and Paraguay, provide their needs in electricity using hydropower engineering almost completely, and export spare electricity to neighboring countries.

Hydropower potential of large rivers in Europe has been used in a great measure. For instance, Italy, Switzerland, Germany, Spain, Austria, Slovakia have exploited their hydropower potential by 85–98%. Ukraine uses its hydropower potential by 60%. The top of European countries, that produce electricity from water resources, includes France (18.38 GW), Sweden (16.32 GW), Italy (14.33 GW) (World Energy Council 2015). Total installed capacity of large HPP and PSPP of Ukraine in 2018 was 6060.1 MW, and electricity production – more than 11 billion kW·hour. Part of capacity hereby produced by hydropower engineering, equals 8% in capacity balance of energy system of Ukraine (Rassovskyi 2018).

3 Tendencies of Hydropower Engineering Development

3.1 List of Main Tendencies

Analysis of accumulated information on development of different parts of hydropower engineering components allowed to identify and formulate a number of objective tendencies of modern hydropower development. There are four main tendencies among them.

1. Intense increasing of hydropower capacity.
2. Expansion of the HPP and PSPP functions.
3. Construction of the high head HPP and PSPP.
4. Enhancing role of small-scale hydropower plants.

3.2 Intense Increasing of Hydropower Capacity

According to data of international power organizations (International Hydropower Association 2015; World Energy Council 2015; Church et al. 2003), the intense development of hydropower engineering takes place around the globe nowadays. The augmentation of HPP and PSPP capacities during 1990s was about 100 GW, however it almost doubled during next decade. In 2014 the hydropower capacity increased by 39 GW. Thus, the installed capacity of hydropower plants in the world was 1036 GW. As reported by forecast of World Energy Council till 2050 the HHP potential may rise to 2000 GW.

The New Energy Strategy of Ukraine till 2030 foresees significant increasing of installed hydropower capacities up to 10300 MW, that will be about 16–20% of total capacity of energy system (Energy Strategy 2013).

3.3 Expansion of the HHP and PSPP Functions

The intense increasing of hydropower capacity for providing ever-growing energy needs of society is very important function of hydropower engineering. Therewith during last decades, functions of HPP and PSPP, which are connected with their operating in energy system of Ukraine, have changed significantly. The additional functions of these plants during their operating in structure of electricity system are regulating the unevenness of electric load curve (i.e. covering the peaks and dips), frequency, capacity and other parameters of energy system.

As far as is known, the daily electric load curve is characterized by large number of unevenness (Fig. 1). Often, regulating this unevenness is made by basis blocks of heat power plants that have capacity about 200–300 MW. As for Ukraine, the number of starts and stops of such blocks is about 3000 per year. Furthermore, for reduction of night dips of daily electric load curve, it has to be stopped more than ten blocks of heat power plants per day (Landau and Stashuk 2018). This regulation is very expensive and can lead to large errors and failures of heat power plants equipment. Such regulation of nuclear power plants is undesirable at all.

Taking into account the mentioned reasons, the regulation of energy system operating has to be made by using high maneuverability generating units of HPP and PSPP. It is obvious, the units of these plants can reach the installed capacity for 1–2 min, furthermore, if units are operated in regime of idle run, this time equals only 15–30 s. For comparison, indicated time for units of heat power plants (after their stop) is 90–180 min, nuclear power plants – 390–600 min, and for "hot" state regimes this time for heat power plants is 20–50 min, nuclear power plants – 60 min (Riabenko et al. 2017; Haddad et al. 2014).

The operating regime of Zahorsk PSPP is good example of significant expansion of PSPP functions. The number of starts of its reverse units reaches 440 per month, and on

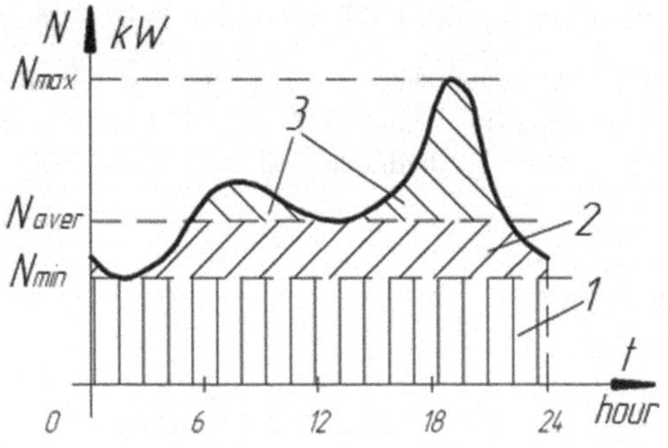

Fig. 1. Daily electric load curve: 1 – base load, 2 – intermediate, 3 – peak

separate days – 30 starts per day without taking into account starts of generating units in the regime of synchronous compensator. The number of changes of HPSPP regimes is assumed to be 4000–8000 per year. As an illustration, this amount for PSPP Drakensberg (South Africa) is 8000, for Gilboa (USA) – 6000, for Dinovik (England) – 5000 (Riabenko et al. 2017).

If there are hydraulic energy losses in turbine and pump regimes, the total efficiency coefficient of such power plants is 0.75–0.80. Economical reason for construction of PSPP can be explained due to difference of electricity price in mentioned regimes, in fact the price of peak energy is in 4–10 times higher than the price of base load energy. One of the main features of pumped-storage power plants is that they are operated using water in revolving regime that depends very little on dryness of the year.

When many perspective locations, that are expedient for constructing the large HHP with good technical and economic parameters, in Europe, North America, Japan have been already used, the building of new PSPP is extremely important. Today about 500 of such power plants are operated and constructed in the world. The total installed capacity of existed PSPP at the end of 2015 equaled 145 GW.

The world leaders of using PSPP are the USA and Japan, which have more than 70 such power plants. The largest pumped-storage power plants in the world are Bath County (the USA, N_i = 3003 MW), Kanagawa (Japan, N_i = 2820 MW), Guangzhou (Chine, N_i = 2448 MW). Among countries of European union, at the end of 2010 the leaders by installed capacity of PSPP were Italy (8895 MW), Germany (7326 MW), Spain (5657 MW), France (5229 MW) (Vasko et al., 2017). PSPP characteristics of founding countries of International Scientific Conference Lviv - Košice - Rzeszów are given in Table 1 (Galat et al. 2018; Hydropower Plants; Krištof et al. 2010).

3.4 Construction of the High Head HPP and PSPP

Analysis of experience of countries such as Austria, France, Norway, etc. allows to assume very important tendency of constructing the PSPP and diversion HPP with pressure derivation in mountain areas. This tendency requires the head increasing of generating unit. For example, the heads of existed and designed hydropower plants

Table 1. Characteristics of PSPP in Ukraine, Poland, and Slovakia

№	Name of PSPP	River	Operating date		Year of start-up	Project data	
			Capacity MW	Number of hydraulic unit		Capacity MW	Number of hydraulic unit
Ukraine							
1	Kyiv	Dnieper	235	6	1970		
2	Tashlyk	Tashlyk	302	2	2007 (2nd unit)	906	6
3	Dniester	Dniester	972	3	2016 (3rd unit)	2268	7
4	Kaniv	Dnieper				1000	4
Poland							
1	Żarnowiec	Piaśnica	716	4	1983		
2	Porąbka-Żar	Sota	500	4	1979		
3	Solina	San	200	4	1968		
4	Żydowo	Radew	157	3	1971		
5	Niedzica	Dunajec	92.75	2	1997		
6	Dychów	Bobr	91.50	3	2005		
Slovakia							
1	Čierny Váh	Čierny Váh	735	6	1981		
2	Dobšiná	Hnilec	24	2	1953		
3	Liptovská Mara	Váh	198	4	1975		
4	Ružín	Hornád	60	2	1972		
5	Mikšová II	Váh	2.6	1	1971		

have next values: Reissecke (Austria, HPP) – 1773 m, Rottau (Austria, HPP) – 1106 m, Tereblynska (Ukraine, PSPP) – 500 m, Ortillon (France, HPP) – 420 m, Tereblya-Rika (Ukraine, HPP) – 200 m, Dniester (Ukraine, PSPP) – 154,9 m (Riabenko et al. 2019). Power plants of such types are very perspective due to providing large capacity with small water discharges by using large heads.

3.5 Enhancing Role of Small-Scale Hydropower Plants (SSHPP)

Last decades are characterized by intense development of small-scale hydropower engineering. In 2016 the total installed capacity of small-scale HPP in the world was 78 GW, that had increased by 4% since 2013. Generally, it is used about 36% of all small hydropower resources in the world. The top 5 countries by using small-scale HPP includes Chine, the USA, Italy, Japan, and Norway, which produce 67% of total installed capacity of SSHPP in the world. According with using of hydropower potential China has 47073, the USA – 1640, Italy – 2304, Japan – 1369, and Norway – 1175 small-scale hydropower plants (United Nations Industrial 2016).

The dynamics of increasing of SSHPP number in Ukraine corresponds to global tendencies of small-scale hydropower engineering development as well. In 2006 the number of SSHPP was 70, in 2010 – 81, in 2016 – 115. According to Register of Power Engineering Objects (Register 2019), at the beginning of 2019 there were about 154 SSHPP in Ukraine. In agreement with New Energy Strategy of Ukraine up to 2030, it is forecasted arising the total capacity of SSHPP to 600 MW (Energy Strategy 2013).

The development of small-scale hydropower engineering in Ukraine has next particularities (Riabenko et al. 2019).

1. Major part of constructed power plants is based on reconstruction of SSHHP that were operated before. It allows to use existed structures at least partially (dams, power houses, water spillways, etc.). Moreover, it becomes easier to solve issue of land set aside for building.
2. The heads of power plants are small and similar to heads of water mills that existed earlier.
3. Water reservoirs of SSHPP is located within river beds and part of nearby bottomland. It excludes the possibility of flooding large areas by these reservoirs. This particularity minimizes the influence of small-scale hydropower plant on environment.
4. The most popular types of hydropower plants are run-of-river and storage types, that have minimum influence on hydrological regimes of river. It is necessary to indicate, river regime during plant operating becomes better due to possibility of regulation by water reservoir which increases the value of minimum river discharge in tail water of hydropower plant.
5. Small-scale HPP of diversion type with open or pressure derivation are used rare than plants of run-of-river and storage types. Main reason of such tendency is ecological complications on the river part from intake to outlet. Mentioned complications are observed during dry periods when plant is operated in forbidden regimes using substantial proportion of river discharge, and water discharge in the river is less than minimum necessary discharge.
6. Ukrainian HPP with small capacity use hydropower equipment of national companies (LTD "Minihidro", PRJSC "Big Size Electric Machine Plant") and also foreign companies – WTW (Poland), Mavel (Czech Republic) and other.

4 Scientific Novelty and Practical Significance

Scientific novelty is to show authors' view on perspective trends of hydropower development and emphasize the four main tendencies. Practical significance of paper is to give received results that can be useful during considering the specific tasks for establishing hydropower facilities and attracting appropriate investors for their implementation. The emphasized recommendations are objective and correspond to global and Ukrainian conditions.

5 Conclusions

1. At the current stage of society development, it is observed the intense increasing of hydropower capacity.
2. Besides function of electricity production, during last decades it was expanded other operated functions of HHP and PSPP connected with regulating the unevenness of electric load curve, frequency, capacity and other parameters of energy system.
3. Last years are characterized by increasing operating heads of PSPP and diversion HPP with pressure derivation in mountain areas.
4. During last decades there is tendency of contribution increasing of small-scale hydropower plants in general electricity production.

References

Church L, Suloway J, Brown S (2003) Hydropower research and development recommendations. J Energy Eng 129(2):33. https://doi.org/10.1061/(ASCE)0733-9402(2003)129:2(33)

Energy strategy of Ukraine for the period up to 2030 (2013) https://de.com.ua/uploads/0/1703-EnergyStratagy2030.pdf (in Ukrainian)

Galat V, Ryzhui V, Riabenko O (2018) Challenges in the design of the upper reservoir for the Kaniv pumped-storage scheme, Ukraine. Int J Hydropower Dams 25(3):80–83

Haddad OB, Ashofteh PS, Rasoulzadeh-Gharibdousti S, Mariño MA (2014) Optimization model for design-operation of pumped-storage and hydropower systems. J Energy Eng 140(2). https://doi.org/10.1061/(asce)ey.1943-7897.0000169

Hydropower Plants (n.d.) https://www.energoprojekt.pl/en/completed-projects/hydro-power-plants

International Hydropower Association (2015) 2015 Key trends in hydropower. https://www.hydropower.org/2015-key-trends-in-hydropower

Krištof V, Kušnír S, Katin M, Csányi Ľ, Marci M (2010) Hydro power plants in Slovak Republic. In: Intensive programme "renewable energy sources", May 2010. Železná Ruda-Špičák, Czech Republic: University of West Bohemia, pp 30–34

Landau YuO, Stashuk IV (2018) Value of hydropower engineering for development of united energy system of Ukraine according to new energy strategy-2035 and environmental issues. Hydropower Eng Ukr 1–2:3–6 (in Ukrainian)

Rassovskyi VL (2018) Hidroenerho 15 years. Hydropower Eng Ukr 3–4:4–7 (in Ukrainian)

Register of objects of power engineering using alternative energy sources (except for blast furnace and coke gases, and with the use of hydropower - only micro and small hydroelectric power stations) (2019) (in Ukrainian)

Riabenko OA, Klyuha OO, Halych OO, Poplavskyi DM (2019) The role of small-scale hydropower engineering in energy balance of Ukraine. In: Proceeding of 10 international conference "unconventional and renewable energy sources as alternative to primary energy sources in region". Lviv Polytechnic National University, Lviv, Ukraine, pp 203–206 (in Ukrainian)

Riabenko OA, Osadchyi SD, Klyuha OO, Tymoshchuk VS (2017) Particularities of PSPP operating under circumstances of translational waves formation. Hydropower Eng Ukr 1–2:45–47 (in Ukrainian)

Sorokina NL, Kucheriavaia IN (2016) Global water resources (state, usage, forecast). Hydropower Eng Ukr 1–2:71–78 (in Russian)

United Nations Industrial Development Organization (2016) World small hydropower development report. https://www.unido.org/sites/default/files/2016-11/WSHPDR_Executive_Summary_2016_0.pdf

Vasko PF, Verbovyi AP, Ibrahimova MP, Pazych ST (2017) Pumped storage power plants as a technological basis for integration of power wind and photovoltaic power plants as part of the electric power system of Ukraine. Hydropower Eng Ukr 1–2:20–25 (in Ukrainian)

World Energy Council (2015) World energy resources: charting the upsurge in hydropower development. https://www.worldenergy.org/wp-content/uploads/2015/05/World-Energy-Resources_Charting-the-Upsurge-in-HydropowerDevelopment_2015_Report2.pdf

Nano-modified Ultra-rapid Hardening Portland Cement Compositions for High Strength Concretes

M. Sanytsky[1(✉)], U. Marushchak[1], Y. Olevych[1], and Y. Novytskyi[2]

[1] Department of Building Production, Lviv Polytechnic National University, S. Bandera Street 12, Lviv 79013, Ukraine
msanytsky@ukr.net

[2] Department of Highways and Bridges, Lviv Polytechnic National University, S. Bandera Street 12, Lviv 79013, Ukraine

Abstract. It was shown that one of the ways for obtaining effective high strength concrete is using ultra-rapid hardening Portland cement compositions. The possibility of development of such compositions on the basis of ordinary Portland cement for high strength concrete with using nanotechnological methods of the modification of cementitious matrix by introduction of nanoscale particles of C-S-H phase and polycarboxylate type superplasticizer is shown. The complex nanomodifier results in high water reduced effect with decreased interparticles distance, accelerated formation of hydrosilicate gel with a more homogeneous distribution of hydrates in a limited intergranular space. Nano-modified Portland cement compositions reaches 54% of standard strength after 24 h. It allows to categorize them as ultra-rapid hardening. According to the indicator of standard strength (R_{c28} = 84.8 MPa), they refer to high strength binder. Effectiveness of nano-modified ultra-rapid hardening Portland cement compositions for high strength concrete under different curing conditions is confirmed.

Keywords: Ultra-rapid hardening Portland cement composition · High strength concrete · Nanomodification · Polycarboxylate superplasticizer · Nanoscale C-S-H phase

1 Introduction

Contemporary building technologies set high requirements concerning functional performance, durability, operational reliability, economic efficiency of building materials, which leads to wider application of high strength concrete (HSC). The use of HSC becomes more common place in the building and transportation industry because of its beneficial economical and material properties. HSC is advantageous since it reduces material requirements in axial or flexural members, permits longer member spans allows for increased member spacing; thereby reducing material and total project costs (Choi et al. 2008; Torres and Burkhart 2016; Solodkyy et al. 2017).

At the same time, an important criterion is the providing of intensive kinetics of the concrete strength development at the early age. Rapid hardening concretes provide the

Z. Blikharskyy et al. (Eds.): CEE 2019, LNCE 47, pp. 392–399, 2020.
https://doi.org/10.1007/978-3-030-27011-7_50

ability of early structures load, increase of formwork turnover and accelerate the construction of buildings with using cast-in-place concrete (Dvorkin et al. 2017). The wider application of precast concrete for civil structures is caused by its advantages of improving the quality of concrete, shortening of construction work, reducing the use of temporary materials and reducing waste. However, steam curing is widely used to achieve early strength (pre-determined, striking and rated strength) of concrete in the production of precast concrete. Thus, significant amount of energy is consumed. The using of rapid hardening high strength concrete allows reducing the production cycle, transition to low- and zero heat curing technologies of precast concrete with significant reduction of energy resources, an increase in turnover of moulds, one day cycle and line productivity (Martí et al. 2013; Min et al. 2014). Therefore, in order to obtain the rapid hardening concrete the introduction of ultra-rapid hardening cements, for which compressive strength after one day or less is normalized, is required. The ultra-rapid hardening cements based on Portland cement clinker have considerable practical interest. Alkaline activation (Krivenko et al. 2018; Savchuk et al. 2018) or calcium-based hardening accelerator (Min et al. 2014) usually use for development of Portland cement compositions with high early strength.

The innovative approach in the technology of rapid hardening high strength concrete is the introduction of nanotechnological techniques based on the directional formation of the structure of the material as a heterogeneous, multiphase system of complex hierarchy from the nanoscale to the macrostructural level by modification with nanosized particles in combination with highly effective polycarboxylate superplasticizers (Kanchanason and Plank 2015). Nanosized particles of additives, which characterize by high values of specific interfacial area and excess surface energy, provide early directed formation of the microstructure of the cement matrix due to seeding, packing effects and pozzolanic reactions in unclinker part (Horszczaruk et al. 2014; John et al. 2018; Marushchak et al. 2016; Sanytsky et al. 2018). The influence of ultrafine additives on strength of high strength cement-based composites at elevated temperatures was investigated (Marushchak et al. 2017; Sikora et al. 2018).

The aim of present study is to investigate ultra-rapid hardening Portland cement compositions, modified with nanosized C-S-H particles and polycarboxylate superplasticizer, and high strength concrete on their base under different curing conditions.

2 Experimental Program

2.1 Raw Materials

The Portland cement CEM I 42.5 R JSC "Ivano-Frankivskcement" (Ukraine) based on the clinker of normalized mineralogical composition (% by mass: C_3S – 60.82; C_2S – 14.62; C_3A – 6.76; C_4AF – 12.32, Na_2O_e – 0.8) was used to prepare nano-modified ultra-rapid hardening Portland cement compositions. This cement is characterized by high strength at early age. Thus, compressive strength after 2 days according to EN 196-1 is 26.9 MPa (68% of minimum standard strength for strength class 42.5); the standard strength of CEM I 42.5 R is 47.8 MPa.

The innovative admixture Master X-SEED 100 (BASF) according to Crystal Speed Hardening concept, which contains synthetically produced C-S-H nanoparticles suspended in a liquid as well as the superplasticizer GLENIUM ACE 430 based on polycarboxylate esters (PCE) with nanodesigned chains were used as nano-modifiers of Portland cement compositions. The admixture Master X-SEED 100 (2.0 wt.%) and the superplasticizer GLENIUM ACE 430 (1.0 wt.%) were added together with mixing water. The natural quartz sand with fineness modulus of 2.1 and coarse aggregate of 5–20 mm fraction were used in researches. The cement: sand ratio of fine grained concrete is 1:2. High-strength concrete based on CEM I 42.5 R and ultra-rapid hardening Portland cement compositions (concrete mixture consistency classes S4 and V4, the Portland cement and Portland cement composition consumption of 360 kg per 1 m^3) was prepared.

2.2 Experimental Process

Compressive strength of ultra-rapid hardening Portland cement compositions at different age (6, 10, 18, 24 h and 2, 7, 28 days) was tested according to EN 196-1 procedure. The consistency of fresh cement mortars and fine grained concrete determined by flow table method according to standard EN 1015-3. Hydration processes of the nano-modified Portland cement compositions were examined by means of X-ray diffractometry (XRD), scanning electron microscopy (SEM). The concrete cubes of 100 × 100 × 100 mm and fine grained concrete prisms of 40 × 40 × 160 mm were used for the compressive strength tests. The concrete specimens were subjected to one of the three curing conditions: moist curing to represent typical curing procedures for concrete (temperature of 20 ± 3 °C and RH of 95 ± 5%); heat curing (the heat cured specimens were placed in a chamber with a heating regime ramping up to a constant temperature of 40 and 60 °C during 8 h); heat curing (the heat cured specimens were placed in a chamber with a heating regime 480 °C · h at constant temperature of 20; 40 and 60 °C). Testing ages for moist cured concrete specimens were 12, 18 h and 1, 2, 7 and 28 days. The heat cured specimens were tested after heat curing and after 28 days.

3 Results and Discussion

The consistency test result of nano-modified compositions at W/C = 0.5 is shown significant plasticization effect (flow F = 320 mm, ΔF = 64.1%). In this case the composition CEM I 42.5 R+PCE+X-SEED is characterized by high early strength (R_{c2} = 30.1 MPa) and the standard strength (R_{c28} = 52.8 MPa), which meets the requirement for strength class 52.5 R (Fig. 1). Due to water-reducing effect (ΔW/C = 37.2%), nano-modified composition CEM I 42.5 R+PCE+X-SEED with workability F = 143 mm is characterized by an intensive development of early strength. So, the strength of the nano-modified Portland cement composition exceeds the strength of the ordinary Portland cement by 3.4 times after 10 h. The strength of nano-modified composition after 1 and 2 days reaches 54 and 67% of the standard strength respectively, which meets requirements for ultra-rapid hardening binders. The strength of composition CEM I 42.5R+PCE+X-SEED is 84.8 MPa after 28 days and

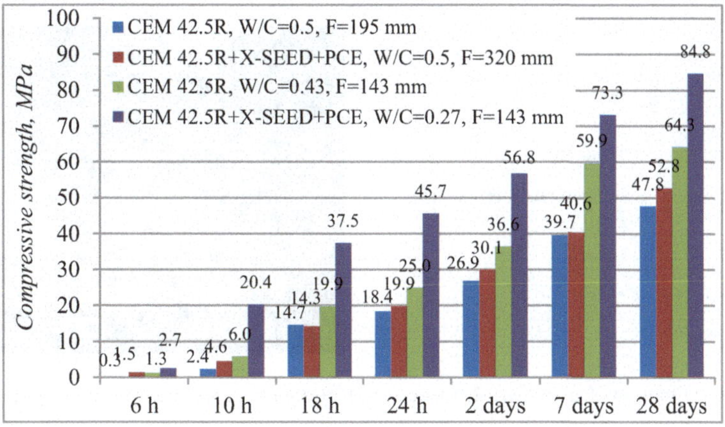

Fig. 1. Compressive strength of Portland cement compositions

these nano-modified Portland cement compositions can be attributed to high strength binders.

The XRD of the hardened pastes after 24 h is shown in Fig. 2a. It can be noticed that the main crystal phases are attributed to ettringite (d/n = 0,973; 0,561 nm) and $Ca(OH)_2$ (d/n = 0,493; 0,263; 0,192 nm). Moreover, the XRD patterns of the nano-modified Portland cement composition paste displays lower intensity of C_3S line (d/n = 0,218 nm), that indicates acceleration of the hydration processes when C-S-H nanoparticles are added. The hydration degree of nano-modified Portland cement composition is 56% after 24 h, at the same time hydration degree of CEM I 42.5 R is 48%. Lower intensity of portlandite lines indicates that the content of amorphous calcium silicate hydrates of type C-S-H(I) is increases after the addition of complex nano-modifiers. Likewise, no phase changes take place after 28 days of curing.

Nanoparticles of complex nano-modifiers play the role of crystallization centers, which accelerate the formation of a hydrosilicate gel, the so-called external product, with a more homogeneous distribution of hydrates in a limited intergranular space. Polycarboxylate superplasticizer with high surface activity due to phenomena of adsorption modification of crystalline products of hydration contributes to the formation of a homogeneous fine crystalline structure of cement matrix.

The intensive formation of calcium silicate hydrates C-S-H(I), which are characterized by fibrous structure at the early ages of hydration of nano-modified Portland cement composition CEM I 42.5R+PCE+X-SEED is observed (Fig. 2b). The calcium silicate hydrates cause stitching together cement grains and fast development of mechanical strength.

The concretes based on nano-modified ultra-rapid hardening Portland cement composition (slump class S4) are characterized by an indicator of specific strength after 1 day (f_{cm1}/f_{cm28}) of 0.46, the compressive strength of concrete after 28 days is 81.5 MPa due to water reducing effect ($\Delta W/C$ = 42.6%), and meet requirements to high strength concrete. The relative strength of concrete based on the ultra-rapid

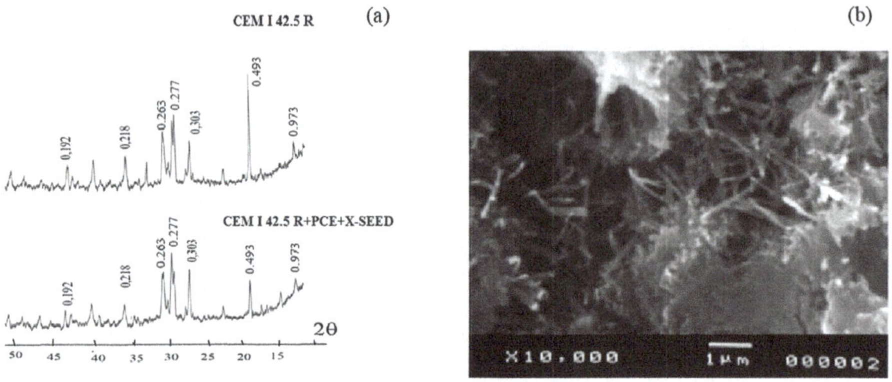

Fig. 2. XRD patterns (a) and SEM image (b) of Portland cement composition hydrated 24 h

hardening Portland cement composition as the ratio of concrete strength for each hardening period and the strength of concrete based on the CEM I 42.5 R after 28 days (f_{cm28} = 33.2 MPa) was determined (Fig. 3). The results of relative strength indicate intensive development of early strength of concrete based on the ultra-rapid hardening Portland cement composition. Thus, the relative strength of concrete based on the ultra-rapid hardening Portland cement composition (f_{cm12h}/f_{cm28} = 0.76) exceeds the required index for concrete with a rapid strength development (f_{cm2}/f_{cm28} = 0.50) after 12 h. The relative strength after 1 and 2 days is 1.13 and 1.62 respectively, which meets requirements to rapid hardening concrete.

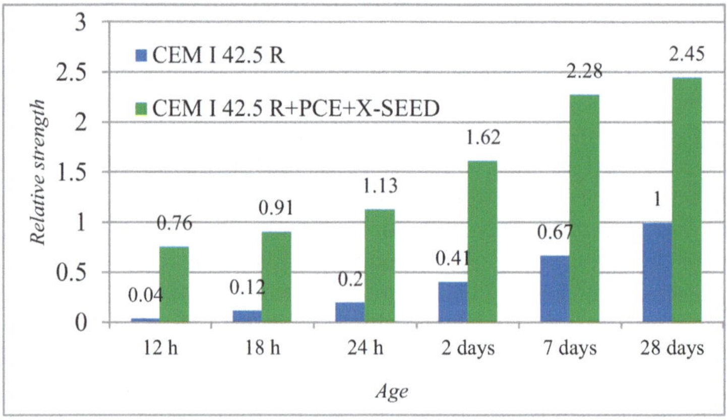

Fig. 3. Relative strength of concretes

The effects of curing procedures on the compressive strength of fine grained concrete are illustrated in Fig. 4. The strength of fine grained concrete based on the CEM I 42.5 R+PCE increases up to 68.3% at 40 °C and up to 32.5% at 60 °C compared with fine grained concrete based on the CEM I 42.5R after 8 h heat curing (Fig. 4a).

The fine grained concrete based on nano-modified Portland cement composition CEM I 42.5 R+PCE+X-SEED characterizes by the highest strength. The strength of nano-modified fine grained concrete increases by 2.5 times at 20 °C compared with fine grained concrete based on CEM I 42.5R and by 1.7 and 1.5 times at 40 °C and 60 °C respectively. The results of strength testing of concrete cured under 480 °C · h regime are shown, that influence of temperature is weak (Fig. 4b). The strength of fine grained concrete based on nano-modified Portland cement composition CEM I 42.5 R+PCE +X-SEED after heat curing at 60 °C exceeds strength after 12 h at 20 °C by 14%.

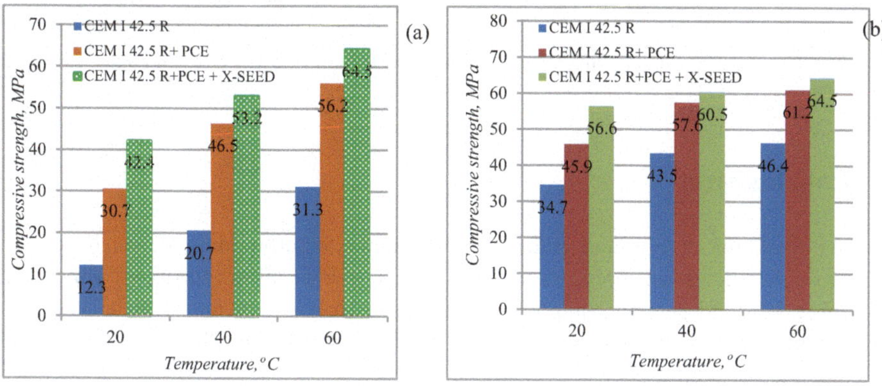

Fig. 4. Compressive strength of fine grained concrete under 8 h heat curing (a) and 480 °C · h heat curing (b)

The using of ultra-rapid hardening Portland cement compositions provide obtaining high early strength (after 24 h) of concrete (Vebe class V4) which reaches 48.9 MPa (an indicator of specific strength $f_{cm1}/f_{cm28} = 0.53$) and strength after 28 days is 91.6 MPa at ambient temperature (20 ± 3 °C), which meets requirements for high strength concrete (Fig. 4).

The 8 h heat cured at 40 °C concrete based on nano-modified ultra-rapid hardening Portland cement composition characterize by strength of 48.5 MPa which is 93% of 28-days strength of concrete based on CEM I 42.5 R and it mets requirement for pre-determined strength of prestressed precast concrete (Fig. 5). The strength of concrete based on ultra-rapid hardening Portland cement composition increases by 44% after 28 days compared with strength after heat curing.

The strength test results of the high strength concretes based on the nano-modified ultra-rapid hardening Portland cement compositions under different curing conditions have scientific novelty and practical significance. They give possibility to provide the required striking strength of concrete and pre-determined strength of concrete at low and zero energy steam curing. This allows reducing consumption of energy, increasing the number of the moulds using, saving of the formwork cost due to reducing striking time and to develop ecotechnologies of precast concrete as well as cast-in-place concrete.

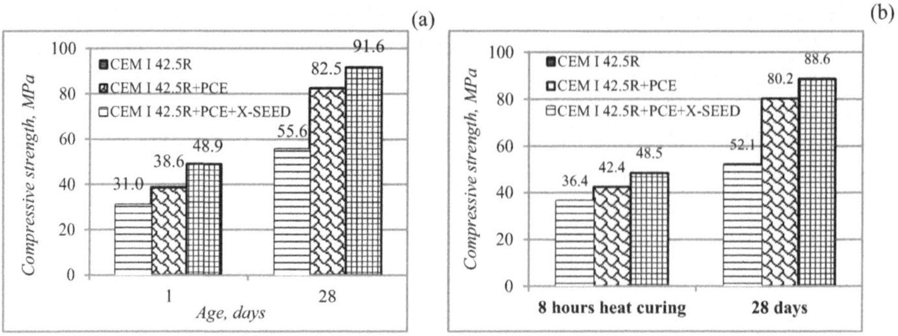

Fig. 5. Compressive strength of concrete under moist condition (a) and after heat curing at 40 °C (b)

4 Conclusions

The modification of Portland cement with complex nano-modifiers, which contain C-S-H nanoparticles and polycarboxylate superplasticizer allows to obtain ultra-rapid hardening Portland cement compositions for high strength concrete, which characterizes by the accelerated early strength gain (R_{c1}/R_{c28} = 0.54 and R_{c2}/R_{c28} = 0.67) and high standard strength (R_{c28} = 84.8 MPa).

Nanoparticles of complex nano-modifiers due to seeding effect accelerate the formation of a hydrosilicate gel with a more homogeneous distribution of hydrates in a limited intergranular space. As a result, concretes based on ultra-rapid hardening Portland cement compositions are characterized by increased rate of early strength development (after 1 day of hardening f_{cm1}/f_{cm28} is 0.46–0.53), which meets the requirements for ultra-rapid hardening concrete. The strength of nano-modified concrete on the basis of ultra-rapid hardening Portland cement compositions is 81.5–91.6 MPa after 28 days and meets the requirements for ultra-rapid hardening concrete. It was shown that concretes based on ultra-rapid hardening Portland cement compositions allowed to provide development of low and zero energy technologies of precast concrete and effective technologies of cast-in-place concrete.

References

Choi W, Rizkalla S, Zia P, Mirmiran A (2008) Behavior and design of high-strength prestressed concrete girders. PCI J 5:54–69

Dvorkin L, Babych Y, Zhytkovsky V, Bordyuzhenko O, Filipchuk S, Kochkarov D, Kovalyk I, Kovalchuk T, Skrypnyk MM (2017) High-strength rapid hardening concretes and fiber reinforced concretes. NUVGP, Rivne (in Ukrainian)

Horszczaruk E, Mijowska E, Cendrowski K, Sikora P (2014) Influence of the new method of nanosilica addition on the mechanical properties of cement mortars. Cem Lime Concr 8: 308–316

John E, Matschei T, Stephan D (2018) Nucleation seeding with calcium silicate hydrate – a review. Cem Concr Res 113:74–85

Kanchanason V, Plank J (2015) C-S-H – PCE Nanocomposites for enhancement of early strength of cement. In: 19th international conference, Weimar, Germany, 16–18 September 2015

Krivenko P, Sanytsky M, Kropyvnytska T (2018) Alkali-sulfate activated blended Porland cements. Solid State Phenom 276:9–14

Martí JV, Gonzalez-Vidosa F, Yepes V, Alcalá J (2013) Design of prestressed concrete precast road bridges with hybrid simulated annealing. Eng Struct 48:342–352

Marushchak U, Sanytsky M, Mazurak T, Olevych Y (2016) Research of nanomodified Portland cement compositions with high early age strength. Eastern-Eur J Enterp Technol 6/6(84): 50–57

Maruchchak U, Sanytsky M, Olevych Y (2017) Effects of elevated temperatures on the properties of nanomodified rapid hardening concretes. MATEC Web of Conf 116:010008

Min T-B, Cho I-S, Park W-J, Choi H-K, Lee H-S (2014) Experimental study on the development of compressive strength of early concrete age using calcium-based hardening accelerator and high early strength cement. Constr Build Mater 64:208–214

Sanytsky M, Kropyvnytska T, Kruts T, Horpynko O, Geviuk I (2018) Design of rapid hardening quaternary zeolite-containing Portland-composite cements. Key Eng Mater 761:193–196

Savchuk Y, Plugin A, Lyuty V, Pluhin O, Borziak O (2018) Study of influence of the alkaline component on the physico-mechanical properties of the low clinker and clinkerless waterproof compositions. Matec Web Conf 230:03018

Sikora P, Abd Elrahman M, Stephan D (2018) The Influence of nanomaterials on the thermal resistance of cement-based composites–a review. Nanomaterials 8:465

Torres A, Burkhart A (2016) Developing sustainable high strength concrete mixtures using local materials and recycled concrete. Mater Sci Appl 7:128–137

Solodkyy S, Markiv T, Sobol K, Hunyak O (2017) Fracture properties of high-strength concrete obtained by direct modification of structure. MATEC Web Conf 116:01016

Calculation of Reinforced Concrete Columns Strengthened by CFRP

Jacek Selejdak[1], Yaroslav Blikharskyy[2(✉)], Roman Khmil[3], and Zinoviy Blikharskyy[3]

[1] Department of Concrete Structures and Geotechnics, Faculty of Civil Enginnering, Czestochowa University of Technology, Akademicka 3, 42200 Czestochowa, Poland

[2] Department of Highways and Bridges, Lviv Politechnic National University, Karpinskogo 6, Lviv 79013, Ukraine
yaroslav.z.blikharskyy@lpnu.ua

[3] Department of Building Constructions and Bridges, Lviv Polytechnic National University, Karpinskogo 6, Lviv 79013, Ukraine

Abstract. In this paper the results of calculation of eccentrically compressed reinforced real scale concrete columns strengthened with carbon fiber reinforced polymers (CFRP) laminates is presented. In order to check the calculation results, four real scale experimental reinforced concrete columns 2200 mm long and with sectional dimensions 180 × 140 mm were manufactured. Reinforced samples were strengthened with a 25 mm CFRP laminates. The calculation was made according to Eurocode 2. The usage of carbon laminates limit strain introduced into the calculation, is proposed. On the basis of normative documents the block diagram of calculation algorithm for such structures is created. As a result of the calculation it was established that according to the presented calculation algorithm the divergence between calculated and experimental values reaches 5 … 10%, while the experimental data exceeds the theoretical ones. It makes it possible to reliably calculate and design strengthening of eccentrically compressed RC structures with CFRP laminates according to the algorithm presented in the work.

Keywords: RC columns · Real scale · Eccentrically compression · Strengthening · CFRP laminates · Load bearing capacity · Calculation method

1 Introduction

Reinforced concrete elements subjected to longitudinal eccentric axial forces or longitudinal axial forces and bending moments work under non-centre compression. Eccentrically compressed RC elements are wide-spread in construction (Krainskyi et al. 2018a, b; Kramarchuk et al. 2019). Examples of such elements could be columns of industrial one-storey buildings, loaded with eccentric force, for example from a bridge crane or other equipment, uttermost columns of public buildings subjected to wind loads, columns of overpasses.

Z. Blikharskyy et al. (Eds.): CEE 2019, LNCE 47, pp. 400–410, 2020.
https://doi.org/10.1007/978-3-030-27011-7_51

The necessity to ensure reliable serviceability of buildings and structures during reconstruction or extension of their service life requires the implementation of works on strengthening, restoration and repair of building structures (Blikhars'kyi and Obukh 2018; Selejdak et al. 2018). Structures` strengthening is rather time-consuming and expensive procedure. In industrialized countries, more than 40% of investment in the construction industry is used for maintenance care and repair of RC structures and less than 60% is used for the construction of new ones (Lee et al. 2006).

In some cases, further reliable exploitation can be ensured without strengthening by changing work conditions: reducing of permanent and variable loads (for example, replacement of heavy insulation, ceiling, roofing decking), reduction of cranes' lifting capacity if it is possible under certain operating conditions, reduction of vibration, etc.

If it is impossible to ensure the reliable structures' performance by changing work conditions for the restoration of design operational properties or for its increase comparatively with the specified ones, as well as when certain defects or damage of the structures are identified their strengthening is avoidless.

When design of the RC structures' strengthening takes place, the choice of reinforcement methods is carried out after the stage of verification calculations, through which a possible failure scheme of the structure is identified (Bobalo et al. 2018). Generally the bending elements failure can take place according to several schemes: in the stretched zone, in the compressed zone, in the cutting zone due to shear forces, in the spatial cut due to torque moment, local wrinkling, tearing, and squashing (Blikharskyy et al. 2018; Khmil et al. 2018; Brozda et al. 2017).

Nowadays according to the valid norms (Eurocode 2, 2004, DBN V.2.6-98, 2009), the load-bearing capacity of reinforced concrete elements is calculated on the basis of the deformation model of the calculated cross section, which is based on the following:

- equilibrium equations of external and internal forces in the normal section are used;
- the strains' distribution along the section height is supposed to be according to the linear law (the appropriateness of the plane cross sections hypothesis);
- the dependency between the axial stresses and the relative strains of concrete and steel bars is assumed in the form of complete diagrams of the state (strain) of concrete and rebar;
- the estimated resistance of the stretched zone concrete is not taken into account;
- the stress diagram in the compressed zone is taken as a simplified rectangular or closer to the actual curvilinear.

Basing on this principles output data is identified for calculating bending, compressed or non-centrically compressed reinforced concrete elements, both non-reinforced and reinforced with glued tapes.

For more widespread use of CFRP laminates and canvas as additional external reinforcement for RC structures, a sufficient amount of experimental-theoretical research and establishment of an optimal calculation methodology basing on valid normative documents is required.

2 Brief Description of Eccentrically Compressed Columns Experimental Studies

In order to achieve this goal, experimental reinforced concrete columns 2200 mm long and with sectional dimensions 180 × 140 mm were manufactured. On both sides of the column cantilever protrusions are arranged in order to create eccentrically load. The rebar of the column sections is symmetrical with the A400C Ø12 mm rebar.

Reinforced samples were strengthened with a 25 mm Sika Carbodur S512 laminate. The laminate was glued on the stretched plane of the column. In order to provide anchoring with tape on the columns' cantilevers SikaWrap fabric 15 cm width was used.

To determine the strength and deformability characteristics of RC columns, they were tested for eccentrically compression until failure. The following columns' marking was accepted: KS-1.1 and KS-1.2, KPS-1.3 and KPS-1.4, where the KS is the standard column, KPS is the column reinforced by a tape. The first digit indicates the series' number and the second one identifies the sample number.

Load on columns was implemented by a concentrated force applied with eccentricity e = 150 mm. Application of the load was created with the use of hydraulic jack on specially prepared equipment. Detailed information about materials, strengthening techniques and experimental testing of experimental samples was at Blikharskyy et al. (2018) and Zhang et al. (2015).

3 Method of Load-Bearing Capacity Calculation of Eccentrically Compressed Reinforced Concrete Columns

Determination of bearing capacity of eccentrically compressed reinforced concrete columns was carried out according to the valid norms (Eurocode 2, 2004, DBN V.2.6-98, 2009) based on the nonlinear deformation section model. This technique uses nonlinear strain diagrams of concrete and rebar and is based on an iterative method.

The calculation is implemented with the assumption of the following prerequisites:

- the hypothesis of plane cross sections is accepted - strains along the section height are distributed according to the linear law;
- the averaged cross section with the greatest strain increment is accepted for calculation;
- in calculations the dependence between stress and relative strain of concrete and rebar as diagrams is accepted, while for concrete this dependence is assumed as nonlinear, and for rebar- as two-linear;
- the resistance of the concrete stretched zone is not taken into account;.

When calculating eccentrically compressed reinforced concrete elements with large eccentricities of rectangular shape, which include eccentrically compressed reinforced concrete columns investigated in this paper, according to the norms following equilibrium equations of external and internal forces in the normal section (Eurocode 2, 2004, DBN V.2.6-98, 2009).

Estimated cross sections of eccentrically compressed columns with strengthening are shown on Fig. 1. It is accepted that the strain of the additional rebar (carbon laminate) are equal to the averaged strain of the stretched concrete fibers $\varepsilon_{c(2)}$, since they are on the same level, as shown on Fig. 1.

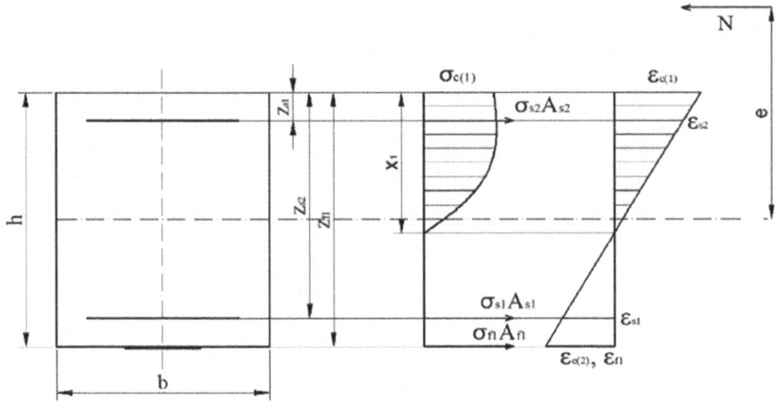

Fig. 1. The calculated section of the eccentrically compressed element of rectangular cross section with strengthening. The cross section of the element, stress diagram and the strain diagram

In DBN V.2.6-98 (2009), the equitation for determining the acting moment is specified:

$$M = N \cdot \eta \cdot (x_1 - y + e_0). \tag{1}$$

where η - coefficient which takes into account the compressed element slenderness; y - distance between the most compressed fibers and sectional center of gravity; e_0 - initial eccentricity; x_1 - compressed concrete zone height.

However, valid norms DBN V.2.6-98 (2009) do not specify how to calculate the coefficient η that takes into account the element slenderness. In order to calculate it, the equitation given in the previous rules could be used:

$$\eta = \frac{1}{1 - \frac{N}{N_{cr}}}, \tag{2}$$

where N – the calculation value of the axial force; N_{cr} – relative critical force, which could be calculated as:

$$N_{cr} = \frac{6.4 E_c I}{\varphi_l l_0^2}\left(\frac{0.11}{0.1 + \delta_e} + 0.1\right). \tag{3}$$

In European norms Eurocode 2 (2004), the calculating moment equals the sum of moments of the first order and the second-order nominal moment, which appears due to the element bending (according to the deformed scheme). Total moment could be calculated by the equitation:

$$M_{Ed} = M_{0Ed} + M_2, \tag{4}$$

where M_{0Ed} - first-order moment; M_2 - nominal second-order moment, calculated as:

$$M_2 = N_{Ed} \cdot e_2, \tag{5}$$

where N_{Ed} - calculating value of the axial force; e_2 - additional eccentricity due to bending, which is equal to

$$e_2 = \frac{1}{r} \cdot \frac{l_0^2}{c}, \tag{6}$$

where $\frac{1}{r}$ - curvature; l_0- estimated length; c - coefficient, which depends on the curvature distribution. It is recommended to accept this coefficient: $c = \pi^2 \approx 10$.

Using the suggestions outlined in Eurocode 2 (2004), an equitation for calculating the slenderness of an element could be derived with the use of equitation given in the DBN V.2.6-98 (2009):

$$\eta = \frac{e_2}{e_0} + 1 = \frac{1}{r} \cdot \frac{l_0^2}{c \cdot e_0} + 1 = \frac{\chi \cdot l_0^2}{c \cdot e_0} + 1, \tag{7}$$

where e_0 - initial eccentricity.

The block diagrams for calculating the deformation model of the stress-strain state of eccentrically compressed RC elements strengthened without the initial load level are shown on Fig. 2.

The sequence of calculation of the stress-strain state parameters of the normal section was accepted according to (DBN V.2.6-98, 2009). It consists in the fact that at each step of the iterative calculation, the values of relative averaged strain of the stretched concrete fibers $\varepsilon_{c(2)}$ were determined basing on the pre-set value of the strain of the compressed concrete fibers $\varepsilon_{c(1)}$. When the specified calculation accuracy for a given value of the strain was reached, the calculation was stopped and the corresponding value of the longitudinal force was determined. Then, at a given value of the compressed fibers' strain $\varepsilon_{c(1)}$, it was increased by the value $\Delta\varepsilon_{c(1)}$ and the calculation was repeated until the requirement $\varepsilon_{c(1)} \geq \varepsilon_{cu1}$ was fulfilled, as it was accepted as a criterion of carrying capacity of reinforced concrete columns exhaustion due to the achievement the limit values by the most compressed concrete fibers. After the calculation is completed, following values were obtained: strains of compressed and stretched fibers of concrete, main rebar, reinforced samples of additional strengthening (carbon laminate), curvature of the normal section bended axis, the value of additional eccentricity due to bending of the reinforced concrete column for the corresponding value of longitudinal axial force.

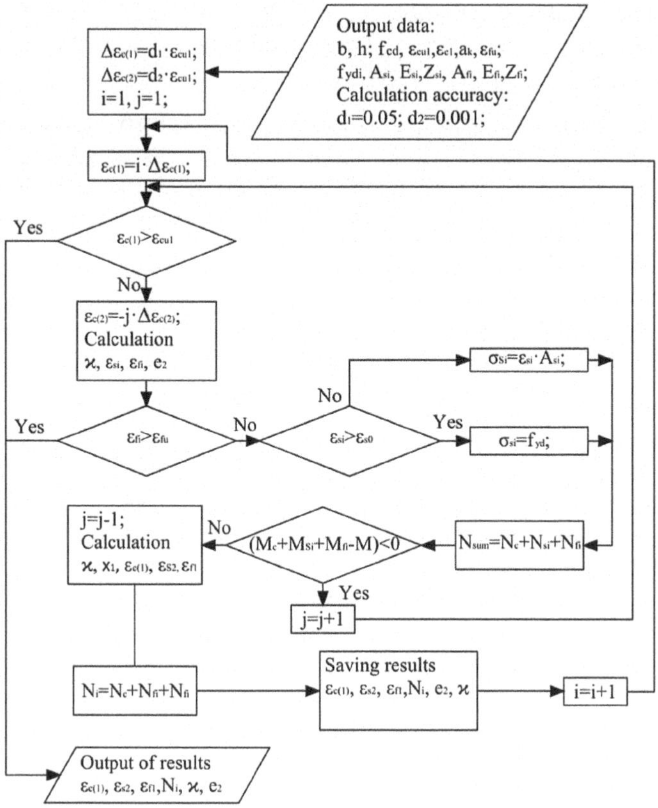

Fig. 2. Block diagram for calculation of stress-strain state of RC eccentrically compressed strengthened element normal section

When the strains $\varepsilon_{c(1)}$ exceeded the limit values ε_{cu1} by iterations, the value of the load, at which $\varepsilon_{c(1)} = \varepsilon_{cu1}$ was redefined. This value is accepted as a load, at which the exhaustion of the reinforced concrete columns` bearing capacity was reached and could be used for further comparison with those determined during experimental studies.

The following output data was accepted for experimental-theoretical studies performed in this paper:

- section parameters (width and height): $b = 140\,\text{mm}$, $h = 180\,\text{mm}$;
- initial eccentricity of a longitudinal force applying from the section center of gravity $e_0 = 150\,\text{mm}$;
  concrete parameters: C30/35: $f_{cd} = 25.5\,\text{MPa}$; $\varepsilon_{cu,1} = 0.0031$; $\varepsilon_{cu,3} = 0.00293$; $\varepsilon_{c1} = 0.00172$;
- for this concrete class the polynomial coefficients:
- $a_1 = 2.6219$, $a_2 = -2.425$, $a_3 = 0.98327$, $a_4 = -0.17908$, $a_5 = -0.00107$;
- rebar parameters 2Ø12 A500C: $f_{yd} = 570\,\text{MPa}$; $E_s = 200000\,\text{MPa}$; $A_s = 226\,\text{mm}^2$, $z_{s1} = 26\ldots36\,\text{mm}$, $z_{s2} = 138\ldots154\,\text{mm}$.

– parameters of additional composite rebar (CFRP laminate 25 mm width): thickness $t = 1.2\,\text{mm}$, $f_{fd} = 2080\,\text{MPa}$; $E_s = 160000\,\text{MPa}$; $A_s = 25 \cdot 1.2 = 30\,\text{mm}^2$, $z_{s1} = 180.6\,\text{mm}$.

According to FIB recommendation (Externally Bonded FRP Reinforcement for RC Structures 2001) the following limit strain for composite laminate are accepted: $500{\cdot}10^{-5}$ for concrete grade C35/45 and lower. By specified parameters the theoretical calculations of reinforced concrete columns were done in order to compare their results with experimentally established ones in this work.

4 Results and Discussion

The comparative analysis of the calculated and experimental strain and bearing capacity values for the columns shows that the use of Eq. (1) for the determination of the bending moment, results in overstatement of theoretical values of the bearing capacity of the columns comparing with the experimental ones, while the main rebar does not reach the yield strength. The use of Eq. (1) and substituting the coefficient of flexibility from Eq. (2), give the results with more close values, however, they are also somehow overstated compared with the experimental ones. The satisfacting similarity could be seen for the results where Eq. (1) and the coefficient of flexibility from Eq. (7) derived with the use of the dependencies of Eurocode 2 (2004) were used, namely

$$M = N \cdot \eta \cdot (x_1 - y + e_0) = N \cdot \left(\frac{\chi \cdot l_0^2}{c \cdot e_0} + 1\right) \cdot (x_1 - y + e_0). \tag{8}$$

The calculation results are presented in the following graphs and tables.

In the columns KS-0.1 and KS-0.2, according to the theoretical calculation (Fig. 3a), the yield strength strain $\left(\varepsilon_y = 280 \cdot 10^{-5}\right)$ of the main steel bars occurred at $N_s^{th} = 118.4\,\text{kN}$ (Fig. 3), further loading was accompanied by significant increase in the strain of the reinforcement and concrete. The load at which the compressed concrete reached the limit strain $\left(\varepsilon_{cu1} = 310 \cdot 10^{-5}\right)$ was $N_c^{th} = 121.8\,\text{kN}$, which is lower than the experimental values on 8.35%.

In the columns KPS-1.3 and KPS-1.4, according to the theoretical calculation (Fig. 3b), the load at which the deformation of the main rebar reached the yield strength $(\varepsilon_y = 280 \cdot 10^{-5})$ was $N_s^{th} = 159.2\,\text{kN}$, which is 9.7 kN less than the experimental values (5.65%). The load at which the compressed concrete reached the limit strain $(\varepsilon_{cu1} = 310 \cdot 10^{-5})$ was $N_c^{th} = 159.3\,\text{kN}$, which is 15.8 kN (9.02%) less than the experimental values.

The strain of additional reinforcement (carbon laminate) at $N_c^{th} = 159.3\,\text{kN}$ in the theoretical calculation reached the value of $\varepsilon_f = 419 \cdot 10^{-5}$ which amounts 83.8% of the limit strain according to (Externally Bonded FRP Reinforcement for RC Structures, 2001) $(500 \cdot 10^{-5})$ and 32.2% $(1300 \cdot 10^{-5})$ according to Sika. The experimental strains of the additional rebar at $N_c^{exp} = 175.1\,\text{kN}$ were equal to $\varepsilon_f = 542 \cdot 10^{-5}$, which amounts 108.4% of the limit strain according to FIB recommendation (Externally Bonded FRP Reinforcement for RC Structures, 2001), and 41.7% according to DBN V.2.6-98 (2009).

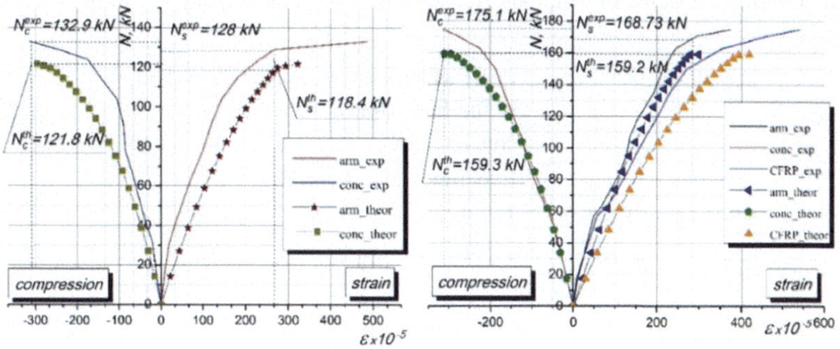

Fig. 3. The graphs of the strains calculated values of the stretched main steel rebar's (arm_theor), the additional reinforcement (CFRP.theor) and the compressed concrete (conc_theor) in comparison with the experimentally determined averaged strains of the stretched main steel rebar's (arm_exp), additional reinforcement (CFRP_exp) and compressed concrete (conc_exp) for columns KP-0.1 and KP-0.2 (left), KPS-1.3 and KPS-1.4 (right)

It should be noted that the theoretical graphs of strain of the stretched main reinforcement, additional reinforcement (in strengthened samples) and compressed concrete show satisfactory consistency with the graphs of the experimental strain values. Herewith, the theoretical strain is less than the experimental one at the same load.

Calculated and experimental values of bearing capacity of columns are presented in Tables 1, 2 and 3.

Table 1. Bearing capacity of columns by the criterion of achieving the limit strain by the main armature

Sample code	Load which corresponds to the limit strain of the main rebars, *kN*		Deviation, %
	Experimental	Theoretical	
KS-0.1 and KS-0.2	128	118.4	7.50
KPS-1.3 and KPS-1.4	168.73	159.2	5.65

By the bearing capacity of the columns, when the yield strain is reached, the deviation between the theoretical and experimental results is 5.65 … 7.5% towards understatement of the calculated values, which ensures the reliability of the proposed calculation method for the reinforced eccentrically compressed columns.

Deviation of the experimental bearing capacity of the columns with the theoretical one by the criterion of achieving the limit compressive strain of the concrete (Table 2) is 8.35 … 9.02% towards the understatement of theoretical values compared with the experimental ones.

Table 2. Bearing capacity of columns by the criterion of achieving the limit strain by the concrete

Sample code	Load which corresponds to the limit strain of the concrete, *kN*		Deviation, %
	Experimental	Theoretical	
KS-0.1 and KS-0.2	132.9	121.8	8.35
KPS-1.3 and KPS-1.4	175.1	159.3	9.02

Therefore, the calculation of the eccentrically compressed columns strength according to the valid norms DBN V.2.6-98 (2009) with the use of the proposed coefficient of slenderness in Eq. (7) which is derived using dependencies proposed in Eurocode 2 (2004) provides satisfactory similarity of results with experimental data, herewith the theoretical strength values are smaller than experimental ones, which ensures the reliability of the proposed calculation method.

Table 3. Deformation of additional reinforcement (CFRP laminate) under load that corresponds to the limit strain of concrete

Sample code	Deformation of CFRP laminate ε_f, $\times 10^{-5}$		Laminate using comparing to ε_{fu} according to codes,%			
	Experimental	Theoretical	FIB		Sika	
			Experimental	Theoretical	Experimental	Theoretical
KPS-1.3 and KPS-1.4	542	419	108.4	83.8	41.7	32.2

According to recommendations of the FIB (Externally Bonded FRP Reinforcement for RC Structures, 2001), the limit strain of the laminate is taken as $\varepsilon_{fu} = 500 \cdot 10^{-5}$, and according to the recommendations of Sika $\varepsilon_{fu} = 1300 \cdot 10^{-5}$. These strain values were used to determine the effectiveness of using the laminate in Table 3.

The deviation between the experimental values of the additional reinforcement (carbon laminate) strain ε_f^{exp} and the theoretical ones ε_f^{th}, recorded at loads N_c^{exp} and N_c^{th} amounts 22% in average, towards the understatement of the theoretical data. If we consider the limit strain according to the recommendations of the FIB (Externally Bonded FRP Reinforcement for RC Structures, 2001) the experimental strain characteristics of the carbon tape in the samples KPS-1.3 and KPS-1.4 are fully used, if the results of the theoretical calculation are considered in such samples the efficiency of using the characteristics of the carbon laminate amounts 84% on average. If Sika's recommendations are used efficiency amounts 42% on average according to experimental data, however the theoretical percentage of carbon laminate characteristics usage in such samples is on average 32%.

5 Conclusions

1. For calculation of eccentrically compressed reinforced concrete columns strengthened with carbon laminate the calculating algorithm on the basis of valid norms for new structures` design has been developed and the recommendations of Sika have been used. In order to determine the additional eccentricity due to bending slenderness coefficient is introduced which is derived using the dependencies of Eurocode 2 (2004).
2. During determination of the reinforced columns bearing capacity when the limit strain of the main reinforcement is reached, the deviation between the theoretical and experimental results is 5.65 … 7.5% towards the understatement of the theoretical values. The difference between the calculated bearing capacity of columns with the experimental criterion for achieving the limit strain of concrete is 8.35 … 9% towards the understatement of theoretical values. Therefore, calculation of the bearing capacity of reinforced eccentrically compressed columns with the use of the proposed method provides satisfactory results, which allows us to recommend the proposed calculation method for further use.
3. The deviation between the experimental values of strain of the additional reinforcement (CFRP laminate) ε_f^{exp} and the theoretical ones ε_f^{th}, recorded at loads N_c^{exp} and N_c^{th} respectively is within 22% towards the understatement of theoretical results. The efficiency carbon laminate characteristics usage amounted 84% on average. If Sika's recommendations are considered, then according to experimental data, efficiency reaches 42%, and the theoretical percentage of carbon laminate characteristics usage in such samples is on average 32%.

References

Blikhars'kyi Z.Ya, Obukh Y (2018) Influence of the mechanical and corrosion defects on the strength of thermally hardened reinforcement of 35GS steel. Mater Sci 54. https://doi.org/10.1007/s11003-018-0183-2

Blikharskyy Y, Khmil R, Blikharskyy Z (2018) Research of RC columns strengthened by carbon FRP under loading. MATEC Web Conf 174:1–8. https://doi.org/10.1051/matecconf/201817404017

Bobalo T, Blikharskyy Y, Vashkevich R, Volynets M (2018) Bearing capacity of RC beams reinforced with high strength rebars and steel plate. MATEC Web Conf 230:02003. https://doi.org/10.1051/matecconf/201823002003

Brózda K, Selejdak J, Koteš P (2017) The analysis of beam reinforced with FRP Bars in bending. Procedia Eng 192:64–68. https://doi.org/10.1016/j.proeng.2017.06.011

DBN V.2.6-98:2009 (2009) Concrete and reinforced concrete structures. National standard of Ukraine [in Ukrainian]

Eurocode 2 (2004) Design of concrete structures – part 1–1: general rules and rules for buildings . EN 1992-1-1, European Committee for Standardisation (CEN) Brussels

Externally Bonded FRP Reinforcement for RC Structures (2001) FIB Bulletin No. 14, Technical report

Khmil R, Tytarenko R, Blikharskyy Y, Vegera P (2018) Development of the procedure for the estimation of reliability of reinforced concrete beams, strengthened by building up the stretched reinforcing bars under load. Eastern-Eur J Enterp Technol 5/7(95). https://doi.org/10.15587/1729-4061.2018.142750

Krainskyi P, Blikharskyy Y, Khmil R, Blikharskyy Z (2018a) Experimental study of the srengthening effect of reinforced concrete columns jacketed under service load level. MATEC Web Conf 183:1–5. https://doi.org/10.1051/matecconf/201818302008

Krainskyi P, Blikharskyy Y, Khmil R, Vegera P (2018b) Influence of loading level on the bearing capacity of RC columns strengthened by jacketing. MATEC Web Conf 230:02013. https://doi.org/10.1051/matecconf/201823002013

Kramarchuk A, Ilnytskyy B, Lytvyniak O, Grabowski A (2019) The increase of seismic stability for existing industrial buildings. In: AIP conference proceedings, p 2077

Lee H, Choi S, Youm K (2006) Seismic performance of repaired RC columns. Mag Concr Res 58:267–276. https://doi.org/10.1680/macr.2006.58.5.267

Selejdak J, Khmil R, Blikharskyy Z (2018) The influence of simultaneous action of the aggressive environment and loading on strength of RC beams. MATEC Web Conf 183:1–6. https://doi.org/10.1051/matecconf/201818302002

Zhang Q, Mol'kov YV, Sobko YM, Blikhars'kyi YZ, Khmil RE (2015) Specific fracture energy of thermally hardened reinforcement. Mater Sci 50(6):824–829. https://doi.org/10.1007/s11003-015-9789-9

Highly Combinatorial Reinforced Concrete Slab System

Valerii Shmukler[1], Olena Petrova[1(✉)], and Valerii Nikulin[2]

[1] Department of Building Structures, O.M. Beketov National University of Urban Economy in Kharkiv, 17 Marshala Bazhanova Street, Kharkiv 61002, Ukraine
petrova.bcd@gmail.com

[2] Additional Liability Company «Zhilstroj-2», 12 Kosmichna Street, Kharkiv 61000, Ukraine

Abstract. Paper represents an architectural and construction system for the erection of civil and residential buildings of monolithic reinforced concrete with a given internal and simple external geometry named "MONOFANT". All structural elements of the frame building of the "MONOFANT" system (foundation, columns, rigid elements, floors and coverings) are made of monolithic reinforced concrete with hollow sections due to the installation inside of them inserts - void formers from organic and inorganic insulating materials. Main advantage of this system is creation of structures with the given stress-strain state. It is possible to design structures with maximum resistance on the fixed material consumption or systems with the lowest material consumption on the fixed resistance. One of the most interesting and yet unsolved questions while designing elements of this system is the cutting of the organic inserts which provides minimum waste of material. In this connection we considered traditional "MONOFANT" slab of civil building and fulfilled the analysis of the features of stress-strain state together with solving the optimal cutting task. Procedure and main results are presented in this paper.

Keywords: Inserts · "MONOFANT" system · Optimal cutting task · Stress-strain state (SSS) · Rationalization

1 Introduction

System "MONOFANT" (Shmukler et al. 2013) is an architectural and construction system intended for erection of monolithic reinforced concrete buildings and structures, which involves the construction of vertical, horizontal or curvilinear structures by shotcreting a fiber concrete or using self-compacting concrete. The formation of a monolithic reinforced concrete frame is carried out with the use of inserts – void formers made of foam polystyrene or mineral wool in all structural elements (columns, walls, etc.) and the formation of a traditional reinforcing frame around them.

The main advantage of "MONOFANT" system compared to other known analogues, such as Cobiax system, Airdec and others, is the possibility of creating systems with a given stress-strain state and material consumption. This circumstance became

Z. Blikharskyy et al. (Eds.): CEE 2019, LNCE 47, pp. 411–419, 2020.
https://doi.org/10.1007/978-3-030-27011-7_52

possible due to the use of a direct approach of design, which is based on the exploitation of energy principles of rationalization of the structure. In particular, the following restrictions are taken as criteria of rationality:

1. the requirement of minimizing the strain energy (SE) of the system:

$$U = inf_{\alpha} U\left(\alpha^{k}\right),\ k = 1, 2, \ldots \infty \tag{1}$$

   where U - strain energy; k - number of compared variant of the topology; $\alpha \in M$; M - set of permissible values of external geometric parameters;
2. requirement of isoenergeticity of the system:

$$e[\{\bar{x}\}] = const \tag{2}$$

where e – the density of the strain energy (DSE); $\{\bar{x}\}$ – vector of internal parameters.

2 Main Features of the Stress-Strain State of the System

Today, there are already available studies on the transformation of the stress-strain state of structures with inserts (Albrecht 2012). We should mention, that in (Steblovsky 2015) the results of the rationalization of the internal parameters of the structure of lightweight overlapping slab with the foam polystyrene inserts are given. These parameters include: the thickness of the flanges, the step and the geometry of the internal ribs. It is noted that the thickness of the flange, as a rule, cannot be calculated due to the technological requirements (EN1992-1-1 2004) and State Building Codes of Ukraine (DBN). Among these restrictions there are a safe transfer of an effort for adhesion between reinforcement and concrete, protection of reinforcing steel from corrosion and the required limit of fire resistance. According to the calculations, the rational step of the internal ribs was adopted as 75 cm for C25/30 concrete class and A500S reinforcement class at given temporary and constant loads. Appointment of the minimum thickness of the ribs was made on the basis of technological conditions and was adopted as 100 mm.

In turn, in (Pomazan 2013) on the technology of lightweight floor slabs, it is noted that the technology of laying concrete in two stages, which is used in Ukrainian construction industry today, can lead to a displacement of inserts in the horizontal plane. Based on these data, authors made an analysis of the effect of linear displacement in space on the SSS of the slab with inserts. It is obtained (Bugaevsky et al. 2016) that during the concreting of floor slabs, special attention should be paid to the position of the inserts over the height of the slab to ensure the accepted thickness of the flanges, since this has a great influence on the operation of this structure under load.

Summarizing what has been said, it is relevant today to analyze the influence of the dimensions of the inserts, and, as a consequence, the step of the internal ribs, on the stress-strain state of the structure, taking into account the concreting process. Thus, based on the study (Shmukler and Babaev 2017) on the choice of the rational rib's topology, let's consider in more detail the location of the internal ribs of the slab which corresponds to the above listed limitations (1, 2). According to the analysis performed by the author, the topology variant with a diagonal arrangement (Fig. 1a) of the ribs among other 13 options satisfies the listed limitations (Fig. 1b), and, moreover, it is distinguished by the maximum value of the natural vibration frequency and the minimum value of the deflection.

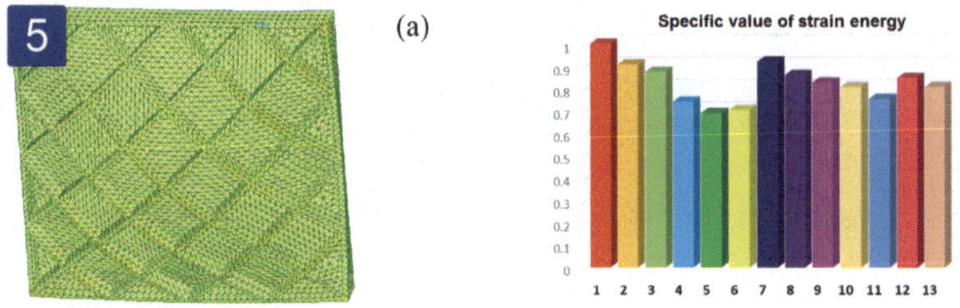

Fig. 1. Fragment of a slab with rational topology of ribs after (Shmukler and Babaev 2017) and graphs of strain energy for all compared options

Based on this, we consider a fragment of the standard plan of a frame building with a cell size of 6 × 6 m (Fig. 2); columns have a square section (300 × 300 mm). Between the columns along the perimeter of the cell, monolithic reinforced concrete beams with a cross section of 200 mm are arranged, they are made flush with the floor slab, and the capitals of a square shape together with the column have a cross section of 500 × 500 mm. The slab is hollow, has a thickness of 200 mm, while the thickness of the upper and lower flanges is 50 mm; the middle part of a slab is filled with lightweight material (polystyrene).

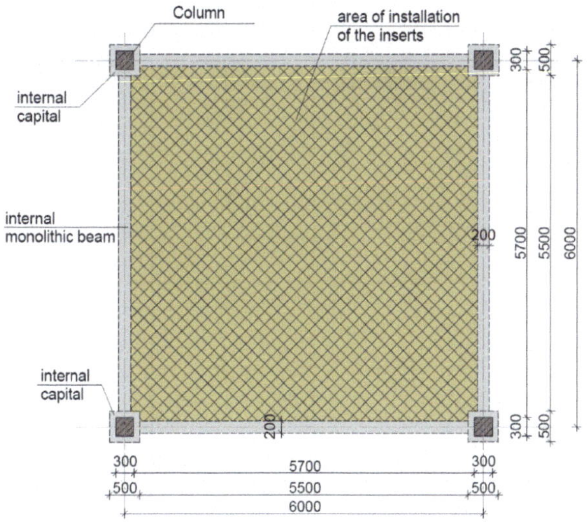

Fig. 2. The considered fragment of the plan of the frame building

Let's analyze the influence of variation of the rib's step on the components of the SSS (value of deflection, bending moments, strain energy) of the system. To begin with, let us set the minimum possible for technological conditions edge width equal to 100 mm and the minimum size of the insert - 500 mm. Thus, under the prevailing conditions, the step of the ribs in the slab will be 600 mm. The cross-section of the slab is shown at Fig. 3.

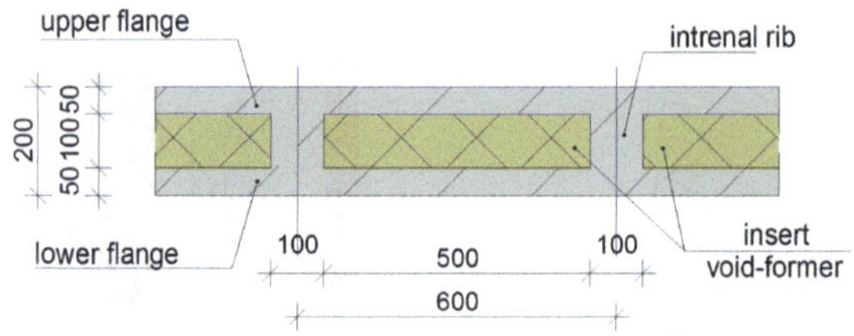

Fig. 3. The cross-section of the considered first option of the slab

Further, on the basis of the obtained data, we will determine the calculated width of the rib, setting the step of the ribs, respectively 800, 1000, 1200 and 1400 mm. At the same time, the total concrete consumption for the ribs obtained in the first iteration for each of the indicated option should remain constant. The consumption of concrete for flanges is also constant, since the thickness of the flanges is unchanged, and is equal to 3.6 m^3.

Table 1. Reduced section for each slab and material consumption

№	Option 1	Option 2	Option 3	Option 4	Option 5
Reduced slab section	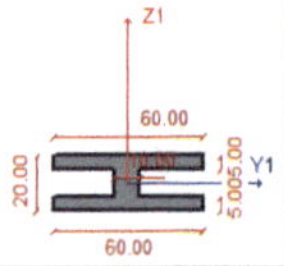	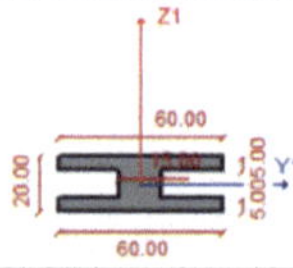	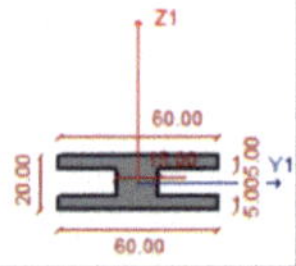	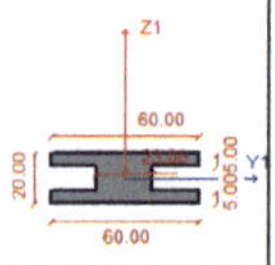	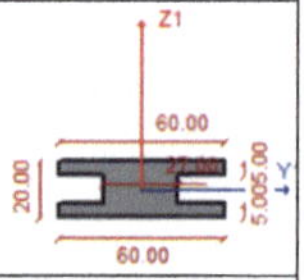
Consumption of concrete for ribs, m^3	1.28	1.37	1.38	1.42	1.43
Consumption of polystyrene, m^3	2.32	2.23	2.22	2.18	2.17
Width of the rib	Set - 100 mm	Estimated – 148 mm, adopted – 150 mm	Estimated – 193 mm, adopted – 190 mm	Estimated – 227 mm, adopted – 230 mm	Estimated – 272 mm, adopted – 270 mm

In order to obtain the components of SSS in all the considered options of the slabs, a finite-element (FE) analysis was performed in the «Lyra 10» software. The created FE models differed not only in the topology of the ribs, but also in the given boundary conditions:

(1) the slab supported in the corners on columns with a height of 1000 mm and a cross-section of 300 × 300 mm; in the lower node of the columns all linear movements and rotation are forbidden;
(2) the slab supported on 4 edges; in all nodes of the extreme faces only vertical movements are forbidden - the hinged support is modeled.

Physicomechanical characteristics are set as for heavy structural concrete C25/30 (E = 25000 MPa, μ = 0.2).

The load for all models is assumed to be 1 t/m^2 (10 kN/m^2). Due to the fact that the slabs are approximated by structures made of rod finite elements, the load diagram for each rod differed depending on the length and loading area (which depends, in turn, on the size of the inserts) and took either a triangular or trapezoidal shape.

The results of determining the main components of the SSS of the considered systems are given in Fig. 4.

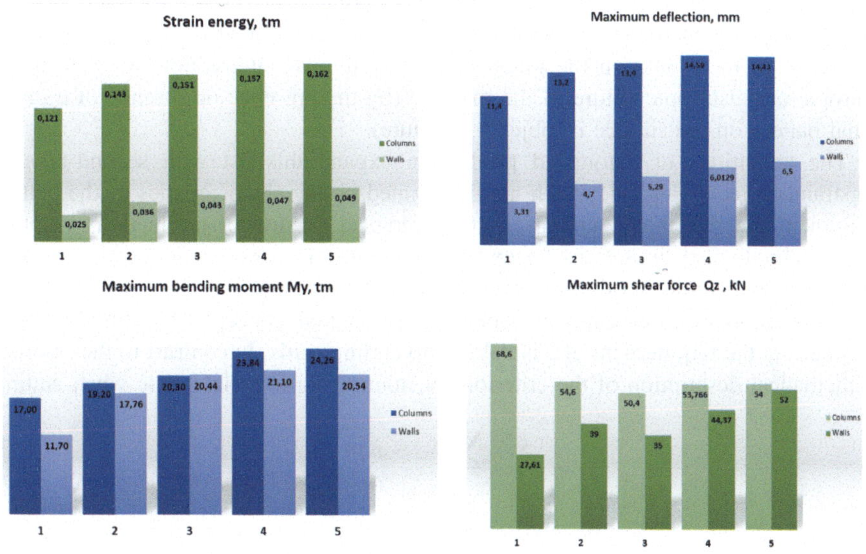

Fig. 4. Results of calculation

As we can see from the results of the study, an increase in the dimensions of the inserts leads to an increase in the values of the strain energy and deflections for both types of support. Thus, the smaller size of inserts and, as a result, a denser grid of ribs, from the point of view of the stress-strain state of the system, are more acceptable as it approaches the traditional solid slab. However, if we go back to the technological side of the issue (Nikulin et al. 2016), in particular, from the point of view of harvesting inserts of various shapes and sizes, we should be sure that the result will be similar when determining the amount of waste after cutting inserts from one blank for each of the considered topology options.

3 Solving of the Optimal Cutting Task

To do this, it is advisable to consider the problem of optimal cutting (Bunakov 2007) of expanded polystyrene inserts, taking into account the topology of the slabs described above (Fig. 4), and the standard blank in the form of a sheet of expanded polystyrene with dimensions of $1.0 \times 3.0 \times 0.2$ m.

The task of rational cutting of plate materials for finished workpieces of various shapes is of great practical importance in designing products for various industries and, in particular, for construction. By its nature, it is a task of a discrete-continuous structure, it belongs to the class of so-called NP-complete problems, which exact solution is possible only by the method of complete enumeration of all possible options.

The mathematical formulation of the problem consists in placing flat geometric objects (initial set of blanks) on sheets of given sizes (full-length sheets) with minimal material waste and taking into account the existing restrictions. The first type of constraints is geometric; they are classical and are determined by the conditions of placement of the blanks in the given area, their mutual intersection, as well as the isotropic or anisotropic nature of the material (by the presence or absence of a directional pattern on the surface of objects – texture).

The conditions of automated production expand this list with second-type of constraints – technological, which are determined by the characteristics of the cutting equipment and the organizational and technological features of production; quantitatively technological limitations far exceed geometric. In addition, they can vary in a wide range depending on the specifics of a particular enterprise.

Automation of production also changes the actual concept of optimal cutting, highlighting the requirement of adaptability to cutting cards. In contrast to the rigorous mathematical description of the criterion for minimizing material waste when cutting

$$S = \sum_{i} S_i \rightarrow min \tag{3}$$

where S_i - the area of the i-th scrap of the material, technological optimization criteria are multiple and often empirical. In general, they can be combined with the notion of "laboriousness of the physical implementation of cutting", which includes such parameters as the total number and total length of cuts, the number of cutting cards, the number of turns of the sheet's pack, the geometric parameters of the scraps, etc.

The high combinatoriality of the discussed "MONOFANT" system practically does not limit the designer in choosing the shape and size of the inserts void formers (Nikulin et al. 2017), which on the one hand is an advantage of the system, and on the other - leads to a qualitative and quantitative complication of the insert itself.

Obviously, in such conditions, with a large number of elements, procedures for processing geometric information will be required. Even with the use of powerful computers time for solving such problems will be unacceptable in real production conditions, therefore different heuristic algorithms are used to solve them, which give an optimal period of time close to the optimal solution.

Thus, when solving the problem of optimal cutting, two stages are distinguished:

- at the first stage, rational methods of cutting material are determined;
- at the second stage, the problem of linear programming is solved to determine the intensity of the use of rational cutting methods.

We now consider in more detail the forms of inserts (Table 2), which are formed during the implementation of the internal topology of slabs, considered in the previous paragraph (Table 1).

So, as noted earlier, we consider the problem of placing the listed blanks on the original sheet of expanded polystyrene with dimensions of 1.0 × 3.0 × 0.2 m. To implement all the possible options for cutting the sheet into inserts of the required size and number, a simulation of the entire set of layout options was performed in the AutoCAD software. Selected the most rational options for cutting presented in Fig. 5.

Baseline data to determine the optimal cutting based on the results of enumeration of possible insertion options as well as results of waste calculation are given in Table 3. The solution to the problem of linear cutting is performed in the environment of Microsoft Excel software.

Table 2. Types and sizes of inserts for each option

№	Option 1	Option 2	Option 3	Option 4	Option 5
Types and sizes of inserts	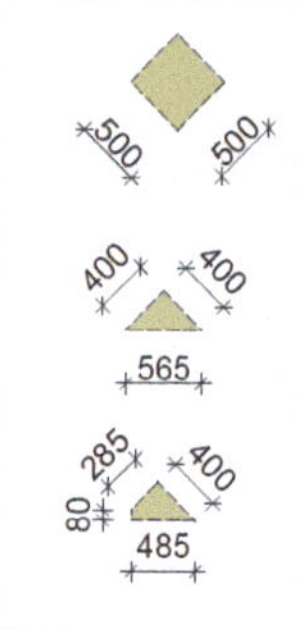	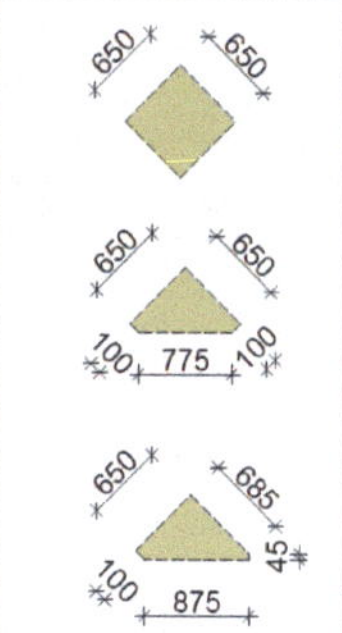	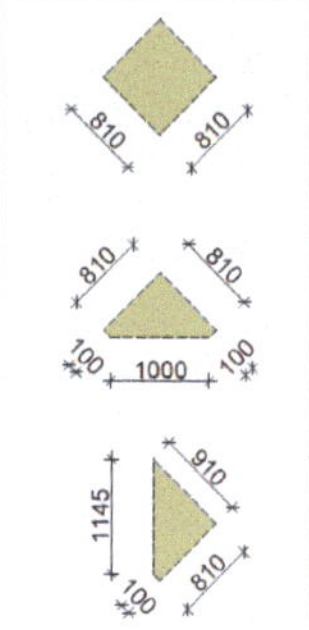	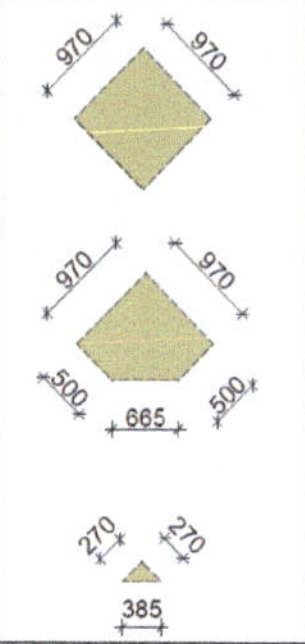	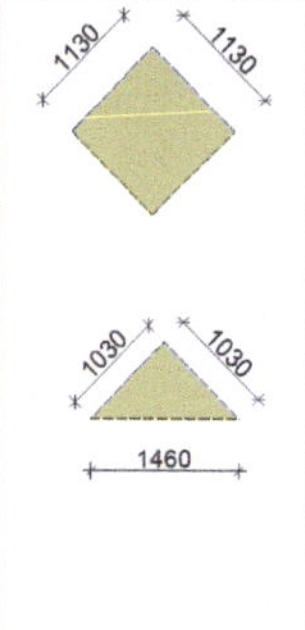

In Fig. 6 a diagram representing the percentage of waste at the optimal cutting case for each of the considered options of slabs is shown.

So, analyzing obtained data, we can see that the smallest size of inserts also is the most justified option together from the point of view of minimum waste of material and taking into account the features of the stress-strain state of the considered structure.

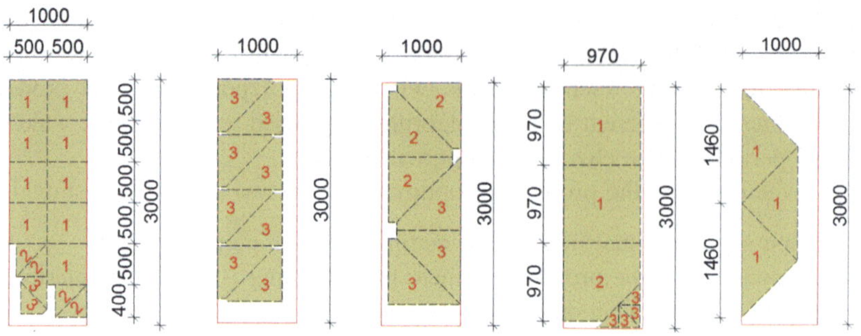

Fig. 5. Rational cutting sets for each slab topology

Table 3. Results of selection of optimal cutting for various plate topologies

№	Option 1			Option 2			Option 3			Option 4			Option 5		
Area of initial workpiece L_w, m^2	3.0														
Number of details, i	1	2	3	1	2	3	1	2	3	1	2	3	1	2	3
Area of details, A, m^2	0,25	0,08	0,07	0,42	0,27	0,28	0,66	0,41	0,40	0,94	0,83	0,04	0,53	0,45	0,37
Amount of details, n	84	20	8	40	12	8	24	8	8	12	12	16	12	24	12
Number of combinations of cut, N	101			30			18			20			12		
Waste share Ωo, %	*3,22*			*42,83*			*32,53*			*8,95*			*39,77*		
Waste area, m^2	1,0			17			11			2			14		

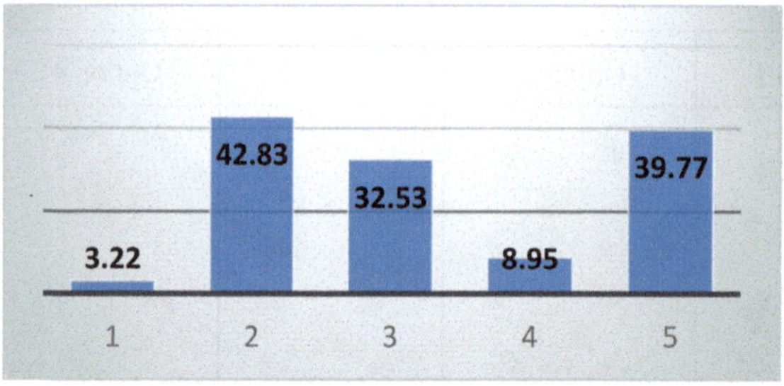

Fig. 6. Waste share for optimal cuttings of considered topologies, % (1…5 – number of option)

4 Conclusions

In the present study, a numerical analysis has been carried out to evaluate the features of the stress-strain state of the hollow slab with rational topology of the ribs according to the energetic principals of the design (Shmukler et al. 2017). It is obtained that the smaller size of the inserts void formers and denser grid of ribs provide the lowest value of deflection, strain energy and bending moments. In addition, the result of solvation of optimal cutting task matched the result of finite-element analysis, so as a conclusion we can admit that first option of the considered slab topologies provided the minimum waste of material at given conditions and restrictions.

References

Shmukler V, Babaev V, Bugaevskiy S, Berezhnha E, Kariakin I, Kondrashchenko E (2013) Patent of Ukraine 89464

Pomazan M (2013) Improving of the technology of lightweight reinforced concrete floors: dis. candidate tech. sciences: 05.23.08/Maksim Dmitrievich Pomazan, Kharkov, 187 p

Steblovsky I (2015) Constructive monolithic reinforced concrete floor with rational parameters: dis. … candidate tech. sciences: 05.23.01/Ilya Archilovich Steblovskiy, Kharkov, 192 p

Bugaevsky S, Berezhnha E, Bormot D (2016) Influence of accuracy of installation of inserts on stress-strain state of the slab of "MONOFANT" system. Naukovyy Visnyk Budivnytstva 3 (85):92–98

Shmukler V, Babaev V (2017) New constructive solutions for building of transport construction facilities. In: MATEC web of conferences, vol 116, p 02004

Bunakov P (2007) Algorithm for optimal cutting of materials for automated production. SAPR i grafika 11:74–77

Shmukler V, Babaev V, Bugaevsky S, Evel S, Evzerov I, Lantukh-Lyashchenko A, Shevetovsky V, Shimanovsky V (2017) Numerical and experimental methods of rational design and erection of constructive systems. "Stal'", Kiev, 404p

Albrecht C (2012) Experimental and theoretical analyses of the load-bearing behaviour of slim biaxial hollow core slabs with flattened void formers. In: Müller HS, Haist M, Acosta F (eds) Proceedings of The 9th fib international PhD symposium in civil engineering. Karlsruhe Institute of Technology (KIT), Karlsruhe, Germany, 22–25 July 2012. KIT Scientific Publishing, Karlsruhe, pp 85–90

Nikulin V, Babaev V, Shmukler V, Bugayevskiy S (2016) Cast reinforced concrete frame of buildings and methods of its erection. J Civ Eng Constr 5(2):143–156

Nikulin V, Bugayevskiy S, Gerasymenko V, Konyukhov A (2017) Basics of modeling and technology of creating reinforced concrete elements of "Monofant" system construction. In: MATEC web of conferences, vol 116, p 02009

EN 1992-1-1 (2004) Eurocode 2: Design of concrete structures - Part 1-1: General rules and rules for buildings, 227p

Cohesion of Slurry Surfacing Mix with Slow Setting Bitumen Emulsions

Iurii Sidun[1(✉)], Oleksiy Vollis[2], Serhiy Solodkyy[1], and Volodymyr Gunka[3]

[1] Institute of Building and Environmental Engineering, Lviv Polytechnic National University, St. Bandery 12, Lviv 79013, Ukraine
siduniurii@gmail.com
[2] Private Enterprise «Laboratory WestRoadServices», Naukova 5a, Lviv 79003, Ukraine
[3] Institute of Chemistry and Chemical Technology, Lviv Polytechnic National University, St. Bandery 12, Lviv 79013, Ukraine

Abstract. In the present scientific work there is done the comparative analysis of two Slurry Surfacing mixes – systems on orthophosphoric and hydrochloric acids. It was determined that application of orthophosphoric acid for preparation of bitumen emulsion on oxidized bitumen leads to the substantial improvement of cohesion strength build-up rate – in comparison with identical system on hydrochloric acid. So as to receive the similar indices of cohesion strength build-up – the orthophosphoric-acid-based system allows using less amount of bitumen emulsion than required amount for the system on hydrochloric acid.

The results received witness about the different natures of breakage and cohesion strength build-up rate for systems on hydrochloric and phosphoric acids. As a confirmation of this fact there can be considered the possibility of using (in the systems with orthophosphoric acid) the aggregates which by methylene blue criterion are unsuitable for such technologies.

The orthophosphoric-acid-based Slurry Surfacing mix systems' properties obtained allow forecasting the substantial advantages of using these mixes at limit-permissible (for bitumen-emulsion technologies) ambient temperatures.

Keywords: Cohesive strength · Slurry-surfacing mix · Slurry Seal · Bitumen emulsion · Hydrochloric and orthophosphoric acids

1 Introduction

In modern road infrastructure it is quite a critical issue of renewing large amounts of road pavement within the shortest time-terms. Unfortunately, while putting forward the problem like that, we forget about maintenance for both the renewed and brand-new-constructed roads, which as the time goes by will be damaged for the reason of increased axis-load and traffic rate. Consequently, there arises a problem of the efficient and quick protection and renewal of road pavement (Demchuk et al. 2018; Nykypanchuk et al. 2013). It is possible to solve this problem by means of arrangement of thin-layer pavements from cast emulsion-mineral mixes, which are known in the world as Slurry Surfacing mix (Slurry Seal, Microsurfacing). These mixes do not

Z. Blikharskyy et al. (Eds.): CEE 2019, LNCE 47, pp. 420–427, 2020.
https://doi.org/10.1007/978-3-030-27011-7_53

increase the bearing capacity of road structure, but eliminate the minor defects of the pavement, allowing saving it from damages for several years. Slurry-surfacing mix can be used also on concrete road surfaces (Sobol et al. 2014; Solodkyy et al. 2017). In general, Slurry Surfacing mix is a technology, which origins from the 1930th, but (as the time was moving on) the technology was reaching the new level and this process is still in progress today (Bhargava et al. 2019; Hou et al. 2018; Poursoltani and Hesami 2018; Gujar and Vakharia 2019). The main task for Slurry Surfacing mix is creation of reliable protection for the existing road pavement by means of arrangement of wear course within the short period of time (practically without restricting the traffic on the road section under repair). This task can be easily fulfilled using high-acid, high-resinous distilled bitumens produced from heavy crude oil. It provides for quick cohesion strength build-up for Slurry Surfacing mix (Pyshyev et al. 2015). Still, the deposits of heavy crude oil are substantially less than those ones of the light crude oil. That is why the application of light-petroleum bitumens is more accessible and less costs-consuming. Therefore, topical today is delineation of potentialities, providing for improved acceleration of cohesion strength build-up rate for Slurry Surfacing mix on the basis of bitumens from light crude oil, obtained by either oxidation or distillation. One of such potentialities is usage of orthophosphoric acid instead of hydrochloric one in the formulations for bitumen emulsions intended for Slurry Surfacing mix. Application of orthophosphoric acid solves the problem of using the unsuitable aggregates (by criterion of "methylene blue"). (Exactly such fillers are dominant on the territory of Ukraine). Besides, the application of orthophosphoric acid allows producers to avoid passing licensing, which is necessary when using hydrochloric acid. To a certain extent it saves for them time and assets. The review of scientists' publications (both of the domestic and the foreign ones) has shown, that the issue of orthophosphoric acid application is not enough investigated. The problems related to this issue are now being fruitfully processed by Alan James (Takamura and James 2015; James 1998; Akzo-Nobel Surface Chemistry 2013). The object of the article is investigation of influence of orthophosphoric acid (along with reactive filler) upon the properties of Slurry Surfacing mix.

2 Materials and Methods

For the Slurry Surfacing mix design there were used Ukrainian aggregates from granite quarries: JSC «Polonskiy Gorniy Combinat» (Khmelnitskiy Region), Public JSC «Ushytskiy Combinat Stroitelnyh Materialov» (Zhytomyr Region), «Rokytnianskiy Granite Quarry» (Kyiv Region) and Klesivskiy Quarry of Non-Ore Minerals Ltd. «Technobud» and «Vyrivskiy Quarry Ltd.» (Rivne Region). For the operative assessment of the aggregate suitability for usage in Slurry Surfacing mix there was used the method of methylene blue adsorption MB (Table 1), which provides for indirect determination of dusty and clayey particles content in aggregate. For the testing there was chosen the aggregate fraction <0.1 mm.

Table 1. Methylene blue adsorption index

Quarry	Methylene blue index, ml
Klesivskiy	9
Vyrivskiy	10
Polonskiy	10
Rokytnianskiy	17
Ushytskiy	20

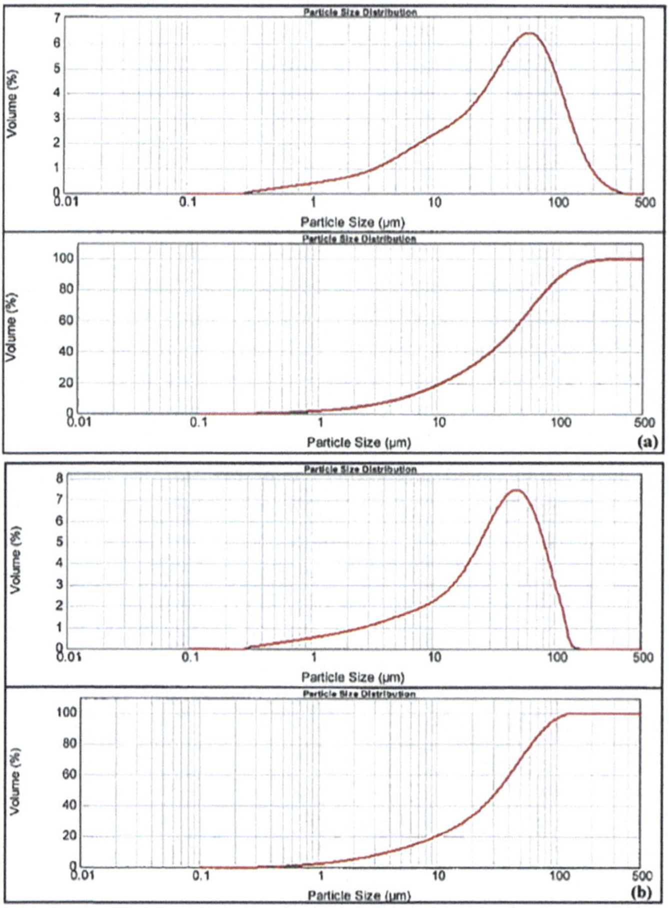

Fig. 1. Particle size for the fine fraction Klesivskiy (a) and Ushytskiy (b) aggregates.

As per data from the Table 1, it may be concluded that the optimum aggregate per MB criterion (as among the tested ones) is Klesivskiy with methylene blue index – 9 ml, while the most reactive one is Ushytskiy with methylene blue index – 20 ml. To find the reason of difference in values of MB-index for the Klesivskiy screenings (optimum by this criterion) and surface active Ushytskiy screenings, there was done laser diffraction of these fillers – so as to see the particles size and distribution. There was determined that the percent ratio of particles sized less than 3 mc – 6.15% (while for the particles sized 3 mc to 90 mc – 77.52%) in Klesivskiy screenings is less in comparison with Ushytskiy ones (correspondingly – 7.67% and 86.33%). As for the particles sized more than 90 mc, the ratio is on contrary larger: Klesivskiy – 16.33%, Ushytskiy – 5.99% (Fig. 1). Blaine value for Klesivskiy screenings (1868 cm^2/g) is also less than for Ushytskiy ones (2175 cm^2/g). By means of laser diffraction done it may be concluded that MB index for Ushytskiy aggregate is higher than for Klesivskiy one by more than two times – due to micro grading of this filler. That is: in Ushytskiy screenings there is present the larger quantity of dusty and clayey particles, while they are characterized by higher specific surface. Accordingly, the high content of such particles in an aggregate in course of Slurry Surfacing mix design will lead to premature breakage of the mix, while that is inadmissible. Therefore, for the mix stabilization there will be a need to add substantial quantity of control additive, and that will negatively influence the Slurry Surfacing mix cohesion strength build-up rate. The Slurry Surfacing mix grading was designed for type 1 with maximum particle size for the coarse filler 5 mm (ISSA A105 2010). In order to avoid influence of grading peculiarities upon the indices of cohesion strength build-up rate the Slurry Surfacing mix design was performed based on the exactly the same narrow fractions. The Slurry Surfacing mix grading selected refers to type 1 Slurry Seal (Table 2) and corresponds to the limit values according to ISSA technical bulletin (ISSA A105 2010).

Table 2. The estimated grading 0–5 for the type 1 mix ISSA

Orifices Ø for round-shaped sieves, mm	Average grading 0–5 for the type 1 mix ISSA, % by mass	
	Partial sieve residues	Complete passing through sieves
0	45	0
0.5	25	45
1	25	70
2	5	95
5	0	100

For the production of bitumen emulsion for Slurry Surfacing there was used BND 70/100 (Petroleum Paving Grade Oxidized Viscous bitumen 70/100) from PJSC "UkrTatNafta" (Kremenchuk Refinery) which corresponds to the requirements of (EN 12591:2009). According to formulation, bitumen emulsions contained Swedish Nouryon (AkzoNobel) emulsifiers Redicote E-11 and Redicote C-320E, as well as co-emulsifier Redicote 540. Each of formulations included Styrene butadiene rubber polymer of Toptex B latex from Algol Chemicals (Finland). Bitumen emulsions were produced on laboratory bitumen emulsion plant SEP-0.3R of Danish company DenimoTech. The emulsions produced were cationic slow-setting (CSS) (Table 3).

Table 3. Bitumen emulsion formulations for Slurry Surfacing

Components	Emulsion formulation (CSS), No, % mass			
	1	2	3	4
Bitumen	62			
Emulsifier	Redicote E-11–1.1			Redicote C-320E – 1.2
Co-emulsifier	–	Redicote 540	–	–
Water phase pH (acid)	pH = 2.5 (HCl)			pH = 2.5 (H_3PO_4)
Water	Till 100			
Latex	–	–	Toptex B – 3	–

3 Results and Discussion

The Slurry Seal mix design was done according to (ISSA A105 2010) with application of Klesivskiy and Ushytskiy aggregates, along with four developed bitumen emulsions formulations (BE), Portland cement (PC) 400 produced by JSC «NicolayevCement» (Ukraine, town of Nikolayev), drinkable water, control additive (CA) – 10% solution of Redicote E-11. The ambient temperature during the testing constituted 25 °C, humidity – 75%. The optimum mixes by breakage criterion (mix time $\geq$180 s) are presented in Table 4.

Table 4. The optimum Slurry Seal mixes by breakage criterion

Mix-design version No	Components content, g, in excess of 100 g of aggregate				Mix time $\geq$180 s
	PC	Water	CA	BE	
1	BE No 1 & Klesivskiy				
	1.0	10	2.0	14	185
2	BE No 1 & Ushytskiy				
	1.5	10	2.5	14	184
3	BE No 2 & Klesivskiy				
	1.5	10	1.5	14	186
4	BE No 2 & Ushytskiy				
	1.5	10	2.0	14	182
5	BE No 3 & Klesivskiy				
	1.0	10	2.0	14	190
6	BE No 3 & Ushytskiy				
	1.0	10	2.5	14	184
7	BE No 4 & Klesivskiy				
	0.75	10	1.5	14	181
8	BE No 4 & Ushytskiy				
	0,75	10	1.5	14	183

The optimum Slurry Seal mix design version No 8 contains lower quantity of control additive for extending the mix breaking time – comparatively to the mix design version No 2, although the both versions of design were done based on unsuitable Ushytskiy filler. Such difference is caused by presence in the emulsion formulation BE No 4 of not hydrochloric but orthophosphoric acid, while the content of Portland cement in these formulations differs. The same situation is regarding the mix design versions based on the suitable Klesivskiy filler. In the system on orthophosphoric acid the cement quantity is the decisive factor: an insignificant deviation of cement content from the optimum one leads to the sharp decrease of the mix breaking time.

The Slurry Seal mixes (versions No 1–8) testing was done according to the regulations (TB139 1990; EN 12274-4 Slurry Surfacing 2018) by means of making the corresponding samples of the mix and testing them after certain time-periods on cohesion-measuring device. The samples tested were classified by destruction character and the corresponding torque value – by four types of destruction: N, NS, S, SS. If Slurry Seal mix sample reaches destruction type NS, the cohesion strength of slurry pavement provides for roller compaction or for starting traffic with speed limit 40 km/h with turning prohibited. If the sample reaches destruction type SS, there is possible traffic start without prohibitions. In Table 5 and on Fig. 2 the results are presented on the cohesion strength build-up rate for the Slurry Seal mix samples made.

Based on Table 5 and Fig. 2, it may be seen that Slurry Seal with application of co-emulsifier Redicote 540 (design version No 3) allows starting traffic with limitations sooner in comparison with other mixes on hydrochloric acid – in 1 h after laying slurry seal pavement. Still, when using in Slurry Seal design version No 4 of unsuitable Ushytskiy screenings the inclusion of Redicote 540 co-emulsifier into the bitumen emulsion proves to be not that efficient. In its turn, traffic start with and without limitations occurs sooner (in 0.5 and 1.0 h accordingly) when applying Slurry Seal based on orthophosphoric acid (design versions No 7 and 8). The modification of bitumen emulsion by Toptex B latex has not led to the substantial growth of cohesion strength build-up rate for Slurry Seal mixes (design versions No 5 and 6). The worst

Table 5. Traffic start time for Slurry Seal pavement done

Mix-design version No	Time period for reaching destruction type, h	
	NS	SS
1	3.5	6.5
2	5.0	7.5
3	1.0	3.0
4	2.0	4.0
5	3.0	5.0
6	4.5	6.5
7	0.5	1.0
8	0.5	1.0

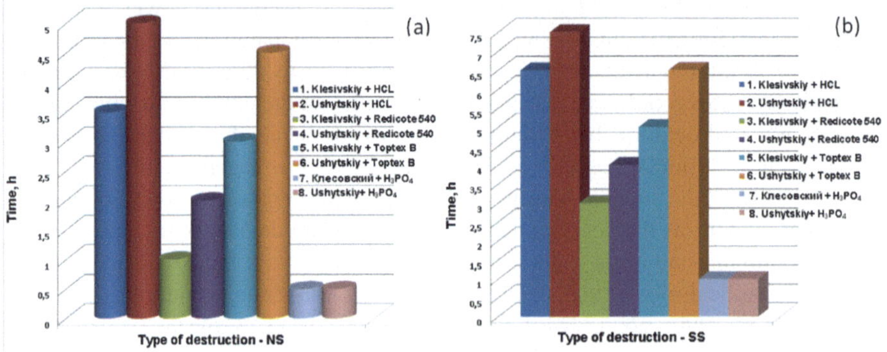

Fig. 2. Time period for traffic start with limitations (a) and without limitations (b)

result as for the cohesion strength build-up rates was demonstrated by the mix with application of «high-reactive» Ushytskiy aggregate and bitumen emulsion on hydrochloric acid (design version No 2).

4 Conclusions

There was determined the suitability of a number of Ukrainian aggregates as for application in Slurry Surfacing mixes by MB criterion. By means of laser diffraction on example of Klesivskiy and Ushytskiy screenings there was analyzed the difference between the suitable and unsuitable aggregates by MB criterion. The unsuitable Ushytskiy screenings are characterized by high content of dusty and clayey particles and larger specific surface. Besides, in their content there is present amphibole mineral – in contradistinction to Klesivskiy ones.

The four CSS bitumen emulsion formulations were developed.

On the basis of the bitumen emulsions, along with Klesivskiy and Ushytskiy aggregates, there were done eight versions of mix design for Slurry Seal mixes – by breakage criterion. The cohesion strength build-up rate was determined for each of the design versions, while it was discovered that the most efficient design versions on hydrochloric acid are the following: No 3 with inclusion of Redicote 540 co-emulsifier (NS = 1 h, SS = 3 h) and No 7 and 8 on orthophosphoric acid (NS = 0.5 h, SS = 1.0 h). Besides, the application of orthophosphoric acid in bitumen emulsion for Slurry Seal mix allows making mix design (No 8) based on unsuitable Ushytskiy screenings – without worsening the cohesion strength build-up rate indices. Above that, the application of Slurry Seal on orthophosphoric acid is advantageous when traffic start without limitations is required the soonest possible.

References

AkzoNobel Surface Chemistry (2013) Asphalt concrete issues. Information Bulletin of Department «Additives for road construction» . Redipave System for rapid-setting Slurry-Seal pavement on arbitrary bitumen. Europe, Middle East, India and Africa 84

Bhargava N, Siddagangaiah A, Ryntathiang TL (2019) State of the art review on design and performance of microsurfacing. Road Mater Pavement Des 1–35. https://doi.org/10.1080/14680629.2019.1607771

Demchuk Y, Sidun I, Gunka V, Pyshyev S, Solodkyy S (2018) Effect of phenol-cresol-formaldehyde resin on adhesive and physic-mechanical properties of road bitumen. Chem Chem Technol 12:456–461. https://doi.org/10.23939/chcht12.04.456

EN 12274-4 Slurry Surfacing (2018) Test methods: European Standard EN 12274-4. Part 4: determination of cohesion of the mix. European Committee for Standardization

EN 12591:2009 (2009) Bitumen and bituminous binders. Specifications for paving fraction bitumens

Gujar R, Vakharia V (2019) Prediction and validation of alternative fillers used in micro surfacing mix-design using machine learning techniques. Constr Build Mater 207:519–527. https://doi.org/10.1016/j.conbuildmat.2019.02.136

Hou S, Chen C, Zhang J, Shen H, Gu F (2018) Thermal and mechanical evaluations of asphalt emulsions and mixtures for microsurfacing. Constr Build Mater 191:1221–1229. https://doi.org/10.1016/j.conbuildmat.2018.10.091

ISSA A105 (2010) Recommended performance guidelines for emulsified asphalt slurry seal. International Slurry Surfacing Association, Annapolis, MD. 2010 Revised

James A (1998) Asphalt emulsions (chemistry and concepts). In: 2nd Asphalt technology conference of the Americas, Austin, Texas

Nykypanchuk M, Hrynchuk Y, Olchovyk M (2013) Effect of modified bitumen on physic-mechanical properties of asphalt concrete. Chem Chem Technol 7(4):467–470. https://doi.org/10.23939/chcht07.04.467

Poursoltani M, Hesami S, (2018) Performance evaluation of microsurfacing mixture containing reclaimed asphalt pavement. Int J Pavement Eng. 1–14. https://doi.org/10.1080/10298436.2018.1551544

Pyshyev S, Grytsenko Y, Solodkyy S, Sidun I, Vollis O (2015) Using bitumen emulsions based on oxidized, distillation and modified oxidized bitumens for slurry seal production. Chem Chem Technol 9(3):359–366. https://doi.org/10.23939/chcht09.03.359

Sobol K, Blikharskyy Z, Petrovska N, Terlyha V (2014) Analysis of structure formation peculiarities during hydration of oil-well cement with zeolitic tuff and metakaolin additives. Chem Chem Technol 8:461–465. https://doi.org/10.23939/chcht08.04.461

Solodkyy S, Markiv T, Sobol K, Hunyak O (2017) Fracture properties of high-strength concrete obtained by direct modification of structure. MATEC Web Conf 116. https://doi.org/10.1051/matecconf/201711601016. Transbud-2017

Takamura K, James A (2015) Paving with asphalt emulsions. Adv Asph Mater Road Pavement Constr 393–426

TB139 Test Method to Classify Emulsified Asphalt (1990) Aggregate mixture systems by modified cohesion tester measurement of set and cure characteristics. International Slurry Surfacing Association

Comparative Study of Fatigue Life Assessment Made by Different Approaches

E. Bernatowska[1], D. Leń[2], and L. Ślęczka[1(✉)]

[1] Department of Building Structures, Rzeszów University of Technology, Powstańców Warszawy 12, 35-959 Rzeszów, Poland
sleczka@prz.edu.pl

[2] State Higher Vocational School in Krosno, Rynek 1, 38-400 Krosno, Poland

Abstract. The fatigue life of steel elements is usually predicted by classical nominal stress method. Such approach is very convenient, when design model of structure is based on bar elements. But the level of analysis in steel structures is still increasing. There are used more and more sophisticated FE packages, which offer shell and 3D elements instead of classical bars modeling. They allow to include real shape of elements and existence of welds, bolts and other joining components. Results obtained from such analysis contain many stress raising effects and can be easily utilized in modern approaches of fatigue life based on local stress values, e.g. structural (hot spot) stress method. Development of modern computational FE packages make prediction of fatigue life using local approaches possible and easier. Major obstacle in wider application of numerical methods for the fatigue assessment of steel structures is apprehension of designers about accuracy of local approaches. The focus of presented research has been on comparison of fatigue life predictions based on nominal stress method and on structural stress method, made for a couple of constructional details. Paper presents influence of geometrical parameters on stress concentration factors and also highlights a wider flexibility of structural stress method.

Keywords: Fatigue life · Nominal stress method · Structural stress method

1 Introduction

Failure caused by fatigue is one of the ultimate limit states, that should be verified during designing of steel structures (EN 1990). The most widely used method to check this conditions is nominal stress method (EN 1993-1-9). It is characterized by a large computational simplicity, because the determination of the stress range at the considered point of the structure (notch) is calculated for nominal stresses, so it can be carried out using elementary formulas. However, simplicity of stress range predictions forces designer to scrupulously determine category of a given constructional detail. A huge variety of constructional details (notches) appearing in the steel structures were pressed into the framework of fourteen fatigue curves for normal stress range and two fatigue curves for shear stress range. In such approach a particular structural detail is assigned to a particular fatigue class with a given fatigue curve. But in many cases, details of real

Z. Blikharskyy et al. (Eds.): CEE 2019, LNCE 47, pp. 428–435, 2020.
https://doi.org/10.1007/978-3-030-27011-7_54

structures are more complicated than basic structural details gathered in standards or recommendations, so it can lead to conservative estimations.

In the recent years the level of analysis in steel structures is increasing. There are used more and more sophisticated FE packages, which offer shell and 3D brick elements instead of classical bars modeling. They allow to include real shape of elements and existence of welds, bolts and other joining components. The stress results obtained from such analysis contain global stress raising effects and can be easily utilized in structural (hot spot) stress method. For this reason only three detail categories are given in standards and recommendations for the application of the hot spot method (Hobbacher 2016; Niemi et al. 2018; EN 1993-1-9). In many cases approach based on local method offers advantage of wider versatility.

2 Aim of the Study

The aim of this study is to compare fatigue life predictions made by nominal and structural stress methods. Structural (hot spot) stress calculations were made by appropriate FE modeling. Range of this study is limited to welded joints with longitudinal attachments and bolted tension flange connections.

3 Method

Two groups of details have been selected to analysis. First one was longitudinal welded attachment (Fig. 1a), with its variable length *L*. Second group was bolted tension flange joint (Fig. 1b), with variable number of bolts *n* and flange thickness t_f. The summary of the study is given in Tables 1 and 2.

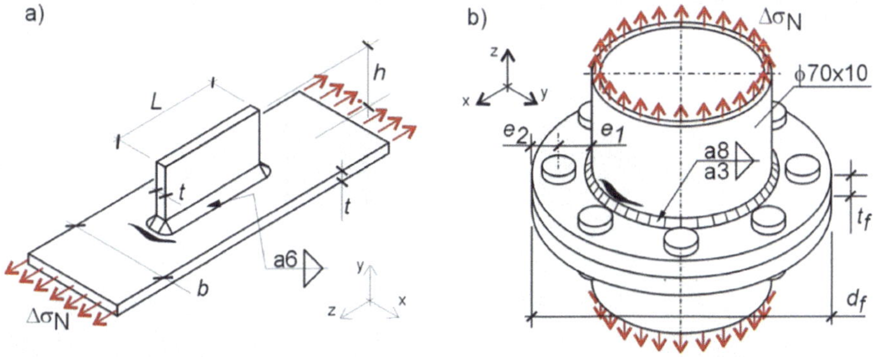

Fig. 1. Investigated structural details; (a) longitudinal welded attachment, (b) bolted flange joints

Table 1. Dimensions of longitudinal welded attachment

Specimen	b [mm]	h [mm]	t [mm]	L [mm]
LA40	100	60	10	40
LA80				80
LA120				120
LA290				290

Table 2. Dimensions of bolted flange joints

Specimen	Flange				Bolts	
	e_1 [mm]	e_2 [mm]	d_f [mm]	t_f [mm]	Diameter	Number [-]
BK 10.4.2	30	30	190	8	M16	4
BK 15.4.2	30	30		15		4
BK 20.4.2	30	30		18		4
BK 10.6.2	35	25		8		6
BK 15.6.2	35	25		13		6
BK 25.6.2	35	25		23		6

It is assumed that both groups of specimens were loaded in the tension range by nominal stress range $\Delta\sigma_N$, with constant amplitude, having pulsating character, i.e. $\sigma_{min} = 0$ and $\sigma_{max} = \Delta\sigma_N$.

Potential crack location was examined in the parent material adjacent to the weld toe of horizontal plate in case of first specimen group, and tube wall in case of second specimen group, Fig. 1. Fatigue checking was done by nominal and structural (hot spot) stress methods. Stress analysis was carried out by FEM. Used numerical models are shown in Fig. 2.

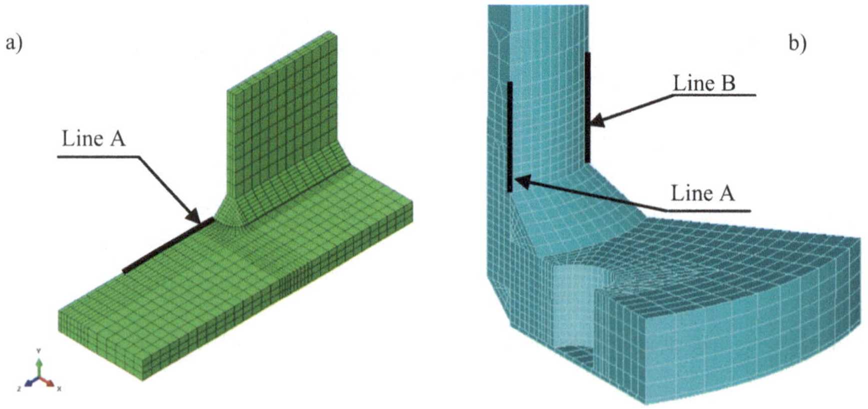

Fig. 2. FEM mesh for structural stress prediction; (a) longitudinal welded attachment LA120, (b) bolted flange joints BK 20.4.2 (the bolt is not shown)

The geometry of the developed numerical model replicated the geometry of the specimens. Due to their shape, symmetry conditions were exploited. Linear elastic analysis was used and no geometric imperfections were applied. Solid model was used and the mesh sizing was chosen according to recommendations (Hobbacher 2016; Niemi et al. 2018).

Two types of stresses were predicted at the potential crack location. Nominal stress range $\Delta\sigma_N$ was directly equal to value of applied load, (Fig. 1). Then hot spot stress σ_{HS} was calculated by extrapolation of the surface longitudinal stress to location of crack site (at the weld toe), Fig. 3. Hot spot is defined as type "a" for both group of specimens, because the potential crack at the weld toe is situated on the plate surface. Extrapolation was done by using linear function. Extrapolating points were chosen according to recommendation (Niemi et al. 2018). In case of bolted flange joints hot spot stresses σ_{HS} were predicted at two locations (Fig. 2). Line A is lying in vertical symmetry plane, which intersect axis of bolt hole, and Line B is situated in vertical symmetry plane, passing along angle bisector between adjacent bolts.

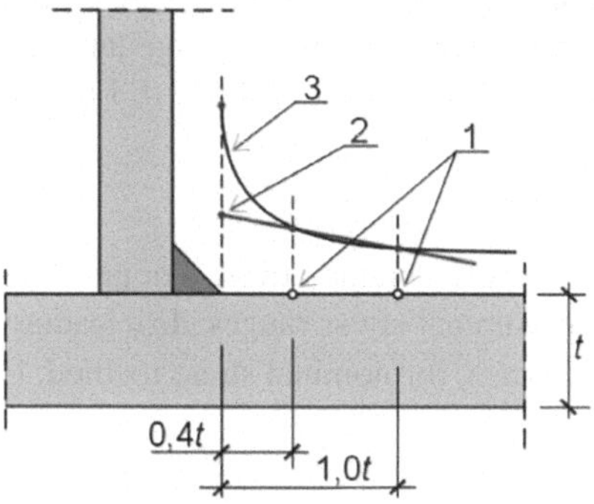

Fig. 3. Linear extrapolation of the structural stress; 1- extrapolating points, 2- hot spot stress; 3-total surface stress

Values of predicted hot spot stresses were used to determine the geometrical stress concentration factor k_f:

$$k_f = \frac{\sigma_{HS}}{\sigma_N} \tag{1}$$

where σ_{HS} is hot spot stress value and σ_N is nominal stress value.

Obtained stress concentration factors are presented in Table 3.

In order to make comparisons, a general criterion, suitable for both approaches, was verified (EN 1993-1-9):

$$\frac{\gamma_{Ff}\sigma_{E,2}}{\Delta\sigma_C/\gamma_{Mf}} \leq 1,0 \tag{2}$$

Table 3. Stress concentration factors k_f

Specimen	Line A	Line B
LA40	1.36	-
LA80	1.43	-
LA120	1.47	-
LA290	1.52	-
BK 10.4.2	5.16	3.11
BK 15.4.2	3.26	2.81
BK 20.4.2	2.54	2.26
BK 10.6.2	4.78	4.50
BK 15.6.2	2.91	2.86
BK 25.6.2	2.20	1.95

where $\sigma_{E,2}$ is equivalent constant amplitude stress range related to $N = 2 \cdot 10^6$ cycles, $\Delta\sigma_C$ is reference fatigue strength at $N_C = 2 \cdot 10^6$ cycles and γ_{Ff} and γ_{Mf} are partial factors for equivalent constant amplitude stress range and for fatigue strength respectively. Assuming, that partial factors $\gamma_{Ff} = \gamma_{Mf} = 1, 0$, Eq. (2) can be written as:

$$\sigma_{E,2} \leq \Delta\sigma_C \tag{3}$$

The results of numerical study obtained for two approaches (nominal and hot spot) were compared in terms of nominal stress ranges $\Delta\sigma_N$ leading to the fatigue failure at $N = 2 \cdot 10^6$ cycles. In such case, for nominal stress method, Eq. (3) can be written as:

$$\sigma_{E,2} = \Delta\sigma_N \leq \Delta\sigma_C \tag{4}$$

where $\Delta\sigma_C$ is reference fatigue strength at $N_C = 2 \cdot 10^6$ cycles, which is equal to detail category according to IIW recommendations (Hobbacher 2016). For first group (Fig. 1a), the detail category varies according to the length of the attachment *L*. For second group (bolted flange joint, Fig. 1b) detail category is constant for each of the studied joints. Reference fatigue strengths for each analyzed joints, considered in nominal method are presented in Table 4.

For structural (hot spot) stress method, Eq. (3) can be written as:

$$\sigma_{E,2} = k_f \Delta\sigma_N \leq \Delta\sigma_C \tag{5}$$

where $\Delta\sigma_C$ is reference fatigue strength at $N_C = 2 \cdot 10^6$ cycles and k_f is geometrical stress concentration factor. Fatigue strengths $\Delta\sigma_C$ using the hot spot stress method depend only on the type of used weld, and were chosen according to EN 1993-1-9 (EN 1993-1-9). Values of $\Delta\sigma_C$ for each considered in this study joints are presented in Table 4.

Table 4. Reference fatigue strengths for nominal and structural stress methods

Specimen	Reference fatigue strength $\Delta\sigma_C$	
	Nominal stress method	Hot spot stress method
LA40	80	100
LA80	71	
LA120	71	
LA290	63	
BK 10.4.2	40	90
BK 15.4.2		
BK 20.4.2		
BK 10.6.2		
BK 15.6.2		
BK 25.6.2		

4 Results

The results of the study, presented in terms of ranges of nominal stress $\Delta\sigma_N$, leading to fatigue at $N = 2 \cdot 10^6$ cycles are shown in Figs. 4 and 5.

Plate with longitudinal attachment is one of the most popular and thus one of the most tested notch, so it can be regarded as a good reference point to compare nominal stress with another approach. It can be noticed, that results obtained from hot spot method in this study are in good agreement with those from nominal stress method. Differences vary between 1÷8% and can be explained by the need to assign $\Delta\sigma_C$ in nominal stress method to certain length ranges of the attachment L. Some allowance is also included in both methods for geometrical imperfections, but levels of stress

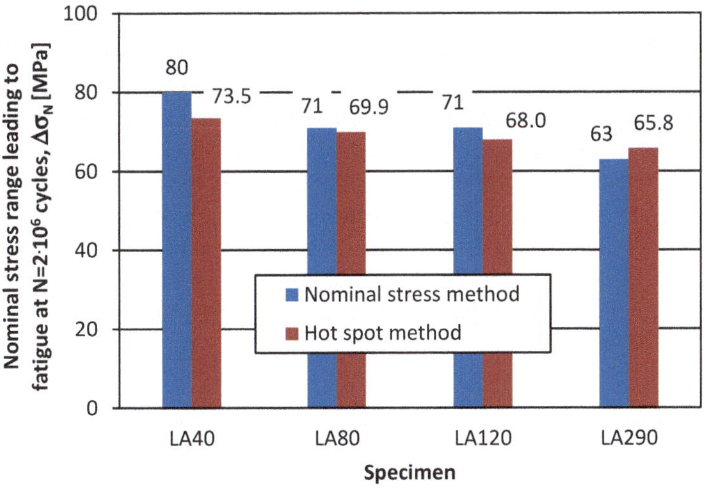

Fig. 4. Longitudinal welded attachments

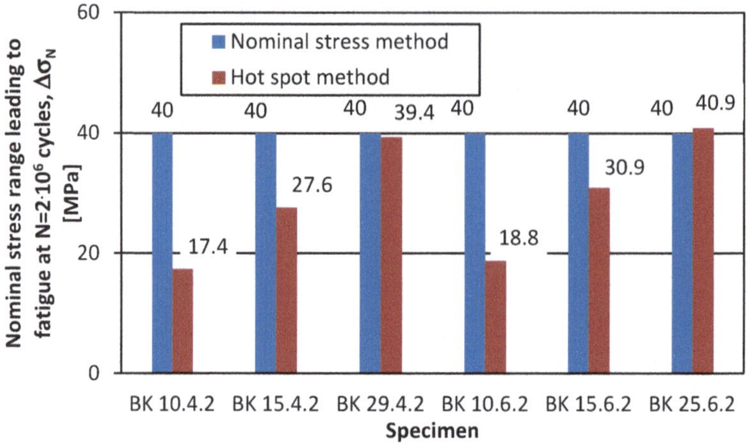

Fig. 5. Bolted flange joints

magnification factors covered in both verification methods are slightly different (Hobbacher 2016).

In case of second group (bolted flange joints) differences between two approaches are significant. However, they probably arise from a noticeable variation of stress concentration factors k_f appearing in the joints (see Table 3), while nominal stress method describe such notch using only one value of reference fatigue strength $\Delta\sigma_C$. An important factors influencing on values of k_f are thickness of flange and number of bolts in joints. So, this example clearly shows a wider flexibility of structural stress method in assessment of fatigue limit state.

5 Practical Significance

In the present study accuracy of hot spot method was tested on two groups of joints, using nominal stress method as a reference point. It has been recognized that the nominal and hot spot methods give consistent results in such cases, where nominal method provides a good representation of the fatigue strength of detail. When nominal method gives rather simplified recommendations, the hot spot method appears to be more appropriate and provide better estimation of fatigue life. Comparisons, using nominal, hot spot and also effective stress method are presented e.g. in (Taras and Unterweger 2017; Aygül et al. 2013; Pettersson and Barsoum 2012) and give information about applicability of different fatigue life estimations and influence of geometrical imperfections.

6 Conclusions

The local approach of fatigue life requires more work in FE modeling. But the level of analysis in steel structures is still increasing and sophisticated FE packages are used to build structural models for the purposes of analysis, design and verification. Obtained results can be easily used to assess fatigue life, by the hot spot method.

References

Hobbacher AF (2016) Recommendations for fatigue design of welded joints and components. Springer

Niemi E, Fricke W, Maddox SJ (2018) Structural hot stress approach to fatigue analysis of welded components. Designer's guide. Springer

EN 1990 Eurocode - Basis of structural design. CEN, Brussels

EN 1993-1-9 Eurocode 3: Design of steel structures – Part 1-9: Fatigue. CEN, Brussels

Taras A, Unterweger H (2017) Numerical methods for the fatigue assessment of welded joints – influence of misalignment and geometric weld imperfections. Eurosteel 13–15 September 2017, Copenhagen, Denemark

Aygül M, Bokesjö M, Heshmati M, Al-Emrani M (2013) A comparative study of different fatigue failure assessments of welded bridge details. Int J Fatigue 49:62–72. https://doi.org/10.1016/j.ijfatigue.2012.12.010

Pettersson G, Barsoum Z (2012) Finite element analysis and fatigue design of a welded construction machinery component using different concepts. Eng Fail Anal 26:274–284. https://doi.org/10.1016/j.engfailanal.2012.04.014

Fire Situation in Case of RC Members by Sika CFRP Strengthening

David Vazquez Cacho[1] and Yuriy Sobko[2(✉)]

[1] Sika Services AG, Calle Aragoneses 17, 28108 Alcobendas, Madrid, Spain
[2] Institute of Building and Environment Engineering (IBEE), Lviv Polytechnic National University, St. Bandery Street 12, Lviv 79000, Ukraine
sobko.yuriy@ua.sika.com

Abstract. The fire situation aspects especially in case of carbon fiber reinforce polymer (CFRP) strengthening of RC elements are one of major design factor in civil engineering today. Here the next options for CRFP strengthening is discussed. Option A1 when CFRP is not necessary under fire scenario and Option B - FRP is necessary under a fire scenario. Two parameters are related to the fire scenario. First parameter is reaction to fire and second parameter is structural members' fire resistance. The need for fire protection and fire resistance of RC structural elements is obtained by means of a calculating following a Eurocode procedures. Simplified expected temperatures profiles calculation process is presented. Structural analysis under fire situation of strengthened unprotected and protected member is shown as well. It is concluded that under fire scenario according to Eurocode calculation methodologies the requested fire resistance (R30–R240) can be fulfilled with no additional measure or protection at more than 80–90% of the real cases.

Keywords: Eurocode calculation · Reaction to fire · Fire resistance · Sika CFRP strengthening · Fire scenario · Structural members

1 Introduction

During the operation of buildings and structures, due to various negative effects, they suffer damage and need to be strengthened and restored (Blikharskyy et al. 2019; Al Sherrawi et al. 2018; Selejdak et al. 2018; Kos et al. 2017). At present, the alternative to traditional methods of strengthening (Bobalo et al. 2018; Krainskyi et al. 2018a, b; Khmil et al. 2018) is the use of various composite materials (Vegera et al. 2018; Brózda et al. 2017). One of the most popular is the use of carbon fiber reinforced polymers (CFRP) (Blikharskyy et al. 2018). The material has several advantages such as quick application, no need for special formwork, neglectable additional weight and high durability. Also the material has a number of disadvantages, such as the negative effects of ultraviolet radiation and insufficient fire resistance. However, there are ways to protect such structures from these negative influences. Therefore, the issue of fire resistance is relevant.

Z. Blikharskyy et al. (Eds.): CEE 2019, LNCE 47, pp. 436–443, 2020.
https://doi.org/10.1007/978-3-030-27011-7_55

Two parameters are related to the fire scenario. Their limits are defined by the local regulations (national/regional/city regulations) in each country.

1-st Parameter - Reaction to fire is the measurement of how a material or system will contribute to the fire development and spread, as well as the emission of smoke/flaming droplets. According to their use, certain quantity and/or type of materials cannot be used for walls/floor/ceiling rendering. Concrete and steel do not contribute to the fire development, and do not generate smoke. In case of an adequate kind of polymer used as saturator/adhesive, the reaction to fire of the strengthening system is moderate.

According to Fire reaction tests (ITB) of multi-layer CFRP Sika systems show more than Euroclass B.

2-nd Parameter - Fire resistance of the structural member: The load bearing capacity of the member can be ensured for a specific period of time (30 to 240 min). The fire resistance is expected to provide time to the building occupants for emergency evacuation before the structure collapses (Fig. 1).

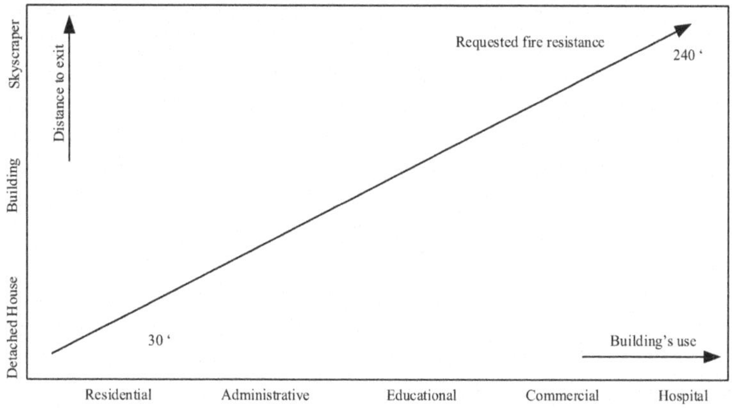

Fig. 1. Requested fire resistance vs distance to exit and building's use

Hence, the requested time to resist is commonly proportional to the quantity of people to evacuate and the distance to the exit. The fire protection for a structural member is therefore not directly oriented to the protection of the structure (e.g. the structure can collapse or be seriously damaged in case of fire, even when protected). In many cases, outdoor structures (e.g. bridges) may not need a satisfy a certain fire resistance as the evacuation is feasible in a few minutes.

2 Options for CRFP Strengthening Needing

Option A1 when CFRP is not necessary under fire scenario: requested fire resistance (R30–R240) can be fulfilled with no additional measure or protection. Option A2 when CFRP is not necessary under fire scenario as well: protection is necessary for the reinforced concrete section to meet a certain fire resistance. Those A1 and A2 options cower > 90% of the real cases. Option B - FRP is necessary under a fire scenario: protection is necessary for the CFRP and the reinforced concrete section to meet a certain fire resistance. This Option cower < 10% of the real cases.

The need for protection and the resulting fire resistance must be obtained by means of a calculation following the Eurocode procedures.

2.1 Step 1. Need of CFRP in Case of Fire

The design of a structure is focused in ensuring the necessary strength under the expected loads. For safety reasons, the different codes take into account additional safety coefficients. Under those circumstances, an appropriate strengthening method must be displayed, so that the structural safety gap required the local regulation is achieved.

For example according to Eurocode (e.g. for residential building) for Design loads the expected loads are magnified by means of safety factors: x 1.5 for imposed loads; x 1.35 for permanent loads.

For Ultimate Strengths the material's strengths are reduced by means of safety factors: y_c = 1.5 for concrete; y_s = 1.15 for steel.

The determination of the anticipated design loads under a fire scenario consists of two Rules. **General Rule** (*Eurocode 1: Actions on structures - Part 1–2: General actions - Actions on structures exposed to fire, section 4.3.1.*) is the fire loads taken as the service, un-factored loads (quasi-permanent combination of loads as usual). **Simplified rule** (*Eurocode 1: Actions on structures - Part 1–2: General actions - Actions on structures exposed to fire, section 2.4.2.*) tell us that fire loads taken as a reduced ratio of the design load (e.g. 70%).

The determination of the anticipated design strengths under a fire scenario the characteristic strengths for concrete and steel to be used (*Eurocode 2: Design of concrete structures - Part 1–2: General rules - Structural fire design, section 2.3.*). Materials safety factors: $Y_{c,fi}$ = 1 for concrete; $Y_{s,fi}$ = 1 for steel.

For example for the design strengths will be (Tables 1 and 2):

Table 1. Strength of concrete and steel for ULS in case of persistent & transient.

ULS, persistent & transient	Characteristic strength	Design strength
Concrete	25 MPA	25/1.5 = 16.6 MPa
Steel	500 MPa	500/1.15 = 434 MPa

Table 2. Strength of concrete and steel for ULS in case of fire situation.

ULS, fire situation	Characteristic strength	Design strength
Concrete	25 MPA	25/1 = 25 MPa
Steel	500 MPa	500/1 = 500 MPa

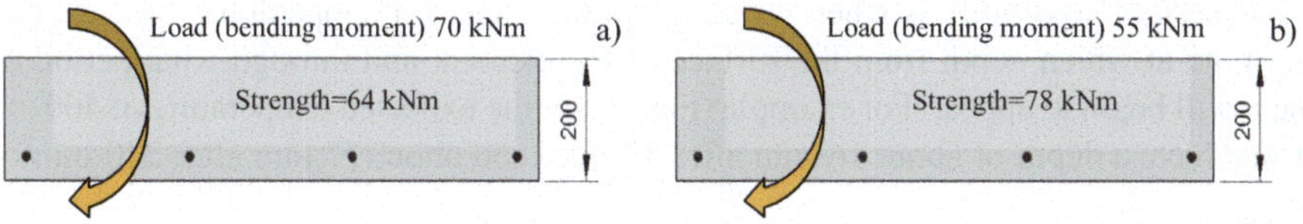

Fig. 2. Load and strength of RC slab member in case of normal design (a) and fire (b) situation

It means that when we have got normal design situation (no CFRP) and for example bending moment 70 kNm and strength of RC slab 200 mm thick equal 64 kNm the design loads exceed the design strength and CFRP is necessary to reach more than 70 kNm (Fig. 2a).

In fire situation for the same slab and load conditions the bending moment will 55 kNm and strength will 78 kNm. Design loads not exceed the design strength and CFRP is not necessary (Fig. 2b).

2.2 Step 2 (FRP not Necessary in Case of Fire). Determination of the Fire Resistance of the Section and Protection of the RC Member

A Concrete. The expected temperatures in a RC member can be evaluated by the simplified temperatures profiles existing in the Eurocode 2: Design of concrete structures - Part 1–2: General rules - Structural fire design (Fig. 3). Using the method of simplified temperature profiles, we have two possibilities. The first possibility is when it is necessary to determine the temperature at the any depth of the section of the reinforced concrete element and the time to reach this temperature. For example, (Fig. 3) for a depth of 30 mm the temperature reaches 500 °C after 90 min, and the temperature is about 760 °C will after 240 min.

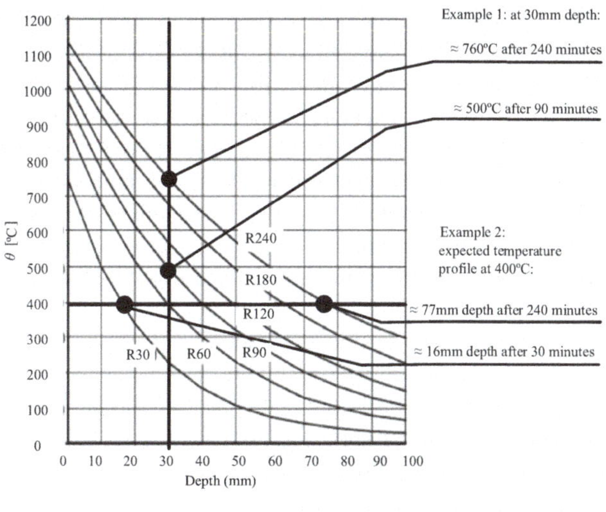

Fig. 3. Simplified temperatures profiles of RC members expexted temperaturas

The second possibility is when we have got the expected temperature, then we can determine at which depth from the surface of the element and through what period of time it will begin to appear. For example (Fig. 3) for the expected temperature of 400 °C we will have a depth of about 16 mm after 30 min, and about 77 mm after 240 min of fire.

The lost of concrete characteristic strength under fire loading is show at the diagrams (Fig. 4). As could be seen from the diagrams in (Fig. 4a), at 200 °C the concrete loses about 5% of its strength, at 400 °C the strength loss is about 20% and at 600 °C the concrete loses are about 50% of its characteristic strength.

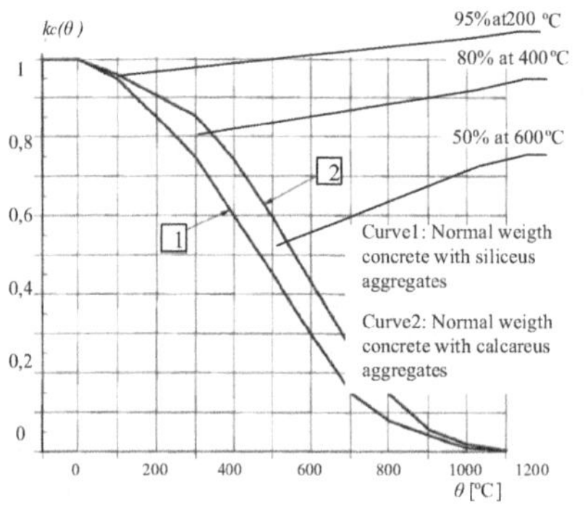

Fig. 4. Coefficient $k_c(\theta)$ allowing for decrease of characteristic strength (f_{ck}) of concrete

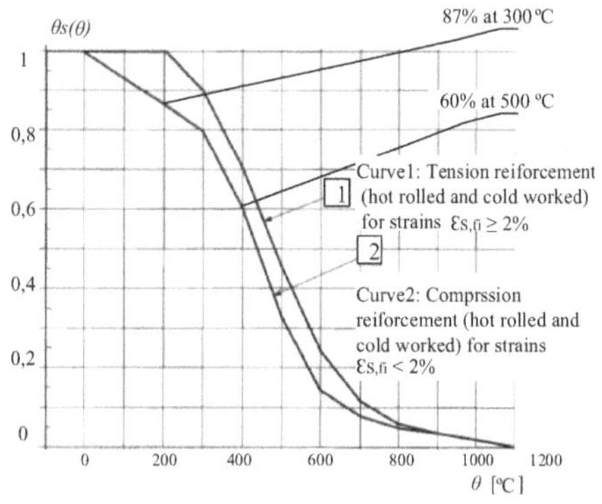

Fig. 5. Coefficient $k_s(\theta)$ allowing for decrease of characteristic strength (f_{yk}) of tension and compression reinforcement (Class X).

B Steel. The design strength is initially increased, as a consequence of the absence of safety factors. From this moment, the strength will decrease according to the fire temperature. As we see from the diagrams in (Fig. 5), at 300 °C the steel loses about 13% of its strength, at 500 °C the strength loss is about 40% of its characteristic value.

The expected temperature profiles calculation process is presented in Fig. 6. For example, first we determine the expected temperature distribution in depth of RC element by simplified temperature profiles (Fig. 3) and build the isotherms for 60 min (Fig. 6a) and for 90 min (Fig. 6b) respectively. Next, we find the losses of concrete strength, depending on the temperature (Fig. 6c, d), as well as the strength loss of the reinforcement. Then we can determine the bearing capacity of the new section of the reinforced concrete element after the fire effects. The simplified model of the 500 °C isotherm method is shown in Fig. 7.

After analysis of calculation results of load and strength development in not strengthened (Fig. 8a), strengthened by CFRP (Fig. 8b) members in case or ULS and load and strength development in protected and unprotected (Fig. 8c) members in case of SLS (fire scenario) show that strength is more than design load for RC member before and after strengthening by CFRP. In case of fire scenario we see the bigger gap

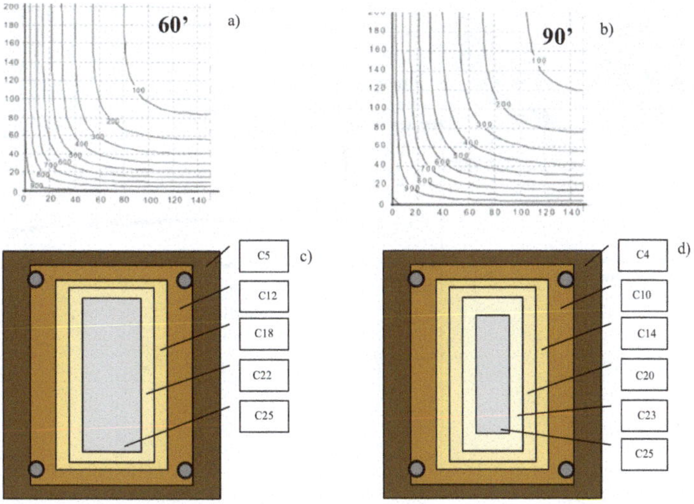

Fig. 6. The expected temperature profiles calculation process

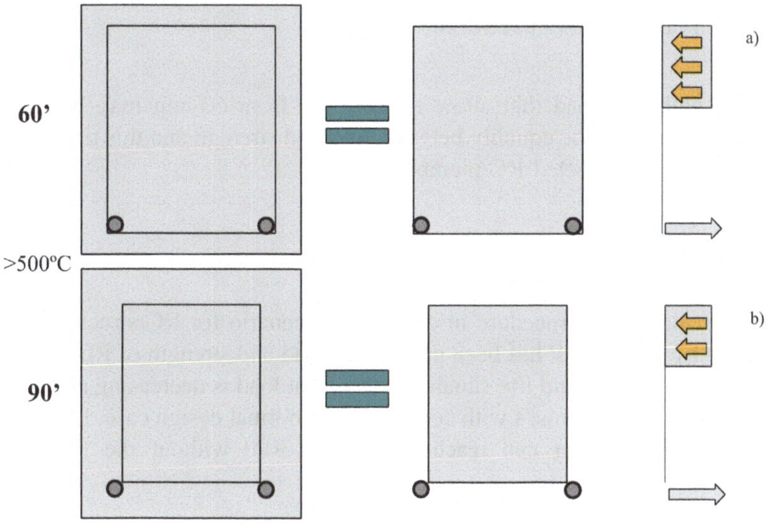

Fig. 7. The simplified 500 °C isotherm method

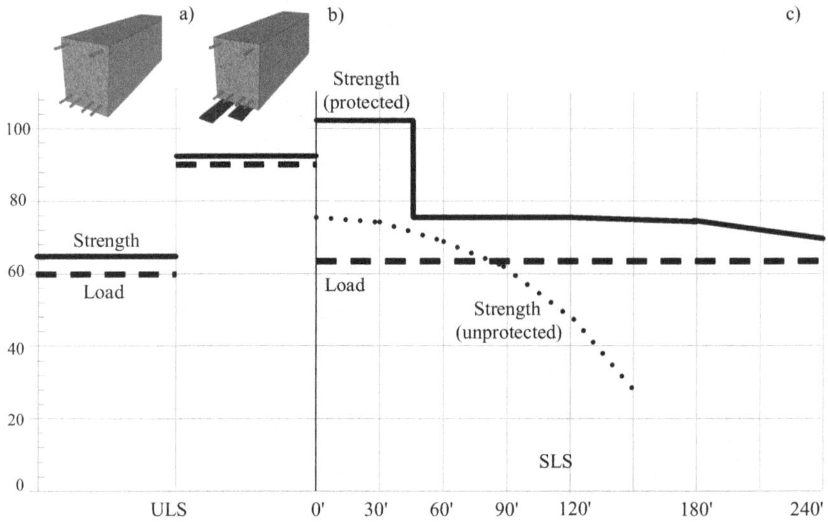

Fig. 8. Load and strength development in not strengthened (a) and strengthened by CFRP (b) members in case or ULS and load and strength development in protected and unprotected (c) members in case of SLS (fire scenario)

between strength and load that allow to have got from 60 min may be up to near 90 min time to achieve the equality between load and strength and this time will be the fire resistance of unprotected RC member.

3 Conclusions

A Eurocode calculation procedure in case of fire scenario for RC structural members strengthened by Sika CFRP has been presented. Load and strength of RC slab member in case of normal design and fire situation shows that load is decreasing and strength is increasing under fire influences with comparison of normal design case. Due to this fire resistance of RC member can reach up to R60–R90 without the fire protection measures.

References

Al Sherrawi M, Lyashenko V, Edaan E, Sotnik L (2018) Corrosion as a source of destruction in construction. Int J Civ Eng Technol 9:306–314

Blikharskyy Z, Selejdak J, Blikharskyy Y, Khmil R (2019) Corrosion of reinforce bars in RC constructions. Syst Saf Hum Tech Facil Environ 1:277–283. https://doi.org/10.2478/czoto-2019-0036

Bobalo T, Blikharskyy Y, Vashkevich R, Volynets M (2018) Bearing capacity of RC beams reinforced with high strength rebars and steel plate. Matec Web Conf 230:02003. https://doi.org/10.1051/matecconf/201823002003

Brózda K, Selejdak J, Koteš P (2017) The analysis of beam reinforced with FRP bars in bending. Procedia Eng 192:64–68. https://doi.org/10.1016/j.proeng.2017.06.011

Khmil R, Tytarenko R, Blikharskyy Y, Vegera P (2018) Development of the procedure for the estimation of reliability of reinforced concrete beams, strengthened by building up the stretched reinforcing bars under load. *East Eur J Enterp Technol* 5/7(95). https://doi.org/10.15587/1729-4061.2018.142750

Kos Ž, Gotal Dmitrović L, Klimenko E (2017) Developing a model of a strain (deformation) of a damaged reinforced concrete pillar in relation to a linear load capacity. Tech J 11(4):150–154 https://hrcak.srce.hr/190990

Krainskyi P, Blikharskyy Y, Khmil R, Blikharskyy Z (2018a) Experimental study of the srengthening effect of reinforced concrete columns jacketed under service load level. Matec Web Conf 183:1–5. https://doi.org/10.1051/matecconf/201818302008

Krainskyi P, Blikharskyy Y, Khmil R, Vegera P (2018b) Influence of loading level on the bearing capacity of RC columns strengthened by jacketing. Matec Web Conf 230:02013. https://doi.org/10.1051/matecconf/201823002013

Selejdak J, Khmil R, Blikharskyy Z (2018) The influence of simultaneous action of the aggressive environment and loading on strength of RC beams. Matec Web Conf 183:1–6. https://doi.org/10.1051/matecconf/201818302002

Vegera P, Vashkevych R, Blikharskyy Z (2018) Fracture toughness of RC beams with different shear span. Matec Web Conf 174:1–8. https://doi.org/10.1051/matecconf/201817402021

Blikharskyy Y, Khmil R, Blikharskyy Z (2018) Research of RC columns strengthened by carbon FRP under loading. Matec Web Conf 174:1–8

Options of Customization in Industrialized Methods of Construction in Terms of Construction 4.0

Marcela Spisakova(✉) and Maria Kozlovska

Faculty of Civil Engineering, Institute of Construction Technology and Management, Technical University of Kosice, Vysokoskolska 4, 042 00 Kosice, Slovakia
marcela.spisakova@tuke.sk

Abstract. Comparing construction to other industries shows that while they have increased productivity, the construction industry lagged behind. One of the way to increased productivity in construction projects is application of industrialized construction which presents a modern method of construction. On the other hand, current consumers require individually manufactured products according an approach "made for me". The connection of individual customer requirements and industrialized construction is considered a challenge for modern construction industry. The solution lies in applying the principles Construction 4.0 which require the transformation of the construction industry towards the 4th industrial revolution, from automated production to a greater level of digitalization. The aim of paper is to analyse the options of customization (customer requirements) in the various stages of construction project - project conception and construction design, construction elements manufacturing and construction.

Keywords: Industry 4.0 · Construction · Productivity · Modern methods of construction · Customization · Construction project stage

1 Introduction

Despite the fact that construction is one of the biggest industries in the world, low level of productivity in construction industry hold true across different countries. Productivity in construction lags far behind productivity improving in other industries. It may seem, the increase in construction productivity is indirectly proportional to the increase in customization in construction industry. The paper provides a new perspective on construction productivity and customization.

1.1 Productivity in Construction Industry

According Changali et al. (2015) there are a variety of reasons why productivity in construction is not at the level of other industries: (i) the construction industry is made up of large-scale and small-scale companies; (ii) paying workers is expensive and given that construction is a labor-intensive industry experiencing a skills shortage, this cost is

Z. Blikharskyy et al. (Eds.): CEE 2019, LNCE 47, pp. 444–451, 2020.
https://doi.org/10.1007/978-3-030-27011-7_56

only likely to increase; (iii) huge amounts of waste in construction, with the power of big data not being utilized; (iv) the level of organization in construction projects is not always at the desired standard; (v) many projects lack the necessary levels of risk management and planning, which can lead to project delays and subsequent financial losses if unforeseen events occur. Increasing productivity in construction is a new challenge for this industry. Using modern methods of construction is possible to improve construction project performance.

1.2 Modern Methods of Construction

Modern Methods of Construction (MMC) presents a wide range of relatively new ways and techniques of construction (Kozlovska et al. 2017). Modern methods of construction can be defined as a method in which less or greater extent building blocks (walls, floor slabs, beams, columns and staircases) are mass-produced in the factory under strict supervision with high quality (Trikha 2000). Other authors (Esa and Nurudin 1998) argued that the MMC is a continuous system utilization craftsman at each construction activity that uses a manufacturing production to minimize wastage of resources and increased value for end users. Managed and process-controlled industrial production in factories is a prerequisite for higher product quality, precision processing, or minimization of waste. The benefits of "off-site" production are demonstrated not only in the increase in production performance (controlled mass production independent of climatic conditions), but also in assembly of construction parts on construction site where is significantly reduced of construction time and the volume of construction works. Compared to traditional construction methods (masonry, reinforced concrete), there are a number of advantages. They are characterized by the production of larger or smaller segments of construction systems of modern construction methods off-site in factories. Application of MMC meet the requirements of Industry 4.0 principles.

1.3 Industry 4.0 Concept in Construction Industry

Implementation of the Industry 4.0 principle is another way to improve construction performance. Industry 4.0 is a strategic initiative of the German government that traditionally heavily supports development of the industrial sector. The main idea is to exploit the potentials of new technologies and concepts such as:

- availability and use of the internet and IoT,
- integration of technical processes and business processes in the companies,
- digital mapping and virtualization of the real world,
- 'Smart' factory including 'smart' means of industrial production and 'smart' products.

According to some sources, Industry 4.0 factory could result in decrease of (Bauernhansl et al. 2016):

- production costs by 10–30%,
- logistic costs by 10–30%,
- quality management costs by 10–20%.

There are also a number of other advantages and reasons for the adoption of this concept including (Rojko 2017): (i) a shorter time-to-market for the new products, (ii) an improved customer responsiveness, (iii) enabling a custom mass production without significantly increasing overall production costs, (iv) more flexible and friendlier working environment, and (v) more efficient use of natural resources and energy.

Compliance with the Industry 4.0 concept in construction industry presents a core of Construction 4.0. Osunsanmi et al. (2018) designed the conceptual framework (Fig. 1). which recognizes Construction 4.0 as the creation of the smart construction site, simulation, and virtualization that will in return lead to improved construction project performance.

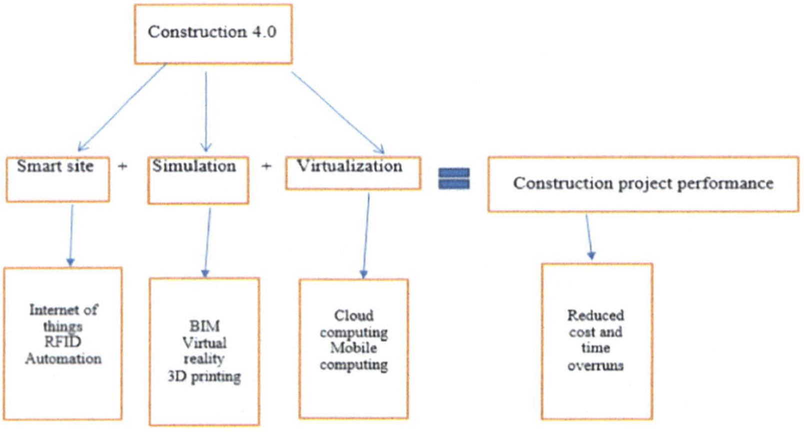

Fig. 1. Conceptual framework of Construction 4.0, Source: Osunsanmi et al. 2018

The gist of this framework is a sharing of information. Mak (2011) referred that the performance level of a construction project is related to the quality of information available to the professional. On the other hand, the Construction 4.0 cannot focus only on the construction phase of construction. It is related to all phases of construction project.

1.4 Stages of Construction Project

A construction project is always a long and demanding process (Landau 2017). The management of construction project can be divided into 5 main project stages (Koutsogiannis 2019):

- project conception - this stage starts with the customer where are specified requirements,

- project design - includes four different steps: programming and feasibility, schematic design, design development, and contract documents. This stage provides information about large the building, how to use the space, a sketch of building, materials, cost for material,
- procurement stage - the project team to order and obtain materials, equipment, and workforce. This stage of the project can be more or less complex and challenging depending on how big the project is, the available resources and the agreed start date,
- construction stage - the goal at this stage is to have planned the construction project (resources of construction project, cost, time, quality) so carefully that everything goes off without a hitch,
- post-construction stage - obtains two part - new construction building commissioning, customer (owner) occupancy.

All phases of the construction project should be in line with the customer's requirements. The completed construction should be carried out at the agreed time, cost and quality required.

2 Customization in Construction Industry

In generally, customization is the process of delivering wide-market goods and services which are modified to satisfy a specific customer's need. Construction is a specific industry. Each building is a unique piece, often made for a pre-know specific customer. Thus, each building is adjusted to a particular customer (investor, user). Due to the low level of construction productivity, it is necessary to look for a suitable approach for increasing construction productivity and increasing the customization of buildings as well as. The goal of construction industry should to be developed the virtual tools for customizing individual stages of construction project, the intelligent models for direct customer involvement in meeting its individual requirements.

2.1 Customization and Industry 4.0

The first industrial revolution is referred to as Industry 1.0, or Customer production (CP), in which products were manufactured based on the requirements of users at a high cost and with a limited number of products. CP is leveraged to depict the production paradigm change from entirely manual production to machine production. In the second industrial revolution, called Industry 2.0 or Mass production (MP), low-cost products were made using large-scale production systems. The variety offered by MP was very small and limited. The third industrial revolution is referred to as Industry 3.0 or Mass customization production (MCP). In the late 1980s, customer demand for a large variety of products led to the development of MP. It was based on the development of information, automation technology, and the computer (Wang et al. 2017).

Most companies have focused on maximizing their value for many years. There is currently a trend of shifting focus from a company's value to customer demand (Wang et al. 2017). Today's consumers demand individually manufactured products and services ("Made-for-Me"). Intelligent articles, products and machines will allow manufacturers to embark on piece production and produce original products without extra cost. The fourth industrial revolution, called Industry 4.0, is approaching. Information and communication technologies (ICTs) are increasing jointly and influencing all aspects of our life and business. Embedded software is integrated into the global communication network, connects and controls devices and systems in our environment. The boundaries between the real world and the virtual world are clearly overlapping. These new technologies will enable wholly new forms of Mass personalization production (MCP). The historic split between cheap mass-produced products, creating value from economies of scale, and more expensive customized products, will be reduced across a wide range of product types.

Figure 2 shows evolution of the production paradigm. There are product variety and productivity in each industrial revolution expressed.

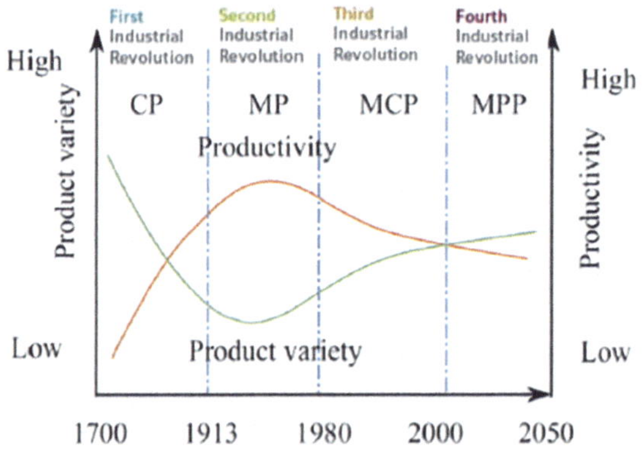

Fig. 2. *Evolution of the production paradigm,* Source: Wang et al. 2017

2.2 Customization and Construction Project Stages

The principles of Industry 4.0 meet the implementation of modern construction methods into construction. Therefore, it is suitable to complete the construction project stages with the manufacturing of the building elements. In this part of paper will analysed three basic stages of MMC project: (i) project conception and construction design, (ii) construction elements manufacturing, (iii) construction stage.

The first step within developing the virtual tools for customizing of construction project is to analyse the customization requirements in each this stage. A support tool for decision-making process should be processed in the form of a knowledge database. The information in the knowledge database will affect the construction project in all analysed stages.

In the stage "project conception and construction design" are defined the building´s requirements (purpose, cost, time, quality, energy efficiency, building material base, etc.) for the first time by customers.

Modern methods of construction present offsite manufacturing or onsite manufacturing. It depends on the rate of completion on site, e.g.: volumetric modern method of construction is a representative of offsite construction. The need for assembly on site is the least (Kozlovska et al. 2016). This stage "construction elements manufacturing" allows a space for implementation Construction 4.0 (smart factory, IT support, (Rojko 2017), Digital Twin, Lean Production and Construction, Design for Manufacturing and Assembly, Low Cost Intelligent Automation, Advanced Planning and Scheduling, (Shin et al. 2008) etc.). Not surprisingly, the customer's requirements have a great impact on it. The customer determines the rate of use Construction 4.0 principles.

The construction stage can meet the customer's requirements for the assembly possibilities of construction elements on the site. It is influenced by the previous two phases and thus by the customer's choice.

Table 1. Impact of customer requirements on construction project phases

Customer's requirement		Construction project stage		
		Project conception and construction design	Construction elements manufacturing	Construction stage
Social	Purpose of construction	✓		
	Spatial possibilities	✓		
	Variability	✓		
	User health (indoor environment quality)	✓		
	Aesthetic aspects	✓		
	Material base	✓	✓	✓
Technological and economic	Construction time	✓	✓	✓
	Construction costs	✓	✓	✓
	Manufacturing time		✓	✓
	Operating costs	✓		
	Disposal costs	✓	✓	
	Life of construction	✓		
	Return on investment	✓		
	Standardization	✓	✓	✓
	Availability of building systems		✓	✓
	Availability of building components		✓	✓
Environmental	Reusable material	✓		
	Material consumption during construction		✓	✓
	Energy consumption	✓		
	Energy efficiency	✓		
	Environmental footprint	✓	✓	

Table 1 shows the impact of customer requirements on construction project phases. The key stage of the construction project that provides information for the next stage. Identification of exact customer requirements provides input information to the decision-making process on the construction method and thus to the measure for application of principles Construction 4.0.

3 Conclusions

Construction, not only in the European but also in the world context, has the lowest productivity of all traditional industries. It is related to the specifics of construction industry final product - the construction (constructive and financially demanding product for a particular customer) and the specifics of the production conditions at the construction site (each product has its own production, in addition to constantly changing). Increasing construction productivity and customization of construction as well as is a challenge for the current construction industry. The solution is provided by Construction 4.0 principles. The paper analysed the options of customization (customer requirements) in the various stages of construction project - project conception and construction design, construction elements manufacturing and construction. The analysis represents the initial information for development of the virtual tools for customizing individual stages of construction project, the intelligent models for direct customer involvement in meeting its individual requirements.

Acknowledgements. The activity presented in the paper is part of the research grant VEGA 1/0557/18 “Research and development of process and product innovations of modern methods of construction in the context of the Industry 4.0 principles”.

References

Bauernhansl T, Krüger J, Reinhart G, Schuh G (2016) Wgp-Standpunkt Industrie 4.0, Wissenschaftliche Gesellschaft für Produktionstechnik Wgp e. v
Changali S, Mohammad A, van Niewland M (2015) The construction productivity imperative. https://www.mckinsey.com/industries/capital-projects-and-infrastructure/our-insights/the-construction-productivity-imperative
Esa H, Nurudin MM (1998) Policy on industrialized building systems. Report on colloquim on industrialise construction system, Kuala Lumpur
Kozlovska M, Spisakova M, Mackova D (2017) Modern methods of construction - towards adoption in Slovakia. 1st edn. Wambeek EuroScietia
Kozlovska M, Spisakova M, Mackova D (2016) Knowledge database of modern methods of construction. Advances and Trends in Engineering Sciences and Technologies II, London, Taylor and Francis
Koutsogiannis A (2019) 6 stages of project in construction. https://geniebelt.com/blog/6-stages-of-a-project-in-construction
Landau P (2017) Quick guide to construction project management. https://www.projectmanager.com/blog/construction-project-management

Mak S (2011) A model of information management for construction using information technology. Autom Constr 21(10):257–263

Osunsanmi T, Aigbavboa C, Oke A (2018) Construction 4.0: the future of the construction industry in South Africa. Int J Civ Environ Eng 12(3):206–212

Rojko A (2017) Industry 4.0 concept: background and overview. Int J Interact Mob Technol 11 (5):77–90

Shin Y, An S-H, Cho H-H, Kim G-H, Kang K-I (2008) Application of information technology for mass customization in the housing construction industry in Korea. Autom Constr 17:831–838

Trikha J (2000) Industrialised building system: prospect in Malaysia. In: Proceeding of world engineering congress, Kuala Lumpur

Wang Y, Ma H-S, Yang J-H, Wang K-S (2017) Industry 4.0: a way from mass customization to mass personalization production. Adv Manuf 5:311–320

Construction Cost Saving Through Adoption of IoT Applications in Concrete Works

R. Bašková, Z. Struková(✉), and M. Kozlovská

Faculty of Civil Engineering, Technical University of Košice, Vysokoškolská 4, 042 00 Košice, Slovakia
zuzana.strukova@tuke.sk

Abstract. Development of digital technologies and processes including Internet of Things (IoT) applications is central to the required transformation of the construction industry within the frame of Construction 4.0 concepts originated from the fourth industrial revolution Industry 4.0. The construction sector has vast potential to improve productivity and efficiency thanks to digitalization, innovative technologies and new construction techniques. The new tool Concremote, offered by multinational expert in solutions of formwork, uses digital sensors to measure the in-situ concrete maturity gradient and with this data it calculates early age strength. Being able to measure in real time the development of in-situ concrete maturity can reduce striking time and thus reduce cost of formwork. The aim of the case study is to demonstrate saving of formwork cost due reducing striking time through intended employment of the tool Concremote in construction of slabs. The cost of slab formwork was estimated in different variants of formwork striking times. For most variants of construction time and striking time, it was verified that the saving of the formwork cost due reducing striking time is higher than cost of the Concremote employment in the referenced construction.

Keywords: Cost of formwork · Slab formwork · Concrete works · Internet of Things (IoT) · Striking time

1 Introduction

Formwork construction for reinforced cast-in-place concrete is costly, often time-consuming and complex to plan and design (Hyun et al. 2018). The cost of formworks, which are generally rented, is one of the highest ones that construction companies have to face when developing concrete structures. It varies averagely between 35–45% of a reinforced concrete structure's cost and for civil engineering structures it can rise up to 60% of the total cost of the project (Rudeli et al. 2015). Considering constructability of formwork items and analyzing efficiency of their utilization at each stage of construction planning seems to be the smartest step to reduce cost of formwork and thus total construction cost (Krawczynska-Piechna 2017). Based on the study of Abdulaziz (2010), the constructability factors that impact formwork productivity the most involve formwork management, repetition, standardization and consistency.

Z. Blikharskyy et al. (Eds.): CEE 2019, LNCE 47, pp. 452–459, 2020.
https://doi.org/10.1007/978-3-030-27011-7_57

Formwork striking time plays a major role in construction scheduling and project managers are always interested in reducing activities along the critical path (formwork striking, curing, pre-stressing …) in order to save time and thus money (Okafor and Ewa 2016). The sooner the formwork is stripped, the more economical and practical it becomes to schedule many reuses of formwork. An economical concrete project requires for formwork planning by principle of maximum reuse of forms however not over the constraints of quality and safety of construction. Several authors have presented models and methodologies which can predict the minimum striking time by continuously measuring the maturity index of concrete at the construction site and methods that can be applied to estimate the compressive strength development in concrete (Galobardes et al. 2015; Yikici and Chen 2015). Benaicha et al. (2016) have studied the temperature evolution of concrete to examine the mechanical strength evolution depending on the age hardening and the cure temperature.

The digital transformation in construction sector, described as Construction 4.0, enable construction companies to improve productivity, reduce project delays and cost overruns, manage complexity and enhance resource-efficiency (Craveiro et al. 2019). Internet of Things (IoT) is one from technological developments having potential to transform the building industry. It can be referred to as the connection of any tool to the Internet, from small accessories to large machines (Atzori et al. 2010). The IoT such as the use of robots, sensors, actuators, electronic signalization and many other Internet enabled physical devices may contribute for new advanced smart applications to be used in construction (Kochovski and Stankovski 2018).

The concrete maturity systems that give accurate real-time assessment of in-situ strength using wireless and smart technology represent the advanced generation of concrete-strength assessment methods (Reddy and Fogarty 2016). Doka, as a multinational expert in solutions of formwork, has recognized impact of IoT in the construction sector and has provided a promisingly excellent example of this. The tool Concremote uses digital sensors to measure the in-situ concrete maturity gradient and based on this data it calibrates early age strength and can send concrete strength assessment to a mobile phone. Automatic notification in-real-time when the concrete reaches the target value enables earlier initiation of critical path activity as formwork striking is. Employment of the mentioned tool in a concrete project has potential to make striking time shorter and to save cost of the formwork.

2 Materials and Methods

The aim of the study is to answer the question "How could adoption of Concremote technology affect the cost of slab formwork in different variants of construction schedules when constructing load bearing structure of a tier building?" The methodology of the study consists of the following steps: 1. Development of variants of construction schedule of the load bearing structure; 2. Development of plan of slab formwork layout inclusive of estimation of unit rental cost of beam slab formwork Dokaflex 1-2-4; 3. Proposal of striking times of slab construction – in case of standard practice and in case of Concremote adoption; 4. Estimating the rental cost of slab formwork – reference of formwork sets for different variants of construction schedules and different striking times.

The referenced apartment building consists of three building sections (BS) – BS 01, BS 02 and BS 03. An elderly apartment building is located between BS 02 and BS 03. The load bearing structure is composed of external masonry walls (size 300 mm), inner masonry walls (size 250 mm) and monolithic concrete slabs (150 thick). The area of one slab in each of the sections (BS 01, BS 02 and BS 03) is around 190 m^2.

3 Results

3.1 Variants of Construction Schedule of the Load Bearing Structure

Six variants of construction schedules were modelled in the submitted study. The building sections were in flow scheduled in construction process – in the first three variants (Fig. 1a) – or they were gradually scheduled in the construction process – in other three variants (Fig. 1b). The variants where building sections are in flow scheduled into construction process enable shorter construction times compared to variants where building sections are scheduled in construction process gradually, but they are more demanding in terms of planning and arranging construction tasks.

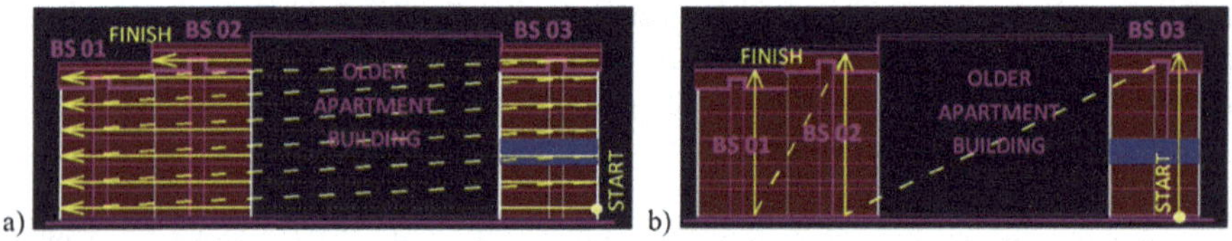

Fig. 1. The scheme of construction schedule (a) in flow and (b) gradual schedule of building sections

As to location parameters of construction, the referenced building consisting of three building sections (BS 01 – 5 floors, BS 02 – 6 floors and BS 03 – 6 floors) is divided into 17 working locations. One working location in all the scheduling variants is represented by one floor (i.e. one slab). Construction schedule of different locations in construction is presented in Fig. 1a for variants with in flow scheduled building sections in construction process and in Fig. 1b for variants with gradually scheduled sections in construction process. In the first case (Fig. 1a) the locations are in horizontally ascendant routing scheduled in construction so that working crews may work continuously and evenly. In other one (Fig. 1b) the vertically ascending routing was scheduled. The working fund is 5 working days per one week in single-shift operation (8.5 h per a shift) and it is the same in all variants of construction schedule. Concrete placing in one slab construction takes no more than one working shift.

Specialized crews for four major tasks of the load bearing structure construction – walling, formworks of slabs, slabs reinforcing, and slabs concreting – were scheduled in the variants with in flow scheduled building sections into construction process. In the variants with gradually scheduled building sections into construction process a requirement to minimize working breaks was taken into consideration. Therefore, one complex crew is scheduled to construct both vertical and horizontal structures. The number of workers in crews in different variants of schedules varies depending on amount of work and on construction time of the tasks in a slab.

Three variants of construction schedule (Variants 1–3) for in flow scheduled crews into construction process, i.e. in horizontally ascending routing, were developed with different duration of takt time. The takt time is seen as the duration of one crew work on one slab (i.e. one location). Three variants of construction schedule (Variants 4–6) for gradually scheduled crews into construction process, i.e. vertically ascending routing, were developed with different construction time of one slab. Resulting construction time in different variants of takt time is presented in Table 1.

Table 1. Construction time of the load bearing structure in the variants of construction schedule

Variant	Takt time (day)	Construction time of one slab – without weekends (day)	Total construction time (day)	Total construction time – without weekends (day)
Variant 1	4	10	104	74
Variant 2	5	13–15	130	93
Variant 3	7	20	182	130
Variant 4	-	10	238	170
Variant 5	-	15	357	255
Variant 6	-	20	476	340

3.2 Unit Cost of Slab Formwork

The beam-type slab formwork Dokaflex 1-2-4 was planned in the presented study. The set of formwork items for a typical slab in one location is the same in all building sections (BS 01, BS 02 and BS 03). The analysis of unit formwork cost estimated based on the formwork layout plan and by offer prices of Doka Group are presented in Table 2. The unrepeated cost involves cost of formwork cuts in case of formwork rental.

Table 2. The initial cost of slab formwork

Unit cost of formwork	The set of formwork items for one slab	A set of floor props and formwork boards in place of floor props
Purchase price (EUR)	25 250	3 060
Rental cost (EUR/day)	65.08	7.08
Unrepeated cost (EUR)	547.69	126.39

3.3 Striking Times of Slab Formwork

Commonly on construction sites, striking time for a slab formwork is 14 days (i.e. 10 days without weekend) and time of floor props removal is 28 days (i.e. 20 days without weekend) after concrete placing. The technology Concremote provides calibration of concrete before concreting at site. By smart monitoring the in-situ concrete maturity, it may reduce striking time to a minimal time which is necessary for reaching sufficient 28-day strength of concrete specified by structural engineer. As far as slab formwork removal, it is usually when concrete in slab has reached 80% of its specified 28-day strength and for removal of all floor props it is when concrete in slab has reached 100% of its specified 28-day strength.

Based on the case studies, published by Doka Group (ConcremoteAnwenderinformation by Doka Industrie GmbH), the Concremote technology may reduce striking time for slab formwork from 14 days to 7 days or even to 3 days. Similarly, the time of floor props removal may be reduced from 28 days to 21 days or even to 14 days.

3.4 Rental Cost of Slab Formwork

From cost of formwork point of view, it is profitable to plan construction process so that the formwork items are required in a working location immediately after their removal in the previous one. Based on the variants of construction schedule and after analysis of striking times, it is possible to determine the number of formwork items which should be rented when constructing the studied apartment building.

The Fig. 2 represents a sample of time-location matrix model where four partial construction processes are scheduled.

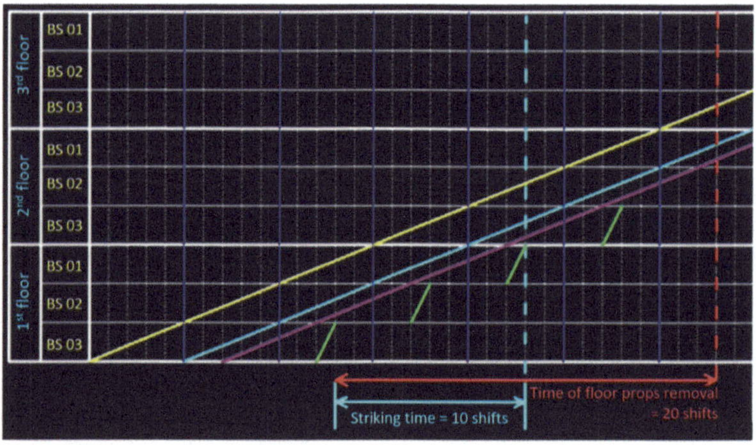

Fig. 2. A sample of time-location matrix model of the studied construction

It involves three locations (i.e. 1st floor, 2nd floor and 3rd floor) of each building section (i.e. BS 01, BS 02 and BS 03), meaning nine slabs from seventeen slabs in total. The work of crews is continual; building sections are in flow scheduled into construction process. The construction time of one location (i.e. one slab of one building section) is 13 days and total construction time is 93 shifts (130 days including weekend). When considering time of slab formwork striking 14 days (i.e. 10 shifts without weekend) and time of floor props removal 28 days (i.e. 20 shifts without weekend), slab formwork items for four slabs and floor props for two slabs should be available all at once. Such construction schedule is optimal in terms of continual work of crews. From formwork cost point of view, the mentioned striking times are not optimal, as formwork items from each slab remain two days unused before their using in following slab. This has indicated an increase of rental cost of formwork by 10%.

The cost of slab formwork in different alternatives of striking time and attributes of cost saving due to adoption of the Concremote technology are presented in Table 3.

Table 3. The cost of slab formwork and representation of the cost saving

CT (day)	ST (day)	ST-FP (day)	X	Y	CSF (EUR)	CFP (EUR)	UC (EUR)	TCF (EUR)	TCFC (EUR)	C-S (%)
Construction method 1 - Flow										
182	14	28	3	2	35 533	2 577	1896	40 007	40 007	
182	7	21	3	1	35 533	1 288	1769	38 590	45 590	−14
182	3	21	2	2	23 689	3 865	1475	29 030	36 030	10
130	14	28	4	2	33 842	1 841	2444	38 127	38 127	
130	7	21	3	2	25 381	1 841	1896	29 118	36 118	5
130	3	21	2	3	16 921	2 762	1601	21 284	28 284	26
104	14	28	5	2	33 842	1 473	2991	38 306	38 306	
104	7	21	3	3	20 305	2 208	2022	24 535	31 535	18
104	3	21	3	3	20 305	2 208	2022	24 535	31 535	18
Construction method 2 - Gradual										
476	14	28	2	0	61956	0	1095	63051	63 051	
476	7	21	1	1	30978	3370	674	35022	42 022	33
476	3	21	1	1	30978	3370	674	35022	42 022	33
357	14	28	2	1	46467	2527	1222	50216	50 216	
357	7	21	1	1	23233	5055	800	29088	36 088	28
357	3	21	1	1	23233	2527	674	26434	33 434	33
238	14	28	2	2	30978	3370	1348	35696	35 696	
238	7	21	2	1	30978	1686	1222	33885	40 885	−15
238	3	21	1	2	15490	3370	800	19659	26 659	25

CT – construction time; ST – striking time of slab formwork; ST-FP – striking time -removal of floor props; X – number of sets of slab formwork for one slab; Y – number of sets of floor props for one slab; CSF – cost of slab formwork; CFP – cost of floor props; UC – Unrepeated cost of formwork (fixed cost); TCF – total cost of formwork; TCFC – total cost of formwork including Concremote; C-S – cost saving

Two from all alternatives represent the common practice at construction sites. It means the striking time of slab formworks - 14 days (i.e. 10 working shifts without weekend) - and time of removal of floor props – 28 days (i.e. 20 working shifts without weekend). The other two alternatives represent the intended striking times provided in adoption of the Concremote technology. The following alternatives were considered: 7 days and 3 days for removal of slab formwork and 21 days for removal of floor props. The cost of the Concremote technology (7 000 EUR) is also included into cost of the formwork.

The Construction method 1 corresponds to construction where building sections are in flow scheduled into construction process and work of several crews, which are at the same time in different locations, is continual. The Construction method 2 corresponds to construction where building sections are gradually scheduled into construction and one crew moves gradually from the first to the last one slab (from 1st to 17th). The calculation of the cost of formwork is based on the initial cost represented in Table 2.

4 Conclusions

Each construction schedule should take into account not only required boundary term of construction, but also demands on construction resources. The presented case study focused on impact of Concremote technology adoption in concrete works on cost of formwork and those on total construction cost. The saving of formwork cost was estimated for different variants of construction time and for different approaches into construction scheduling. The first approach was characterized by construction method where building sections are in flow scheduled into construction and work of several crews, which are at the same time in different slabs, is continual. The second one was represented by construction method where building sections are gradually scheduled into construction and one crew moves gradually from the first to the last one building section. The continual work of crews in both construction methods had the biggest priority in construction scheduling. If the main emphasis was on optimizing/ minimizing the cost of formwork, striking times would have a decisive impact on construction scheduling and on amount of workers in crews. However, each construction schedule is risky either from construction cost or quality point of view or in terms of meeting the deadlines. The Concremote technology may cause some risk in terms of construction cost, quality or time. In terms of concrete structure quality, short striking times may cause risks of failure to comply with the conformity of the concrete, as the sensors only monitor the concrete temperature and don't respond to deviations in the quality of concrete compared to concrete tested during calibration before construction. The differences between digital concrete representative and the real concrete at the site are quite likely. Despite additional cost of Concremote technology adoption and other possible risks of its deployment, such IoT application should be certainly considered to be progressive in concrete works from productivity and efficiency point of view. However, from the study presented, it could be deduced that the technology is preferable for more expensive formwork systems with lower labor difficulty.

Acknowledgements. The activity presented in the paper is part of the research grant - Grant No. *VEGA – 1/0557/18 Research and development process and product innovations of modern methods of construction in the context of the Industry 4.0 principles.*

References

Abdulaziz MJ (2010) Buildability factors affecting formwork labour productivity of building floors. Can J Civ Eng 37(10):1383–1394

Atzori L, Iera A, Morabito G (2010) The internet of things: a survey. Comput Netw 54:2787–2805

Benaicha M, Burtschell Y, Alaoui AH (2016) Prediction of compressive strength at early age of concrete – application of maturity. J Build Eng 6:119–125

Craveiro F, Duarte JP, Bartolo H, Bartolo PJ (2019) Additive manufacturing as an enabling technology for digital construction: a perspective on Construction 4.0. Autom Constr 103:251–267

Doka Industrie GmbH (2014) ConcremoteAnwenderinformation. Amstetten Germany, p 32

Galobardes I, Cavalaro SH, Goodier ChI, Austin S, Rueda A (2015) Maturity method to predict the evolution of the properties of sprayed concrete. Constr Build Mater 79:357–369

Hyun Ch, Jin Ch, Shen Z, Kim H (2018) Automated optimization of formwork design through spatial analysis in building information modeling. Autom Constr 95:193–205

Kochovski P, Stankovski V (2018) Supporting smart construction with dependable edge computing infrastructures and applications. Autom Constr 85:182–192

Krawczynska-Piechna A (2017) Comprehensive approach to efficient planning of formwork utilization on the construction site. Proc Eng 182:366–372

Okafor FO, Ewa DE (2016) Estimating formwork striking time for concrete mixes. Niger J Technol 35(1):1–7

Reddy J, Fogarty A (2016) Assessing concrete strength and the use of temperature matched curing. http://www.engineersjournal.ie/2016/12/13/temperature-matched-curing-concrete-strength/. Accessed 7 May 2019

Rudeli N, Santilli A, Arrambide F (2015) Striking of vertical concrete elements: an analysis using the maturity method. Eng Struct 95:40–48

Yikici TA, Chen H-L (2015) use of maturity method to estimate compressive strength of mass concrete. Constr Build Mater 95:802–812

Analysis of Cost of Building Equipment Used for Removal of Water Conduits Failure

Andrzej Studzinski(✉)

Department of Water Supply and Sewage Systems,
Rzeszow University of Technology, Powstancow Warszawy 12,
35-959 Rzeszow, Poland
astud@prz.edu.pl

Abstract. The costs of removing water pipe breakdowns can be divided into the costs of materials, equipment and labor. The study concerns the analysis of the costs of the equipment used to remove water pipelines failures. The basis of the work are operational data collected during the removal of 302 pipe breakdowns in the city with a population exceeding 100,000. The dependence of these costs on the diameter and material of the pipe was noticed, they increase with increasing diameter. The differentiation of costs is also the result of the location of the pipeline - where a high degree of soil compaction is required, its replacement is carried out, which results in an increase in the costs of both materials and equipment used. In the water supply system under study, the average cost of equipment used to repair the failure was an average of PLN 2178 (price level 2017), was higher than other costs, which amounted to a total of PLN 1317.

Keywords: Water network · Failure rate · Costs of failure removal · Construction equipment

1 Introduction

Waterworks belong to technical systems, which require special reliability of operation (Kwietniewski and Rak 2010; Pietrucha-Urbanik 2015; Tchórzewska-Cieślak et al. 2018). The continuity of water supply to recipients is mainly the result of failure-free operation of the water network, hence water supply companies place particular emphasis on the efficient removal of emergencies. This requires maintenance teams, construction equipment and stock levels of parts necessary to carry out repairs. This involves providing funds to cover repair costs.

Among the costs of removing pipe breaks, the significant component is the cost of equipment used to repair the failures. They are so-called direct costs of equipment, this term is understood as the sum of the product of the working hours and its unit price rates. Unit costs are costs applied by a water supply company to the calculation of the price of the building equipment that is usually owned by the company. The exemplary unit cost of a backhoe loader was 96 PLN/h (2017). The aim of this work is to analyze

Z. Blikharskyy et al. (Eds.): CEE 2019, LNCE 47, pp. 460–467, 2020.
https://doi.org/10.1007/978-3-030-27011-7_58

these costs for one of the south-eastern Poland waterworks, the basis for the development are operational data collected during pipe failures. Data are presented in division to material and diameter of pipes for nominal diameter in the range of 25–800 mm.

2 Object and Methodology of Research

The research was carried out on a water supply system located in southern Poland. Water network supplies water to over 100,000 residents of the city and neighboring towns. The water supply network had a length exceeding 1000 km, 5.9% of the pipelines were water mains, 53.3% distributing pipes and 40.8% of length were water house service lines (at the end of 2015). The material structure was dominated by polyethylene pipes 57.4% cast iron - 22.5%, the remaining materials had the following share in the network length: steel 8.6%, PVC 7.9%, asbestos AC 3.3% and ductile iron 0.3%. The age of the oldest operated water pipelines exceeded 100 years. The reliability of water lines measured by the failure rate showed values lower than those recommended in literature (Herbert 1994; Kwietniewski and Rak 2010).

The method of conducting the tests was presented in (Studzinski and Pietrucha-Urbanik 2017), they were kept while recording data on the developed failure charts filled in by the employees of the water supply company. They contained information about failures (time, place, pipe data such as age, material or diameter, consequences of failure, scope of actions necessary for repair, equipment and materials used). The research was carried out in 2014–15.

The cost of the equipment was calculated by multiplying the time of its use by the unit cost set by the water supply company, and was updated at the end of 2017.

The Hellwig's information capacity indicator method was used for analyzes (Hellwig 1968). The information capacity indicator method is based on basic formulas from combinatorics and correlation coefficients of the analyzed features. On the basis of the Hellwig's method, explanatory variables for equipment costs were determined and then the relationship between the cost variable and these variables was developed. The importance of the issues of methods like prediction analysis, regression analysis and fuzzy inference in the technical system are presented in numerous publications (Skrzypczak et al. 2018).

3 Failures

Operational data from pipeline failures in the tested water supply are shown in Table 1. In total, 302 failures were recorded, of which 298 presented below concerned water lines, only in 4 cases the fittings were repaired. The percentage share of breakdowns of individual types of water pipes was as follows: 7.4% of failures were recorded on water mains, 31.7% on distributing pipes, 59.6% on house service lines, and 1.3% on fittings.

Table 1. Failure list (Studzinski and Pietrucha-Urbanik 2017)

Diameter [mm]	Number of failures [-]	Number of failures for pipe materials					
		PE	PVC	steel	c. ir.	d. ir.	AC
25	143	132	-	11	-	-	-
32	22	19	-	3	-	-	-
40	2	-	-	2	-	-	-
50	8	5	-	2	1	-	-
65	3	2	-	1	-	-	-
80	47	2	5	3	33	-	4
100	28	3	6	1	13	-	5
125	1	-	-	-	1	-	-
150	16	3	6	-	6	1	-
200	5	-	5	1	-	-	-
250	3	3	-	-	-	-	-
300	4	-	1	-	2	-	1
400	7	2	-	1	4	-	-
450	3	2	-	-	-	1	-
500	2	-	-	-	2	-	-
600	3	-	-	3	-	-	-
800	1	-	-	-	1	-	-

The pipelines are grouped for their nominal diameters (ND) in relation to the inside diameter. An increased number of failures has been observed for pipes having the largest share in the length of the water network - for house service lines of 25 and 32 mm, and for 80 and 100 mm distributing pipelines, house service lines diameters are typical for single-family buildings, and distributing pipes diameters are based on fire regulations - hydrants recommended for use and the minimum diameters of pipes on which hydrants are installed. The share of particular materials building the water network is typical - PE and steel in the case of house service lines, all materials for distributing pipes, steel and cast iron for water mains - which in the last case results from their age.

4 Cost of Building Equipment

Table 2 shows the obtained range of equipment costs and their median for individual nominal diameters as well as the range and median time of failure removal.

The data presented in the table indicate an increase in equipment costs accompanied by an increase in the diameter of the pipeline being repaired. Particular consideration should be given to the analysis of data for diameters for which the number of failures is significant, i.e. 25, 32, 80, 100 and 150 mm. The presented trends are better visible by grouping the pipes according to their functions: house service lines, distributing pipes and water mains. The diameters are 25–65 mm, 80–200 mm and 150–800 mm respectively. The annual costs of equipment and the average cost for a given type of pipelines are summarized in Table 3.

Table 2. Costs of materials, time of failure by pipe diameter

Diameter [mm]	Cost range [PLN]	Median of cost [PLN]	Time range [h]	Median of time [h]
25	111–15247	858	1–225	4
32	220,6–1439	762	0,5–52	3,25
40	820–3426	2123	1–4	2,5
50	571–3325	1360	1–15	5
65	505–1862	637	2–4	3
80	107–7789	2344	1–22	4
100	97–7713	2053	1,5–16	5
125	1392	1392	2	2
150	222–8134	2323	1–12	4
200	707–2817	2319	1–10	2
250	1794–5291	3023	4–10	6
300	1777–14724	4918	7–18	8
400	2416–21548	4212	4–182	7
450	778–19833	7731	2–119	32
500	990–1430	1210	2–3	2,5
600	2359–5007	3540	4–11	8
800	22735	22735	21	21

Table 3. Costs of materials divided into pipeline function

Pipeline function	Total cost of materials [PLN]	Average cost [PLN]	Percentage [%]
Water mains	234905	6340	21,0%
Distributing pipes	255684	2691	41,5%
House service lines	158495	1320	37,5%

Table 3 shows the total costs in division to pipeline function and their share in total costs. The growing cost of materials with the increase of diameters of pipelines is noteworthy, the largest values concern the water mains with the nominal diameter range 150–800 mm, lower for distributing pipelines with diameters of 80–200 mm and the lowest for house service lines.

The presented relation should be explained with diameters of pipelines and not with their function. With the increase in diameter, the volume of excavations increases, and as a consequence, the time of failure removal increases. The lowest average cost of equipment for house service lines is also due to the fact that works are usually carried out manually.

The Hellwig's method was used to analyze the dependency of costs (Dziechciarz 2002). The dependence of the dependent variable of equipment costs Y [PLN] on the basis of potential explanatory variables (predictors) was considered:

- X1 – diameter [mm],
- X2 – pipeline function [-],

- X3 – pipeline material [-],
- X4 – time of failure [h].

Values of predictors X2 and X3 are included in Table 4.

Table 4. Values of X2 and X3

X2		X3	
House service line	1	steel	1
Distributing pipeline	2	PE	2
Water main	3	c. iron	3
		AC	4
		PVC	5
		d. iron	6

The method of information capacity indicators is based on the determination of individual and integral indicators for individual information carriers.

The basic problem associated with the practical use of the Hellwig's method is the need to calculate for all possible $1, 2, \ldots, k-1$, k element combinations k of potential control variables explaining the so-called individual indicators of information capacity. This means a very fast increase in computational complexity with an increase in the number of explanatory variables. In addition, these formulas have a "heterogeneous" form, which prevents rapid data analysis.

The number of all combinations K that can enter the model in the analyzed case is equal to:

$$K = 2^m - 1 \tag{1}$$

where m is the number of exogenous variables – m = 4, so the number of combinations is 15.

The individual capacities of individual information carriers appearing in the specified combinations were calculated in accordance with the Eq. (2):

$$h_{ji} = \frac{r_i^2}{1 + \sum_{i=1}^{n} |r_{ij}|} \tag{2}$$

where: r_i - coefficient of correlation of the i parameter and of the explained function of y, r_{ij} - correlation coefficient of explanatory variables i and j.

The next step was to determine the maximum integral capacity H according to the Eq. (3):

$$H_j = \sum_{j,i=1}^{n} h_{ji} \tag{3}$$

which corresponds to the combination of C = {X1, X2, X4} = 0.4933, the results for individual combinations of random variables are presented in the Table 5 below.

Table 5. Values of the maximum integral capacity

A combination of explanatory variables	H_j
{X1}	0,2710
{X2}	0,2068
{X3}	0,0230
{X4}	0,2888
{X1, X2}	0,2619
{X1, X3}	0,2345
{X1, X4}	0,4811
{X2, X3}	0,1536
{X2, X4}	0,4599
{X3, X4}	0,2823
{X1, X2, X3}	0,2327
{X1, X2, X4}	0,4933
{X1, X3, X4}	0,4358
{X2, X3, X4}	0,3901
{X1, X2, X3, X4}	0,4341

This means that the variables X1, X2 and X4 (diameter, time of failure and type of network) should enter the model of assessment of factors determining the costs of equipment for repairing water lines. The model Eq. (4) was obtained:

$$Y = -438{,}589 + 6{,}89\,X1 + 946{,}02\,X2 + 66{,}00\,X4 \quad (4)$$

The value of the determination coefficient R^2 was 0.50. The significance value of the F test is much less than 0.05, so the F test confirms the significance of the relationship between the variable Y and the variables X1, X2 and X4. All model parameters are statistically significant at the 0.05 statistical significance. It can be concluded that the estimated model is a good tool for inference and forecasting (Zak and Walek 2014).

5 Conclusions

During the year, the water supply company spent almost PLN 650,000 on the costs of equipment used during repairs of water pipes, over 4 times more than on materials. The average cost of equipment is PLN 2178 (price level 2017), while materials amounted to PLN 517 (Studziński and Pietrucha-Urbanik 2018), labor costs PLN 800 (Studziński and Pietrucha-Urbanik 2017) (price level 2015) (price level 2015), so equipment costs account for more than half of the cost of removing failures.

On the basis of the presented data, it can be concluded that the equipment costs are mainly dependent on the diameter, duration of the failure and the type of a pipeline. Dependence on the type of the water conduit can be explained by the fact that the most important pipelines are repaired in the first place, which results in the involvement of considerable resources, including equipment.

There is a clear differentiation of costs for three groups of pipelines resulting from their functions, for house service lines 1320 PLN, distributing pipes 2691 PLN and for the water mains 6340 PLN. Observed differences mainly result from the diameters of these pipelines, so it can be concluded that with the increase in diameter, the average cost of building equipment increases (Hotloś 2007; Studzinski 2014). Analogous relationships were observed for the costs of materials and labor.

There were no significant differences in equipment costs for the material of the pipeline, as observed for the materials.

Due to the small number of failures, especially in the case of pipes with a diameter of DN 250 mm and larger, it is necessary to continue the research.

The dependence of equipment costs determined by Eq. (4), is a good tool for cost prediction, the presented results can be helpful in managing water supply systems.

References

Dziechciar J (2002) Ekonometria. Metody, przykłady, zadania. Wrocław, Wydawnictwo Akademii Ekonomicznej we Wrocławiu

Hellwig Z (1968) Toward a system of quantitative indicators of components of human resources development. UNESCO

Herbert H (1994) Technical and economic criteria determining the rehabilitation and for renewal of drinking water pipelines. In: Proceedings of IWSA regional conference, Zurich, Switzerland

Hotloś H (2007) Quantitative assessment of the impact of selected factors on the parameters and operating costs of water supply networks. Wrocław University of Technology Publishing House, Wrocław (in Polish)

Kwietniewski M, Rak J (2010) Reliability of water supply and sewage infrastructure in Poland. Polish Academy of Sciences, Warsaw

Pietrucha-Urbanik K (2015) Failure prediction in water supply system - current issues. In: Theory and engineering of complex systems and dependability, vol 365, pp 351–358. https://doi.org/10.1007/978-3-319-19216-1_33

Skrzypczak I, Kokoszka W, Buda-Ożóg L, Kogut J, Słowik M (2018) Environmental aspects and renewable energy sources in the production of construction aggregate. In: International conference on advances in energy systems and environmental engineering, (ASEE), Wroclaw, Poland

Studziński A, Pietrucha-Urbanik K (2017) Analiza kosztów robocizny usuwania awarii sieci wodociągowych. Technologia Wody 1:24–28

Studziński A, Pietrucha-Urbanik K (2018) Analiza kosztów materiałów użytych do usuwania awarii przewodów wodociągowych. Rocznik Ochrona Srodowiska 20:1453–1464

Studzinski A (2014) Amount of labour of water conduit repair. In: Safety, reliability and risk analysis: beyond the horizon, London, England

Tchorzewska-Cieslak B, Pietrucha-Urbanik K, Urbanik M, Rak J (2018) Approaches for safety analysis of gas-pipeline functionality in terms of failure occurrence: a case study. Energies, 11 (6):1589. https://doi.org/10.3390/en11061589

Zak L, Walek A (2014) Selected mathematical functions used for operation data information. In: Safety, reliability and risk analysis, pp 1303–1308

Effect of Silicate Additives on the Impact Strength of Cement Pastes

W. Szewczenko(✉) and G. Kotsay

Faculty of Civil Engineering, Mechanics and Petrochemistry,
Warsaw University of Technology, Łukasiewicza 17, 09-400 Płock, Poland
wiktor.szewczenko@pw.edu.pl

Abstract. In construction, the usage of silicate additives is constantly expanding. Depending on the type and amount of mineral additives and the time of cement hydration, the structure of basic cement hydration products varies and it affects the properties of cement pastes. To ensure general-purpose cements of high quality, it is necessary to monitor constantly of activities of used mineral additives. To investigation of activities of additives used physical and chemical methods. However, during continuous production, such methods are time consuming, so, it is necessary to use some quicker way to check activities of mineral additives in cement. Therefore, in the paper, impact bending strength method is proposed for evaluation of mineral additives' activity. However impact strength of cement pastes depended on curing conditions. Unbound water in capillary and gel pores works as shock-absorbent during impact bending tests. Therefore, impact bending tests, should be performed on samples stored in air-dry conditions.

Keywords: Paste · Mineral additives · Pozzolanic activity · Impact bending strength

1 Introduction

Nowadays, when it is getting more and more difficult to obtain high-quality raw materials, and when we face necessity of efficient usage of them, it raises the issue of replacing clinker elements of general-purpose cements with mineral additives of natural or artificial origins. In order to save energy and reduce CO_2 emission during cement production, the use of mineral additives is getting more and more popular. In many papers, additives not listed in PN-EN 197-1 norm were suggested, like, for example, metakaolin, fine-ground waste glass, fluidized fly-ash or ashes from biomass combustion (Shi et al. 2011; Puertas and Torres-Carrasco 2014; Pavlu 2018; Kosior-Kazberuk 2011; Dhirendra et al. 2019; Hendi et al. 2019; Kurda et al. 2019; Siddique et al. 2018; Sanytskyy et al. 2010). To ensure general-purpose cements of high quality, it is necessary to monitor constantly, activities of used mineral additives, thus, to perform such thorough control, physical and chemical methods has been used. According to (ASTM C311-02; Tkaczewska 2011; Bobrowski et al. 2015), physical methods are preferred, because they help to predict compressive strength after 7, 28 and 90 days of hardening. However, during continuous production, such time consuming

Z. Blikharskyy et al. (Eds.): CEE 2019, LNCE 47, pp. 468–473, 2020.
https://doi.org/10.1007/978-3-030-27011-7_59

methods are useless, so, it is necessary to use some quicker way to check activities of mineral additives in cement.

It is known that calcium silicate hydrate (C-S-H) is the main product of the hydration of alite and belite, as well as pozzolanic reaction, so, it has crucial role in determining properties of hardened cement paste. Depending on the time of cement hydration, quantity and type of mineral additives, structure of calcium silicate hydrate gel (C-S-H-gel) varies.

The basis of C-S-H-gel structure is a net of CaO octahedrons encircled with ionized silanol groups. Particles of C-S-H-gel have high surface charge density, according to (Nonat 2010; Kurdowski 2010), therefore, cohesive attraction between C-S-H packages is so strong that entire system exhibits rigid structure. At low concentration of calcium ions, in the paste occurs $\equiv SiOH\text{-}H_2O\text{-}HOSi\equiv$ binding group. However, with increase of CH concentration, where negative charges are balanced with Ca^{2+}, the distance between layers shortens and it results in higher density of C-S-H phase and tensile module increases tenfold.

In cement pastes without mineral additives, ratio CaO/SiO_2 equalled 1,75–1,85. Application of mineral additives reduced the ratio and lengthened the chain structure of C-S-H-gel phase thereby altering its morphology. At the same time, H_2O/CaO ratio also, affected the gel properties. At full saturation, it equalled 2.3–2.5, while after drying at 100 °C the ratio reduced to 0.8–0.9.

Assessment of the activity of mineral admixtures by determining compressive strength has not solved a number of problems related to the changes in microstructure of cement paste. Thus, it is crucial to evaluate the strength of the gel bonding in order to assess properly, the activity of mineral admixtures. Therefore, it is very important to use a method that allows to obtain information on properties of additives after short-term hardening, like 1 day, in order to shorten the time for determination of admixture's activity. After the analysis of the strength of mortar with CEM I 32.5R cement, according to (Wójkowska et al. 2016), it was found that the difference in compressive strength after 1 and 28 days of hardening was over tenfold, in bending tensile strength – sixfold, impact bending – only 1.3 times, and it means that the difference in impact bending after 1 and 28 days was minimum. Therefore, in order to determine the activity of additives, the method of impact bending strength has been proposed.

2 Materials and Research Methods

CEM I 32.5R and mineral additives were used for these research purposes. Chemical composition and specific area of additives are presented in Table 1.

Pozzolanic activity detection was carried out according to (Butt et al. 1973) chemical method. Research involved studying the amount of calcium hydroxide bonded by 1 g of additive that was added for 30 days.

For the research on impact bending strength, cement paste with W/C = 0.30 was used. To determine the strength with impact bending method, the Charpy impact test was used, with scale 0.05 kg * m, where the distance between pillars was 40 mm, and the speed of the hammer at the moment of impact was ~5.25 m/s. Dedicated research revealed that samples must be cuboids of specific size – 10 × 10 × 80 mm. Only such

Table 1. Chemical composition (%) and specific area of additives

Oxides	Ground glass		Fluidized fly-ash	Microsilica	Synthetic silica
	Soda-lime	Borosilicate			
SiO_2	72,20	81,00	59,50	96,40	93,52
$Al_2 O_3$	1,80	2,00	18,90	2,10	-
CaO+MgO	11,75	0,50	13,40	1,06	-
Na_2Oeq	13,80	4,50	0,70	1,14	0,30
Fe_2O_3	0,04	-	5,40	1,10	-
B_2O_3	-	12,00	-	-	-
Cr_2O_3	0,01	-	-	-	-
SO_3	0,40	-	1,60	-	0,40
Specific area (m^2/g)	0.32	0.31	0.43	15	173

prepared samples were ripped into two parts with surface of rupture circa about 1 cm^2. When this method measures the amount of work lost to damage the sample with cross-section of 1 cm^2, the unit of impact bending strength is J/cm^2. Results for impact strength were statistically analysed as average from 10 samples with 95% confidence.

3 Results Summary

Impact bending strength depends also, on moisture content in the sample, according to (Mielczarek et al. 2017). Water in pores of the material works as shock-absorber during impact tests. In cement paste, water may occur in capillary pores and gel pores, between layers of hydrated calcium silicate. With moisture at the level of <50% the paste properties are determined by gel surface energy. Therefore, impact tests were performed on cement pastes stored outdoor after 1, 2, 7, 14 and 28 days of hardening, on pastes stored above and under the water surface after 7, 14 and 28 days of hardening (Table 2).

Table 2. The influence of storage conditions of cement pastes on impact bending strength

Hardening (days)	Impact bending strength (J/cm^2):		
	Outdoor	Above the water	In water
1	0,115	-	-
2	0,141	-	-
7	0,142	0,147	0,162
14	0,143	0,158	0,163
28	0,135	0,145	0,145

Based on the data located in Table 2, impact bending strength of cement cuboids stored outdoor increased by 23% during 14 days of hardening, however, after 28 days it decreased to 17%. Increase in the impact strength after 7 and 14 days may be explained by increase in the amount of hydration products, while decrease after 28 days is most likely related to reduction of water amount in gel and capillary pores. 7-day storing of samples above and under the water surface resulted in increase of impact bending strength by 4 and 14%, respectively, when comparing to samples stored outdoor. After 28 days, increase in impact bending strength equalled 7 and 15%, respectively. Conducted research showed that impact bending tests, used to determine bonding forces of gel, should be performed on samples stored in air-dry conditions.

Increased impact strength of samples stored above and under the water surface may be related to water in capillary and gel pores, which played as well a shock-absorber role at high impact velocity of 5–6 m/s, according to (Mielczarek et al. 2017).

It is known that pozzolanic additives containing active silicon and aluminium oxides, due to pozzolanic reaction, increase the amount of C-S-H or C-S-A-H-gel and, at the same time, decrease ratio of CaO/SiO_2, according to (Kurdowski 2010). Gel structure in pastes with pozzolanic additives is fibrous and tight, but also, it is much brittle. In order to evaluate changes in gel structure, impact tests were performed on pastes with silicate additives in amount of 0.5 and 5% of adhesive mass and samples were stored in air-dry conditions. Pozzolanic activity of additives and impact bending tests are presented in Table 3.

Table 3. Pozzolanic activity of additives and impact bending tests of cement pastes with additives

N	Cement pastes with additives	Pozzolanic activity additives (mg CaO/1 g of additive)	Amount (%)	Impact bending strength (J/cm^2)		
				1 day	2 days	28 days
1	Soda-lime glass	36	0,5	0,119	0,143	0,136
			5,0	0,129	0,143	0,140
2	Borosilicate glass	98	0,5	0,128	0,152	0,157
			5,0	0,146	0,163	0,140
3	Fluidized fly-ash	149[a]	0,5	0,120	0,152	0,145
			5,0	0,120	0,156	0,131
4	Microsilica	216	0,5	0,130	0,152	0,135
			5,0	0,140	0,154	0,138
5	Synthetic silica	454	0,5	0,106	0,140	0,134
			5,0	0,103	0,134	0,121

[a]According to (Kotsay 2016)

Increase of the amount of mineral additives in cement paste from 0.5 up to 5% resulted in increase in impact strength after 24 h of hardening for almost all of them, excluding synthetic silica. Decrease in impact strength of pastes with synthetic silica has been linked to its high absorbency, because it resulted in lower amount of water in the paste, and, as a consequence, shock-absorbency effect was smaller during impact test. Low impact tests results for pastes with fluidized fly-ash were related to absorbent properties of ashes due to unburned carbon residues. It is worth to notice that each combination led to increase in impact strength but only after 2 days, while after 28 days, it dropped due to smaller amount of water in pores.

Comparing impact tests results with pozzolanic activity, the most significant correlation R was observed for pastes with 5% of additives. This relation is negative, so, after 24 h of hardening correlation coefficient R = −0,7, and after 28 days R = −0,9. Decrease in impact strength, when pozzolanic activity was increasing, most likely was related to the change in structure of hydrated calcium silicate by reducing CaO/SiO_2 ratio, therefore, it was brittle.

4 Summary

Based on conducted research, following conclusions can be drawn:

1. Impact strength of cement pastes depended on curing conditions. Unbound water in capillary and gel pores resulted in impact strength increase by 7–15%, depending on time of hardening. At the same time, water worked as shock-absorbent during impact bending tests.
2. Greater amount of fine-ground glass, from 0.5 to 5%, resulted in impact strength increase by 8–14%, depending on the type of glass. In case of lime-soda glass, it may be related to its high alkaline activity, while in case of borosilicate glass – with boron coordinate change.
3. Impact bending strength of cement pastes with additives depended on amount, pozzolanic activity and absorptive properties of those additives. Comparing impact tests with pozzolanic activity, the greatest correlation R was notices for pastes with 5% of additives.
4. Decrease in impact strength when pozzolanic activity of additives increased, most possibly was related to the change in C-H-S-gel structure, where, due to smaller CaO/SiO_2 ratio, greater brittleness was observed.
5. When comparing to methods that has been used currently, the impact bending test allows to asses activity of mineral additives in cement in significantly less time.

References

Shi C, Fernández-Jiménez A, Palomo A (2011) New cements for the 21st century. The pursuit of an alternative to Portland cement. Cem Concr Res 41(7):750–763

Puertas F, Torres-Carrasco M (2014) Use of glass waste as an activator in the preparation of alkali-activated slag. Mechanical strength and paste characterisation. Cem Concr Res 57:95–104

Pavlů T (2018) The utilization of recycled materials for concrete and cement production. In: FIB Conference: Sustainable Concrete: Materials and Structures IOP Conference Series: Materials Science and Engineering, vol 442, pp 1–11

Kosior-Kazberuk M (2011) Nowe Dodatki mineralne do betonu. Civ Environ Eng/Budownictwo i Inżynieria Środowiska 48(2):47–55

Dhirendra P, Tiwari RP, Shrivastava R, Yadav RK (2019) Effective utilization of waste glass powder as the substitution of cement in making paste and mortar. Constr Build Mater 199:406–415

Hendi A, Mostofinejad D, Sedaghatdoost A, Zohrabi M, Naeimi N, Tavakolinia A (2019) Mix design of the green self-consolidating concrete: Incorporating the waste glass powder. Constr Build Mater 199:369–384

Kurda R, Brito J, Silvestre JD (2019) Water absorption and electrical resistivity of concrete with recycled concrete aggregates and fly ash. Cem Concr Compos 95:169–182

Siddique I, Rauf M, Khayam SU, Alamgeer M, Faisal H (2018) Effect of tempered glass fines in concrete at elevated temperature. In: ICAET-2018, IOP Conference Series: Materials Science and Engineering, vol 414, pp 1–8

Sanytskyy M, Sobol Ch, Markiv T (2010) Modified composite cements. Lvivska Politechnika, Lviv (in Ukrainian)

ASTM C311–02 standard test methods for sampling and testing fly ash or natural pozzolans for use in Portland-cement concrete

Tkaczewska E (2011) Methods of testing pozzolanic activity of mineral additives. Materiały ceramiczne 63(3):536–541

Bobrowski A, Gawlicki M, Jagosz M, Łój G, Nocuń-Wczelik W (2015) Cement. Metody badań. Wybrane kierunki stosowania. AGH, Cracow

Nonat A (2010) C-S-H i właściwości betonu. Cement Wapno Beton 6:315–326

Kurdowski W (2010) Chemia cementu i betonu. SPC, Cracow; PWN, Warsaw (2010)

Wójkowska J, Szewczenko W, Papieżyński M (2016) Badanie udarności standardowych zapraw cementowych. Aktualne problemy naukowo - techniczne budownictwa 201–205

Butt J, Sychev M, Timashev V (1973) Workshop on chemical technology binders, Moscow

Mielczarek D, Wojkowska J, Papieżyński M, Szewczenko W (2017) Optymalizacja parametrów oznaczania udarności dla gipsu budowlanego. Problemy techniki 32–35

Kotsay G (2016) Pozzolanic activity diagnostics of fly ash for Portland cement. Ch. Ch. Techn. 10(3):354–360

Numerical Analysis of the Landslide Geohazards - Case Study with Gabions and Piles Solutions

Krzysztof Trojnar(✉)

Faculty of Civil and Environmental Engineering and Architecture, Rzeszow University of Technology, Poznańska 2 Street, 35-084 Rzeszów, Poland
ktrojnar@prz.edu.pl

Abstract. The paper describes experiences from the stabilization of landslide slope. Evaluation of the landslide stability is one of the most important problems in geotechnical engineering because failure could cause catastrophic environmental. Analysis of the landslide slope was performed using FE method. It was possible to evaluate two alternative stabilizing techniques and to choose the optimal one. Methods of stabilization consisted in the use of gabions wall and stabilizing piles, as the supports of the slope. Although the gabions wall were able to stabilize the slope, increasing the number of gabions and their weight was not effective in increasing the safety factor. The proposed pile technique was more effective. The numerical analysis provided a good prospect to optimize the design of stabilizing landslide slope. Using piles in the base of the slope with their anchoring in soil can improve the stability of the analyzed area with an increase in the stability factor to a safe value FS = 1.6. The resulting improvement in stability of slope will be sufficient in accordance with the technical requirements. The numerical simulations of the landslide showed that such analysis could be a useful tool in the design process landslide stabilization.

Keywords: Stabilization · Landslide · Gabions · Piles · Numerical analysis

1 Characteristics of the Landslide Problem

The landslide is localized in a sub-mountain area of southern Poland, near a stream runs (PIG 2012). The landslide developed in the lower part of the slope range - Fig. 1. Periodic activity was found in the south-western part, below the secondary slope and above the road. Erosion along the banks of the stream was the most probable cause of landslide activation, endangering the road crossing it (Sroczyńska 1997). Signs of massive movements were observed, both as the deterioration of the road surface and terrain deformation. In order to protect the road, and to fix the problem of stream bank strengthening, a gabions retaining wall or piles is proposed. Also ecological and landscape preservation issues were taken into account. The whole landslide has a length of about 300 m, but the length of its lowest part, protection of which is considered here, is of 110 m. The design and the possible construction variants of the stabilization (especially the final one) analyzed are described. Special attention is paid

Z. Blikharskyy et al. (Eds.): CEE 2019, LNCE 47, pp. 474–479, 2020.
https://doi.org/10.1007/978-3-030-27011-7_60

to numerical analysis, which was used to static calculations performing (in order to check stability conditions of the landslide and estimate internal forces in structural elements). The height of the slope above the road reaches 7.5 m, so it looks quite possible to built a retaining wall as a protective structure. On the basis of data from geological documentation, the land configuration, arrangement of soil layers and values of geotechnical parameters in the area of landslide were adopted. For the calculation analysis, the area of the slope was 70 m length. Averaged geotechnical layers, resulting from the geotechnical cross-section, were adopted. Figure 2 show the calculation model of the landslide area and location of the slip surface.

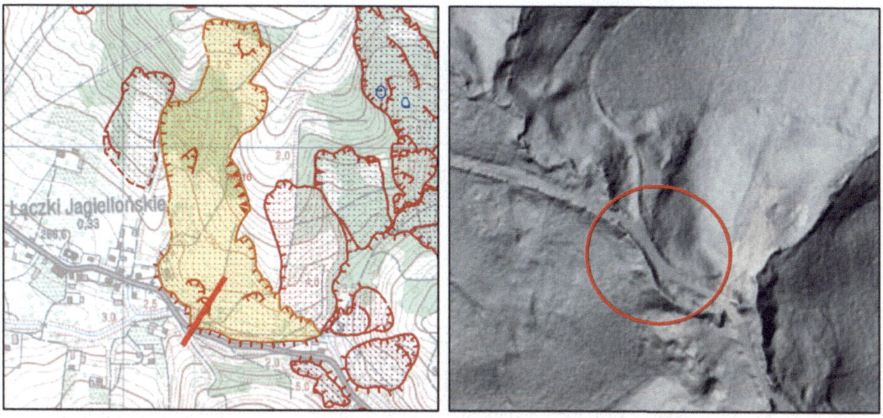

Fig. 1. Location of the landslide area

2 Numerical Simulations

Numerical simulations were performed using FEM (Cała et al. 2004). According to geotechnical data, the existing slope is built mostly from weak silty clay (cohesion c = 7–13 kPa, internal friction angle Φ = 8–13°). Soft rocks with compressive strength Rc = 0.36 MPa are found at about 4 m below the terrain level, stronger rock with Rc = 0.78 MPa is 10 m below the terrain level. Table 1 show the values of geotechnical parameters used in MES calculations. All simulations were performed in the plane strain conditions. The Coulomb – Mohr elastic-plastic model was used for the soil. The soil-water composite was treated as a single-phase media, mainly because the geotechnical documentation, only contains data allowing for this kind of approach. Two variants of protection techniques were adopted: i. the gabions retaining wall, ii. the one row of piles placed every 2 m with RC cap and soil anchors in 5 rows above piles. For gabions and piles, the elastic model of material were used. Interface elements between gabions and soil and between piles an soil were used. For modelling of the joints between gabions resistance to gap opening were used, with tensile strength of the joints of 20 kN/m (Grodecki 2017). The stability calculations were performed using the procedure of c - Φ parameters reduction (Cała et al. 2004). The effect of numerical analyses is to determine the most unfavorable location of the slip surface in the slope and to determine the minimum value of the stability factor (FS). The calculations were performed, assuming that the soil features deteriorate due to the occurrence of particularly unfavorable conditions affecting stability e.g. due to long-term moisture caused by precipitation, increase in the level of groundwater table or weakening of soil layers.

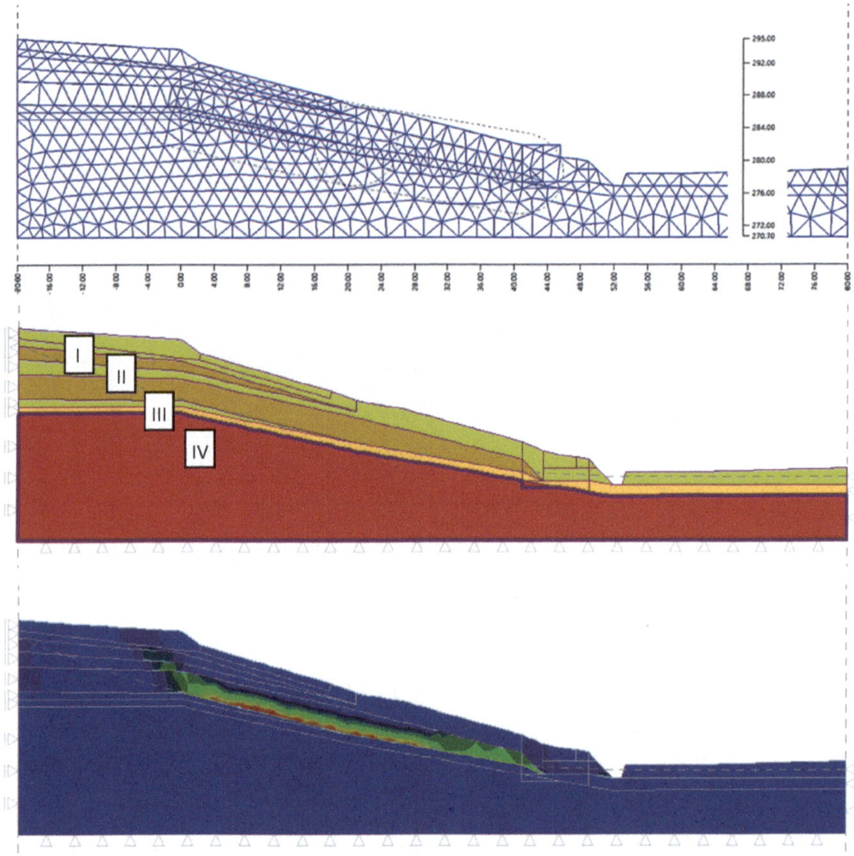

Fig. 2. Numerical model FE analysis and critical slip surface location (FS_{min} = 1.08)

First of all, a stability analysis of the existing "in situ" state of the slope was performed. The value of the stability factor SF = 1.08 obtained shows that slope is practically in the limit state of the stability. So design of the stabilizing structure was really necessary. The location of the slip surface on the basis of calculations corresponds to the real area of the landslide in the actual terrain. This state can be considered as a verification of the applied calculation model (Matsui and San 1992). The zone of the highest soil plasticization occurs in the central part of the landslide, on the length of about 12 m, in the layer II (clay). The minimum values of internal friction angle and cohesion for clay determined in the MES calculations confirm the formation of the slip surface for the values of geotechnical parameters: $\phi = 8.4°$, c = 6.9 kPa (Table 1). This corresponds approximately to the values of the internal friction angle $\phi = 8.9°$, cohesion c = 6.9 kPa and consistency from laboratory tests in the direct shear apparatus for dusty clay samples taken from the borehole R1 (Dokumentacja 2014) from the depth corresponding to the position of the slip zone - (Figs. 2 and 3).

Table 1. Geotechnical parameters

Soil layer	FEM stability analysis with c-ϕ reduction					Laboratory test		SF_{min}
	γ [kN/m3]	ν [-]	E [MPa]	Φ_{min} [°]	c_{min} [kPa]	Φ [st]	c [kPa]	
I Silt	20.0	0.30	15	13.2	13.2	14.8	16.9	1.08
II Sensitive clay	19.50	0.30	8	8.4	6.9	8.9	6.9	
III Soft rock	21.0	0.30	8	14.8	16.9	16	9.5	
IV Rock	23.0	0.25	1000	10	100	-	-	

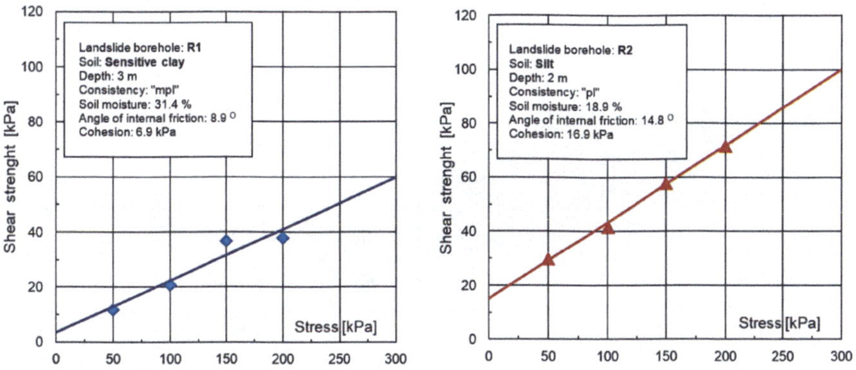

Fig. 3. Results of laboratory strength tests of silt and clay

3 Methods of Stabilizing the Landslide

The variant of the gabions retaining wall consists three blocks levels of 1.5 m high (Grodecki 2017). Numerical simulation shows that such a construction does not improve the stability of the landslide area (SF = 1.3). In order to obtain a satisfactory value of SF > 1.50, piles use in one row was analyzed. Figure 4 show the CFA piles 0.5 m in diameter with Titan 40/20 type soil anchors solution (Ashour and Ardalan 2012). An anchors length is 12 m (including 5 m active zone), distance between anchors is 3 m, prestressing N = 20 MPa was applied. Five anchor level was proposed and an SF = 1.60 was achieved. So anchoring of the pile wall leads to stability of the landslide and the calculation result obtained were satisfactory. This variant was chosen as the optimal and selected for realization (Fig. 5).

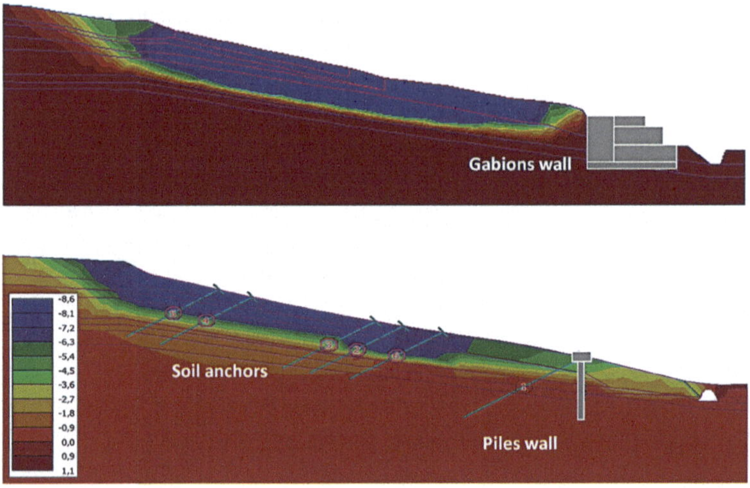

Fig. 4. Displacement zones in the stabilized slope of the landslide; a. Gabions wall (FS = 1.3), b. Piles wall with anchors (FS = 1.6).

Fig. 5. Realization of the landslide stability solution with piles and anchors.

4 Conclusion

Numerical modeling is an important tools to estimate slope stability. Numerical calculation showed positive application of a FEM for evaluation of stability of the Łączki Jagiellońskie landslide. The slide surfaces, which were indentified in borehole R1, were used to compute factors of safety (SF). Calculations of the unprotected landslide confirmed the loss of stability and showed the location of the slip surface within the sensitive clay (layer II). The data from the geotechnical investigation were used to verify the numerical model of the landslide. On the FEM calculations the value of stability (safety) coefficient for the existing landslide SF = 1.08 was obtained. This results in the danger of further landslides in the area of the damaged road, if there are no safety works. Calculations of slope stability for the landslide were carried out for two scenarios. First with gabions wall, second scenario with piles wall and soil anchors. As the result for each scenario the factors of safety were estimated. It was found that stabilization of the landslide using piles in one row in the base of the slope with their anchoring in soil can improve the stability of the analyzed area with an increase in the stability factor to a safe value of SF = 1.6. The resulting improvement in stability will be sufficient and in accordance with the technical requirements (SF > 1.5). The alternative numerical simulations of the landslide show that such analyses could be a useful tool in the design process. It was possible to evaluation two variants of the stabilizing structure and to choose the optimal one. Based on the assessment it can be concluded that the method adopted and the extent to which the landslide is protected with piles in rows was better than gabions.

References

Ashour M, Ardalan H (2012) Analysis of pile stabilized slopes based on soil–pile interaction. Comput Geotech 39:85–97

Cała M, Flisiak J, Tajdus A (2004) Numeryczne metody analizy stateczności skarp i zboczy. Warsztaty górnicze 7:37–50

Grodecki M (2017) Numerical modelling of gabion joints. Tech Trans 2:83–89

Matsui T, San K-C (1992) Finite element slope stability analysis by shear strength reduction technique. Soils Found 32:59–70

Sroczyńska A (1997) Osuwiska karpackie zagrożenie dla dróg. Drogownictwo 2:62–66

Dokumentacja (2014) geologiczno-inżynierska osuwiska w c.d. nr 115170R w m. Łączki Jagiellońskie, pow. krośnieński (in polish)

PIG (2012) Karta dokumentacyjna osuwiska nr 1807092B1, Państwowy Instytut Geologiczny

Application Features for the Activator of Mineral Materials "RENA-Aquador" for Asphalt Concrete Pavings

Serhiy Y. Solodkyy[1(✉)], Zinovii M. Ilnytskyi[2], Ihor P. Poliuzhyn[3], and Fedir I. Tsiupko[3]

[1] Department of Roads and Bridges, Institute of Building and Environmental Engineering, National University "Lviv Polytechnic", Karpinskoho Street 6, Lviv 79013, Ukraine
s.solodkyy@ukr.net
[2] Scientific and Production Company "Halychyna" Ltd., Stryiska Street 443, Drohobych 82106, Ukraine
[3] Department of Physical, Analytical and General Chemistry, Institute of Chemistry and Chemical Technologies, National University "Lviv Polytechnic", Sviatoho Yura Square 9, Lviv 79013, Ukraine

Abstract. The use of activated mineral powder (mineral filler) obtained with the aid of the RENA-Aquador activator for mineral materials reduces the consumption of bitumen by 15% while simultaneously improving the physical and mechanical properties of asphalt concrete pavings. This indicates the high efficiency of activating the mineral filler by means of this activator.

Keywords: Activator for mineral powder · Mineral filler · Asphalt · Bitumen · Filler - binder interaction · Road pavements

1 Introduction

Mineral filler aggregates play an important role in asphalt mixtures because they fill voids in paving mix and improve the cohesion of asphalt binder. Limestone powder containing over 90% of $CaCO_3$ is the most frequently used type of filler. According to standard DSTU B.V.2.7-121-2003, the main purpose of mineral fillers (MF) for asphalt mixtures is the formation of a disperse system together with organic binder (bitumen), moving from a bulk state to a film. This disperse system acts as asphalt-binding agent in asphalt concrete pavements. The list of products used for the activation of mineral filler include: (1) anionic surfactants such as of higher carboxylic acids, for example gossypol resin (cotton tarpaulin), fat tar, synthetic fatty acids C_{17}-C_{20}, refining residues of synthetic fatty acids, oxidized petrolatum); (2) cationic surfactants of the amine type for example higher aliphatic amines or their salts, quaternary ammonium salts of organic compounds); (3) coal tar resin; (4) viscous road petroleum bitumen; (5) plasticizers such as oil fuel and diesel fuel.

It is known (Gezencvej et al. 1985) that MF particles act as carriers for thin orientated layers of bitumen. Since MF has a high specific surface area compared to

Z. Blikharskyy et al. (Eds.): CEE 2019, LNCE 47, pp. 480–487, 2020.
https://doi.org/10.1007/978-3-030-27011-7_61

other mineral constituents, it has a significant effect on the properties of asphalt concrete. The activation technology for MF involves the creation on the beads of a primary contact layer of highly structured bitumen, which changes the properties of MF. In order to improve the uniformity of bitumen distribution, increase the efficiency of crushing and MF activation, its processing with bitumen is carried out in the presence of appropriate surfactants.

The effect of activated MF on the performance of asphalt-concrete pavements, which determines their durability, is carried out in such directions (Gezencvej et al., 1985): (1) strengthening of the structured dispersion system "bitumen-MF"; (2) increase of density and decrease of water permeability of asphalt concrete; (3) spontaneous liberation of aging processes for asphalt concrete; (4) increase asphalt concrete resistance against water and frost. According to one of the technologies developed before 1985, to activate the MF surface during the milling process bitumen and surfactant in a mass ratio of 1:1 are included in the activation mixture. These activation mixtures were in the amount of 1,5–2,5% relatively the MF mass. (Gezencvej et al. 1985).

Amines, poly-amines, and amido-amine adhesion promoters (AMAPs) represent one of the most common classes of organic additives, which are basic compounds (Rossi et al. 2017). The optimum adhesion power is the result of a compromise between the length of the nonpolar hydrocarbon chain and the number of amine polar groups. The best performance can be reached with 14–18 carbon chains, linked to one or even two amine groups, with at least one primary amine group.

Other compounds such as naphthenates, metal salts of organic acids obtained from the alkali washes of petroleum fractions, have also been demonstrated to play the role of adhesion promoters (Rossi et al. 2017). Organosilanes-based Adhesion Promoters (OSAPs) adhesion promoters with the general formula $R_1Si(OR)_3$ have hydrocarbon chain with affinity for bitumen and a polar silane end group with affinity for inorganic surface (Rossi et al. 2016). Polymer materials, including styrene butadiene styrene copolymer (SBS), have also been shown to be effective as anti-stripping additives for bitumen (Mirzaiyan et al. 2019). The bitumen modified with SBS in conjunction with polyphosphoric acid (PPA) positively affected wettability performance and increased binder stability during long-time storage at elevated temperatures (Xiao et al. 2014). The hardening effect played by PPA has also been compared to bitumen containing different commercial types of waxes (Edwards et al. 2006).

The comparison between amido-amine adhesion promoters (AMAPs), organosilanes-based adhesion promoters (OSAPs), and Styrene Butadiene Styrene copolymer (SBS) has recently been reported by (Cui et al. 2014) where a fracture mechanics approach was applied to quantify the adhesion between the modified bitumens and the aggregates in asphalt mixtures. The best result was achieved for modification with OSAPs.

A class of compounds that contain both basic (amine) and acid (phosphate) polar residues can be represented by natural phospholipids such as, for example, phosphatidyl choline, phosphatidyl inositol, phosphatidyl ethanolamine, and phosphatidic acid (Liu and Ma 2011).

Since the use of activated MFs reduces the bitumen amount for the asphalt concrete preparation, therefore the search and testing of new activated MFs are relevant studies in this area. The research results are presented in (Solodkyi et al. 2017) on the effectiveness for products from Scientific & Production Company "Halychyna" Ltd. for the manufacture of activated mineral powder. According to (Solodkyi et al. 2017) the use of activator for MF such as "RENA-CEMERIN" grade PM can reduce the amount of bitumen by 10–15% in preparation of asphalt concrete when activated MF is obtained at the activator amount as 1% of the limestone mass.

The aim of this work is to evaluate the effectiveness of new industrial mineral powder activators for asphalt concrete mixtures on the basis of researching the properties of activated mineral powder and such physical and mechanical properties as water saturation, residual porosity and compression strength for asphalt concrete pavements obtained by using this activated mineral powder as MF.

2 Experimental

The objects of the study were two industrial prototypes of new mineral filler activators which are produced by Scientific & Production Company "Halychyna" Ltd. (http://rena.ua/about_us): (1) prototype RENA-PM and (2) prototype RENA-PT.

For mineral fillers obtaining the limestone was used from production by "Mineral" Ltd (Novyi Rozdil).

The bitumen 70/100 (Open Joint Stock Company "Mozyr Oil Refinery", Republic of Belarus) was used for the preparation of laboratory samples of asphalt concrete mixtures. This bitumen is close for the properties to distillation bitumen as brand DB 60/90 in accordance to standard DSTU 4044-2001.

Experimental conditions are described in (Solodkyi et al. 2017). The asphalt concrete testing was carried out in accordance to the requirements of standard DSTU B V.2.7-89-99 for the following parameters: (1) average density; (2) water saturation; (3) residual porosity; (4) compressive strength limits at temperatures of 20 °C and 50 °C.

3 Results and Discussion

The dependence of the MF specific surface area on the amount is shown in Fig. 1. For both samples of activator prototypes it is evident (Fig. 1) that when the activator amount increases, then the MF specific surface area decreases significantly after the activator content of 0.5%. On the interval of 0.5–1.5% for activator amount this decrease is almost linear with an average increment of 2800–4800 cm^2/g on 1% for activator amount. This decrease in the specific surface area of the limestone MF can be explained by certain aggregation of particles, the surface of which is covered with organic matter.

With an increase in the amount of MF activator, the bitumen capacity of the MF decreases, as shown in Fig. 2. For identical amounts of the activator, there is always a greater decrease in bitumen capacity for the "RENA-PM" sample. The maximum of this difference (6 g) relates to 0.5% of activator and decreases to one gram of industrial oil at activator amount as 1.5%. An increase in the activator amount from 1% to 1.5%

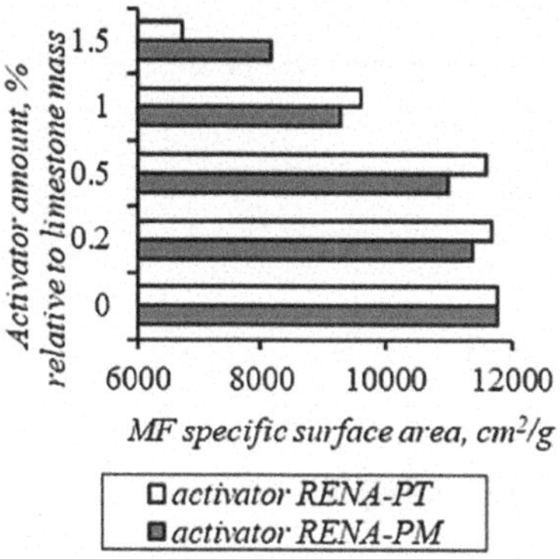

Fig. 1. Dependence of the specific surface area for the mineral filler (MF) on the activator amount (grinding time is 3 h). Specific surface area is air permeability specific surface according to Lea and Nurse method)

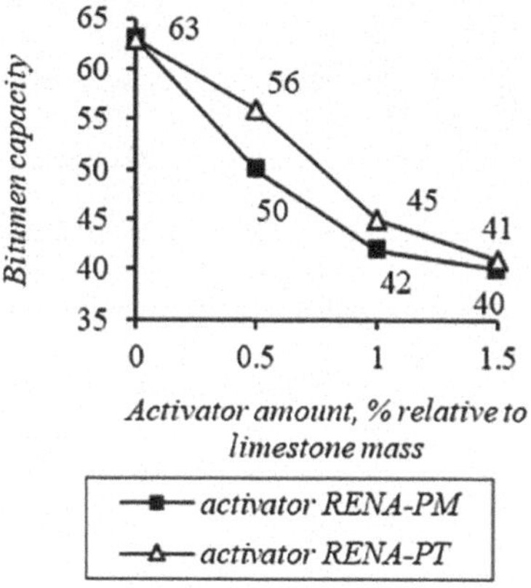

Fig. 2. Dependence of MF bitumen capacity on the activator amount (grinding time is 3 h). Bitumen capacity is expressed in grams of petroleum oil per 100 cm^3 of MF

reduces the bitumen capacity for RENA-PM activator by only 2 g. Therefore, in order to investigate the effectiveness of activated MF in asphalt concrete mixtures, it is expedient to use RENA-PM as MF activator in the amount of 1.0% relative to the limestone mass. No MF samples using RENA-PT as activator failed to pass the hydrophobic test. Reducing the amount of RENA-PM as MF activator below 0.5% resulted in loss of hydrophobic MF.

The efficiency of activated MF was determined in sandy asphalt concrete mixtures as type G and fine-grained as type A, which were made in the laboratory of the Department of roads and bridges of the National University "Lviv Polytechnic" and the composition of which is given in Fig. 3. The first series asphalt concrete mixtures contains non-active limestone mineral powder, and on the other side for the prepared by RENA-PM as the activator with the amount of 1% relative to the limestone weight. For the batch series with numbers 2, 3 and 4, the bitumen content was less by 10%, 15% and 20%, respectively, than the bitumen content for the series number 1.

It was found that when decreasing of bitumen content in asphalt concrete mixture from 7 to 6% for type G and from 6 to 5% for type A, then the asphalt concrete mixture density increases almost linearly, since the density of bitumen (0.95–1.5 g/cm^3) is less than the density of the mineral part. However, if the bitumen content in the series number 4 is less than the content for the series number 1 by 20%, then a sharp decrease in the asphalt concrete mixture density occurs with a decrease in the content of bitumen in asphalt concrete mixture from 6 to 5.6% for type G and from 5.1 to 4.8% for type A (Fig. 4). The asphalt concrete mixture porosity decreases almost linearly with decreasing bitumen amount. For porosity a phenomenon similar to density change is observed, that is, if the bitumen content in the series number 4 is less than the content

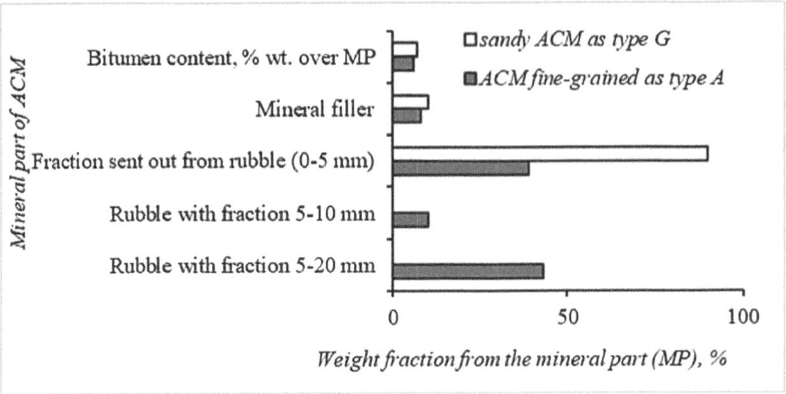

Fig. 3. The composition of the mineral part (MP) and the bitumen content in the laboratory samples of the asphalt concrete mixtures (ACM) of the first series for the investigation of performance of such activator as RENA-PM

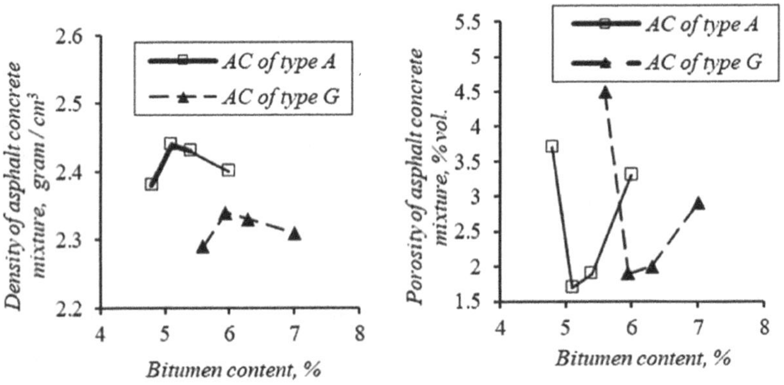

Fig. 4. Dependence of density and porosity for asphalt concrete (AC) mixture on various bitumen contents (%). Bitumen content is bitumen content in asphalt concrete mixture, % wt. over the mineral part

for the series number 1 by 20%, then a sharp increase was obtained in the asphalt concrete mixture porosity with a decrease for the bitumen content in the asphalt concrete mixture (Fig. 4).

It is obvious that with increasing asphalt concrete mixture porosity then its water saturation will increase. There are linear correlations between the porosity and water saturation as well as the porosity and density of asphalt concrete mixture that is shown in Fig. 5.

According to Fig. 6, at a temperature of 20 °C, when the bitumen amount is reduced, but with the activated mineral filler there is a noticeable increase in the compressive strength limits for both types of asphalt concrete pavement. At the same time, for all prepared series of asphalt concrete mixtures (ACMs), the compressive

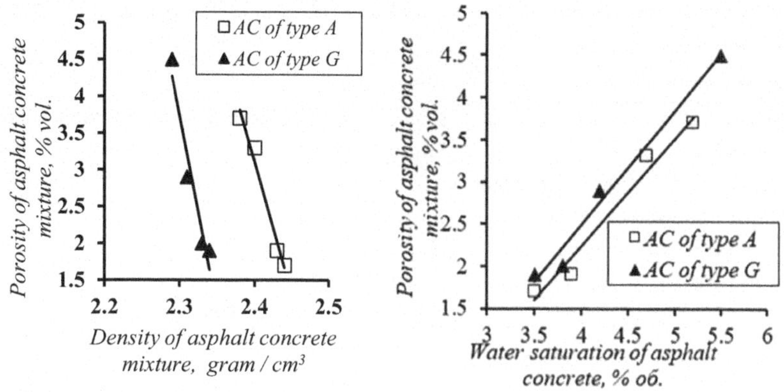

Fig. 5. The relationship between water saturation and porosity as well as the porosity and density of asphalt concrete (AC) for various bitumen contents

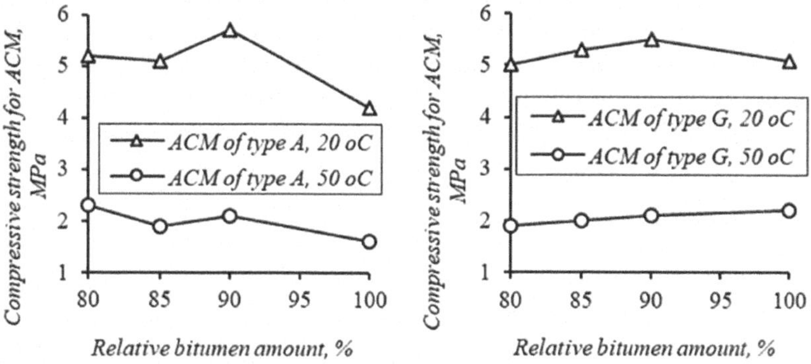

Fig. 6. Compressive strength (MPa) for asphalt concrete mixtures (ACM) types A and G with various bitumen amounts expressed as relative bitumen amount (%). Relative bitumen amount (%) is relation of bitumen amount to bitumen weight for ACM series №1 (100%) where nonactivated limestone mineral filler was used, %

strength limit of ACM is 1.6–2.1 times greater than the standard value of 2.7 MPa at a temperature of 20 °C. With increasing temperature the strength limit decreases. For temperatures of 50 °C for all prepared series of asphalt concrete pavings, the compression strength limit is 1.1–1.6 times larger than the maximum from two values in regulatory value of 1.4 MP for asphalt concrete of type A (1.2 MPa) and of type G according to standard DSTU B V.2.7-119:2011.

Consequently, for the investigated series of the asphalt concrete mixtures with activated mineral filler, the main factors are the water absorption and porosity of ACM taking into account limiting the decrease in the bitumen amount by no more than 15%.

According to the results of experimental research for the selected best sample among two prototypes of the mineral filler activators the technical specifications

Table 1. Characteristics of the activator for mineral materials "RENA-Aquador" (http://rena.ua/dobavky.html)

Characteristic	Norm
Appearance	Liquid with color from light brown to dark brown
Density at 20 °C, g/cm^3, within	From 0.9 to 1.1
Mass fraction of moisture and volatile substances, %, not more than	4.0
Warranty period of storage	12 months from the date of manufacture
Safety of toxicity	Grade 4 for hazard (small dangerous substances)

were developed for the industrial product from Scientific & Production Company "Halychyna" Ltd. (Drohobych city) with the product name "RENA-Aquador" as an activator of mineral materials. Table 1 gives the basic parameters of the product "RENA-Aquador". The activator of mineral materials "RENA-Aquador" is produced according to the technical specifications as TU U 20.4-30084964-018:2018 and is used to intensify the grinding process and modification of mineral materials, for example limestone. The dosage of the activator is individually selected for each type of raw material and technological conditions. For using the activator amount is determined in the range from 0.1% to 2.5% relatively to the mass of mineral materials.

4 Conclusions

The assessment of the activator "RENA- Aquador" use in the industrial conditions for limestone milling in the mineral filler production, as well as the use of mineral filler (MF) for asphalt concrete pavement when performing road repair works, shows that this mineral materials activator effectively intensifies the grinding process and modifies the mineral filler giving hydrophobic properties for MF. The use of activated MF allows to reduce the consumption of bitumen by 15% with a one-time improvement of the physical and mechanical properties of asphalt concrete pavement, and that indicates the high efficiency of activation. Reducing the bitumen amount by 20% leads to a decrease in the quality of asphalt concrete, and that indicates a lack of bulk bitumen quantity.

References

Aman MY, Shahadan Z, Noh MZ (2014) A Comparative study of anti-stripping additives in porous asphalt mixtures. Jurnal Teknologi 70:139–145

Cui S, Blackman BR, Kinloch AJ, Taylor AC (2014) Durability of asphalt mixtures: Effect of aggregate type and adhesion promoters. Int J Adhes Adhes 54:100–111

Edwards Y, Tasdemir Y, Isacsson U (2006) Rheological effects of commercial waxes and polyphosphoric acid in bitumen 160/220—low temperature performance. Fuel 85(7–8):989–997

Gezencvej LB, Gorelyshev NV, Boguslavskij AM, Korolev IV (1985) Dorozhnyj asfal'tobeton, Gezencvej LB (ed) Moskva, USSR: «Transport»

Goh SW, Hasan MR, You Z (2013) Performances evaluation of Cecabase® RT in warm mix asphalt technology. Procedia Soc Behav Sci 96:2782–2790

Haritonovs V, Tihonovs J, Smirnovs J (2016) High modulus asphalt concrete with dolomite aggregates. Transpo Res Procedia 14:3485–3492

Kakar MR, Hamzah MO, Akhtar MN, Woodward D (2016) Surface free energy and moisture susceptibility evaluation of asphalt binders modified with surfactant-based chemical additive. J Clean Prod 112:2342–2353

Liu D, Ma F (2011) Soybean phospholipids (Chap 22). In: Krezhova D (ed) Recent trends for enhancing the diversity and quality of soybean products, pp 483–500

Mirzaiyan D, Ameri M, Amini A, Sabouri M, Norouzi A (2019) Evaluation of the performance and temperature susceptibility of gilsonite- and SBS-modified asphalt binders. Constr Build Mater 207:679–692

Rossi CO, Caputo P, Baldino N, Lupi FR, Miriello D, Angelico R (2016) Effects of adhesion promoters on the contact angle of bitumen-aggregate interface. Int J Adhes Adhes 70:297–303

Rossi CO, Teltayev B, Angelico R (2017) Adhesion promoters in bituminous road materials: a review. Appl Sci 7(5):524

Solodkyi SY, Hrymak OYa, Sidun YuV, Ilnytskyi ZM (2017) Doslidzhennia efektyvnosti produktsii PAT « NVK-HALYChYNA » dlia vyhotovlennia aktyvovanoho mineralnoho poroshku. Visnyk Kharkivskoho natsionalnoho avtomobilno-dorozhnoho universytetu 79:147–150

Xiao F, Amirkhanian S, Wang H, Hao P (2014) Rheological property investigations for polymer and polyphosphoric acid modified asphalt binders at high temperatures. Constr Build Mater 64:316–323

Strength and Fracture Toughness of Cement Concrete, Dispersedly Reinforced by Combination of Polypropylene Fibers of Two Types

Yurii Turba[1], Sergii Solodkyy[1(✉)], and Taras Markiv[2]

[1] Department of Highways and Bridges, Lviv Polytechnic National University, Bandera Street 12, Lviv 79000, Ukraine
s.solodkyy@ukr.net
[2] Department of Building Production, Lviv Polytechnic National University, Bandera Street 12, Lviv 79000, Ukraine

Abstract. The results of the study of the impact of the reinforcement of cement concrete with polypropylene fiber of various types on the strength and crack-resistance characteristics are presented in the article. Consumption of the polypropylene fiber of two types (ordinary and deformed) was taken as the factors. The methods and criteria of fracture mechanics, which are based on the fracture toughness and fracture energy, allowed to carry out a quantitative and qualitative assessment of the effectiveness of addition of two types of polypropylene fiber. The optimum ratio of contents of two types of polypropylene fibers is determined, at which the maximum values of strength and crack resistance of concrete are reached.

Keywords: Fiber reinforced concrete · Fiber · Crack · Fracture mechanics

1 Introduction

Concrete is widely used and the most important construction material all over the world (Krainskyi et al. 2018, Bobalo et al. 2018). However, its relatively low tensile strength, resistance to cracking, and brittle character of destruction limit its use under high strain-rate loadings (Brandt 2008, Khmil et al. 2018).

According to Xu et al. (2012), the addition of fibers to concrete prevents effectively the creation and propagation of cracks in it by gaining bridging effect, as well as increase the toughness under high strain-rate loading. Different types of fibers such as short discrete steel, glass, polyester, and polypropylene fibers can be used to reinforce concrete. Diameter, specific gravity, Young's modulus, tensile strength of used fibers influence on the properties of concrete on the level of meso- and macrostructure (Marushchak et al. 2018; Solodkyy and Turba 2015a, 2015b). Polypropylene fiber is attributed to flexible fibers. They improve the crack resistance of concrete due to their high ductility (Qiang et al. 2018). As a result, the advantages of concrete incorporating

Z. Blikharskyy et al. (Eds.): CEE 2019, LNCE 47, pp. 488–494, 2020.
https://doi.org/10.1007/978-3-030-27011-7_62

polypropylene fiber are its homogenous tensile properties and better micro cracking behavior under load.

The mechanical properties of polypropylene fiber reinforced concrete (PFRC) have been studying for last several decades. The influence of polypropylene fiber on the compressive strength of concrete has been studied by some researchers and they showed that its effect on the compressive strength of concrete is negligible in some cases. Vairagade et al. (2012) reported that polypropylene fiber in small volume fraction (0.25%) has no or very small effect on the compressive strength of fiber reinforced concrete. Ne'ma et al. (2011) have shown a significant increase in the compressive strength of fiber reinforced concrete. They noted that the addition of polypropylene fiber in the amount of 0.4, 0.8, 1.0, and 1.5% increased significantly the compressive strength up to 64% and flexural strength up to 55.5%. Mazaheripour et al. (2011); Kakooei et al. (2012); Karahan and Atis (2011) concluded also that, when polypropylene fibers are added into concrete, the tensile and flexure strength, as well as durability of concrete are significantly improved.

The aim of this study is to determine the strength and fracture toughness properties of concrete reinforced by combination of polypropylene fibers of two types (ordinary and fiber with deformed shape).

2 Experimental Program

2.1 Raw Materials

Portland-slag cement CEM II/A-S 42.5 as cementitious materials, granite gravel (5.0–20.0 mm) as coarse aggregate, quartz sand with fineness modulus of 1.29 and crushed granite sand (1.25–5.0 mm) as fine aggregates were used for this program.

W/C ratio was 0.44, consistence class of concrete mixture was S1 (Slump = 20–30 mm). Magnesium lignosulphonate based plasticizer was used in the concrete mixtures to obtain the required slump value.

Ordinary polypropylene (PP) fiber and polypropylene fiber with deformed shape were used to improve deformation characteristics of concrete.

Characteristics of the ordinary fiber: length is 12 mm, diameter is 18–20 μm and consumption is 0.7–1.4 kg/m^3.

Polypropylene fiber with deformed shape were used to improve the polymer-concrete bond, because it is well known, that mechanical bond of ordinary polypropylene (PP) fiber with the cement matrix is rather low.

Characteristics of the deformed fiber: length is 45 mm, cross section is 1.0 × 0.5 mm, consumption is 8.0–12.0 kg/m^3. The wavy form provides better adhesion to concrete.

PP fiber have been carefully scattered to provide a uniform distribution in the mixture.

The mixture proportions of concretes are described in Table 1.

Table 1. Concrete mixture proportions

Mixture identification	Content of ingredients (kg/m^3)						
	Cement (CEM II/A-S 42.5)	Aggregates			Plasticizer	Fiber	
		Fine		Coarse		Ordinary	With deformed shape
		Quartz sand	Crushed granite sand				
1	350	435	255	1247	2.45	1.4	12
2	350	435	255	1247	2.45	0.7	12
3	350	435	255	1247	2.45	1.4	8
4	350	435	255	1247	2.45	0.7	8
5	350	435	255	1247	2.45	1.05	12
6	350	435	255	1247	2.45	1.05	8
7	350	435	255	1247	2.45	1.4	10
8	350	435	255	1247	2.45	0.7	10
9	350	435	255	1247	2.45	1.05	10
Reference	350	435	255	1247	2.45	0	0

2.2 Experimental Process

The specimens have been cured for 3, 7 and 28 days in the ambient climatic conditions such as temperature of 20 ± 2 °C and relative humidity of 95 ± 5%. The specimens were taken out from the curing chamber after 3, 7 and 28 days.

Samples have been tested according to three-point bend scheme using 200-ton hydraulic press. Arrangement of flexural tests with crack opening mode for the prismoidal specimen is given in Fig. 1.

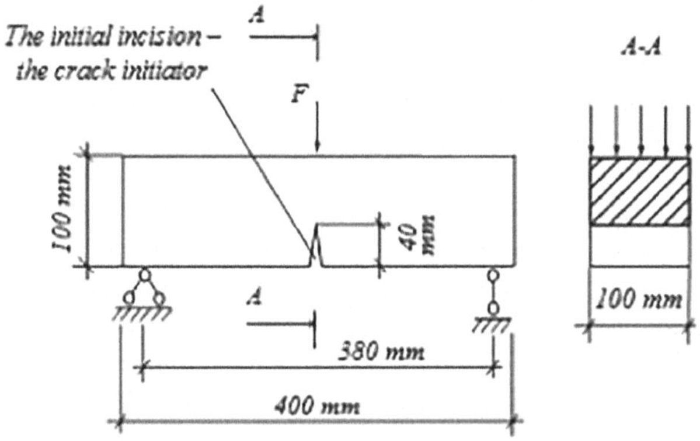

Fig. 1. Arrangement of flexural tests with crack opening mode for the prismoidal specimen

Samples with size of 100 × 100 × 400 mm have been used for testing with initial notch of 40 mm in height and 2 mm in width. Initial notch have been formed, using diamond saw. Samples have been loaded continuously until their fracture with the recording of load-deflection curves by means of computer-aided data acquisition system. An estimate of fracture energy was obtained from the L–d diagrams according to the RILEM method (work-of-fracture). All series consisted of four samples. The size of cube specimens for determination of compressive strength was 100 × 100 × 100 mm.

3 Results and Discussion

The impact of the complex combination of ordinary and deformed fiber on the strength of concrete was studied. The results of testing of fiber reinforced concrete at the age of 3, 7 and 28 days are shown in Fig. 2. Concrete with no reinforcement (base series) was tested at the same time in order to compare the effectiveness of hybrid dispersed reinforcement.

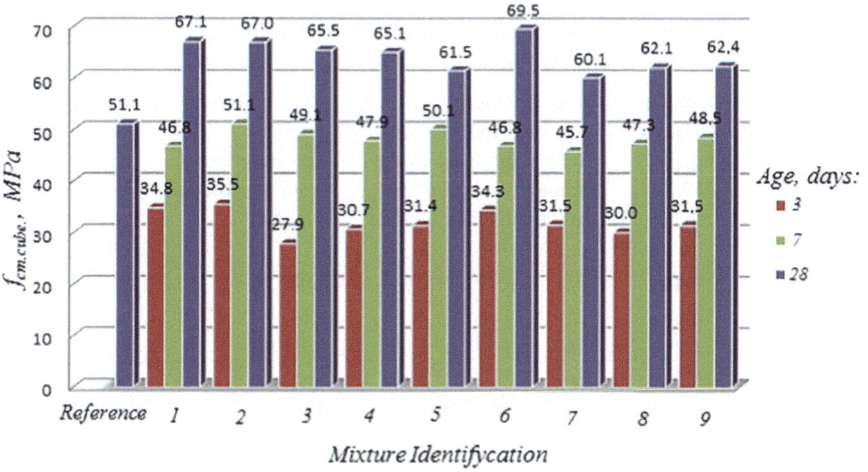

Fig. 2. Compressive strength of concretes

Dispersed reinforcement provided increase in the compressive strength within 9–18 MPa after 28 days regarding the base unreinforced series. The undeniable benefits of a particular fiber combination were not found. At the same time, the compressive strength of concrete that is reinforced by various types of fibers differs insignificantly and ranges within 4–16%.

According to the obtained results, the L-d diagrams were created (Fig. 3), the energy and load bearing characteristics of crack resistance of dispersed reinforced concretes (Table 2) were calculated together with the flexurol tensile strength.

The configuration of L-d diagrams allows to carry out the qualitatively assessment of the impact of the hybrid dispersed reinforcement of cement concretes with two types

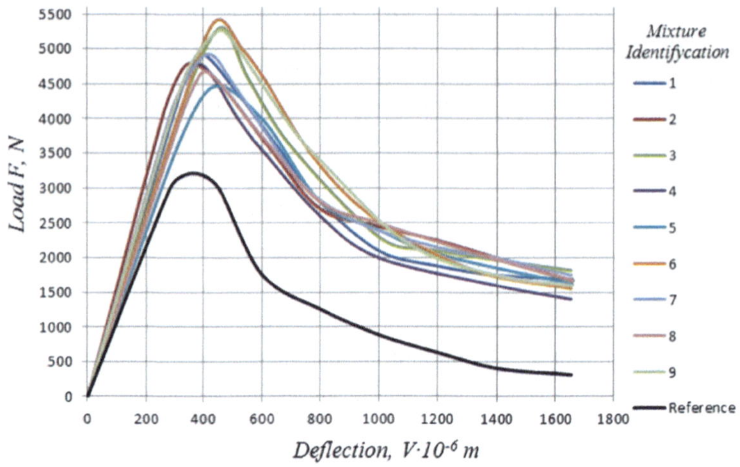

Fig. 3. Load – deflection diagrams of concrete

Table 2. Physical and mechanical characteristics, fracture and energy characteristics of the hybrid fiber reinforced concrete with various types of polypropylene fiber

Mixture identification	Flexural strength, f_{ctfm}, MPa	Specific energy consumption of static deformations up to the moment of movement of a main crack G_i, J/m^2	Specific effective energy consumption of static destruction G_F, J/m^2	Stress intensity factor (fracture toughness) K_i, MPa$^{-1/2}$
1	10.44	171	698	0.80
2	10.15	166	714	0.85
3	11.21	219	728	0.88
4	10.08	183	631	0.85
5	9.43	177	672	0.76
6	11.40	207	760	0.86
7	10.42	199	696	0.89
8	9.87	162	715	0.75
9	11.12	225	727	0.93
Reference	6.79	112	327	0.57

of polypropylene fibers and with various consumption thereof. The reinforced concrete samples resist internal stress caused by external load more effectively than the unreinforced samples. Such concretes begin to fracture at critical loads that are 39–68% higher than for unreinforced concretes. After macrocracks are formed (actual beginning of the fracture of concrete), reinforced concretes continue to slow down complete fracture of the samples due to the used polypropylene fiber. At this moment the unreinforced samples are being defragmented until splitting into pieces.

It is arguable that the efficiency of hybrid dispersed reinforcement according to energy fracture resistance criteria is high on the indicator of areas of L-d diagrams of studied concretes, which complies with the overall fracture energy consumption.

Concrete series No. 3, 6, and 9 demonstrate the best results. The highest rates of crack resistance and strength are correlated with each other in these concrete series.

According to the obtained results, the response surfaces of the studied criteria were constructed as per combinations of consumption of various types of fiber (Fig. 4).

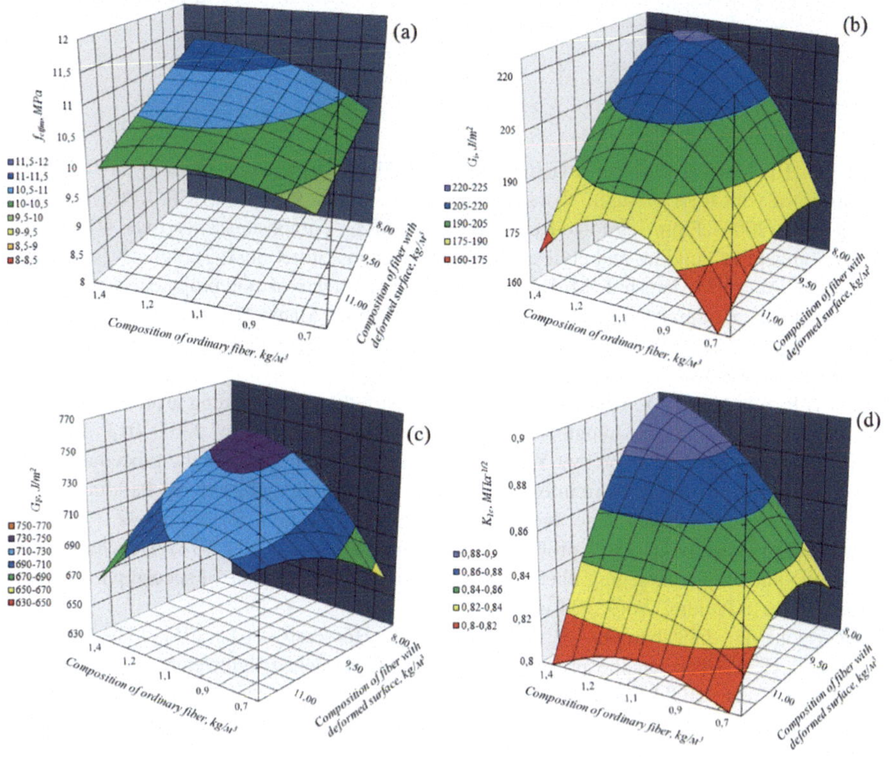

Fig. 4. The response surfaces of the studied characteristics of fiber reinforced concrete with various types of polypropylene fibers: the flexural tensile strength, f_{ctfm}, MPa (a); the specific energy consumption of static deformations up to the moment of movement of a main crack G_i, J/m^2 (b); the specific effective energy consumption of static destruction G_F, J/m^2 (c); the stress intensity factor (fracture toughness) K_i, MPa$^{-1/2}$ (d)

The obtained results show that when various types of fiber are applied in hybrid combination, their optimal composition is 1.0–1.2 kg/m^3 of ordinary fiber and 8–9 kg/m^3 of deformed fiber. This optimal composition allows to reach maximum studied characteristics.

4 Conclusions

1. Dispersed reinforcement of cement concrete with a combination of polypropylene ordinary and deformed fiber increases the indicator of compressive strength by 9–18 MPa and the flexural strength by 2.6–4.6 MPa.
2. Dispersed reinforcement causes a significant increase in the fracture resistance of concrete according to the load bearing and energy criteria of fracture mechanics.
3. The maximum values of strength and fracture characteristics are achieved when ordinary fiber (1.05 kg/m^3) and deformed fiber (8 kg/m^3) are applied together.

References

Bobalo T, Blikharskyy Y, Vashkevich R, Volynets M (2018) Bearing capacity of RC beams reinforced with high strength rebars and steel plate. Matec Web Conf 230:02003

Brandt AM (2008) Fibre reinforced cement-based (FRC) composites after over 40 years of development in building and civil engineering. Compos Struct 86(1–3):3–9

Kakooei S, Akil HM, Jamshidi M, Rouhi J (2012) The effects of polypropylene fibers on the properties of reinforced concrete structures. Constr Build Mater 27(1):73–77

Karahan O, Atis CD (2011) The durability properties of polypropylene fiber reinforced fly ash concrete. Mater Des 32(2):1044–1049

Krainskyi P, Blikharskyy Y, Khmil R, Vegera P (2018) Influence of loading level on the bearing capacity of RC columns strengthened by jacketing. Matec Web Conf 230:02013

Khmil R, Tytarenko R, Blikharskyy Y, Vegera P (2018) Development of the procedure for the estimation of reliability of reinforced concrete beams, strengthened by building up the stretched reinforcing bars under load. East Eur J Enterp Technol 5/7(95):32–42

Marushchak U, Sanytsky M, Korolko S, Shabatura Y, Sydor N (2018) Development of nanomodified rapid hardening fiber reinforced concretes for structure of special purpose. East Eur J Enterp Technol 2/6(92):34–41

Mazaheripour H, Ghanbarpour S, Mirmoradi SH, Hosseinpour I (2011) The effect of polypropylene fibers on the properties of fresh and hardened lightweight self-compacting concrete. Constr Build Mater 25(1):351–358

Ne'ma NH, Abbas AM, Mtasher RA (2011) Strength prediction of polypropylene fiber reinforced concrete. Eng Technol J 29(2):305–311

Solodkyy S, Turba Y (2015a) Improvement of the fracture crack resistance of the dispersion-reinforced concrete. In: 19th Internationale Baustofftagung, vol 2. IBAUSIL. Weimar Band, pp 1075–1082

Solodkyy SY, Turba YuV (2015b) Experimental-statistical modeling of fracture resistance of concrete reinforced with polypropylene fibers. Bull Natl Univ Lviv Polytech Theory Pract Constr 823:301–305

Vairagade VS, Kene KS, Deshpande NV (2012) Investigation on compressive and tensile behaviour of fibrillated polypropylene fibers reinforced concrete. Int J Eng Res Appl 2(3):1111–1115

Qiang F, Ditao N, Jian Z, Daguan H, Mengshu H (2018) Impact response of concrete reinforced with hybrid basalt-polypropylene fibers. Powder Technol 326:411–424

Xu Z, Hao H, Li HN (2012) Experimental study of dynamic compressive properties of fibre reinforced concrete materials with different fibres. Mater Des 33:42–45

Problems in the Implementation of the Lean Concept in the Construction Industries

R. Ulewicz[1(✉)] and M. Ulewicz[2]

[1] Department of Production Engineering and Safety, Czestochowa University of Technology, Armii Krajowej 19B, 42-201 Czestochowa, Poland
robert.ulewicz@wz.pcz.pl

[2] Department of Civil Engineering, Czestochowa University of Technology, Akademicka 3, 42-201 Czestochowam, Poland

Abstract. Many companies are implementing Lean management in order to be more competitive. The goal of management in the construction industry is the reasonable allocation of resources in order to complete the project in accordance with the approved budget, time allowance and the set quality level. Lean management tools can significantly contribute to increasing the efficiency of resource use at a construction site through their rational use. In the article, based on surveys, expanded with an expert interview, the results of using lean tools in the construction industry are presented, as well as their impact on improving the construction process from the management's point of view when executing a construction project. According to the results of the survey, it is clear that lean management is still unknown and not widely used in the construction industry outside large construction companies. Based on observations of the implementation of construction processes, the areas of implementation possibilities were identified and barriers to Lean implementation in the construction sector were identified.

Keywords: Lean management · Construction industry · Tools · Improvement process

1 Introduction

The lean management concept – LM, developed in the 1990s, based on the Toyota Production System. It is a set of practices, tools and organization solutions implemented since 1948 at Japan's Toyota Motor Company (Holweg 2007). The main goal of TPS is the reduction of costs (limits in expenditures) by eliminating processes that don't bring any value, identified in Japanese by the word *muda*, or waste (Womack and Jones 2003). The construction sector does not differ so much from other industries, and it is also possible to apply the tools and methods of improvement known, for example, in the automotive industry. The definition of construction management was explained by Clough and Sears (1994), in which construction management was defined as "The judicious allocation of resources to complete a project at budget, on time, and at desired quality". From the point of view of this definition, construction is a production process

Z. Blikharskyy et al. (Eds.): CEE 2019, LNCE 47, pp. 495–500, 2020.
https://doi.org/10.1007/978-3-030-27011-7_63

subject to all the rules applicable during the implementation of production processes. The storage of materials is one of the key elements of the construction process. Very often investment is carried out in a place with space constraints, where there is no possibility to directly store materials. Such a situation triggers the entire logistics process based on the just-in-time and the pull systems. Standard methods of management on construction sites are usually ineffective (Nowotarski and Paslawski 2015) and can increase work hours or overall project costs (Sacks and Goldin 2007). One way to help manage such projects are the principles and tools of lean management, which enable the comprehensive management of a whole construction process and the gradual improvement of a situation related to construction management in order to eliminate waste (Nahmens and Mullens 2011). The introduction of the concept to construction began when the traditional theories of production were questioned by researchers, and Koskela (1992) proposed the new integrated view of Transformation, Flow, and Value (TFV) theory of production. However, compared to other industries, lean tools are not so widely known, and their use is not so common (Shah and Ward 2003). Very often it is limited only to improving planning, making semi-finished products and delivering them for construction in the just-in-time system (Stasiak-Betlejewska 2012).

2 Barriers to Implementing Lean Management in the Construction Industry

The concept of lean management is very simple and is implemented directly for the elimination of waste (Hasle et al. 2012). However, in its simplicity acceptance of the concept is difficult not only on a managerial level but also on an operational one (Mazur and Momeni 2018). This is not only due to a lack of knowledge of lean tools but also a lack of comprehensive knowledge of the processes and their mutual connection. We are often faced with this phenomenon, where despite implementing the ISO 9001 system, which aims to use the process approach, we still have to deal with the functional approach (Klimecka-Tatar 2018). Experienced authors (Ulewicz and Kucęba 2016, Maszke et al. 2018) from other industries show very similar barriers (including the lack of developed standards and good practices) in implementing lean in the construction industry. Implementing a lean approach has not been standardized (Bicheno and Holweg 2016). Another mistake made by enterprises is limiting themselves to the implementation of individual tools that solve only single problems. The aim of lean management is to build a culture of continuous improvement and not temporarily solving individual problems, which is the domain of the construction industry (Mazur 2016). One of the most important barriers in the implementation of lean management is the lack of knowledge, support and involvement of top management (Yang et al. 2011). This concept can be implemented only from the board or company owners. Formal authority is indispensable not only to change culture but also to overcome resistance and solve conflicts that may arise in optimizing horizontal level processes. Managers must understand and accept the principles of the lean concept as well as be competent in change management skills. Focusing on interim goals and outcomes in the construction industry is not the only big obstacle in the implementation of Lean management. Frequent changes in leadership as well as a high employee churn rate, as well as seasonal work are also substantial barriers causing inconsistencies in performed tasks related to process improvement. Another mistake is to set very ambitious goals without first preparing the organization, which causes general discouragement and a lack of faith in the concept of lean management. Another barrier is the need to extend lean to sub-suppliers in order to achieve positive synergy (Melton 2005). This is not a simple task from the viewpoint of costs incurred as well as the distribution of benefits obtained from a joint implementation of the lean concept. The aim of joint activities within lean management is to create the highest value for the end customer. Lean management cannot be implemented in organizations that do not want to change their course of action. There are good examples of the use of Lean in construction, particularly in the implementation of large projects such as airports, hospitals, roads and hotels where a good organization of teams at the right time with the appropriate equipment enables proper execution of construction works. Analyzing the mentioned examples, the success of a lean implementation occurs when the same repeated activities are performed. In other cases in construction, people often wait and circle around the so-called problem, because someone did something incorrectly or equipment was not delivered, or work was delayed. This is a notorious difficulty in construction.

3 Survey Research

The surveys were conducted in southern Poland. The CAWI (Computer-Assisted Web Interview) and PAPI (Paper & Pen Personal Interview) methods were used. Only 23% of respondents representing the small- and medium-sized construction companies sector qualified for the study. 77% of company respondents did not qualify. This is due mostly to incorrect answers or a lack of knowledge about process improvement based on the lean concept or the execution of construction works as a subcontractor based on the standards of the main contractor. This indicates a gap in knowledge management in the area of change management and the improvement of management in the construction industry. In construction companies performing subcontracting work for large corporations there is great interest in the applicable standards in the field of transport logistics, storage and the rational use of machinery and equipment. Talks conducted with management staff in this case showed the need to transfer over a model from the company to the parent sub-contracting company. There is a barrier in the way which was previously identified as a lack of knowledge, training, and the expectation of immediate results; while in the lean approach results are achieved in the long term. The results of the survey on the extent of the impression of the lean concept is presented in Fig. 1.

The practical use of lean requires the use of numerous management methods and tools. Respondents were asked to identify known and used lean tools and methods in the construction industry and their impact on the elimination of waste (MUDA), overburden of employees and machinery on construction sites (MURI), and the elimination of inconsistency (MURA).

Respondents pointed to the following instruments: 5S, value stream mapping (VSM), Kaizen, just-in-time, pull system, Total Productive Maintenance (TPM) and

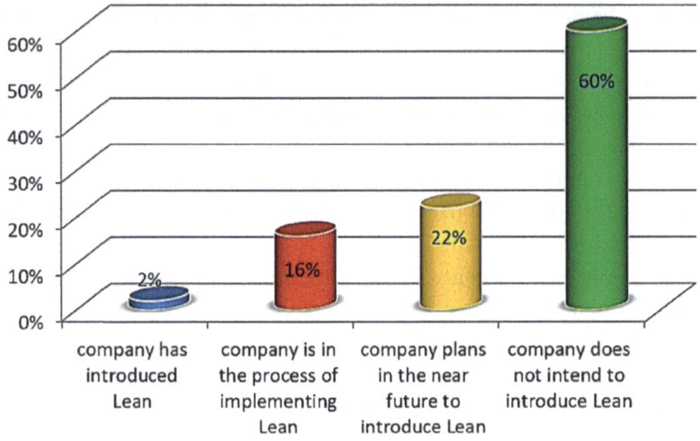

Fig. 1. The percentage share of companies participating in the survey responding to the question about the level of involvement of companies in the implementation of the lean concept

Table 1. Lean instruments used in construction in specific areas for the elimination of waste.

Instruments	Elimination of waste MUDA	Elimination of overburden MURI	Elimination of inconsistency MURA
5S	++++	+	+++
VSM	+++	+++	+++
Kaizen	++	++	+
Just-in-time	++++	+	+++
Pull system	+++	+	++
TPM	+	++	+
Poka-Yoke	++	+	++

Poka-Yoke. Table 1 shows the allocation of specific instruments to particular areas for improvement, i.e. the elimination of waste.

In order to get to know the opinions of construction companies about the impact of using lean instruments to improve the functioning of a company and attain a more efficient use of resources, respondents were asked to indicate the importance of each of the seven instruments by positioning them in order of importance. Then the significance index W was calculated Eq. (1) for all of the instruments previously identified by respondents.

$$W = \frac{\sum_{i=1}^{k} n_i \cdot w_i}{k \sum_{i=1}^{k} n_i} \tag{1}$$

where W – indicator of importance, i – an indication of the place of the method (tool), k – the max weight (the indicated order of methods meant assigning weights in reverse order), n_i – number of indications of the bottom method (technique) at the i-th place, w_i – weight corresponding to the place of the technique i (Fig. 2).

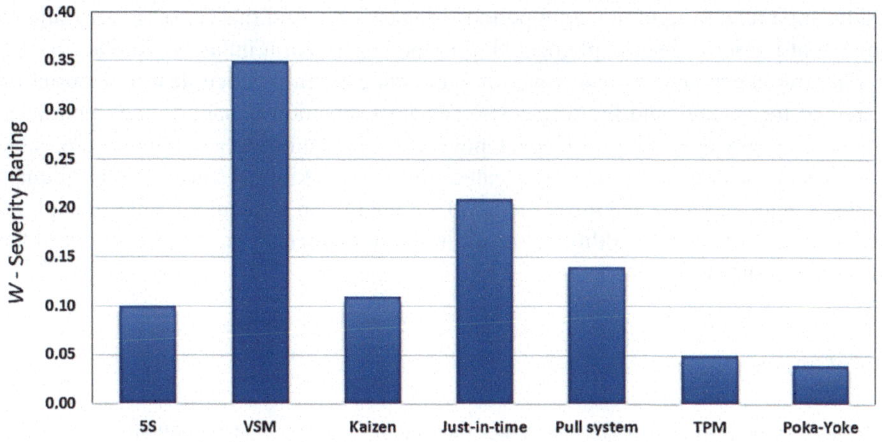

Fig. 2. The results of the selected lean instruments on the effectiveness of company functioning by eliminating waste in construction companies

Obtained results of research clearly indicate the emphasis on activities that are important in the area of transport logistics. This is indicated by the very high position of value stream mapping as well as the high position of the just-in-time system and the pull system. It has great justification in construction practice because a lot of new companies have a deficit in terms of space for storage of materials and equipment; in such cases, the just-in-time and pull systems are invaluable. What is surprising is the low assessment of comprehensive maintenance of machinery when the W coefficient is equal to 0.05. Based on the conducted interview it can be said that it is not a universal solution for companies, especially for the construction industry. It is most accurately described by Nakajima (1988): "The mode and details of use of the TPM system in order to maximize efficiency in machinery and equipment must be adjusted in practice to the individual possibilities of companies". Unfortunately, according to the respondents the possibilities of the surveyed businesses are not sufficient to implement TPM.

4 Conclusions

Based on the results of the survey, we can conclude that lean management is not known and used in the construction industry outside large construction companies, whose lean approach and battle with waste is treated as standard. This approach is lacking in the sector of small- and medium-sized construction companies to which this survey was directed. Among respondents, only 23% had a correct knowledge of the lean concept. Lean implementation can have both a positive and negative impact on the work environment. On the one hand it gives employees greater influence on the way work is performed. On the other hand, in "lean" organizations there is a requirement for broad specialization and acquiring qualifications in many fields (multiskilling), readiness to work in various positions, as well as being involved in continued improvement. What's more is that there is also a measurement of the results of work and its intensity, which results, inter alia, from eliminating periods of inactivity. And this aspect is very often a significant barrier to lean implementation in the construction industry. Another barrier is a narrow understanding and usage of lean concepts. In practice, lean is focused on searching for savings and limiting costs. Too little attention is paid on the importance of lean concepts as a means of developing an organization. Lean supports profitable growth by providing customers with values they are seeking in construction buildings or roads with required features, the level of quality in a required time and place. Reducing costs is just an additional result in the implementation of this concept in the construction industry.

References

Bicheno J, Holweg M (2016) The lean toolbox a handbook for lean transformation. Picsie Books, (S.l.)

Clough RH, Sears GA (1994) Construction Contracting. Wiley, New York

Hasle P, Bojesen A, Langaa Jensen P, Bramming P (2012) Lean and the working environment: a review of the literature. Int J Oper Prod Manag 32(7):829–849

Holweg M (2007) The genealogy of lean production. J Oper Manag 25(2):420–437

Klimecka-Tatar D (2018) Context of production engineering in management model of Value Stream Flow according to manufacturing industry. Prod Eng Arch 21(21):32–35

Koskela L (1992) Application of the new production philosophy to construction, CIFE Technical report: 72. Stanford University, 75

Mazur M, Momeni H (2018) LEAN production issues in the organization of the company - the first stage. Prod Eng Arch 21(21):36–39

Mazur M (2016) Assumptions concept of LEAN processes in the organization of the work on example the production of building components. Prod Eng Arch 13:41–43

Maszke A, Dwornicka R, Ulewicz R (2018) Problems in the implementation of the lean concept at a steel works - case study. MATEC Web Conf 183:01014

Melton T (2005) The benefits of lean manufacturing: what lean thinking has to offer the process industries. Chem Eng Res Des 83(6A):662–673

Nahmens I, Mullens MA (2011) Lean homebuilding: lessons learned from a precast concrete panelizer. J Arch Eng 17(4):155–161

Nakajima S (1988) Introduction to TPM: total productive maintenance. Productivity Press, (S.l.)

Nowotarski P, Paslawski J (2015) Barriers in running construction SME - case study on introduction of agile methodology to electrical subcontractor. Procedia Eng 122:47–56

Sacks R, Goldin M (2007) Lean management model for construction of high-rise apartment buildings. J Constr Eng Manag 133(5):374–384

Shah R, Ward PT (2003) Lean manufacturing: context, practice bundles, and performance. J Oper Manag 21(2):129–149

Stasiak-Betlejewska R (2012) Value engineering as the way of quality problems solving in the steel construction management. Manuf Technol 12(13):242–247

Ulewicz R, Kucęba R (2016) Identification of problems of implementation of lean concept in the SME sector. Eng Manag Prod Serv 8(1):19–25

Womack JP, Jones DT (2003) Lean thinking: banish waste and create wealth in your corporation. Free Press, New York

Yang MG, Hong P, Modi SB (2011) Impact of lean manufacturing and environmental management on business performance: an empirical study of manufacturing firms. Int J Prod Econ 129(2):251–261

Optimal Design of a Three-Hinged Arch with Given Topology Under Constant Load

Glib L. Vatulia(✉), Sophia D. Komagorova, Olena V. Opanasenko, and Oleksii V. Lobiak

Ukrainian State University of Railway Transport (UkrSURT), Feuerbach Sq. 7, Kharkiv 61050, Ukraine
glebvatulya@gmail.com

Abstract. The article describes the optimal design method for three-hinged arches with given topology. The examples of optimization of the three-hinged arch with constant and uniform-strength cross section subjected to the distributed load are considered. The optimization technique is based on the geometrical parameters and design features of the systems. The calculations of volume material of three-hinged arches for the uniaxial compression stress strength conditions are shown. The results of comparison of the arch volume of material s depending on the axis shape are presented.

Keywords: Three-hinged arch · Optimal design · Optimality criterion · Materials consumption · Equal strength design · Arch axis equation

1 Introduction

From the very beginning of the use of arched structures, there has been constant search for their rational parameters. Numerous researchers (Ermakova; (2007)) were engaged in investigations in this area (Hayrullin; (2013)).

Solving the problem of optimal design has been the subject of multiple studies in various fields (Mironenko; (2011), Yurev et al.; (2007), Fairbairn, et al.; (2004)). E.g., in works by Perelmuter; (Lyahovich and Perelmuter 2014), the principle of solving optimal design problems was formulated which is applied to a wide range of structures. The optimization problem was formulated as a nonlinear programming problem (NLP) for design variables, and a finite element model of the structure was used as the design scheme. Fiacco and McCormick proposed the barrier method which they were called the sequential unconstrained minimization technique with constraints for the inequality type (Marano; (2018)), that is $\min f(x), x \in R = \{g_s(x) \geq 0, s = 1,p\}$. In problems of optimal design of structures, genetic algorithms (GA) are becoming ever more commonly used (Klein, et al.; (2018), Belevičius, et al.; (2018), Klyuev, et al.; (2007)).

Let us consider a subset of systems with given topology. A subset of systems with given topology is a set of structures that have the same structural design (Du; (2004)), but different arrangement of supports and intermediate ties (Qizhi He (2014)) and equation of axes in arches. Each subset of systems is described by its parameters which depend on the distribution of forces.

Z. Blikharskyy et al. (Eds.): CEE 2019, LNCE 47, pp. 501–509, 2020.
https://doi.org/10.1007/978-3-030-27011-7_64

However, the practical implementation of their studies revealed a number of shortcomings. The most significant of them include: 1. Computational process is complicated with regard to representation of the mathematical model. 2. The complete range of loads affecting the structures cannot be taken into consideration. This fact prevents them from being commonly used in the commercial design of arched structures. Therefore, a new improved method for determining rational constructive design shapes is necessary, which would take into account the entire range of possible loads and action on the structure (Chihladze, et al.; (2006), Kitov, et al.; (2017)). Thus, the aim of the study is reducing the materials consumption of arched structures by giving them a rational shape.

2 Three-Hinged Arch Rational Shape

In this subset of arches, the optimal design is influenced by equation of axis $y = \psi(x)$ and rise of the arch f. We will conditionally call them topological variables.

In the general case, in the arbitrarily shaped arch, internal forces M, Q and N occur from the given load. The area of the arch sections and, consequently, the materials consumption for its construction are functions of these forces and the mechanical characteristics of the material. The experience of practical design and the results of the calculations show that when the axis of the arch changes, while the span and rise are constant, the length of the axis of the arch and longitudinal force N in it change slightly, while bending moment M changes significantly. Therefore, the minimum theoretical volume of the arch should be expected with such a shape of its axis with the zero bending moments M and shear forces Q from the constant load in all its sections. Such axis of the arch is called rational (Vatulia, et al.; (2018)). Thus, the differential equation of the axis of the arch under the constant load varying according to arbitrary law $q(x)$ along the span should be composed (Fig. 1), provided only internal forces in the form of longitudinal forces N will occur in all sections of the arch.

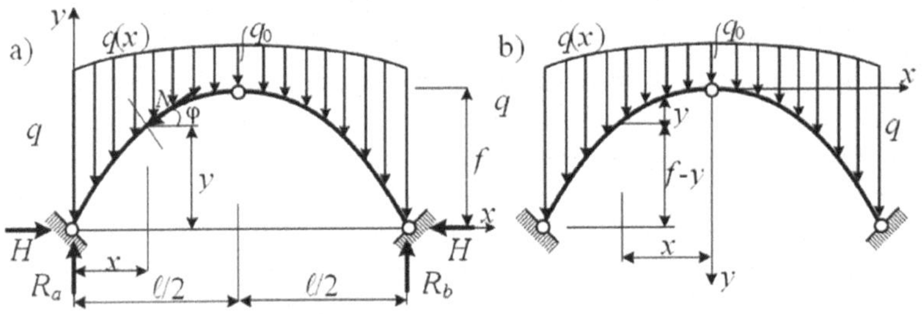

Fig. 1. Three-hinged arch: arrangement of axes considering: (a) distributed load which does not explicitly depend on the shape of axis of the arch y; (b) load $q(x) = q_0 + \gamma y$

Let us consider the equilibrium of the part of the rationally shaped arch (Lyahovich and Perelmuter; (2014)) to the left of section x (Fig. 1a). Let us compose the equilibrium condition as a projection of forces to axis x:

$$\sum X = 0,\ H - N\cos\varphi = 0,\ N = \frac{H}{\cos\varphi}. \tag{1}$$

From the equilibrium condition to the vertical axis we obtain

$$R_a - \int\limits_0^x q(\xi)\,d\xi - N\sin\varphi = 0, N = \frac{R_a - \int\limits_0^x q(\xi)\,d\xi}{\sin\varphi}; \mathrm{tg}\,\varphi = \frac{dy}{dx} = \frac{R_a - \int\limits_0^x q(\xi)\,d\xi}{H} \tag{2}$$

When we take derivative from Eq. (1), we get the differential equation of the rational axis:

$$\frac{d^2y}{dx^2} = -\frac{q(x)}{H}. \tag{3}$$

The constant load acting on the arch may differently depend on the shape and size of the arch span. Its intensity can be uniformly distributed along the arch span:

$$q(x) = q_0 = \text{const}, \tag{4}$$

or change according to the shape of the axis of the arch, for example,

$$q(x) = q_0 + \gamma(f - y). \tag{5}$$

Let us first find a rational equation of the axis of the arch under constant load *uniformly distributed along the span* (the arrangement of the axes is shown in Fig. 1a). Substitute in differential equation Eq. (3) $q(x) = q = \text{const}$ and integrate it:

$$\frac{d^2y}{dx^2} = -\frac{q}{H}, \frac{dy}{dx} = -\frac{qx}{H} + C, y = -\frac{qx^2}{2H} + Cx + D$$

Boundary conditions for arbitrary constants C and D are:

$$x = 0, y = 0; \quad x = \frac{\ell}{2}, \frac{dy}{dx} = 0.$$

When we insert in the boundary conditions expressions for y and $\frac{dy}{dx}$, we get

$$D = 0, C = \frac{q\ell}{2H}, y = -\frac{qx^2}{2H} + \frac{q\ell x}{2H} = \frac{qx(\ell - x)}{2H}. \tag{6}$$

If q=const, we obtain the equation of the rationally shaped arch with constant load

$$y = \frac{4fx\,(\ell - x)}{\ell^2}. \tag{7}$$

Let us find the distribution function for axial forces in arch using Eqs. (1), (4) and (5) and expressing cos φ through y:

$$\cos\,\varphi = \frac{1}{\sqrt{1+\mathrm{tg}^2\varphi}} = \frac{1}{\sqrt{1+\left(\frac{dy}{dx}\right)^2}} = \frac{1}{\sqrt{1+\left[\frac{4f\,(\ell - x)}{\ell^2}\right]^2}} = \frac{\frac{4f}{\ell^2}}{\sqrt{(\ell - 2x^2) + \frac{\ell^4}{16f^2}}}, \tag{8}$$

Let us now find the equation of the axis of the rationally shaped arch *with variable load* (5). To simplify, the axes are first arranged as shown in Fig. 1b. Taking into account the sign of the curvature, we obtain the differential equation

$$\frac{d^2y}{dx^2} = \frac{q_0 + \gamma y}{H}, k^2 = \frac{\gamma}{H}, \frac{d^2y}{dx^2} - k^2 y = \frac{q_0}{H} \tag{9}$$

where γ – specific weight of the load above the axis of the arch.

This is a linear nonhomogeneous second order differential equation. The integral of this equation consists of the homogeneous and the particular solutions:

$$y(x) = C\,\mathrm{sh}(kx) + D\,\mathrm{ch}(kx) - \frac{q_0}{k^2 H}. \tag{10}$$

Constants C and D can be found from the boundary conditions:
when $x = 0$ $y = 0$; when $x = 0$ $y' = kC\,\mathrm{ch}(kx) + kD\mathrm{sh}(kx) = 0$.
It follows from the above condition that

$$D = \frac{q_0}{k^2 H}, kC = 0, C = 0. \tag{11}$$

When we substitute Eqs. (11) in Eq. (10), we get

$$y(x) = \frac{q_0}{k^2 H}\mathrm{ch}(kx) - \frac{q_0}{k^2 H} = \frac{q_0}{\gamma}[\mathrm{ch}(kx) - 1], \tag{12}$$

from which subject to Eq. (11) it follows that the rational shape of the axis under load N_x has the form known as Legay's catenoid:

$$y(x) = \frac{q_0}{\gamma}[\mathrm{ch}(kx) - 1]. \tag{13}$$

In the co-ordinate system shown in Fig. 1a, Eq. (13) has the form

$$y(x) = f - \frac{q_0}{\gamma}\left[\operatorname{ch} k\left(x - \frac{\ell}{2}\right) - 1\right]. \tag{14}$$

Let us make some transformations. Let us denote the intensity of the distributed load in the abutment as q_{abut} and write down the following equality according to N_x:

$$q_{\text{abut}} = q_0 + \gamma f, \gamma = \frac{q_{\text{abut}} - q_0}{f} = \frac{q_0(m-1)}{f}, m = \frac{q_{\text{abut}}}{q_0}, \frac{\gamma f}{q_0} = m - 1$$

from

$$\operatorname{ch}\left(\frac{k\ell}{2}\right) = m, k = \frac{2}{\ell}\operatorname{ar\,ch}(m), H = \frac{\gamma}{k^2} = \frac{q_0(m-1)}{fk^2}. \tag{15}$$

For comparison, Table 1 shows ordinates of axes of arches when $q(x) = q_0$ Eq. (6) and $q(x) = q_0 + \gamma(f - y)$ Eq. (14) with dimensions $\ell = 30\,\text{m}$, $f = 7.5\,\text{m}$ and intensity values $q_0 = 60\,\text{kN/m}$, $q_{\text{abut}} = 120\,\text{kN/m}$.

Table 1. Ordinates of axes of arches

#	x, m	y, m		$\Delta y = y_2 - y_1$, m
		y_1 when $q = q_0$	y_2 when $q = q_0 + \gamma(f - y)$	
1	0	0	0	0
2	1.5	1.42	1.59	0.17
3	3.0	2.70	2.94	0.24
4	4.5	3.82	4.08	0.26
5	6.0	4.80	5.03	0.23
6	7.5	5.62	5.81	0.19
7	9.0	6.30	6.44	0.14
8	10.5	6.82	6.91	0.09
9	12.0	7.20	7.24	0.04
10	13.5	7.42	7.43	0.01
11	15.0	7.5	7.5	0

The axis delineated in the square parabola and rational when $q = \text{const}$ has a slightly larger radius of curvature (is less steep).

3 Three-Hinged Arch of Constant Cross-section Subjected to a Uniformly Distributed Load

We get the cross-sectional area from the strength condition

$$\sigma_x^{\max} = \frac{N_x}{A_x} \leq mR_y \tag{16}$$

when $N_x = N_{\max}$, which will be the case when $x = 0$. Then from Eq. (16) we find

$$N_{\max} = \frac{q\ell}{2}\sqrt{1 + \frac{\ell^2}{16f^2}},\ A_{\max} = \frac{N_{\max}}{mR_y} = \frac{q\ell}{2mR_y}\sqrt{1 + \frac{\ell^2}{16f^2}} \tag{17}$$

Let us find the dependence to determine it:

$$\begin{aligned} L(y,f) &= \int_L dL = \int_L \sqrt{dx^2 + dy^2} = \int_L \sqrt{1 + \left(\frac{dy}{dx}\right)^2} dx \\ &= \frac{8f}{\ell^2}\int_0^{\ell/2} \sqrt{(\ell - 2x)^2 + \frac{\ell^4}{16f^2}} dx. \end{aligned} \tag{18}$$

Let us use the substitution $u = \ell - 2x$ to compute the integral, then $du = -2dx$ and

$$L(y,f) = -\frac{4f}{\ell^2}\int_\ell^0 \left(u^2 + \frac{\ell^4}{16f^2}\right) du = -\frac{4f}{\ell^2}\left[\frac{u}{2}\sqrt{u^2 + \frac{\ell^4}{16f^2}} + \frac{\ell^4}{32f^2}\left(u + \sqrt{u^2 + \frac{\ell^4}{16f^2}}\right)\right]_\ell^0$$

Let us substitute the limits and perform backward substitution. We finally get

$$L(y,f) = 2f\left[\sqrt{1 + \frac{\ell^2}{16f^2}} + \frac{\ell^2}{16f^2}\ln\frac{4f}{\ell}\left(1 + \frac{\ell^2}{16f^2}\right)\right]. \tag{19}$$

To analyze the influence of function y_{opt} Eq. (14) on the change of the volume of the material of the arch, let us find it when y_{opt} and rise $f = 7.5$ m, the same as in the round arch. Subject to Eqs. (17), (19), we get

$$V_c(y_{\text{opt}}, f) = A_{\max}L = \frac{q\ell f}{mR_y} \times \left\{1 + \frac{\ell^2}{16f^2}\left[1 + \sqrt{1 + \frac{\ell^2}{16f^2}}\ln\frac{4f}{\ell}\left(1 + \sqrt{1 + \frac{\ell^2}{16f^2}}\right)\right]\right\}. \tag{20}$$

When we substitute values $q = 60\,\text{kN/m}$, $\ell = 30\,\text{m}, f = 7.5\,\text{m}$ and $mR_y = 200\,\text{MPa}$ in (22), we can find $V_c(y_{opt}, 7.5) = 219.113\ \text{cm}^3$

This is about 85% less than in the round arch, since the shape of the axis plays an important role in optimal arch designing.

Now let us find the optimal value of the second design variable – rise of the arch. To determine f_{opt}, for which the function $V_c(y_{\text{opt}}, f_{opt})$ will be minimum, equation $dV(y,f)/df = 0$ can be obtained and solved. However, considering the form of function $V(y,f)$, optimality equation will be transcendental, and it has to be solved using approximate approaches. Therefore, numerical investigation of function $V(y,t)$ allows to find the value of the minimum volume of arch $V_c(y_{opt}, f_{opt}) = 210.104\ \text{cm}^3$ for $f_{opt} = 10.25\text{m}$. This is 4% less than the volume of the arch with rise $f = 7.5\,\text{m}$. Hence, changing the rise of the arch has less impact on obtaining the optimal design than changing the shape of the axis.

4 Three-hinged Arch of Uniform-strength Cross Section

In this problem, the strength condition

$$\sigma_x = \frac{N_x}{A_x} = mR_y \tag{21}$$

should be observed as the equality for all sections of the arch. In view of this, let us write the function of the volume of the material of the arch

$$V_e(y_{opt}, f) = \int_V A_x dL = \frac{4qf}{mR_y\ell^2} \int_0^{\ell/2} \left[(\ell - 2x)^2 + \frac{\ell^4}{16f^2}\right] dx = \frac{2q\ell f}{mR_y}\left(\frac{\ell^2}{16f^2} + \frac{1}{3}\right) \tag{22}$$

Let us find f_{opt} with which the minimum volume is achieved:

$$\frac{dV_e(y_{\text{opt}}, f)}{df} = 0, -\frac{\ell^2}{16f^2} + \frac{1}{3} = 0,$$

From which $f_{\text{opt}} = \frac{\ell}{4}\sqrt{3}$. With $q = 60\,\text{kN}$, $\ell = 30\,\text{m}$, $mR_y = 200\,\text{MPa}$, $f_{\text{opt}} = 12.99\,\text{m}$

$$V_e(y_{\text{opt}}, f_{\text{opt}}) = 155.885\,\text{cm}^3.$$

This is 26% less than the volume of the optimal arch of the constant cross-section $V_e(y_{\text{opt}}, f_{\text{opt}}) = 210.104\,\text{cm}^3$.

5 Conclusions

The dependencies of the shapes of the axes of the three-hinged arch on the applied load are given. The optimal volume of the arch of a constant cross section is about 7 times less than in a round arch, i.e. the shape of the axis plays a significant role in the optimal design of arches. Numerical investigation of function $V(y,t)$ allowed to find the value of the minimum volume of the arch $V_c(y_{opt},f_{opt}) = 210.104\ \text{cm}^3$ when $f_{\text{opt}} = 10.25\,\text{m}$. This is 4% less than the volume of the arch with rise $f = 7.5\,\text{m}$. Hence, changing the rise of the arch has less impact on obtaining the optimal design than changing the shape of the axis. The optimal value of the volume of material of a three-hinged arch of continuously equal strength under the action of constant load was also obtained $V_c(y_{opt},f_{opt}) = 155.885\ \text{cm}^3$, which is 26% less than the volume of the optimal arch of constant cross section $V_c(y_{opt},f_{opt}) = 210.104\ \text{cm}^3$ by, which is a fairly good effect.

References

Mironenko, I (2011): Analiz shodimosti evolyutsionnoy optimizatsii zhelezo-betonnyih konstruktsiy. Sovremennyie problemyi nauki i obrazovaniya 4

Yurev A, Klyuev C, Klyuev A (2007) Optimizatsiya stroitelnyih konstruktsiy na osnove geneticheskogo algoritma. Izvestiya Tomskogo politehnicheskogo universiteta 310(1):61–64

Fairbairn EM, Silvoso MM, Toledo Filho RD, Alves JLD, Ebecken NFF (2004) Optimization of mass concrete construction using genetic algorithms. Comput Struct 2–3:281–299

Chihladze E, Vatulia G, Kitov Yu et al (2006) Osnovyi rascheta i proektirovaniya kombinirovannyih i stalebetonnyih konstruktsiy. Transport of Ukraine, Kiev

Du J (2004) Topological optimization of continuum structures with design–dependent surface loading – Part II: algorithm and examples for 3D problems. Struct Multi Optim 3:166–177

Vatulia G, Komagorova S, Pavliuchenkov M (2018) Optimization of the truss beam, verification of the calculation results. Matec Web of Conf 230:02037. https://doi.org/10.1051/matecconf/201823002037

Abd Elrehim, MZ, Eid, MA, Sayed, MG (2019): Structural optimization of concrete arch bridges using Genetic Algorithms. Ain Shams Engineering Journal

Marano GC, Trentadue F, Greco R, Vanzi I, Briseghella B (2018) Volume/thrust optimal shape criteria for arches under static vertical loads. J Traffic Transp Eng (English Edition) 5(6):503–509

Dos Santos Coelho L, Klein, CE, Mariani, VC, Do Nascimento, CAR, Askarzadeh, A (2018): Electromagnetic optimization based on Gaussian crow search approach. SPEEDAM. In: Proceedings: International Symposium on Power Electronics, Electrical Drives, Automation and Motion 8445209. Pp. 1107–1112

Belevičius R, Juozapaitis A, Rusakevičius D (2018) Parameter study on weight minimization of network arch bridges. Periodica Polytechnica Civ Eng 62(1):48–55

Ermakova, AV (2007): Metod dopolnitelnyih konechnyih elementov dlya rascheta zhelezobetonnyih konstruktsiy po predelnyim sostoyaniyam. ASV

Lyahovich LS, Perelmuter AV (2014) Nekotoryie voprosy optimalnogo proektirovaniya stroitelnyih konstruktsiy. Int J Comput Civ Struct Eng 2:14–23

He, Q, Kang, Z, Wang, Y (2014): A topology optimization method for geometrically nonlinear structures with meshless analysis and independent density field interpolation. Comput Mech. https://doi.org/10.1007/s00466-014-1011-7

Kitov Y, Verevicheva M, Vatulia G, Orel Y, Deryzemlia S (2017) Design solutions for structures with optimal internal stress distribution. BulTrans'2017 Matec Web of Conf 133:03001. https://doi.org/10.1051/matecconf/201713303001

Simulation of Thermal Processes in the Solar Collector Which Is Combined with External Fence of an Energy Efficient House

Stepan Shapoval, Vasyl Zhelykh, Iryna Venhryn(✉), and Khrystyna Kozak

Department 'Heat and Gas Supply and Ventilation', Institute of Building and Environmental Engineering, National University "Lviv Polytechnic", S. Bandery Street 12, Lviv 79013, Ukraine
iryna.venhryn@gmail.com

Abstract. Efficient and rational use of energy carriers is one of the important issues of energy policy of the EU and Ukraine. In this regard, it is important to design systems that will be based on the use of renewable or inexhaustible energy for efficient and rational use of the fuel and energy complex of Ukraine at the international level and heat supply of houses at local levels. This solution can be achieved by using solar systems to generate electricity and heat. In general, territory of Ukraine belongs to the average zone with solar radiation intensity. The value of annual solar radiation per 1 m^2 on the earth's surface has a static distribution. The work considers energy efficient solar system combined with external fence of energy-efficient building. The paper describes the data of functioning mechanism of the solar collector using mathematical modeling. The article analyzes the heat technical characteristics of the proposed solar collector for the possibility of their subsequent installation and application by consumers.

Keywords: Computer simulation · Solar fence · Solar roof · Heat carrier · Efficiency · Energy-efficient building

1 Introduction

Alternative fuels are becoming more important than other fuels. For example, the main advantage of solar energy over conventional fuels are the inexhaustibility of this energy source. As well as the possibility of using solar energy in almost all areas of the earth's surface. Such alternative energy source is environmentally friendly but geographically differentiated.

In general, Ukraine territory belongs to the zone with average solar radiation intensity. The annual amount receipt of solar radiation per 1 m^2 on the earth's surface has a static distribution pattern and changing throughout all territory of Ukraine.

It is known that only about 3% of the solar flux which falls on the earth's surface could be used from the total solar energy from cosmos without damage on the

Z. Blikharskyy et al. (Eds.): CEE 2019, LNCE 47, pp. 510–517, 2020.
https://doi.org/10.1007/978-3-030-27011-7_65

biosphere. This is 1000 billion kW of energy in terms of mankind needs. This capacity is 100 times higher than the current capacity of energy production on the planet.

Improving energy efficiency in buildings is the one of the most cost-effective ways in all sectors for reducing energy consumption and reduce greenhouse gas emissions as the consequence. Certification and additional measures can solve these problems (International Energy Agency 2010).

In 2008 the number of passive houses were ranged from 15 000 to 20 000 units worldwide. The vast majority of them were built in German-speaking countries and Scandinavia. The first certified passive house in the Antwerp region from Belgium was built in 2010. In 2011 the government of Heidelberg in Germany initiated the Bahnstadt project which was marked as the world's largest passive house. The company in Qatar established the first passive house in the country in 2013 (Kaklauskas 2015).

In the computer industry with regard to heat supply systems the TRNSYS 17 software is used to evaluate the efficiency of several designs of hybrid systems. Such hybrid systems which are consisting of solar thermal collectors, photovoltaic panels and natural gas combustion engines. Studies are conducted in Spain in five different places with different climatic characteristics. Moreover, there is analyzing the behavior of the designed systems for the one house in all five cases (Rodríguez et al. 2016).

A promising solution in Smart Energy Networks (SEN) is to combine renewable energy sources with cogeneration systems that are highly efficient. This, in turn, can help to integrate alternative technologies in widespread use (Mathiesen et al. 2015; Sig Chai et al. 2013; Lund et al. 2012).

2 Objectives the Formulation of the Problem

On the one hand, promising development direction of the construction industry and architecture is use of combined solar heating systems with the purpose to solve set tasks which are exist before Ukraine in area to improve the energy and environmental situation. On the other hand, such systems are based on solar collectors combined with architectural constructions of energy-efficient buildings (Shapoval 2019).

In connection with the above, modeling of thermal processes in solar collectors combined with an external fence of an energy-efficient house is an important study for further large-scale introduction of solar collectors by consumers.

Therefore, the purpose of this work was to simulate external solar fences for the house with taking into account using the useful area of energy-efficient facade and the implementation such fences experimentally in the future.

3 The Analysis of Recent Research and Publications

Exploitation of solar energy in construction is described in (Chwieduk 2014). In addition, the work contains the scientific basis for the use of solar energy and practical recommendations to this.

In particular, a number of studies are described in (Yudong 2012) and there was dedicated to green buildings. However, this thesis focused on the management of microclimate systems and the result forecasting of their work. On the basis of this work was developed the predictive control model (MPC). Such model was as a unified management methodology that could be systematized and take into account future projections during the design phase while adhering to system constraints.

Studies of the use of solar energy was engaged by Shchukyna (2011), which offered scientific and methodical approaches to search of ways to increase the efficiency of passive/active solar systems for achievement of the maximum possible power supply in buildings.

Scientists have developed recommendations for system analysis and modeling of active/passive methods of radiation capture with using absorbing materials. The system analysis includes thermal insulation materials and materials for the screening of solar protection devices under unstable weather conditions and according to seasonal changes in loads. Studies confirm the prospects of alternative energy supply studies through the use of flat collectors with corrugated translucent protection for houses of only a certain type. However, there is no comprehensive approach to the energy supply of passive houses.

In modern studies, particularly in Italy (Buonomano et al. 2018) has been developed the dynamic model of the simulations codes which are written in MatLab. This model allowed to investigate the energy, economic and environmental performance of new solar systems pageneral. This is based on adsorbing or absorption technology chillers, which are served in concentrated or flat photovoltaic/thermal collector.

Christoph Maurer of the Fraunhofer Institute of Solar Energy from Germany with co-authors from Spain, Ireland argue that with few system studies BIST (building-integrated solar thermal systems), most studies related to the BI configurations, namely PVT, PV. Thus, there is a need for additional research related to BIST installations, especially for active configurations that can provide thermal (or electrical/thermal) energy for energy consumption needs. Taking into account previous studies where BIST modeling focuses on the system itself authors draw conclusions about the necessary additional studies of the system in conjunction with the building (Lamnatou 2015).

4 The Main Material

In the current conditions, it is necessary to perform a preliminary computer analysis of solar heating systems, since in real conditions it is difficult to explore in details more factors affecting their operation.

To determine the feasibility of exploitation solar collectors integrated with the external fencing of the building it is necessary to conduct a preliminary simulation. Figure 1 shows a house containing a solar roof, a solar wall and a solar window. To determine the effectiveness of the solar wall, solar window and solar roof, it is advisable to consider these elements separately and simulate their work with the help of

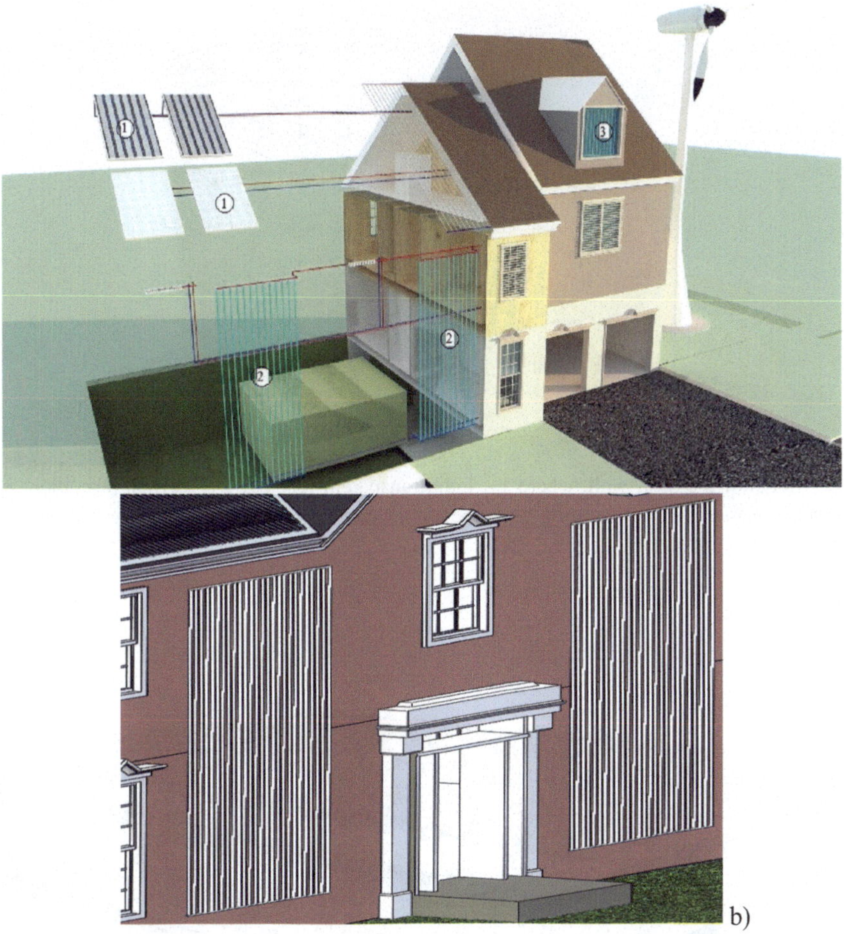

a)

b)

Fig. 1. Model (a) energy-efficient building where the solar roof (1), solar wall (2), solar window (3) and (b) enlarged image of the solar wall in the SolidWorks software

computer technology, in particular in the SolidWorks program. In all studies, the physical parameters of water were taken as a heat carrier.

In the article was considered the modeling of the solar wall and roof as a solar collector. Figure 2 shows a computer model of the solar wall, and Fig. 3 - the solar roof. With this design, the solar collector is structurally combined with the wall of the house or with the roof. It is carried out for the purpose of useful use of the protections area of the building.

Figure 4 shows the model of the solar wall where the heat carrier is heated by programmed solar radiation.

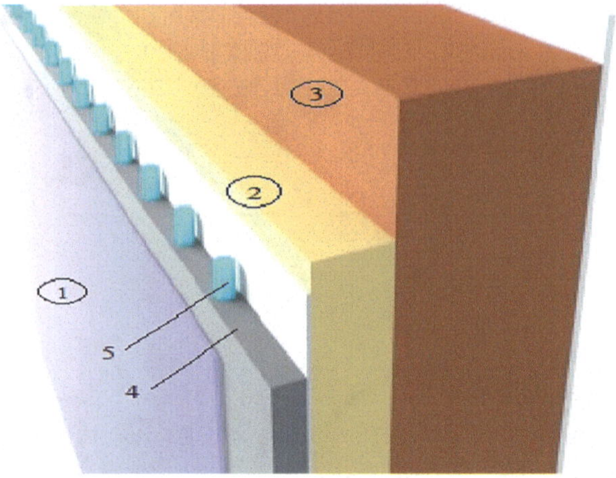

Fig. 2. Model of the solar wall, where 1 – light transparent fencing, 2 – thermal insulation, 3 - load-bearing wall design, 4 – screed, 5 – tube circuit circulation

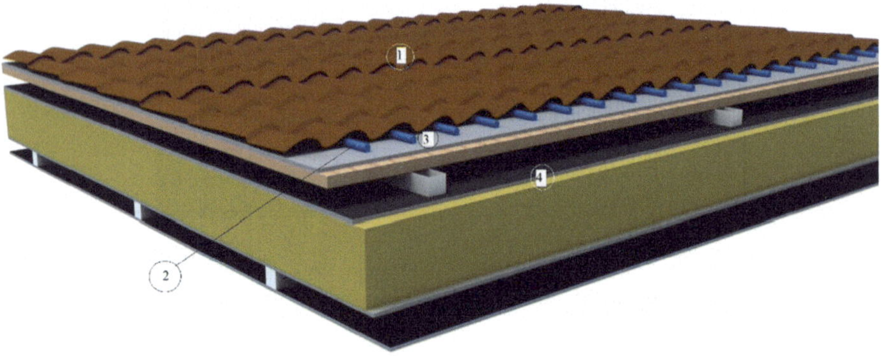

Fig. 3. Model of the solar roof 1- profiled decking, 2 – tube circuit circulation, 3 – thermal insulation; 4 – supporting structure of the roof

The angle of the solar wall installation was taken into account in the program code since it is a vertical structure. In Fig. 4, the solar wall is drawn from this angle in order to better illustrate the heat carrier movement to the storage tank.

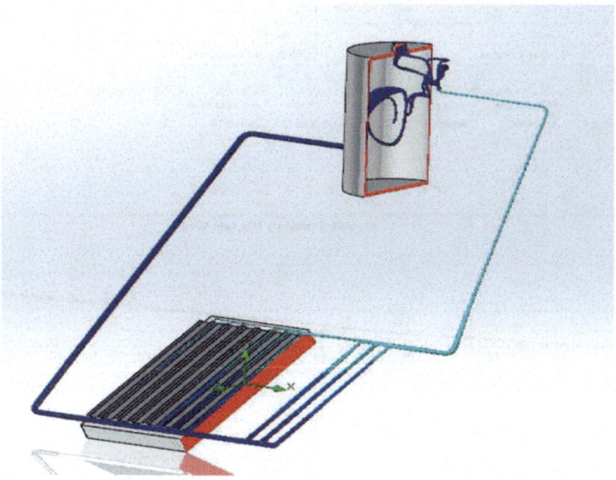

Fig. 4. The heat carrier heating due to pre-programmed solar radiation in the solar wall model

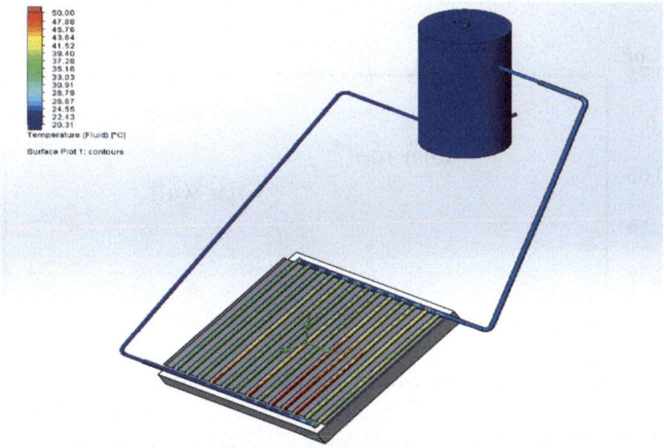

Fig. 5. Model of the solar roof

The solar roof is shown in Fig. 5, where the heat carrier is heated under the heat absorber (under profiled decking) due to the gravitational motion of the heat carrier in the solar heat supply system. The maximum temperature reached in this computer model was amounted 50 °C.

The result of the heat supply system simulation of the energy-efficient home with solar fences is shown in Fig. 6.

The average coefficient of efficiency (CoE) are obtained on the basis of the computer simulation of solar fences separately and in the solar heating system.

As a result of the simulation of the accumulation of solar radiation by the solar roof and the solar wall it was found that the CoE of the solar wall reached 0.65 and the solar roof – 0.69 (Fig. 7). Such data shows about efficiency and prospects of application of these designs.

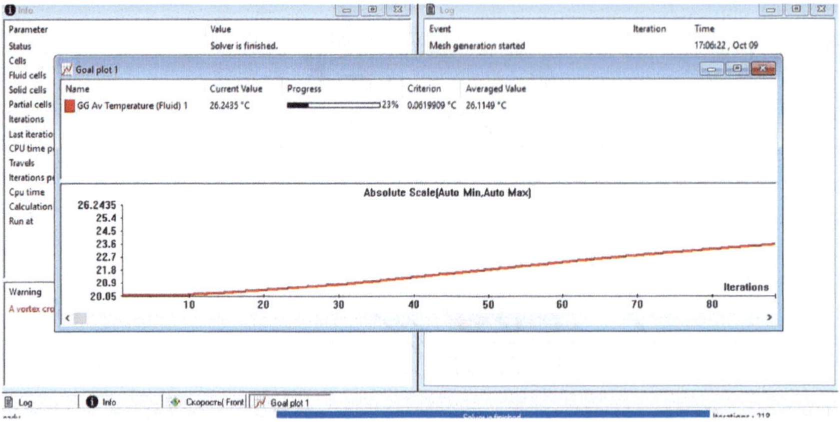

Fig. 6. Modeling of the heat carrier heating in the storage tank when simultaneous operation of solar fences

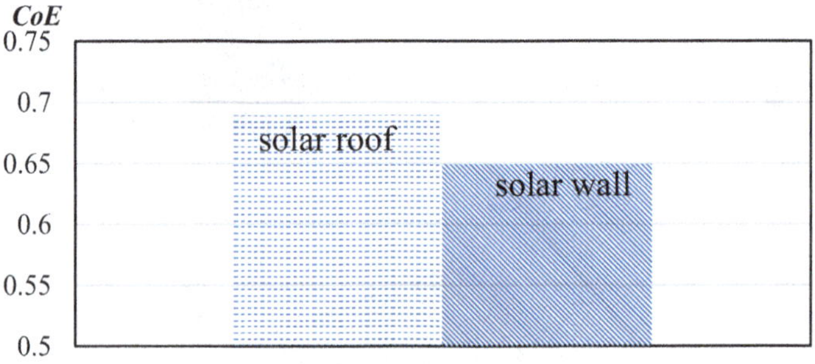

Fig. 7. The average coefficient of efficiency for solar fances

5 Conclusions

Modeling of the processes work in structures for their subsequent application in practice is the important step for the development and preliminary improvement before using by consumers. Computer simulation in the software Solidworks has improved the design of combined solar wall or roof of the building. Software simulation is carried out for these circuit solutions allows for engineers to place the solar structure correctly on the facade of the energy-efficient house and choose cost-effective materials for the construction of such fence by the consumer. The efficiency calculated by the software was 69% for the solar roof and 65% for the solar wall.

In light of further research, it is planned to improve computer models and operations of the solar roof and wall.

References

Energy Performance Certification of Buildings (2010) International Energy Agency, Paris, p 64

Kaklauskas A (2015) Passive house model for quantitative and qualitative analyses and its intelligent system. Biom Intell Decis Mak Support 81:87–112

Rodríguez LR, Lissén JS, Ramos JS (2016) Analysis of the economic feasibility and reduction of a building's energy consumption and emissions when integrating hybrid solar thermal/PV/ micro-CHP systems. Appl Energy 165:828–838

Mathiesen BV, Lund H, Connolly D, Wenzel H (2015) Smart energy systems for coherent 100% renewable energy and transport solutions. Appl Energy 145:139–154

Sig Chai D, Wena JZ, Nathwani J (2013) Simulation of cogeneration within the concept of smart energy networks. Energy Convers Manag 75:453–465

Lund H, Andersen AN, Ostergaard PA, Mathiesen BV, Connolly D (2012) From electricity smart grids to smart energy systems – a market operation based approach and understanding. Energy 42:96–102

Shapoval S, Zhelykh V, Venhryn I (2019) Theoretical and experimental analysis of solar enclosure as part of energyefficient house. East Eur J Enterp Technol 8:38–46

Chwieduk D (2014) Solar energy in buildings. In: Thermal balance for efficient heating and cooling. Elsevier Inc., Boston, p 362

Yudong, M. (2012). Model predictive control for energy efficient buildings. Doctor of Philosophy in Control and Design, University of California, p 108

Shchukyna T (2011) Scientific methodological bases of solar energy use in the replacement of buildings thermal loads. Dissertation of D.Sc.: 05.23.03 Voronezh, p 293 (in Russian)

Buonomano A, Calise F, Palombo A (2018) Solar heating and cooling systems by absorption and adsorption chillers driven by stationary and concentrating photovoltaic/thermal solar collectors: modelling and simulation. Renew Sustain Energy Rev 82(2):1874–1908

Lamnatou C, Mondol J, Chemisana D, Maurer C (2015) Modelling and simulation of Building-Integrated solar thermal systems: behaviour of the system. Renew Sustain Energy Rev 45:36–51

The Influence of Chairs and Passengers on Air Velocity in Bus Passenger Compartment

Orest Voznyak(✉), Yuriy Yurkevych, Oleksandr Dovbush, and Yaroslav Serediuk

National University "Lviv Polytechnic", Bandery Str. 12, 79013 Lviv, Ukraine
orest.voznyak@i.ua

Abstract. The aim of the investigation is determination of compressing coefficient and air jet's initial velocity under the condition, when bus passenger compartment is filled to a great extent with seats and passengers (sitting and standing).

Air distribution experimental research was carried out in passenger compartment of different LAZ (Lviv Bus Plant) bus production models both at presence and absence of the passengers (in filled and empty bus passenger compartment). Air was supplied at the various flow rate through the different air distribution devices. Air exhaust was realized through the driver window. Air jet's initial velocity of outlet from inflow hole and axial velocities in the bus passenger compartment service area has been experimentally determined. Measurement of the air jet's velocity was made by thermal electrical anemometer Testo-405. Universal calculating equations of air jet's velocities and determining factors have been expressed as dimensionless. Velocity of air jet's outlet from inflow holes of different size and axial velocities in empty and filled bus passenger compartment of different production models have been measured. Dependence of compressing coefficient from three factors is determined and presented as a chart. Obtained chart has been approximated by equation.

Keywords: Air distribution · Compressed air jet · Air velocity · Jet compression coefficient

1 Introduction

The bus passenger compartment is filled largely with seats and passengers (sitting and standing). That is why, it is necessary to take into account dependence of jet's compression coefficient k_c from several determinative factors (Lorin et al. 2007; Rumsey and Spalart 2009; Srebric and Chen 2002; Voznyak et al. 2015; Grimitlin 2004). The compression coefficient k_c should be considered as the relation of the jet's axial velocity v_{xc} in a limited space to the axial velocity v_x in a free space on the same distance (running coordinate) from inflow hole. The resolving task of the air distribution in the passenger compartment is subject to compliance with the internal air standard parameters (Gumen 2016; Dovgaliuk and Milejkovski 2002; Dovgaliuk and Milejkovski 2013; Dovgaliuk and Milejkovski 2008). This article presents generalized

Z. Blikharskyy et al. (Eds.): CEE 2019, LNCE 47, pp. 518–525, 2020.
https://doi.org/10.1007/978-3-030-27011-7_66

relation between the compression coefficient k_c and three arguments: a running coordinate, a passenger compartment filling degree by chairs and persons, a jet compressing degree.

2 Analysis of Literary Data and Problem Statement

The air velocity in the room working area should be within the normative limits (Kapalo et al. 2014; Korbut 2017). It is possible due to velocity attenuation (Voznyak et al. 2015). The initial intensity of turbulence significantly influences on the velocity attenuation and temperature coefficients (Allmaras 1999; Allmaras et al. 2012; Zhao et al. 2003; Chen and Srebric 2001; Holyoake 2006; Huo et al. 2000). It is known, that the one of the main turbulence generators is the local resistance sets of air moving, particularly twisting plates, bends, change of moving direction, etc. (Grimitlin 2004). When air is going through these devices there will be significant gradient of average velocities, which causes the formation of vortices. As the consequence, increases the intensity of air flow turbulence (an average of 15–20%) (Grimitlin 2004). An impulse of external forces is created due to the difference in static pressure at the formation site that is directed towards to the flow and causes loss of the motion in it. Thus, the loss of the motion amount of air jet (that flows from the outlet) causes the decrease in the velocity attenuation coefficient (Grimitlin 2004).

3 The Purpose and Objectives of the Study

The purpose of the investigations is determination of compression coefficient k_c and, finally, air jet's initial velocity v_0.

The objectives of the study are to perform experimental studies of air distribution by the transverse flat spread compressed air jets and swirl compressed air jets.

4 Methods, Materials and Research Results

In this regard, it is proposed to apply a two-flow air distributor (TFAD) (Fig. 1) and inflow-exhaust hatch combined (IEHC) (Fig. 2). It should be noted that there will also be an uneven velocity field in the outlet of the air distributor and the lateral raising of the air flow. This means that when the air is emitted from the holes of IEHC the intensity of turbulence will be higher than with direct airflow. As the result, the value of the velocity attenuation coefficient will be lower. Based on the foregoing, it can be stated that the transverse air distribution implementation by the IEHC is effective in terms of the velocity and temperature attenuation.

Based on the review of inflow jets development characteristics (Allmaras 1999; Allmaras et al. 2012; Zhao et al. 2003; Chen and Srebric 2001; Holyoake 2006), we can make the following generalizations:

1. There are methods for calculating air distribution systems with the free and compressed jets, the distribution of air parameters affects the air jet compression degree.

2. Engineering methods for the calculation of air distribution systems are based on the study about jet flows regularities or based on research statistical methods and it is not excluded to carry out calculations according to air jet regularities.
3. There is no method for calculating the air flow generated by a TFAD and by flat jets from IEHC.

To solve this problem it is needed:

1. To conduct the natural experimental research on the air distribution of the TFAD and IEHC. To establish calculation dependences for the theoretical solution of the air distribution in the bus passenger compartment.
2. To establish interconnection of air environment parameters in order to provide an optimal thermal condition of the bus.
3. To compare theoretical results with experimental data and to establish correction coefficients.

The flow long-range of inflow jets is directly related to the attenuation coefficient of the velocity m as it can be seen from (Allmaras 1999). Proceeding from this, for a concentrated air distribution, it is advisable to use as an air distributor the outlet having a direct section of the air line and supplying jets that are applied to the surface of the ceiling, the coefficient $m > 1$.

The coefficient of velocity attenuation of the air distributor is determined by equation:

$$m = \frac{v_x}{v_0} \frac{x}{\sqrt{F}} \tag{1}$$

where v_x and v_0 are accordingly air velocity on the axis of the stream and initial air velocity of a stream, m/s.

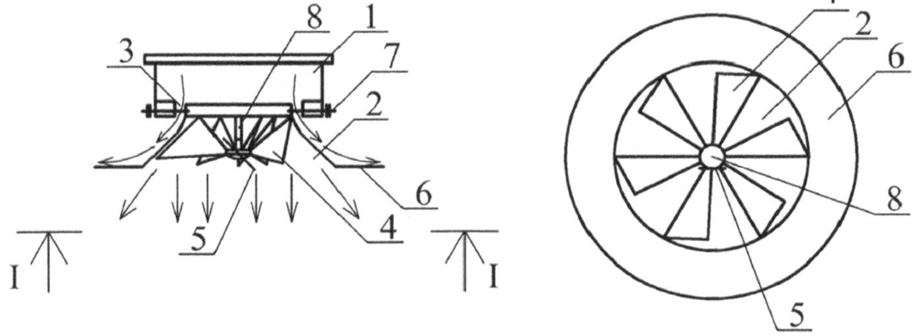

Fig. 1. Two-flow air distributor with the formation of a swirl and flat spread jet: 1 – tidal nozzle; 2 – diffuser; 3 – annular regulated crack; 4 – plates; 5 – control knob; 6 – shield; 7 – screw adjustment slit, 8 – rod

To keep the required (rated) air velocity in the working area is one of the important question that appear in designing the microclimate maintaining systems in small volume rooms. It is natural to consider that bus passenger compartment as the small volume room, which is characterized by small volume due to the chairs and the passengers (Fig. 3). This factor results in increasing compressing degree of the supply inflow jet and causes more intensive falling of the jet velocity and the temperature.

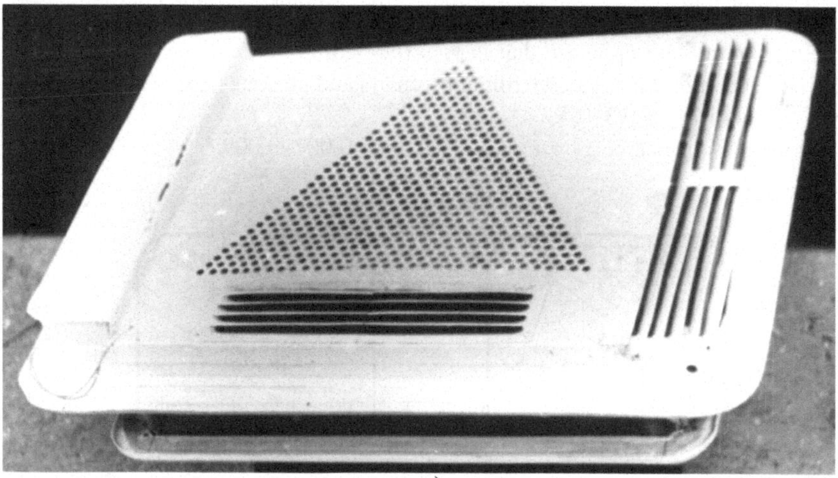

a)

b)

Fig. 2. Ventilation hatch combined

The transverse flat compressed air jets and swirl compressed air jets that flow respectively from IEHC and from TFAD in bus passenger compartment, which is considered to be as a limited space, are characterized by coefficient of compression k_c. It was assumed, that following parameters influence on this coefficient (Fig. 3):

- the aperture square $F_{0,}$ m$^{2;}$
- the jet's space crossing section $F_{j,}$ m$^{2;}$

- the running coordinate x, m;
- the total crossing section of the bus passenger compartment F, m;
- the distance to the serviced area of bus passenger compartment l, m.

There was carried out multifactor experiment with 3 factors, where physical variable values were changed into the dimensionless during the investigations. For experimental research was compiled an experiment planning matrix (Table 1) with taking into account factors interaction. That means was adopted a nonlinear mathematical model. There are taken such values as determinative factors:

- running coordinate $x_1 = x/l$, that is $x_1 = 0{,}1–1{,}0$;
- passenger compartment degree filling by chairs and persons (sitting and standing) $x_2 = F_j/F$, that is $x_2 = 0{,}3–0{,}5$;
- jet compressing degree $x_3 = F_0/F$, that is $x_3 = 0{,}001–0{,}009$.

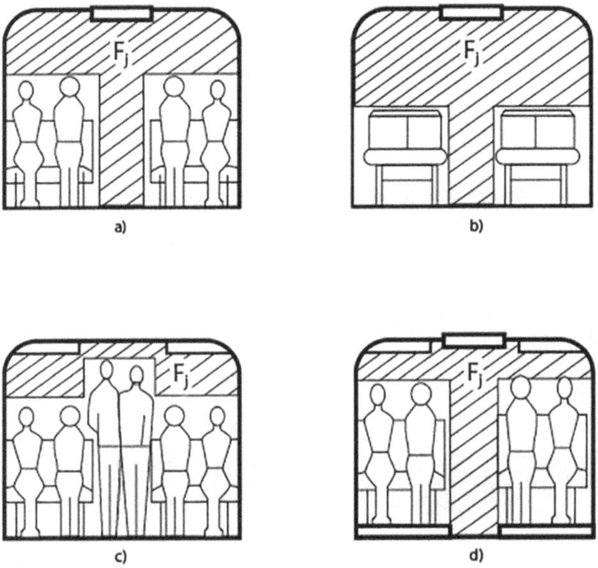

Fig. 3. Calculating cross sections of the bus passenger compartment for the air jets: (a) intercity bus with only the sitting passengers $F_j \approx 0{,}5\ F$ (b) bus without passengers $F_j \approx 0{,}65\ F$ (c) city bus with the sitting and standing passengers $F_j \approx 0{,}3\ F$ (d) turist bus with only the sitting passengers $F_j \approx 0{,}4\ F$.

Measurements of the air velocity v were carried out by thermal electrical anemometer Testo-405 using the coordinator with a grid of points 10 × 10 cm.

Measurements were carried out in cycles for at least two times; when the results differed more than 10% they were performed for the third time, while the randomization was maintained. In addition, the velocity measurement at each point was made for at least 120 s with a discrete (after 10 s) recording of results and their subsequent averaging.

Air distribution experimental research was carried out in passenger compartments of different LAZ (Lviv Bus Plant) bus production models (LAZ-695, LAZ-4202, LAZ-42021, LAZ-4204). Various air distributors provide changing for aperture square F_0 and as the result changing of factor $x_3 = F_0/F$. Presence and absence of passengers (filled and empty bus salon) as well as using different bus production models (Fig. 3) gives a possibility to vary factor $x_2 = F_j/F$. Value x_1 is varied as any coordinate on a scale from 0 to l as complete length to the serviced area of bus passenger compartment. Dependency of compression coefficient k_c from these 3 factors is shown in Fig. 4 as a chart, which is approximated by Eq. (2):

$$k_c = 0,4 - 52 \cdot \frac{F_0}{F} + 0,8 \cdot \left(1 - \left(2,87 - 5 \cdot \frac{F_j}{F}\right) \cdot \frac{x}{l}\right)^2. \tag{2}$$

Table 1. The matrix of the 3-factor experiment planning

N	x_0	$x_1 = x/l$	$x_2 = F_j/F$	$x_3 = F_0/F$	$x_1\ x_2$	$x_1\ x_3$	$x_2\ x_3$	$x_1 x_2\ x_3$	$\overline{v}_{xc}$
1	+	–	–	–	+	+	+	–	0,80
2	+	+	–	–	–	–	+	+	0,23
3	+	–	+	–	–	+	–	+	0,57
4	+	+	+	–	+	–	–	–	0,05
5	+	–	–	+	+	–	–	+	1,00
6	+	+	–	+	–	+	–	–	0,37
7	+	–	+	+	–	–	+	–	0,76
8	+	+	+	+	+	+	+	+	0,12

Due to this chart or Eq. (2) the compression coefficient k_c is determined for the different kinds of buses. The initial velocity v_0 of the air supply jet is calculated from Eq. (3):

$$v_0 = \frac{v_r}{\left(1 - 0,5 \cdot \overline{R}^{1,4}\right) \cdot m \cdot k_c} \cdot \sqrt{\frac{x}{b}} \tag{3}$$

where v_r – rated air velocity in the serviced area for the specific kind of the bus; m/s; b – height of the air supply aperture; m; $\overline{R}$ – dimensionless distance from the considered point to the jet axis in the definite cross section; m – coefficient of the velocity attenuation (Allmaras 1999).

Equation (3), which includes coefficient of jet's compression k_c that can be determined from chart (Fig. 4) or Eq. (2), is final for calculation of air jets in bus passenger compartment.

The data presented gives the possibility to calculate initial velocity v_0 of leakage of the flat spread compressed transverse and swirl air jets at condition of the known rated velocity maintaining v_r in the serviced area of the LAZ bus passenger compartment.

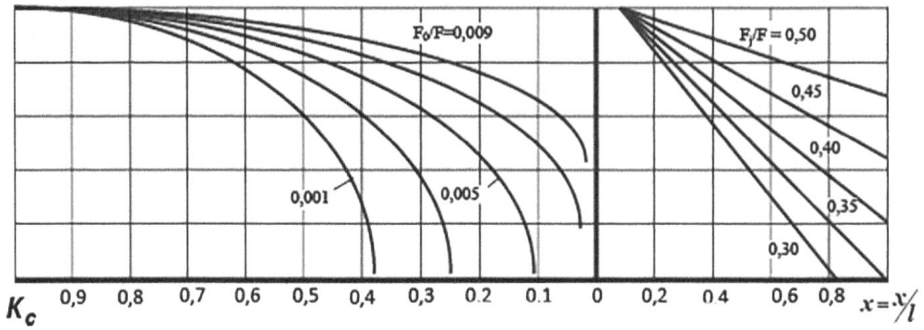

Fig. 4. Chart for obtaining compressing coefficient k_c of the flat jets.

5 Conclusions

1. Theoretical and experimental studies of air distribution in the bus passenger compartments have confirmed that in terms of velocity attenuation the IEHC and TFAD with air supply, respectively, with transverse and swirl air jets are effective devices.
2. A chart has been designed and the calculated dependences for determining the current parameters of the inflow jets in the bus passenger compartment working area were obtained at different air distribution schemes. On this basis an engineering method of their calculation was created.

References

Allmaras SR (1999) Multigrid for the 2-D compressible navier-stokes equations. In: 14th computational fluid dynamics conference, American Institute of Aeronautics and Astronautics, Norfolk, USA

Allmaras SR, Johnson FT, Spalart PR (2012) Modifications and clarifications for the implementation of the spalart-allmaras turbulence model ICCFD7-1902. In: 7th International Conference on Computational Fluid Dynamics, Hawaii

Zhao B, Li X, Zhao QY (2003) A simplified system for indoor airflow simulation. Build Environ 38:543–552

Chen Q, Srebric J (2001) Simplified diffuser boundary conditions for numerical room airflow models. Final Report for ASHRAE RP-1009, Department of Architecture, Massachusetts Institute of Technology, Cambridge, MA, p 181

Holyoake (2006) Diffuser Performance Data Sheet, Ceiling Fixed Pattern Radial Swirl Diffuser, Model CFP Radial Induction Swirl Diffuser

Huo Y, Haghighat F, Zhang J, Shaw C (2000) A systematic approach to describe the air terminal device CFD simulation for room air distribution analysis. Build Environ 35(6):563–576

Kapalo P, Vilcekova S, Domnita F, Voznyak O (2014). Determine a methodology for calculating the needed fresh air. In: The 9th International Conference "Envi-ronmental Engineering", Vilnius, Lithuania Selected Papers eISSN 2029-7092/eISBN 978-609-457-640-9

Korbut V, Voznyak O, Myroniuk K, Sukholova I, Kapalo P (2017) Examining a device for air distribution by the interaction of counter non-coaxial jets under alternating mode. Eastern Eur J Enterp Technol 2(8(86)):30–38 ISSN 1729-3774

Lorin E, Benhajali A, Soulaimani AA (2007) Positivity preserving finite element-finite volume solver for the Spalart-Allmaras turbulence model. Comput Meth Appl Mech Eng 196(17–20):2097–2116

Rumsey CL, Spalart PR (2009) Turbulence model behavior in low reynolds number regions of aerodynamic flowfields. AIAA J 47(4):982–993

Srebric J, Chen Q (2002) Simplified numerical models for complex air supply diffusers. HVAC&R Res 8(3):277–294

Voznyak O, Sukholova I, Myroniuk K (2015) Research of device for air distribution with swirl and spread air jets at variable mode. Eastern Eur J Enterp Technol 6/7(78):15–23

Grimitlin M (2004) Indoor air distribution. S.Petersburg, p 320 (in Russian)

Gumen M, Dovgaliuk V, Milejkovski V (2016) Determination of the streams turbulence intensity with large-scale vortices on the basis of geometric and kinematic analysis of macrostructure. In Proceedings of Lviv Polytechnic National University: The theory and practice of construction, No 844, pp 76–83 (in Ukrainian)

Dovgaliuk V, Milejkovski V (2002) The calculation model of the stream, which is laid on the curvilinear surface. In: Proceedings of kyiv construction and architecture national university: ventilation, lighting and heat supply, vol 5, pp 26–46 (in Ukrainian)

Dovgaliuk V, Milejkovski V (2013) Analytical studies of the macrostructure of jet flows for the calculation of energy efficient air distribution systems. Energy Effi Constr Architect 4:73–81 (in Ukrainian)

Dovgaliuk V, Milejkovski V (2008) Estimated model of nonisothermal stream, which is laid out on a convex cylindrical surface. In: Proceedings of Kyiv Construction and Architecture National University: Ventilation, lighting and heat supply, vol 12, pp 11–32 (in Ukrainian)

Air Distribution Efficiency in a Room by a Two-Flow Device

Orest Voznyak[1(✉)], Vadym Korbut[2], Borys Davydenko[3], and Iryna Sukholova[1]

[1] Lviv Polytechnic National University, Bandery Street 12, Lviv 79013, Ukraine
orest.voznyak@i.ua
[2] Kyiv National University of Building and Architecture, Povitroflotsky Avenue 37, Kiev 00037, Ukraine
[3] Institute of Technical Heat Physics of National Academy of Science of Ukraine, Bulakhovskyy Street 2, Kiev 03164, Ukraine

Abstract. The article is devoted to decision of actual task of air distribution efficiency increasing due to swirl and spread air jets to provide normative air parameters in the production apartments. The mathematical model of air supply with swirl and spread jets in that type of apartments is improved. It is shown that for reaching of air distribution maximal efficiency it is necessary to supply air by jets, that intensively extinct before entering into a working area. Results of air flow simulation are presented. The graphical and analytical dependences are shown. Dynamic parameters of air flow that is created due to swirl and spread jets has been determined. Results of experimental investigations of air supply into the room by swirl jets are presented. The results of the theoretical and experimental investigations of dependence of the attenuation coefficients and coefficient of device local resistance from the angle of inclination of the twisting plates are presented. It was established that an increase in the angle results in increase of the attenuation coefficients and in decrease of the resistance coefficient of two-flow air distributor (TFAD). The optimum angle of the plates is determined considering aerodynamic and energy aspects.

Keywords: Air distribution · Swirl jet · Spread jet · Alternating mode · Velocity

1 Introduction

The effectiveness of human work depends largely on how the sanitary-hygienic parameters of the microclimate meet the physiological needs (Kapalo et al. 2014, 2018, 2019, Kapalo and Siroczki 2014). In small-sized industrial premises, where permanent staff is in operation, there is a need to supply a significant amount of inflow air. For this purpose, a two-flow air distributor is proposed which provides a high intensity of the velocity *v* and temperature *t* attenuation of the inflow stream with the formation of a swirling and spread air jet. Studies in the premises of both public and industrial buildings indicate that the human's warmth is favorably influenced by the variables of the stimulus (Korbut et al. 2017, Voznyak et al. 2015). The variable mode of inflow air

Z. Blikharskyy et al. (Eds.): CEE 2019, LNCE 47, pp. 526–533, 2020.
https://doi.org/10.1007/978-3-030-27011-7_67

jets means the creation of a dynamic microclimate and the thermal regulation of the human body is reflected positively. In essence, a person is exposed to a dynamic microclimate, being constantly in the natural environment, and therefore accustomed to it with constant fluctuations in air temperature, velocity and humidity. Thus, it is obvious that the periodic creation of a dynamic microclimate is also advisable in the closed spaces.

2 Analysis of Literary Data and Problem Statement

Analysis of literary data has shown that the effect of air velocity on a comfortable state of a person must be considered together with the temperature, humidity of the air environment of the room, temperature of the surfaces and thermal resistance of clothing. The influence of the alternating air supply regime on the state of health of a person and the efficiency of his/her labor (Korbut et al. 2017, Voznyak et al. 2015) is analyzed. In industrial premises, for the adaptation of the apparatus of thermal regulation, improvement of the state of health and reduction of workers' fatigue, especially for the monotonous nature of work, it is hygienically justified to change according to the periodic order one of the parameters, for example, the temperature or air velocity, that is, to create a dynamic microclimate (Voznyak et al. 2015). Along with the periodic change in the ventilation efficiency, the CO_2 concentration in the room is changed according to the periodic order (Kapalo et al. 2014, Kapalo and Siroczki 2014). Air distribution schemes are considered and analyzed for being most appropriate to be used indoors, taking into account the attenuation of inflow non-isothermal air jets. It is substantiated that the maximum efficiency can be achieved using horizontal flat jets with the effect of their laying, as well as vertical twisted streams (Korbut et al. 2017, Lorin et al. 2007).

Fig. 1. Natural model of TFAD device with Belimo automation

The TFAD was developed, which creates a swirl and flat jet stream flowing in analternating mode (Korbut et al. 2017, Lorin et al. 2007) (Fig. 1). The main task of this system is to create a dynamic microclimate in the room, providing a comfortable state of health of people in the working area, as well as achieving energy and investment savings. In the application of TFAD, the variation of the flow rate and the angle of inclination of the swirling plates is provided by Belimo automation (Fig. 1).

It is needed to consider indoor airflow simulation (Lorin et al. 2007, Rumsey and Spalart 2009, Srebric and Chen 2002) and mathematical treatment of results should be done by Zorych (1981), Grimitlin (2004), Gumen et al. (2016).

3 The Purpose and Objectives of the Study

The purpose is to determine the characteristics of the air flow that is created by swirl and flat spread air jets during their leakage and obtaining analytical dependencies for determining the required dynamic parameters of air jets; to determine the relationship between the coefficient of local resistance and the rate of attenuation of the velocity and temperature of the air flow; optimize the angle of inclination of the swirling plates of the air distributor.

The objectives of the study are to perform experimental studies of air distribution in stationary and alternating modes by the TFAD.

4 Methods, Materials and Research Results

The flow of air in a non-stationary regime from a TFAD in the form of non-isothermal air jets, which determine its parameters (Fig. 1) (Dovhaliuk and Mileikovskyi 2007, 2008, 2013) was considered.

The axial velocity v_x and the excess temperature Δt_x at the calculated point A with the coordinate x_A in the case of steady motion (partial case of non-stationary mode) was determined from the known formulas for calculating the axial velocity v_x and the excess temperature Δt_x at the known initial values:

$$v_x = v_0 \cdot m \frac{\sqrt{F_0}}{x}; \tag{1}$$

$$\Delta t_x = \Delta t_0 \cdot m \frac{\sqrt{F_0}}{x}. \tag{2}$$

where v_0, v_x – respectively, the initial and axial air velocity, m/s;
m, n, – coefficients of attenuation according to velocity and temperature;
Δt_0, Δt_x – respectively, the initial and axial excess temperatures, °C;
F_0, – air outlet space area, m^2;
x – current coordinate, m.

As it can be seen from (1) and (2), in order to provide normative values of air velocity and temperature in the working area it should be possible to reduce the velocity and temperature attenuation coefficients. This is achieved by decreasing of the twisting plates angle, but at the same time it significantly increases both devices resistance and the system as a whole. Consequently, it is necessary to carry out experimental studies and to establish numerical relationship between these values.

Experimental studies were carried out under such conditions and simplifications:

- the air jets are non-isothermal;
- the initial air velocity in the air distributor was within the limits: v_0 = 5–15 m/s;
- the initial excess air temperature in the air distributor was: Δt_0 = 5–8 °C;
- the air flow rate in the experiments was within: L = 200–500 m^3/h.

Measurements of air velocity and temperature were carried out by thermal electrical anemometer Testo-405 using a coordinate system with a grid of points 5 × 5 cm.

Based on the results, graphs (Fig. 2) are designed.

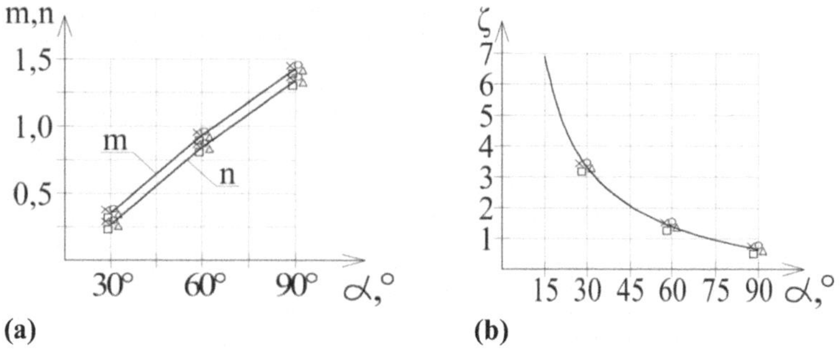

Fig. 2. Charts for determination (a) of attenuation coefficients of the velocity m and the temperature n of the air jet (b) as well the coefficient of local resistance ζ depending from the angle of inclination α of the twisting plates

In Fig. 2(a) the results of attenuation coefficients research of velocity m and temperature n of an air distributor are presented, and in Fig. 2(b) – coefficient of TFAD local resistance ζ depending from the angle of inclination α of the twisting plates. These graphs (Fig. 2) are approximated by Eqs. (3) (4) for determining of the attenuation coefficients m and n depending from the angle of inclination α of the twisting plates:

$$m = 0{,}02 \cdot \alpha - 0{,}24; \quad (3)$$

$$n = 0{,}013 \cdot \alpha - 0{,}115, \quad (4)$$

and for determining of the local resistance coefficient ζ of TFAD – hyperbolic (5) and power (6) equations to achieve higher accuracy of the results and greater reliability:

$$\zeta = \frac{90}{\alpha}; \quad (5)$$

$$\zeta = 2{,}9 - 0{,}25\sqrt{\alpha - 30}. \quad (6)$$

It is advisable to verify the results obtained and to propose a combined formula (7) for determining the local resistance coefficient ζ:

$$\zeta = 0{,}5\left(2{,}9 - 0{,}25\sqrt{\alpha - 30} + \frac{90}{\alpha}\right). \quad (7)$$

In order to conduct an analytical study of the relationship between the values of *m*, *n* and ζ, it is necessary to set these functions explicitly and graphically represent them in Fig. 3.

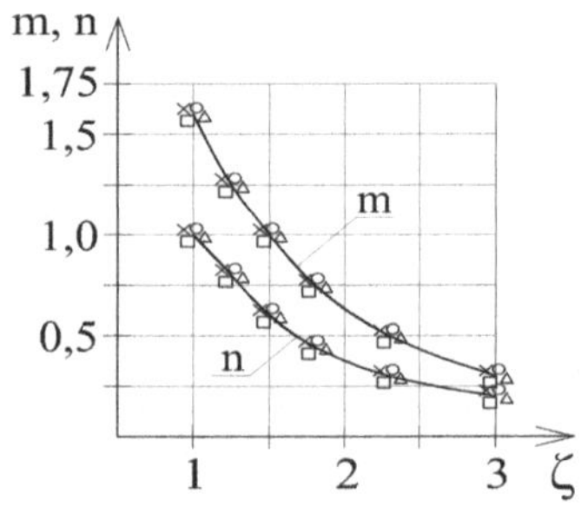

Fig. 3. Charts of attenuation coefficients dependence from the coefficient of TFAD local resistance

The graphs (Fig. 3) are approximated by dependences (8) and (9):

$$m = \frac{0{,}9}{\zeta - 0{,}5}; \quad (8)$$

$$n = \frac{0{,}7}{\zeta - 0{,}5}. \quad (9)$$

Numerical values of the attenuation coefficients of the velocity $m = 0.36 - 1.56$ and temperature $n = 0.27 - 1.04$ indicate the high efficiency of the application of twisted jets. However, the coefficients of attenuation substantially increase at an angle of inclination of plates 90°, since the inflow stream is close to the straight-line. This phenomenon is negative because the aim is to maximize the attenuation coefficient. The aerodynamic resistance of the air distributor $\zeta = 2{,}9$ is determined, which varies depending from the angle of inclination of the twisting plates. The highest value of the aerodynamic resistance is $\zeta = 2{,}9$ for the angle of inclination of the twisting plates $\alpha = 30°$ and with the increase of this angle the coefficient ζ decreases, which is positive fact.

Consequently, in terms of velocity and temperature attenuation, it is effective to use the device at smaller angles of the torsion plates, where the attenuation coefficients are minimal. But at the same time, in terms of aerodynamic resistance, the situation is completely opposite. Therefore, it is quite logical to see the optimization problem of determining the optimal angle of inclination of the twisting plates, which would satisfy the appropriate requirements, such as the velocity attenuation m of the air jet (material content of the system) and the aerodynamic resistance ζ (energy intensity of the system).

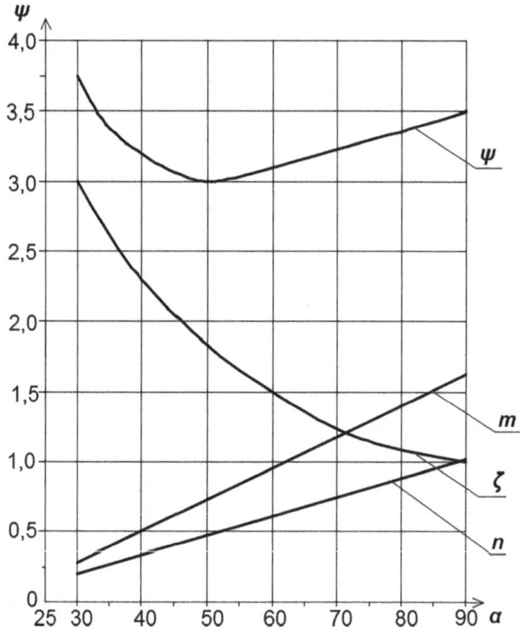

Fig. 4. Graph for determining of the twisting plates inclination optimal angle α taking into account coefficient of attenuation of temperature n.

To solve this problem, it is proposed to introduce an additional function ψ, which combines these values and represents their sum: $\psi(\alpha) = m(\alpha) + \zeta(\alpha) + n(\alpha)$. This representation is correct, since all of these quantities are dimensionless, depend on the same argument (the angle of inclination of the twisting plates α) and are given by analytical expressions (3–7).

Graphically, it is presented in Fig. 4, but in analytical form – in expression (10):

$$\psi = 0{,}02\alpha - 0{,}24 + 0{,}5\left(2{,}9 - 0{,}25\sqrt{\alpha - 30} + \frac{90}{\alpha}\right) + 0{,}013\alpha - 0{,}115. \qquad (10)$$

Along with the graphical method of determining the optimal angle of inclination also the analytical method has been applied. To do this, we differentiate the expression (10), equate the derivative to zero, and obtain the equation of the fifth degree (11), which is unreasonable to solve by an algebraic method.

$$0{,}1 \cdot \alpha^5 - 4 \cdot \alpha^4 - 46{,}08 \cdot \alpha^3 - 1382{,}4 \cdot \alpha^2 + 51840 \cdot \alpha - 1{,}55 \cdot 10^6 = 0. \qquad (11)$$

Instead, an iteration method is proposed, for which we obtain the solution $\alpha_0 = 52°$. Consequently, graphically, we obtain $\alpha_0 = 50°$, and the analytical Eqs. (10) and (11) – $\alpha_0 = 52°$.

5 Conclusions

1. The two-flow air distributor is proposed, which provides increasing of intensity of air jet attenuation and the possibility of air supply in an alternating mode, that is, the creation of a dynamic microclimate in the room; TFAD's geometric characteristics are determined and optimized.
2. It has been determined that in order to achieve an increase in the velocity attenuation of the air flow, it is necessary to reduce the angle of inclination of the twisting plates and apply the effect of the jet laying; its optimum value is $\alpha_0 = 52°$.
3. Application of TFAD substantially increases the quality of air distribution for supplying a significant amount of air into small-scale industrial premises.

References

Kapalo P, Domnita F, Bacotiu C, Podolak M (2018) The influence of occupants' body mass on carbon dioxide mass flow rate inside a university classroom - case study/Peter Kapalo... [et al.] - 2018. Int J Environ Health Res 28(4):432–447. https://doi.org/10.1080/09603123.2018.1483010. ISSN 0960-3123

Kapalo P, Meciarova L, Vilcekova S, Burdova E, Domnita F, Bacotiu C, Peterfi K (2019) Investigation of CO_2 production depending on physical activity of students. Int J Environ Health Res 29(1):31–44. https://doi.org/10.1080/09603123.2018.1506570. ISSN 09603123

Kapalo P, Siroczki P (2014) Calculating the intensity of ventilation in classrooms on the basis of measured concentrations of carbon dioxide in slovakia - case study. Int J Vent 13(3):247–257. ISSN 1473-3315. WOS:000348585900004

Kapalo P, Vilcekova S, Domnita F, Voznyak O (2014) Determine a methodology for calculating the needed fresh air. In: The 9th international conference "environmental engineering", Vilnius, Lithuania selected papers. eISSN 2029-7092/eISBN 978-609-457-640-9

Korbut V, Voznyak O, Myroniuk K, Sukholova I, Kapalo P (2017) Examining a device for air distribution by the interaction of counter non-coaxial jets under alternating mode. Eastern Eur J Enterp Technol 2(8)(86):30–38. ISSN 1729-3774

Lorin E, Benhajali A, Soulaimani AA (2007) Positivity preserving finite element-finite volume solver for the Spalart-Allmaras turbulence model. Comput Meth Appl Mech Eng 196(17–20):2097–2116

Rumsey CL, Spalart PR (2009) Turbulence model behavior in low reynolds number regions of aerodynamic flowfields. AIAA J 47(4):982–993

Srebric J, Chen Q (2002) Simplified numerical models for complex air supply diffusers. HVAC&R Res 8(3):277–294

Voznyak O, Sukholova I, Myroniuk K (2015) Research of device for air distribution with swirl and spread air jets at variable mode. East Eur J Enterp Technol 6/7(78):15–23

Grimitlin MI (2004) Air distribution in the room. Issue 3, adapted and supplemented, publication AVOK North-West, 320 p (in Ukrainian)

Gumen OM, Dovhaliuk VB, Mileikovskyi VO (2016) Determination of the intensity of turbulence of streams with large-scale vortices on the basis of geometric and kinematic analysis of macrostructure. J Lviv Polytech Natl Univ Ser Theory Build Pract 844:76–83 (in Ukrainian)

Dovhaliuk VB, Mileikovskyi VO (2007) Efficiency of organization of air exchange in heat-stressed premises in compressed conditions. J Build Ukraine 3:36 (in Ukrainian)

Dovhaliuk VB, Mileikovskyi VO (2008) Estimated model of non-isothermal stream, which is laid out on a convex cylindrical surface. In: Ventilation, illumination and heat and gas supply: scientific and technical collection, Issue 12, Kyiv, KNUBA, pp 11–32 (in Ukrainian)

Dovhaliuk VB, Mileikovskyi VO (2013) Analytical studies of the macrostructure of jet currents for calculating energy-efficient systems of air distribution. Energy Effic Constr Arch (4):11–32 (in Ukrainian)

Zorych VA (1981) Mathematical analysis. Science 427 (in Russian)

The Impact of Air Flows on the Environment

Orest Voznyak[1(✉)], Khrystyna Myroniuk[1], Iryna Sukholova[1], and Peter Kapalo[2]

[1] Lviv Polytechnic National University, Bandery str 12, Lviv 79013, Ukraine
orest.voznyak@i.ua
[2] Faculty of Civil and Environmental Engineering, Technical University of Kosice, Košice, Slovakia
peter.kapalo@tuke.sk

Abstract. The aim of this work is the investigation of different outlets acoustic properties from the different air jets leakage, which flow out from outlets under varying conditions. Such holes as: rectangular chink for a compact air jet; rectangular chink with a bend for a flat air jet; round outlet for a swirl air jet; round outlet for air jets interaction; rectangular chink for the flat air jets interaction. In this article has been determined relationship between such factors as a noise level, air flow rate and inflow hole size at air jet leakage. There are considered some sources of noise: fan, vibration of ventilation piping metal walls and aerodynamic noise of the air jet flowing out from outlet. There has been designed the chart of sound level's dependence from the initial air velocity and dimensionless outlet square. Obtained curves was approximated by equation. Comparing chart for the round, compact, swirl, flat air jets and their interaction, which flow out from a rectangular aperture with a bend, rectangular chink's aperture and cylindrical pipe with a bend, we shall note that noise level is the highest at flowing out of flat jet in cause of a rectangular chink with a bend.

Keywords: Air jet · Acoustic properties · Noise level · Air distribution · Air velocity

1 Introduction

The factor of noise appearance at jet's flowing out from air distributing devices is necessary to take into account in designing ventilation and air conditioning systems (Lau and Huang 2018, Zhao et al. 2018). Basically, noise generation at air jet's leakage from outlet depends from its constructive realization and air jet flow velocity (Lai et al. 2018, Kobayashi et al. 2017, Choi and Pate 2017). As a result of noise emissions influences on the environment (Zhang et al. 2012, Katinas et al. 2016), solution of this problem is very significant.

The aim of the paper is investigation of flat and round jet's acoustic properties, which flow out from outlets at the different conditions. They could have different forms as flat and compact air jet leakage at its direction change or without any change. Task of this article is relationship determination between such factors as the noise level, air flow rate and incoming hole size at air jet leakage.

To solve this problem it should be carried out experimental investigations.

Z. Blikharskyy et al. (Eds.): CEE 2019, LNCE 47, pp. 534–540, 2020.
https://doi.org/10.1007/978-3-030-27011-7_68

The first part of this paper addresses to state of the problem, conditions of the experimental research and the measurement equipment.

The next one addresses to the analysis obtained experimental results and to the chart creation has been realized. Then, this diagram has been approximated by analytical equation.

The conclusion of this paper focuses on the assessment of the proposal data in design of ventilation and air conditioning systems.

2 Analysis of Literary Data and Problem Statement

The noise factor is important under leakage from the air supply devices, when designing ventilation and air conditioning systems in rooms. The generation of noise by air-forming nozzles depends largely on its design performance and the flow rate of the jets. There are known acoustic characteristics of air distributors such as cylindrical pipe with a bend, rectangular slit openings (Voznyak and Kovalchuk 2002). However, it is possible to use air-distributing units in ventilation and air conditioning systems with the use the interaction effect of counter-non-coaxial round jets, opposing non-uniform slits or an air distributor, which creates a swirl inflow jet in an accompanying flow.

In this work there is proposed the two-flow air distributor (TFAD) (Fig. 1), which provides normative velocity and temperature in the working area of the room, creates dynamic microclimate in the room due to automation.

Fig. 1. Natural model of TFAD with Belimo automation.

3 The Purpose and Objectives of the Study

The purpose of this work was to study and research acoustic properties of different air jets, which flow out from the different outlets at conditions like this.

To achieve the goal, it should be doing the following research objectives: theoretical and experimental research, concerning noise level determining, measurement of air jet velocity and flow rate, measurement of vibration of metal pipes' walls.

4 Methods, Materials and Research Results

There is known acoustic characteristics of air distributors like cylinder pipe with fitting, rectangular openings with chinks (Ye 1985, 1974). Nevertheless, we often use these air distribution devices with change of leakage direction (Fig. 2) due to flow twist in Ventilation and Air Conditioning Systems (Korbut et al. 2017, Lee et al. 2009, Voznyak and Kovalchuk 2002, Voznyak et al. 2015).

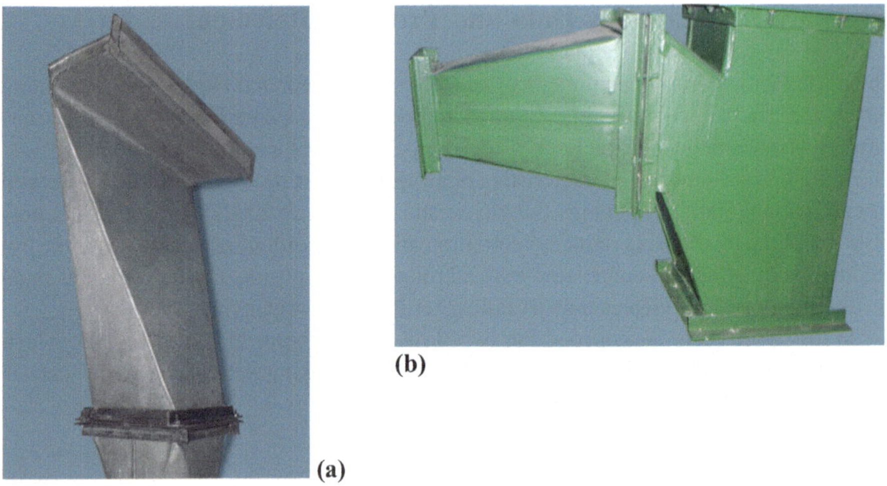

(b)

(a)

Fig. 2. (a) Rectangular chink with a bend for a flat air jet; (b) rectangular chink for a compact air jet.

The experimental research of air supply devices with interaction of opposed non-coaxial round and flat air jets as well device that creates swirl air jet have been carried out on the installations, that are presented respectively on Fig. 3a, b and c, under such conditions and simplifications:

- jets are isothermal;
- supply nozzles – cylindrical nozzles and rectangular openings with coefficients of the velocity attenuation accordingly $m = 6, 8$ and $m = 2, 5$;
- linear dimension of nozzles didn't change: diameter of cylindrical nozzles $d_o = 50$ mm, width of rectangular opening $b_o = 20$ mm;
- linear dimension of air ducts didn't change and was equaled to $H = 1, 5$ m;
- relation of air flow rates that interact remained unchanged, with $L_L = L_R$;
- the diameter of the opening for swirl supply air jet remained constant and was $d_{sw} = 46$ mm;
- the diameter of the opening for spread supply air jet remained constant and was $d_{sp} = 98$ mm;

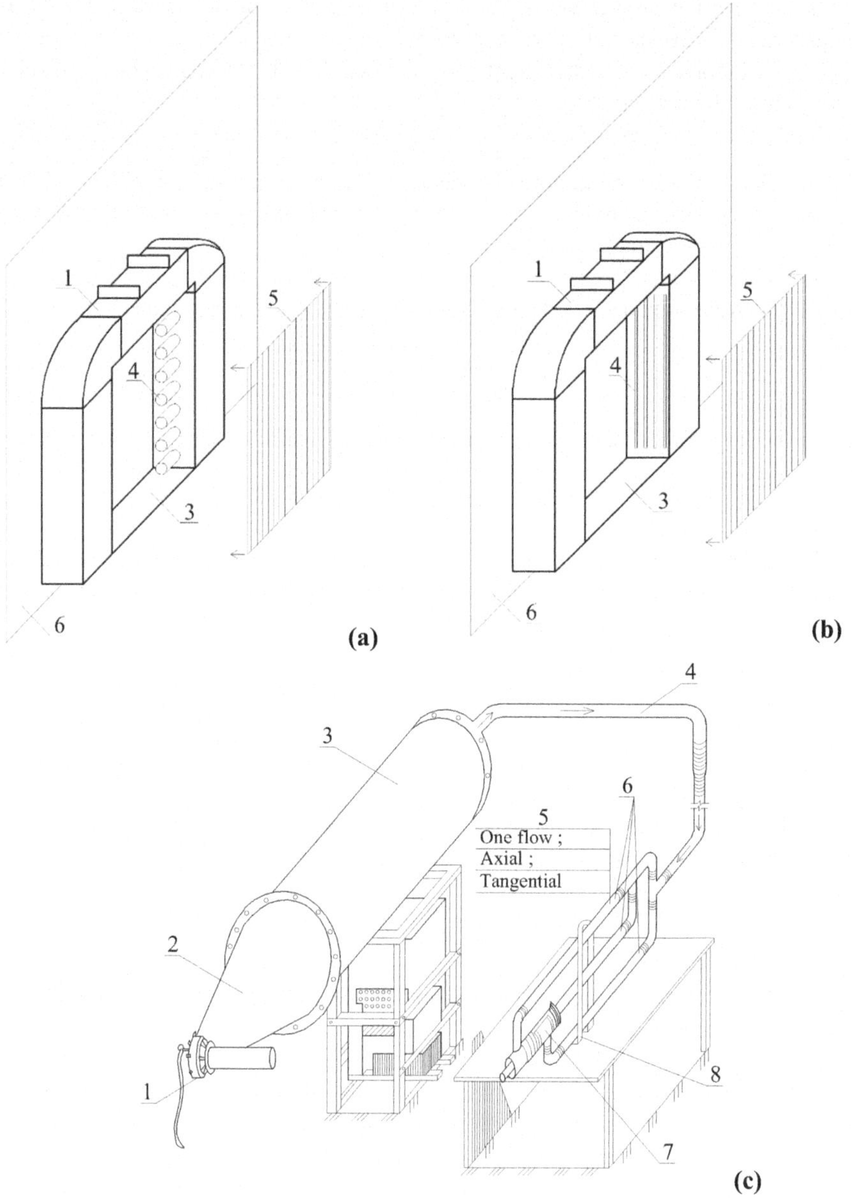

Fig. 3. Schemes of experimental installations (a) air supply device with interaction of opposed non-coaxial circular air jets; (b) air supply device with interaction of opposed non-coaxial flat openings: 1 – air duct; 2 – constructive part of air supply device; 3 – constructive part of air supply device; 4 – supply cylindrical nozzles $d_o = 50$ мм; supply openings $b_o = 20$ мм; 5 – netting; 6 – basis; (c) air supply device that creates swirl supply jet in one flow: 1 – ventilator; 2 – contractor; 3 – chamber of static pressure; 4 – collector (flexible duct); 5 – air supply device; 6 – valve; 7 – nozzle; 8 – stand.

- the hole for the production of the swirl jet beetle out of the aperture for the production of an spread jet on X_0 = 20 mm;
- rectangular outlet size was: height – 40 mm, width – 200, 400 and 600 mm; amount of inflow air was variable;
- initial velocity of air in nozzles for supply was in the range of: v = 3–10 m/s.

Air velocity has been measured by electrical anemometer Testo-405, level of sound capacity has been measured by noise meter ШМ-1-M1 and vibration of air pipes' metal walls – by vibration-meter AV-160A.

As follows from (Ye 1985), vibration's quantity at spectrum measurement are carried out in decibels (V – velocity), than meaning of V is determined:

$$V = 10^{\frac{L_V}{20}} \cdot 5 \cdot 10^{-5}\mathrm{mm/s} \tag{1}$$

On the basis of dependence (1) there is level of sound capacity in decibels:

$$L_v = 20(\lg V + 5 - \lg 5) \tag{2}$$

There are some sources of noise: fan, vibration of ventilation piping metal walls and aerodynamic noise of the air jet flowing out from outlet. Summary level of sound capacity is determined according to formula (3):

$$L = 10\lg\left(10^{\frac{L_1}{10}} + 10^{\frac{L_2}{10}} + 10^{\frac{L_3}{10}}\right) \tag{3}$$

where L_1, L_2, L_3 – are levels of sound capacity, which are created (according to) by fan, vibration and jet's air flow.

On the basis of (3) there is obtained a value of sound capacity, that is created by flat and compact air jet flowing out, taken separately:

$$L_3 = 10\lg\left(10^{\frac{L}{10}} - 10^{\frac{L_1}{10}} - 10^{\frac{L_2}{10}}\right) \tag{4}$$

where L_2 and L – values, that were measured by experimental method accordingly by vibration-meter and noise-meter.

The dependence (4) gives the possibility to estimate the acoustic characteristics of the air flat jet, which flows out from outlet – chink with height 40 mm with the bend.

According to experimental results obtained there has been designed the chart (Fig. 4) of sound level's dependence L on initial air velocity v_0 (m/s) and dimensionless square $\overline{S}$ as relation of outlet's free section S to $S_0 = 0,01\,\mathrm{m}^2$ (Ye 1974), that is $\overline{S} = S/S_0$.

These curves are approximated by Eq. (5):

$$L = \left(1,875\overline{S} + 42,5\right)V^{0,18} \tag{5}$$

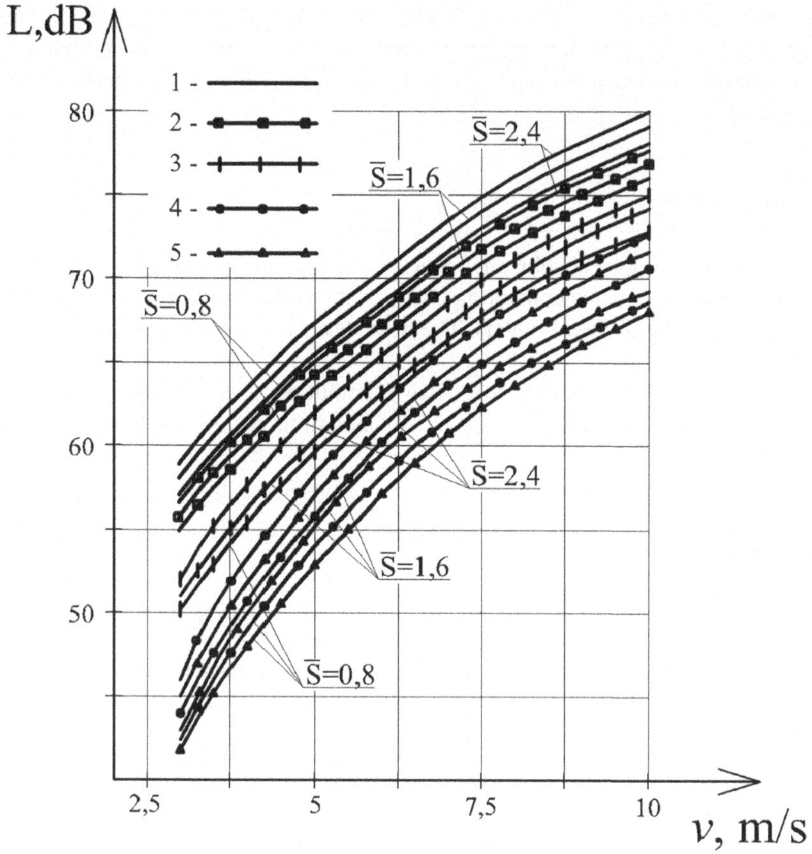

Fig. 4. Chart of noise level L dependence from air jet velocity v and dimensionless square $\overline{S}$ of the different outlets at the different air jets leakage: 1 – rectangular chink for a compact air jet; 2 – rectangular chink with a bend for a flat air jet; 3 – round outlet for a swirl air jet; 4 – round outlet forair jets interaction; 5 – rectangular chink for the flat air jets interaction

In Fig. 4 the similar curves for incoming air jet from round outlet with a bend (Fig. 2a) and rectangular chink (Fig. 2b) are shown. They are constructed on the basis of known analytical dependence:

$$L_p = 60\lg V + 30\lg \xi + 10\lg F + A \tag{6}$$

Where

V – average initial air jet's velocity, (m/s);
ζ – dimensionless coefficient of element's local hydraulic resistance, (-);
F – section's square of outlet, (m^2);
A – parameter, which depends from kind of the element.

A numerical simulation such as ANSYS_CFX (Voznyak et al. 2015) can be used to solve pipeline design and subsequent pressure loss solutions. These simulation tools make it possible to clearly visualize the real conditions in the air duct before it will be implemented.

5 Conclusions

We should note that noise level is the highest at flowing out of compact air jet in cause of a rectangular chink when comparing chart for the round, swirl, flat and compact jets flowing out from the different outlets. The noise level is the lowest for the device with flat air jets interaction.

It is possible to carry out control of air distribution selection taking into account limited noise level for particular capacity of incoming outlet based on the obtained dependences.

References

Lau ASH, Huang X (2018) The control of aerodynamic sound due to boundary layer pressure gust scattering by trailing edge serrations. J Sound Vib 432:133–154

Lai D, Qi Y, Liu J, Dai X, Zhao L, Wei S (2018) Ventilation behavior in residential buildings with mechanical ventilation systems across different climate zones in China. Build Environ 143:679–690

Korbut V, Voznyak O, Myroniuk K, Sukholova I, Kapalo P (2017) Examining a device for air distribution by the interaction of counter non-coaxial jets under alternating mode. East Eur J Enterp Technologies 2(8):30–38 (86). ISSN 1729-3774

Zhao K, Alimohammadi S, Okolo PN, Kennedy J, Bennett GJ (2018) Aerodynamic noise reduction using dual-jet planar air curtains. J Sound Vib 432:192–212

Lee K, Jiang Z, Chen Q (2009) Air distribution effectiveness with stratified air distribution systems. ASHRAE Trans 115(2):1–16

Zhang M, Kang J, Jiao F (2012) A social survey on the noise impact in open-plan working environments in China. Sci Total Environ 438:517–526

Kobayashi T, Sugita K, Umemiya N, Kishimoto T, Sandberg M (2017) Numerical investigation and accuracy verification of indoor environment for an impinging jet ventilated room using computational fluid dynamics. Build Environ 115:251–268

Katinas V, Marciukaitis M, Tamasauskiene M (2016) Analysis of the wind turbine noise emissions and impact on the environment. Renew Sustain Energy Rev 58:825–831

Voznyak O, Kovalchuk A (2002) Air distribution by opposite non-coaxial air jets. In: Proceedings. Zbornik prednasok: VII Vedecka konferencia s medzinarodnou ucastou Kosicko-Lvovsko-Rzeszowska, pp 173–178

Voznyak O, Sukholova I, Myroniuk K (2015) Research of device for air distribution with swirl and spread air jets at variable mode. East Eur J Enterp Technol 6/7(78):15–23

Choi W, Pate MB (2017) An evaluation and comparison of two psychoacoustic loudness models used in low-noise ventilation fan testing. Build Environ 120:41–52

Ye Y (1985) Noise control at work. Directory, Moscow, p 400 (in Russian)

Ye Y (1974) Noise protection. Moscow, p 111 (in Russian)

Influence of Heat Source Choice on Building Energy Certification Process and CO_2 Emissions

F. Vranay and Z. Vranayova(✉)

Department of Building Services, Faculty of Civil Engineering, Technical University of Kosice, Vysokoskolska Street, No. 4, 04002 Kosice, Slovakia
{frantisek.vranay, zuzana.vranayova}@tuke.sk

Abstract. In pursuit of energy savings, reducing production costs and securing heat to buildings, we are trying to find new suitable solutions. The intention is also supported by the requirements for energy efficiency of buildings. The requirements are defined by the value of the global indicator – Primary energy. In the article we focus on the category of residential buildings. From January 1st 2016, buildings will be refurbished to meet category A1. In cases justified by high investment intensity without return, it is possible to accept the reconstructed building with classification in category B. Measures are usually implemented on building structures that generally have the greatest impact on reducing heat demand. However, heat is "produced" in various sources to which energy is supplied to produce it. The total energy delivered for global assessment is driven by the efficiency of production, and the quantity of "primary energy factor," which is different for different types of energy. It is the primary energy factor that fundamentally influences the category to which the building under assessment gets. In the article, we will make a comparison on a real certified object, with the impact of different primary energy sources on the overall classification of the object.

Keywords: Energy certificate · Primary energy · CO_2 emissions · Thermal insulation · Heat pump

1 Introduction

Buildings are responsible for approximately 40% of energy consumption and 36% of CO_2 emissions in the EU, making them the single largest energy consumer in Europe. At present, about 35% of the EU's buildings are over 50 years old and almost 75% of the building stock is energy inefficient. At the same time, only 0.4–1.2% (depending on the country) of the building stock is renovated each year. Renovation of existing buildings can therefore lead to significant energy savings and play a key role in the clean energy transition, as it could reduce the EU's total energy consumption by 5–6% and lower CO_2 emissions by about 5% (EC 2015). Many authors deal with this appeal e.g. (Thiers and Peuportier 2012; Aste et al. 2013; Kerdan et al. 2016; Jouhara and Yang 2018; Guelpa et al. 2018). We deal with this area for Slovak republic conditions

Z. Blikharskyy et al. (Eds.): CEE 2019, LNCE 47, pp. 541–548, 2020.
https://doi.org/10.1007/978-3-030-27011-7_69

(Stefanco et al. 2014), but similar study are described for different countries as US, China, Germany, Poland (Wu and Skye 2018; Su et al. 2018; Peán et al. 2019; Deng et al. 2019; Stec et al. 2017).

2 Aim

Our paper is intended as deciding process what measures to use when reconstructing an object or when designing a new building project. For demonstration we used a real residential house, which has undergone thermal insulation of building structures.

3 Methods

3.1 Computational Assessment

We performed the computational assessment of the building before and after insulation. The methodology of environmental certification was used according to valid standards and regulations valid for the Slovakia. The criterion is the determination of total energy for heating and hot water. Primary energy is determined by the type of fuel used. We used different types of fuel for comparison. CO_2 emissions were calculated from the fuels used.

3.2 Measurement of Energy Consumptions

Measurements were performed in the operation of the current heat source - gas boiler. From the conclusions of the previous calculations, the use of a heat source heat pump appears to be most advantageous. In calculations, we used a replacement for water/water heat pump and an air/water heat pump. In our faculty laboratories we have installed these heat sources, where we monitored their operation. Also many years of experience in designing and operating other objects are taken into account in the calculation. We have developed mathematical models to simulate heating and hot water systems with these devices.

3.3 Description of Assessed Object

The building is a residential self-standing house with 5 above-standard apartments (Fig. 1). Facade orientation is south-north. The total heated floor area is 712 m^2 and the volume of 2 136 m^3. The source of heat is the gas hot water boiler (Fig. 2). The measurement is provided at the boiler inlet by a gas meter, the heat supplied to the heating at the outlet of the boiler and the heat at the hot water heating. The amount of hot water supplied is also measured.

Fig. 1. Object - south side

Fig. 2. Boiler room with heat consumption measurement for heating and hot water heating

4 Results

Table 1 lists the values and categories according to consumption sites. In terms of energy certification, the reconstructed object must meet class A1 according to primary energy. Variant "A" is an assessment of the object before insulation. The gas boiler is in the category "C" = unsatisfactory. The only fuel that will ensure a satisfactory condition (A0 or A1) is wood.

Table 1. Results of heat needs for the building before and after insulation

	A BEFORE insulation		B AFTER insulation	
HEAT HEATING NEEDED	92.0	kWh/m2	48.7	kWh/m2
		class		class
HEATING ENERGY	119.6	C	63.3	B
WATER HEATING ENERGY	27.8	C	27.8	C
TOTAL ENERGY	**147.4**	C	**91.1**	B

PRIMARY ENERGY = (TOT.ENERGY / FACT.TRANSFORMATION x FACT.PRIMARY ENERGY)

After thermal insulation the object gets into the category "B" and it is still unsatisfactory. Results for other types of fuel are shown in Tables 2 and 3. In addition to wood, a heat pump (A1) in Table 3 is used in the case of using a heat source. The next step was a comparison of measured consumption – object operation. The measurement was applied to the building in 2015. The whole year with the measurement before insulation is from 2016. During 2017, the object was insulated, so we do not use the data in the balances. The year 2018 is entering the comparison.

Table 2. Results of the energy certificate for the uninsulated object

DETERMINING THE GLOBAL PRIMARY ENERGY INDICATOR
A - VARIANT BEFORE INSULATION

TOTAL ENERGY =	147.5	kWh/(m2.a)	PRIMARY ENERGY	LIMIT kWh/(m2.a) FOR INDIVIDUAL CLASSES								EMISSIONS CO2	
	factor (v.324/2016)			A0	A1	B	C	D	E	F		factor emmisions	production
				32	63	126	189	252	315	378			
	transformation	primary energy	kWh/m2	X exceeds the limit								kg/kWh	kg/m2
natural gas	1.00	1.10	162	5.1	2.6	1.3	0.9	0.6	0.5	0.4	C	0.220	32.5
gas heat pump	1.60	1.10	101	3.2	1.6	0.8	0.5	0.4	0.3	0.3	B	0.220	20.3
LPG	1.00	1.35	199	6.2	3.2	1.6	1.1	0.8	0.6	0.5	D	0.220	32.5
coal	0.73	1.10	222	6.9	3.5	1.8	1.2	0.9	0.7	0.6	D	0.360	72.7
black coal	0.74	1.10	219	6.9	3.5	1.7	1.2	0.9	0.7	0.6	D	0.360	71.8
brown coal	0.70	1.10	232	7.2	3.7	1.8	1.2	0.9	0.7	0.6	D	0.360	75.9
heating oil	0.90	1.10	180	5.6	2.9	1.4	1.0	0.7	0.6	0.5	C	0.290	47.5
wood pellets	0.86	0.20	34	1.1	0.5	0.3	0.2	0.1	0.1	0.1	A1	0.020	3.4
wood chip	0.78	0.15	28	0.9	0.5	0.2	0.2	0.1	0.1	0.1	A0	0.020	3.8
wood piece	0.70	0.10	21	0.7	0.3	0.2	0.1	0.1	0.1	0.1	A0	0.020	4.2
wood gasification	0.83	0.10	18	0.6	0.3	0.1	0.1	0.1	0.1	0.0	A0	0.020	3.6
electricity	0.99	2.20	328	10.2	5.2	2.6	1.7	1.3	1.0	0.9	F	0.167	24.9
electric heat pump	3.40	2.20	95	3.0	1.5	0.8	0.5	0.4	0.3	0.3	B	0.167	7.2
centralized source	0.76	0.70	136	4.2	2.2	1.1	0.7	0.5	0.4	0.4	C	0.290	56.3

Table 3. Results of the energy certificate for the insulated object

DETERMINING THE GLOBAL PRIMARY ENERGY INDICATOR
B - VARIANT AFTER INSULATION

TOTAL ENERGY =	91.1	kWh/(m2.a)	PRIMARY ENERGY	LIMIT kWh/(m2.a) FOR INDIVIDUAL CLASSES								EMISSIONS CO2	
	factor (v.324/2016)			A0	A1	B	C	D	E	F		factor emmisions	production
				32	63	126	189	252	315	378			
	transformation	primary energy	kWh/m2	X exceeds the limit								kg/kWh	kg/m2
natural gas	1.00	1.10	100	3.1	1.6	0.8	0.5	0.4	0.3	0.3	B	0.220	20.0
gas heat pump	1.60	1.10	63	2.0	1.0	0.5	0.3	0.2	0.2	0.2	A1	0.220	12.5
LPG	1.00	1.35	123	3.8	2.0	1.0	0.7	0.5	0.4	0.3	B	0.220	20.0
coal	0.73	1.10	137	4.3	2.2	1.1	0.7	0.5	0.4	0.4	C	0.360	44.9
black coal	0.74	1.10	135	4.2	2.1	1.1	0.7	0.5	0.4	0.4	C	0.360	44.3
brown coal	0.70	1.10	143	4.5	2.3	1.1	0.8	0.6	0.5	0.4	C	0.360	46.9
heating oil	0.90	1.10	111	3.5	1.8	0.9	0.6	0.4	0.4	0.3	B	0.290	29.4
wood pellets	0.86	0.20	21	0.7	0.3	0.2	0.1	0.1	0.1	0.1	A0	0.020	2.1
wood chip	0.78	0.15	18	0.5	0.3	0.1	0.1	0.1	0.1	0.0	A0	0.020	2.3
wood piece	0.70	0.10	13	0.4	0.2	0.1	0.1	0.1	0.0	0.0	A0	0.020	2.6
wood gasification	0.83	0.10	11	0.3	0.2	0.1	0.1	0.0	0.0	0.0	A0	0.020	2.2
electricity	0.99	2.20	202	6.3	3.2	1.6	1.1	0.8	0.6	0.5	D	0.167	15.4
electric heat pump	3.40	2.20	59	1.8	0.9	0.5	0.3	0.2	0.2	0.2	A1	0.167	4.5
centralized source	0.76	0.70	84	2.6	1.3	0.7	0.4	0.3	0.3	0.2	B	0.290	34.8

The energy consumption values measured directly at the heat source enter into the comparison. The deductions were made at weekly intervals and are shown in the Fig. 3.

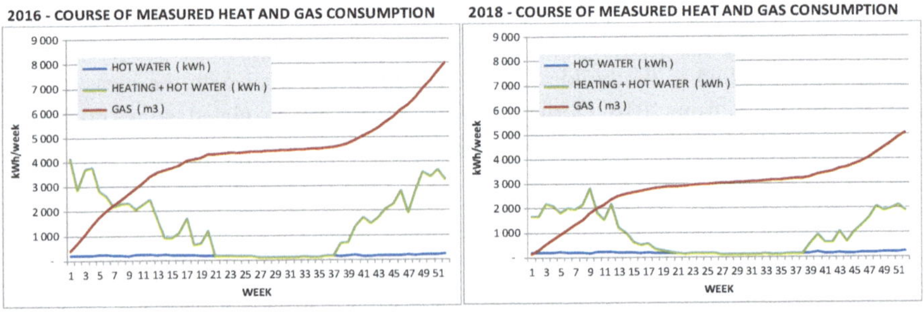

Fig. 3. Consumption for heating, hot water and gas for years 2016 and 2018

The values of energy consumption measurements were recalculated according to the daily rates, so that it is possible to compare the years 2016 and 2018 and to determine the energy benefit of the insulation of the object (Fig. 4).

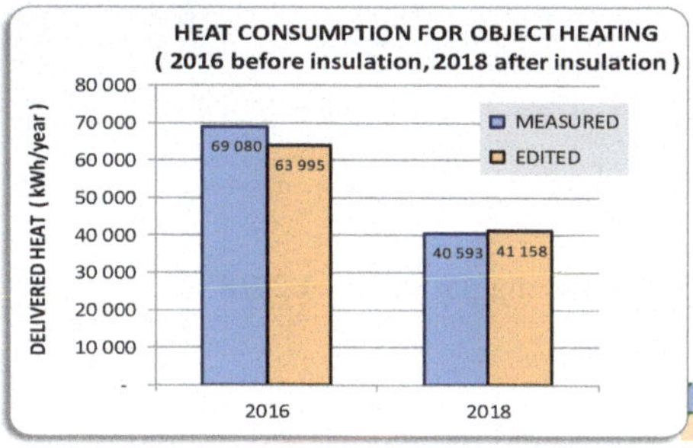

		2016		2018		DIFFERENCE
		1	2	1	2	
DAYS	-	237	212	213	212	
Te	°C	4.414	3.86	4.155	3.86	
Ti	°C	20	20	20	20	
D°	-	3 694	3 422	3 375	3 422	
HEATING	kWh	69 080	63 995	40 593	41 158	22 837
						35.7%

DAYS = number of heating days per year
Te = average outdoor air temperature at heating time
Ti = average interior temperature at the time of heating
D° = number of daily rates for the year considered
HEATING = consumed energy for heating
1 = measured values
2 = values recalculated by daily rates

Fig. 4. Conversion of measured consumption 2016 and 2018 by daily rates

By calculation it is possible to declare that the heat consumption has decreased by **35.7%.**

5 Discussion

5.1 Heat Source Designs

The first result of the paper is to determine the most suitable heat source according to the energy certificate. Most preferably wood fuel is unjustified in the present case. Therefore we propose heat pump, in the first case, water/water in the air/water species. The comparison is with respect to the existing gas-fired boiler room (Figs. 5 and 6).

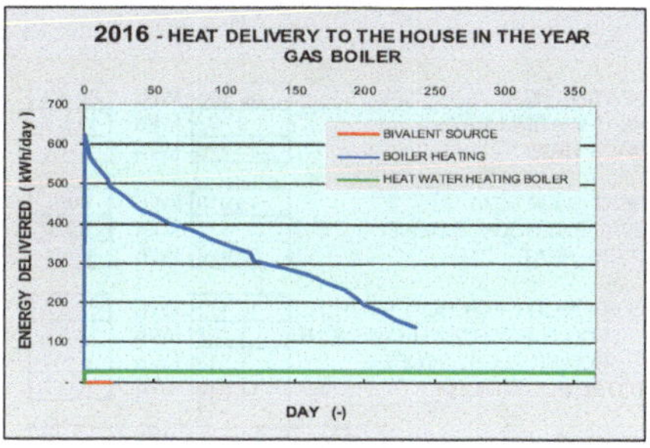

HEATING HEAT	69 076	kWh	87.0%
HEAT WATER HEAT	10 309	kWh	13.0%
MADE HEAT	79 385	kWh	100%
HEAT MADE FROM BOILER	79 385	kWh	93.6%
LOSS IN BOILER PRODUCTION	5 385	kWh	6.4%
BOILER ENERGY DELIVERED IN GAS	84 770	kWh	100%
ELECTRICITY FOR HEAT PUMP	0	kWh	0.0%
. . . FOR SUPPORTING EQUIPMENT	654	kWh	100%
. . . FOR BIVALENT SOURCE	0	kWh	0.0%
TOTAL ELECTRICITY	654	kWh	100%
HEATING	EFFECTIVENESS		93.8%
HOT WATER	EFFECTIVENESS		92.4%
SYSTEM	EFFECTIVENESS		93.6%

Fig. 5. Measured parameters of the existing gas boiler in 2016

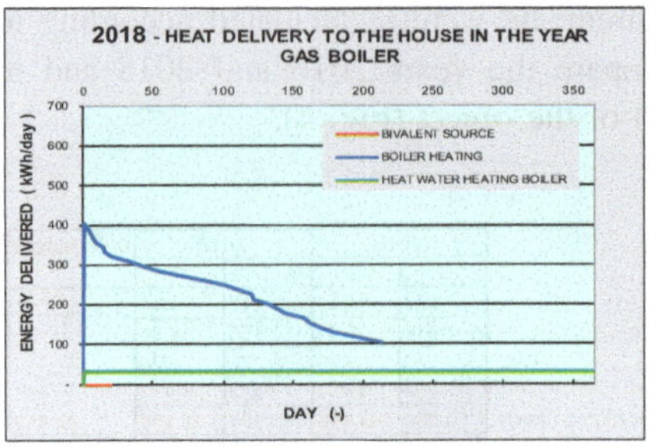

HEATING HEAT	40 593	kWh	77.2%
HEAT WATER HEAT	11 996	kWh	22.8%
MADE HEAT	52 589	kWh	100%
HEAT MADE FROM BOILER	52 589	kWh	94.1%
LOSS IN BOILER PRODUCTION	3 317	kWh	5.9%
BOILER ENERGY DELIVERED IN GAS	55 906	kWh	100%
ELECTRICITY FOR HEAT PUMP	0	kWh	0.0%
. . . FOR SUPPORTING EQUIPMENT	569	kWh	100%
. . . FOR BIVALENT SOURCE	0	kWh	0.0%
TOTAL ELECTRICITY	569	kWh	100%
HEATING	EFFECTIVENESS		94.4%
HOT WATER	EFFECTIVENESS		93.0%
SYSTEM	EFFECTIVENESS		94.1%

Fig. 6. Measured parameters of the existing gas boiler at 2018

5.2 Heat Pumps Use

The simulation was realized for the operation of the object with water/water heat pump (HP w/w) under the same conditions for 2016 and 2018 (Figs. 7 and 8).

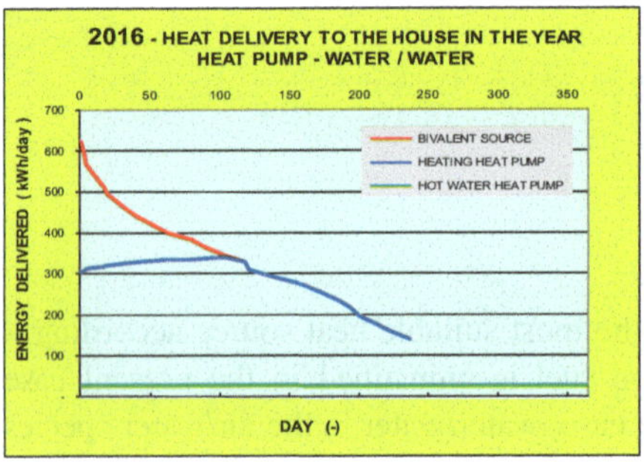

HEATING HEAT	69 076	kWh		87.0%
HEAT WATER HEAT	10 309	kWh		13.0%
MADE HEAT	79 385	kWh		100%
HEAT PUMP HEAT	69 493	kWh		87.5%
BIVALENT SOURCE HEAT	9 892	kWh		12.5%
TOTAL HEAT	79 385	kWh		100%
ELECTRICITY FOR HEAT PUMP	15 230	kWh		55.3%
. . . FOR SUPPORTING EQUIPMENT	2 413	kWh		8.8%
. . . FOR BIVALENT SOURCE	9 892	kWh		35.9%
TOTAL ELECTRICITY	27 535	kWh		100%
HEATING	COP =	4.71	SPF =	2.81
HOT WATER	COP =	3.86	SPF =	3.45
SYSTEM	COP =	4.56	SPF =	2.88

Fig. 7. Water/water heat pump parameters simulated in 2016

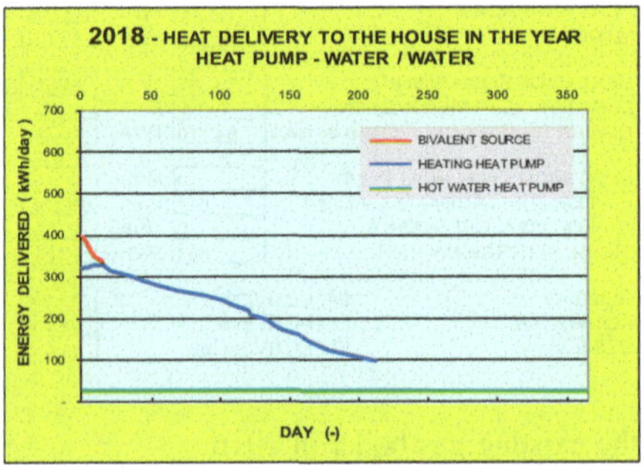

HEATING HEAT	40 593	kWh		80.5%
HEAT WATER HEAT	9 833	kWh		19.5%
MADE HEAT	50 426	kWh		100%
HEAT PUMP HEAT	49 913	kWh		99%
BIVALENT SOURCE HEAT	513	kWh		1.0%
TOTAL HEAT	50 426	kWh		100%
ELECTRICITY FOR HEAT PUMP	10 413	kWh		82.5%
. . . FOR SUPPORTING EQUIPMENT	1 702	kWh		13.5%
. . . FOR BIVALENT SOURCE	513	kWh		4.1%
TOTAL ELECTRICITY	12 628	kWh		100%
HEATING	COP =	4.96	SPF =	4.06
HOT WATER	COP =	4.21	SPF =	3.74
SYSTEM	COP =	4.79	SPF =	3.99

Fig. 8. Water/water heat pump parameters simulated in 2018

Due to the significant reduction in heat loss of the building due to thermal insulation, the efficiency of the heat pump is increased. The heating water temperature is lower, which has a beneficial effect on the heat pump efficiency. The reduced power requirement will greatly displace the need for a bivalent heat source. This was followed by air/water heat pump operation simulation. As in the previous case, we modeled an air/water heat pump (HP a/w) system. The balances are indicated in the Table 4.

Table 4. Balances of energies and CO_2 productions for studied variants

	BALANCE FOR REAL YEARS					
	ENERGY CONSUMED			EMISSIONS PRODUCED CO_2		
	GAS	HP w/w	HP a/w	GAS	HP w/w	HP a/w
	kWh	kWh	kWh	kg	kg	kg
2016	84 770	27 535	33 497	18 649	4 598	5 594
	100%	32.5%	39.5%	100%	24.7%	30.0%
2018	55 906	12 628	17 153	12 299	2 109	2 865
	100%	22.6%	30.7%	100%	17.1%	23.3%

	ADJUSTED BY D°					
	ENERGY CONSUMED			EMISSIONS PRODUCED CO_2		
	GAS	HP w/w	HP a/w	GAS	HP w/w	HP a/w
	kWh	kWh	kWh	kg	kg	kg
2016	78 530	25 508	31 031	17 277	4 260	5 182
	93%	30.1%	36.6%	93%	22.8%	27.8%
2018	56 684	12 804	17 392	12 471	2 138	2 904
	101%	22.9%	31.1%	101%	17.4%	23.6%

It is possible to save money by replacing a source **before** insulation:

- energy consumption from 84 770 kWh (gas) to 27 535 (electricity) - savings 67,5%
- CO_2 emissions from 18 649 (gas) to 4 598 (electricity) - savings 75,3%

Thermal insulation can save on the sources:

- energy consumption from 55 906 kWh (gas) to 12 628 (electricity) - savings 77,4%
- CO_2 emissions from 12 299 (gas) to 2 109 (electricity) - savings 82,9%

Daily Rate Conversion (source replacement + insulation)

- gas before insulation = 78 530 kWh, after insulation (electricity) 12 804 kWh - savings 83,7%
- CO_2 before insulation (gas) - 17 277 kg, CO_2 after insulation (electricity) 2 138 kg - savings 87,6%

6 Conclusions

Measurements and balances were converted to a residential house. Balance programs used are the result of long-term measurements and practical experiences. We present only outputs here. These results can be generalized. They can help in deciding on renovations or energy concepts when designing new buildings. The building was assessed on the terms of the Slovak Republic with European standards application. We did not include the price in the balance sheets because there are significant differences between countries in this commodity. From an energy certification point of view, wood is the best fuel for primary energy. Its potential and future accessibility must be considered. We do not consider it a prospective fuel mainly in urban areas. The advantage of heat pumps in a given balance is also supported by meeting the A1 criterion. Replacing the original heat source with a heat pump is also possible because the heat demand is significantly reduced after the insulation. Reducing heating system performance can be achieved by lower heating water temperature. This has a beneficial effect on the higher operating efficiency of the heat pump. If it is difficult to reach category A1, resp. B is possible to install renewable sources - solar systems, in the case of hot water systems with higher efficiency for preheating hot water and heating support. When using photovoltaic systems, it is possible to multiply the electric power generated by the heat generation by using heat pumps.

Acknowledgements. The paper is part of the project ITMS: 26220220064 under the title: Research Center for the Integration of Combined Renewable Energy Systems', and ITMS: 26220220182 under the title TECHNICOM University Science Park for Innovative Applications with Knowledge Technology Support - Phase I.

References

Act No. 86/2019 Coll. 555/2005 Coll. on energy performance of buildings and on amendments and supplements

Aste N, Adhikari RS, Manfren M (2013) Cost optimal analysis of heat pump technology adoption in residential reference buildings. Renew Energy 60:615–624

Deng J, Wei Q, Liang M, He S, Zhang H (2019) Does heat pumps perform energy efficiently as we expected: field tests and evaluations on various kinds of heat pump systems for space heating. Energy Build 182:172–186

European Commision (2015) Energy performance certificates across the EU. Mapping of national approaches. https://ec.europa.eu/energy/en/topics/energy-efficiency/energy-performance-of-buildings/certificates-and-inspections

Guelpa E, Mutani G, Todeschi V, Verda V (2018) Reduction of CO2 emissions in urban areas through optimal expansion of existing district heating networks. J Clean Prod 204:117–129

Jouhara H, Yang J (2018) Energy efficient HVAC systems. Energy Build 179:83–85

Kerdan IG, Raslan R, Ruyssevelt P, Gálvez DM (2016) An exergoeconomic-based parametric study to examine the effects of active and passive energy retrofit strategies for buildings. Energy Build 133:155–171

Péan TQ, Salom J, Costa-Castelló R (2019) Review of control strategies for improving the energy flexibility provided by heat pump systems in buildings. J Process Control 74:35–49

Stec A, Kordana S, Słyś D (2017) Analysing the financial efficiency of use of water and energy saving systems in single-family homes. J Clean Prod 151:193–205

Stefanco M, Kosicanova D, Vranay F, Kusnir M, Lojkovics J, Stone C (2014) Heat pumps as a means of efficiency using renewable energy sources. In: 14th International geoconference on energy and clean technologies. SGEM, Albena, Bulgaria

Su Ch, Madani H, Palm B (2018) Heating solutions for residential buildings in China: current status and future outlook. Energy Convers Manag 177:493–510

Thiers S, Peuportier B (2012) Energy and environmental assessment of two high energy performance residential buildings. Build Environ 51:276–284

Wu W, Skye HM (2018) Net-zero nation: HVAC and PV systems for residential net-zero energy buildings across the United States. Energy Convers Manag 177:605–614

Computer Aided Design - Optimisation of Cross-Sections of Steel Profiles of Industrial Sliding Gate

A. Wojnar(✉) and K. Sieńkowska

Department of Building Structures, Rzeszow University of Technology, ul. Poznańska 2, 35-084 Rzeszów, Poland
awojnar@prz.edu.pl

Abstract. The paper presents optimisations of industrial sliding gate structure of 6 m wide. The optimisation process was carried out in two steps. In the first step, the structure of the gate was optimized due to its weight and the value of vertical displacement of the corner node of the gate. In the second step, the cross-section of the ground track of the gate was optimized. It was assumed that the gate is loaded with its self-weight and wind pressure. The calculations were made using computer-aided design. The following programs were used: SHAPE-THIN, Autodesk Robot Structural Analysis and ConSteel. The structures of the gate was made of SHS profiles. The ground track of the gate was made of cold formed profile, of sheet thickness 4 mm. All elements were made of S235 steel grade.

Keywords: Metal structures · Industrial gates · Cold formed profiles · Ground track of the industrial sliding gate

1 Introduction

Sliding gates are perfect for most types of private and industrial business activity. The gates are installed in fences around industrial buildings, guarded car parks, schools, private property, etc. Industrial sliding gates usually are made of steel SHS sections. This modular design has no welds at the joints of gate elements, which made it easy assemble. Industrial sliding gates travel along the fence. For its installation and proper operation, free space is required to enable free movement of the leaf. A structure of the cantilever slim industrial gates lets to close off really big openings – up to 16 m wide with a single sliding gate. The structure of the gate consist (Fig. 1):

- ground track, (1)
- cantilever structure of the gate, (2)
- balancing back frame to prevent the gate from tilting when fully opened, (3)
- columns of supporting structure. (4)

Z. Blikharskyy et al. (Eds.): CEE 2019, LNCE 47, pp. 549–557, 2020.
https://doi.org/10.1007/978-3-030-27011-7_70

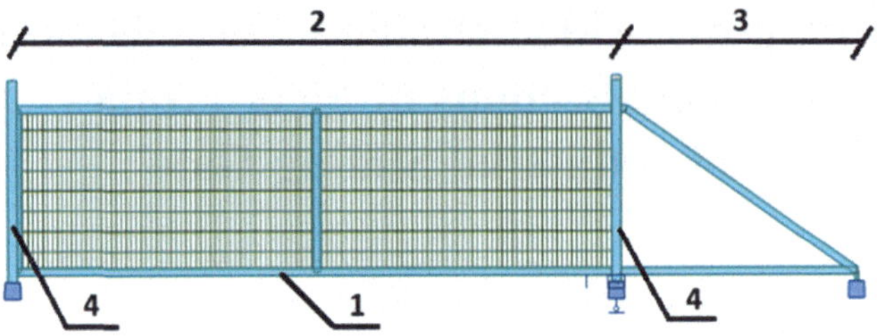

Fig. 1. View of the structure of typical industrial sliding gate. Description in the text.

2 Aim of the Paper

The optimisation of the structure of 6 m wide and 1,6 m high industrial sliding gate was presented in this paper. The following parameters were taken into account during optimisation:

– the shape of the gate – the type of structure of the sliding gate (truss structure/frame structure), and dimensions of cross-sections its elements and profiles, Fig. 2,

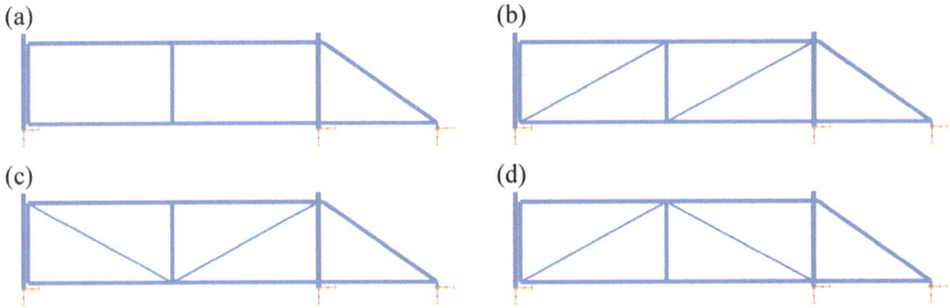

Fig. 2. View of the structure of 6 m wide industrial sliding gate: (a) frame structure, (b) truss structure type 1, (c) truss structure type 2, (d) truss structure type 3

– the shape of the gate ground track, Fig. 3.

All elements were made using S235 steel grade.

3 Numerical Models of the Structure of Sliding Gates

Numerical models of structure of industrial sliding gates used to checking ULS and SLS were made using Autodesk AutoCAD, Autodesk Robot Structural Analysis Professional and ConSteel software. Autodesk software were used to modelling of geometry of the structure and obtaining wind load. Value and distribution of wind load pressure were obtained using wind load simulation. This method of calculations generating the wind load on the structure in dependence on the direction of wind and its

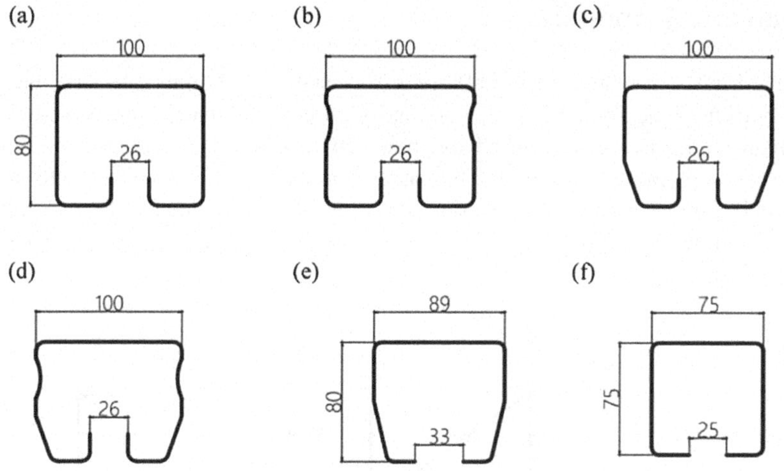

Fig. 3. View types of cross-sections of track ground; from a to d – shapes designed by Rzeszow University of Technology design team, e, f – shapes design by another firms

velocity (pressure). Self-weight of the structure was calculated automatically by the software. ConSteel software was used to checking of Ultimate Limit State of the structure (using the global method according to EN-1991-1-1) and Serviceability Limit State (vertical displacement of the free-end of the gate, in calculation case No. 1). Two calculation cases were taken into account during calculations:

1. case No. 1 – gate are open,
2. case No. 2 – gate are closed.

In the first case (No. 1) the cantilever beam was taken as a main static schema of the gate. It was assumed that the "cantilever part" of the gate structure was fixed in columns supporting structure located on one side of the gate. The ULS and SLS of the structure were checked in this case. In the second case the simply supported beam was taken as a main static schema of the gate. It was assumed that the "beam part" of the gate structure was supported in columns supporting structure located on both sides of the gate. In this case only ULS of gate was checked. Additionally in this case the gate's plane was modelled using panels, which caused an increasing the wind load value. The direction of wind load was taken as perpendicular to the gates plane, wind velocity was establish as 22 m/s.

According to recommendations from the gates manufacturers the following profiles were used in the structure of 6 m wide industrial sliding gates:

- ground track – design as a cold-formed profile,
- cantilever truss (frame) structure of the gate – columns, upper and lower chords, balancing back frame from SHS 80 × 80 × 3 profile, rods from SHS 40 × 40 × 3 profile,
- columns of supporting structure from SHS 100 × 100 × 4 profile.

4 Numerical Analysis

Ultimate Limit State and Serviceability Limit State of sliding gate structure were checked using Fine Elements Analysis and ConSteel software. Values vertical displacements industrial gates caused by their self-weight (in dependence of types of structures) were shown on Fig. 4. The calculation case No. 1 was taken into account during calculations. Values of utilization of profiles were presented on Figs. 5 and 6. In this calculations case No. 1 (see Fig. 5) and case No. 2 (see Fig. 6) were taken into consideration.

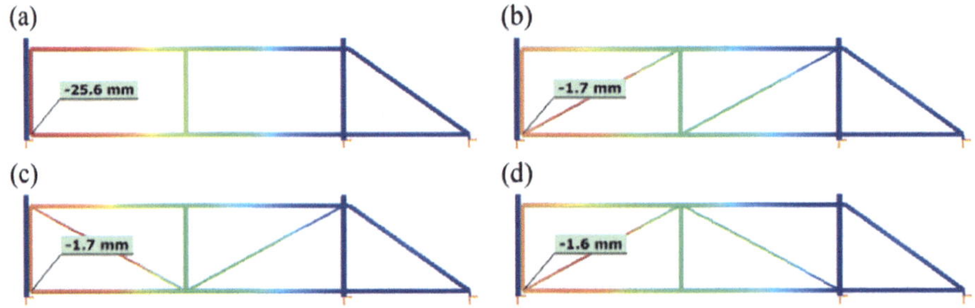

Fig. 4. Values of vertical displacements of industrial sliding gate: (a) frame structure, (b) truss structure type 1, (c) truss structure type 2, (d) truss structure type 3

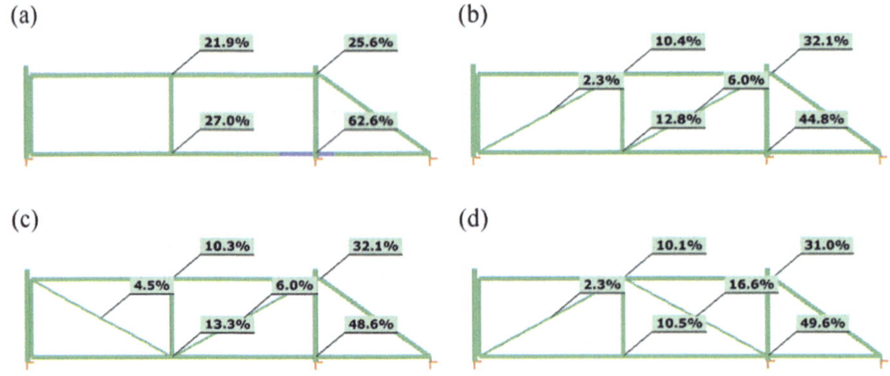

Fig. 5. Values of utilization (global check) of industrial sliding gates (gates were opened, without panels): (a) frame structure, (b) truss structure type 1, (c) truss structure type 2, (d) truss structure type 3

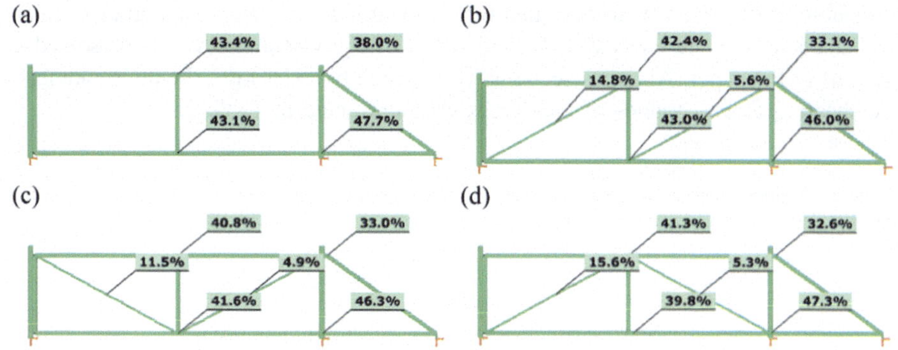

Fig. 6. Values of utilization (global check) of industrial sliding gates (gates were closed, panels): (a) frame structure, (b) truss structure type 1, (c) truss structure type 2, (d) truss structure type 3

5 Conclusions

Based on the results of analyses, the following conclusions were drawn:

1. The frame cantilever frame structure of industrial sliding gate (Fig. 4a) has the smallest stiffness in the vertical plane. Its vertical displacement of free-end of the gate is equal to 26 mm and it is about 15 times more than in case of gates structure shown on Fig. 4b, c and d (truss cantilever structure of the gates).
2. Type of the truss structure used as a cantilever part of the gate (Fig. 4b, c, d) has no influence on gate stiffness in the vertical plane. Vertical displacements of free-end gates are almost the same.
3. In case No. 1 (gate is open) the static schema of the structure is a cantilever beam. Utilization of profiles is about 63% in case structure modelled as a frame-structure or about 45–50% in case structure modelled as truss-structure.
4. In case No. 2 (gate are closed) the static schema of the structure is self-supported beam. Utilization of profiles is about 47% (lower chord) and 43% (upper chord).
5. Taking above into consideration it can be seen that case No. 1 is more unfavourable and determines the load capacity of the gates structure.

6 Optimisation of the Shape of the Track Ground Cross-Sections

In the second part of this paper, optimisation of the shape of the track ground cross-sections was made. Based on results of above calculations, frame-structure of the gate taken into account, Fig. 5a. Shapes of analysed cross-sections were presented on Fig. 3. Shapes presented on Fig. 3a, b, c, d were created by a design team of the Rzeszow University of Technology in cooperation with an industrial company. Shapes presented on Fig. 3e, f were designed by competing companies. Comparing geometrical characteristics all cross-sections and values of equivalent stress were presented in this part of the paper. All cross-sections were loaded by axial force, bending moments and shearing forces. Values of this forces were presented in Table 1.

Table 1. Values of internal forces in analysed track ground cross-sections of industrial sliding gate

$N_{c,Ed}$ [kN]	$M_{y,Ed}$ [kNm]	$M_{z.Ed}$ [kNm]	V_y [kN]	V_z [kN]
3,87	1,51	2,70	1,37	1,22

$N_{c,Ed}$ – axial compressive force, $M_{y,Ed}$ – bending moment in vertical plane, $M_{z,Ed}$ – bending moment in horizontal plane, V_y – shear force in horizontal plane, V_z – shear force in vertical plane

All calculations were made using SHAPE-THIN computer software. All geometrical parameters, values of internal forces and steel grade were introduced in to the numerical model of cross-sections. All results were calculated using Fine Element Analysis and presented on Figs. 7 and 8 (Fig. 9).

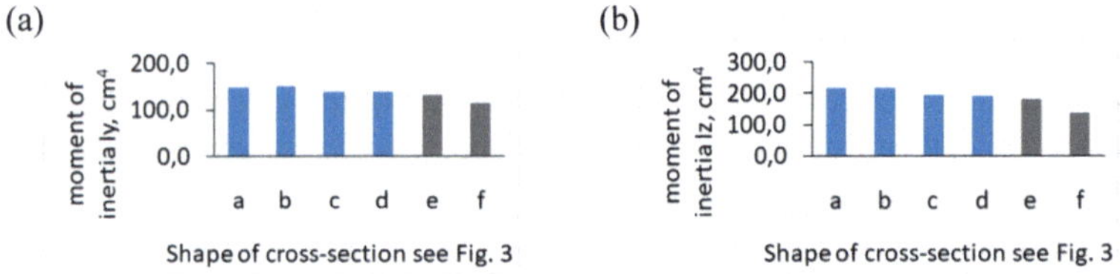

Fig. 7. Results of calculations: (a) moment of inertia I_y (in vertical plane) of cross-sections, (b) moment of inertia I_z (in horizontal plane) of cross-sections

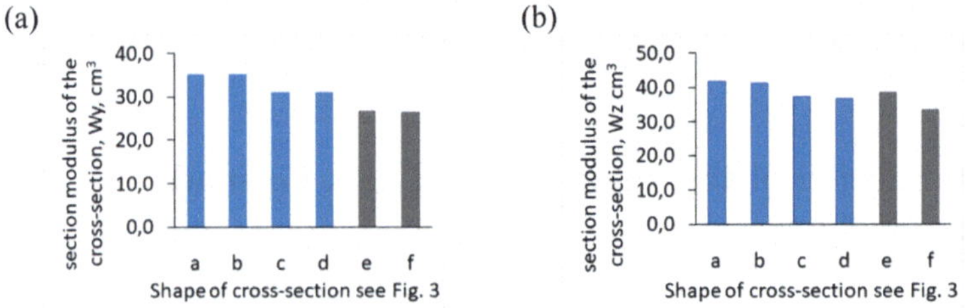

Fig. 8. Results of calculations: (a) section modulus of the cross-section, W_y (in vertical plane), (b) section modulus of the cross-section, W_z (in horizontal plane)

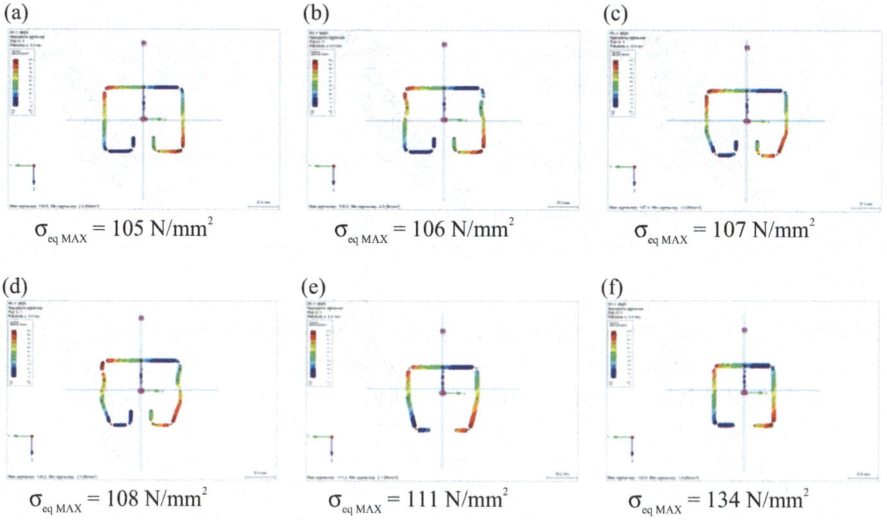

Fig. 9. Results of calculations, values of equivalent stress in ground track cross-sections, cross-sections according to: (a) Fig. 3a, (b) Fig. 3b, (c) Fig. 3c, (d) Fig. 3d, (e) Fig. 3e, (f) Fig. 3f

7 Conclusions

Based on the results of analyses, following conclusions were drawn:

1. Cross-section shown on Fig. 3f has the smallest self-weight. Self-weight of designed profiles (cross-sections from a to d, Fig. 3) is more then this one. In case profiles 3c, 3d (Fig. 3) increasing self-weight is small – up to 2%. In case profiles 3a, 3b is more up to 7%, Fig. 10.

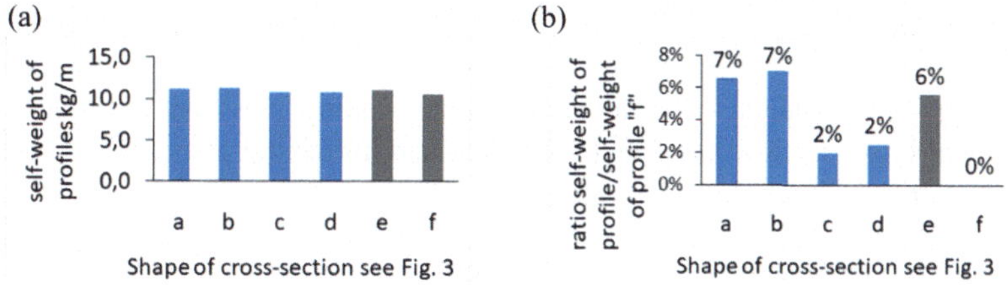

Fig. 10. Results of calculations: (a) self-weight of profiles, (b) changing values of self-weight

2. All types of designed profiles have higher moment of inertia (in vertical and horizontal plane) than existing profiles, especially cross-section 3f. In case of stiffness in vertical plane the increase is up to 30%, in case horizontal plane up to 63%, Fig. 11.

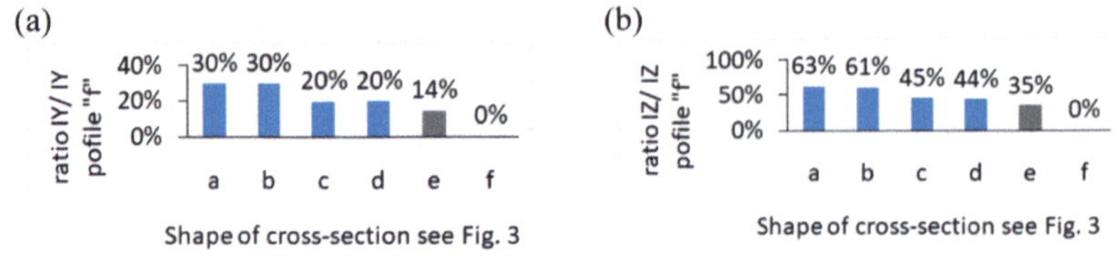

Fig. 11. Results of calculations, changing values of moment of inertia in relation to cross-section 3f: (a) changing moment of inertia in vertical plane, (b) changing moment of inertia in horizontal plane

3. Resistance of designed cross-sections is more than existing profiles (especially in case 3f cross-section, Fig. 3). Value of equivalent stress in designed profiles is about 105–108 N/mm^2 and is about 20% less than in case cross-section 3f (about 134 N/mm^2).
4. Next, the ratio: values of moment inertia I_y and I_z to self-weight profiles was calculated. Results were presented on Fig. 12. It can be seen that the most advantageous value of ratio is for cross-section 3a, Fig. 3, the worst for existing profiles 3e, 3f.

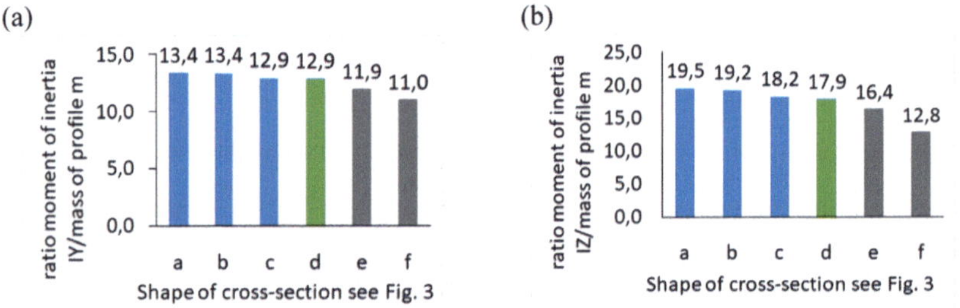

Fig. 12. Results of calculations, ratio moment of inertia to mass profile: (a) ratio moment of inertia I_y – vertical plane to self-weight profile, (b) ratio moment of inertia I_z – vertical plane to self-weight profile

5. The same types analysis like above was made taking into consideration section module of cross-section calculated in vertical and horizontal plane, Fig. 13. Results of calculations are similarly like above.

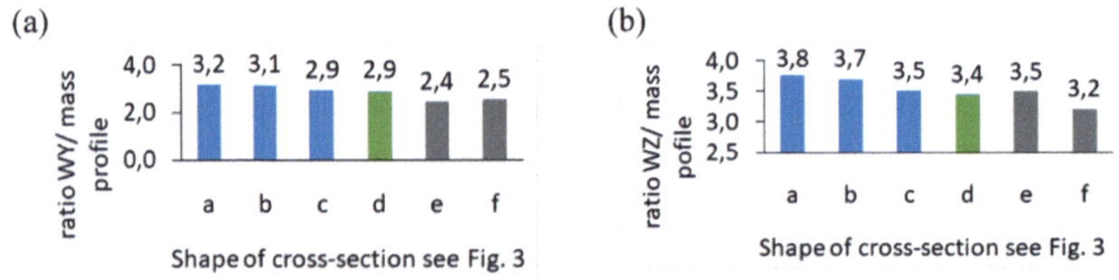

Fig. 13. Results of calculations, ratio section module of cross-section to mass profile: (a) ratio section module of cross-section W_y – vertical plane to self-weight profile, (b) ratio section module of cross-section W_z – vertical plane to self-weight profile

Final conclusion: Taking account all types of calculations described above and its results, it can be said that cross-section 3d – Fig. 3 – has optimal shape. This is proved by favourable values of ratio presented of Figs. 11, 12 and 13 and self-weight profile 3d to 3f – Fig. 10.

References

EN 1991-1-4 (2005) Eurocode 1: actions on structures - Part 1-4: general actions - wind actions. European Committee for Standardization, Brussels

EN 1993-1-1 (2005) Eurocode 3: Design of steel structures – Part 1-1: general rules for buildings. European Committee for Standardization, Brussels

Surface Reinforcement in Concrete Beams to Ensure Controlled Cracking

Szczepan Woliński(✉)

Department of Building Structures, Faculty of Civil and Environmental Engineering and Architecture, Rzeszow University of Technology, ul. Poznańska. 2, 35-084 Rzeszow, Poland
szwolkkb@prz.edu.pl

Abstract. Principles and methods for design of surface reinforcement required to control cracking on side faces of reinforced concrete beams are presented and discussed. On the basis of the modified concept for assessing the effective tension area in the beam's cross-section, which dimensions was assessed using the fracture mechanics of concrete, a method was proposed to calculate cracks width over the entire depth of beam provided with surface reinforcement. Proposed method allows to determine the quantity and distribution of surface reinforcement above the main reinforcement axis necessary for the acceptable width of cracks.

Keywords: Surface reinforcement · Effective tension area · Controlled cracking

1 Introduction

According to fib Model Code 2010 and Eurocode EN 1992-1-1 the surface reinforcement in RC beams should be used to resist spalling in the case of the main reinforcement is made up of bars with diameter greater than 32 mm or bundled bars with equivalent diameter greater than 32 mm. Similarly, it should be used for enhanced durability where the cover of reinforcement is greater than 70 mm. In the previous generation of Polish Standards, the use of surface reinforcement or additional longitudinal reinforcement bars placed between the main reinforcement axis and the neutral axis of the beam cross section was recommended in beams of the depth over 600–700 mm. The role of this reinforcement was to limit the width of the cracks above the main reinforcement and to protect the vertical external surfaces of the beam against cracking due to concrete shrinkage and thermal actions. The methods of calculating and detailing of the surface reinforcement recommended by various standards and authors of relevant research lead to significantly different results (Basteskår et al. 2019; Knauff 2018; Subramanian 2005; Ziari and Kianoush 2009). This is mainly due to taking into account only one function of this reinforcement, most often treated as anti-shrinking or limiting the width of cracks at the depth of the cross-section of the beam.

The paper presents a brief review of methods for design the longitudinal surface reinforcement in the RC beams and for calculation the crack width at their depth based on the concept of effective tensile area of concrete, determined by means of the of

Z. Blikharskyy et al. (Eds.): CEE 2019, LNCE 47, pp. 558–565, 2020.
https://doi.org/10.1007/978-3-030-27011-7_71

nonlinear fracture mechanics, that allowing to determinate of the necessary reinforcement sectional area, numbers, diameters and spacing of reinforcement bars for the assumed allowable width of cracks.

2 Formulas for Surface Reinforcement

Conventional criterion applied in designing of the surface reinforcement for RC beams is to maintain its integrity in case of distortion actions. It leads to the requirement that the tensile force in the surface reinforcement should be not greater than the cracking force. The solutions of the problem formulated in this way differ quite a lot but the minimum value of reinforcement ratio for surface longitudinal reinforcement $\rho_{l,min}$ is always proportional to the ratio of the tensile strength of concrete to the yield stress of reinforcement. For example, according to (Rüsch and Jungwirth 1979):

$$\rho_{l,min} = (1 + nv)\frac{f_{ctm}}{f_y} \approx 1.05\frac{f_{ctm}}{f_y} \tag{1}$$

and (Flaga 2015):

$$0.55\frac{f_{ctk0.05}}{f_{yk}} \le \rho_{l,min} \le 1.10\frac{f_{ctk0.95}}{f_{yk}} \tag{2}$$

where f_{ctm},$f_{ctk0.05}$,$f_{ctk0.95}$ are the mean, lower and upper characteristic values of the tensile strength of concrete, f_y, f_{yk} are the mean and lower characteristic value of the yield stress of surface reinforcement, v is the Poisson's ratio.

According to the recommendations of the withdrawn Polish Standard PN-B-03264:2002, in beams with cross-section depth greater than h $\geq$ 0.7 m, longitudinal bars with diameters of $\emptyset \geq$ 8 mm, treated as reinforcement needed due to shrinkage stresses, should be placed at the side surfaces, spacing s $\leq$ 0.40 m.

Recently, the opinion prevails that the basic criterion for designing the surface reinforcement should be related to limiting the width of cracks on the side surfaces of beams in the persistent situation. Based on a study by Breen and Frantz (1978), confirmed by many other investigators, in Fig. 1 the representative distribution of the mean crack widths at the depth of web of the reinforced concrete beam is presented, depending on the spanning and diameters of the surface reinforcement bars. The results of tests and observations of distribution and widths of cracks on the side surfaces of reinforced concrete beams with significant cross-section depths allow to formulate the following conclusions:

- For beams with cross-sectional depth h $\geq$ 0.6–0.7 m, without surface reinforcement or additional longitudinal bars on the side surfaces of the beam, the maximum width of cracks above the axis $w_{max}(x)$ of the main tensile reinforcement is usually greater than cracks width at the level of main reinforcement $w_{max}(x = 0)$.

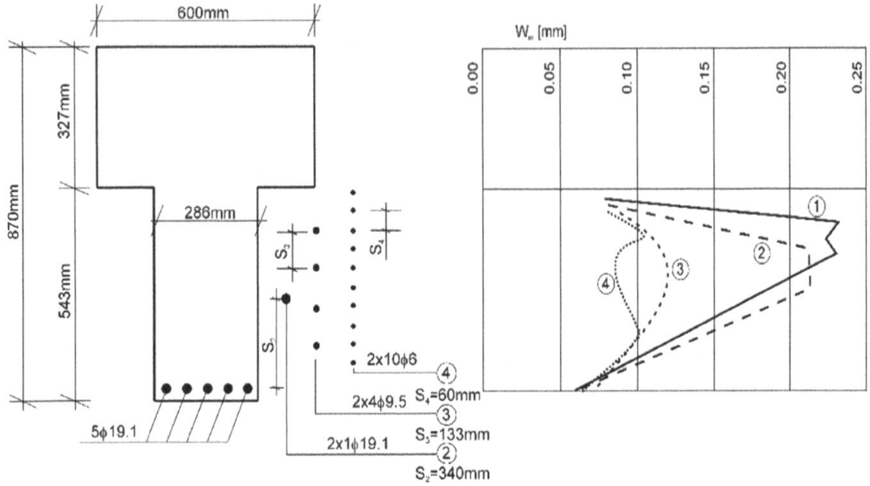

Fig. 1. Extent and width of cracks at the depth of the side faces of RC beams with different surface reinforcement according to (Breen and Frantz 1978).

- For beams without surface reinforcement or additional longitudinal reinforcement bars on the side surfaces of the beam, the ratio $w_{max}(x)/w_{max}(x = 0) \cong 1.5 \div 5$ and depends mainly on the depth of the of beam's cross-section.
- The use of additional longitudinal reinforcement bars at the side surfaces of the beam results in reduction of the crack width above the reinforcement axis which depend mainly on: spacing of reinforcement bars and their diameter, bond stresses, thickness of concrete cover, ultimate tensile strain of concrete, tensile strength and fracture energy of concrete and the width of beam's cross-section.

In the calculation of the crack width in elements with reinforcement bars concentrated in the part of tensile zone of the cross-section area, the concept of effective tensile area of concrete $A_{ct,eff}$ is used. This concept was proposed by A. Beeby in 1971 and used in Eurocode 2. It consists in replacing the whole tension area of the cross-section, in which the tensile reinforcement is concentrated in 1–3 layers by "effective tension area" $A_{ct,eff}$ in which the reinforcement axis is located in the gravity center of $A_{ct,eff}$.

Figure 2 presents several ways to determine the range and size of $A_{ct,eff}$ in cross-sections of concrete beams with reinforcement concentrated in one and two levels: (a), (b), (c), (e) and distributed at the depth of the tension area A_{ct}: (e), (f) and (g). Because the formulas for determining $A_{ct,eff}$ are usually empirical, they have a limited applications and lead to different results. The crack width w_m be determined using the formula recommended in Eurocode 2:

$$w_m = s_{rm}(\varepsilon_{sm} - \varepsilon_{cm}) \tag{3}$$

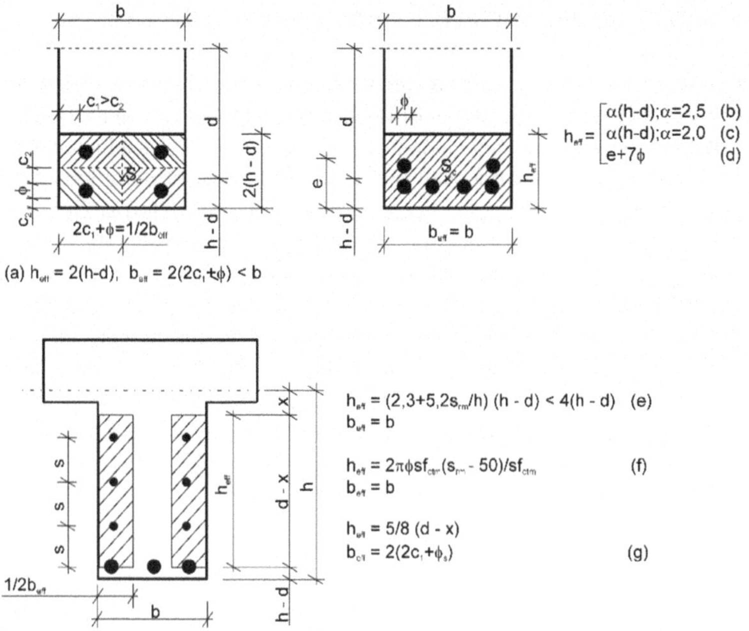

Fig. 2. Recommendations for calculating the effective area of the tensile concrete $A_{ct,eff}$ in the cross-section of RC beams according to different sources.

where s_{rm} is the maximum cracks spacing, ε_{sm} and ε_{cm} are the mean strain in the reinforcement (taking into account the effect of tension stiffening) and the mean strain in concrete between cracks. The strain difference $(\varepsilon_{sm} - \varepsilon_{cm})$ is calculated using a simplified formula which is exact for pure tension but not for bending where the concept of effective tension area $A_{ct,eff}$ has been used.

According to the proposal presented in Eurocode 2 the surface reinforcement needed to limit the crack width on the beam's side faces in the form of longitudinal bars should be evenly distributed between the main reinforcement and the neutral axis of the beam's cross-section and located inside stirrups. The cross sectional area of that reinforcement $A_{s,l}$ can be determined using the following expression:

$$A_{s,l} = k_c k f_{ct,eff} \frac{A_{ct}}{f_{yk}} \tag{4}$$

where k_c, k are empirical coefficients which takes account of the form of strain distribution, shape and dimensions of concrete tension area A_{ct} and the effective tensile strength of concrete $f_{ct,eff} \cong 3\,\text{MPa}$.

3 Evaluation of the Surface Reinforcement Effectiveness

As a measure of the surface reinforcement effectiveness, the ratio of the maximum width of cracks measured at the depth of the beam's side surfaces to the crack width at the level of the main reinforcement axis $w_{max}(t)/w_{max}\,(t = 0)$ was adopted. Increase of the cracks width at the side surfaces of the beam without surface reinforcement is due to increase of the cracks spacing and decrease of the ultimate deformability of concrete with the distance from reinforcement concentrated near the tensile edge of the beam's cross-section. So, the maximum width of cracks in between the axis of main reinforcement and the neutral axis can be observed in cases of deep beams without surface reinforcement. In order to calculate the crack opening width, the concept of effective tension area was used. Application of the nonlinear fracture mechanics method for concrete and the crack band model (Bazant 2002) enabled author of this paper to obtain relatively simple, analytical expression to assess the range of the advantageous effect of reinforcement on the deformability of concrete g_w (Wolinski 1991, 2005):

$$g_w \approx \frac{1}{6}\left[l_{ch} + (l_{ch}(l_{ch} + 12\emptyset))^{\frac{1}{2}}\right] \tag{5}$$

where $l_{ch} = G_F E_c / f_{ctm}^2$ is characteristic length of concrete (material's parameter), G_F is the fracture energy of concrete, E_c is modulus of elasticity of concrete, $\emptyset$ is diameter of reinforcing bar. According to (fib MC 2010) the fracture energy G_F [N/m] for normal concrete can be calculated from the relation:

$$G_F = 73 f_{cm}^{0.18} \tag{6}$$

where f_{ctm} is the mean value of axial tensile strength of concrete.

Assuming that the g_w-value determines the range of the effective tension area $A_{ct,eff}$ in the beam's cross-section, it is possible to calculate the average width of the crack, which will not be exceeded within this zone. Outside the $A_{ct,eff}$ area, the width of the cracks can exceed the value of $w_{max}(x = 0)$ if the mean value of crack spacing s_{rm} at the depth of the beam sides increases faster than the mean value of strain $\varepsilon_{rm} = (\varepsilon_{sm} - \varepsilon_{cm})$ decreases.

The value of $s_{rm}(x)$ can be determined by means of the semi-empirical relation formulated by (Braam 1990):

$$s_{rm(x)} = s_{rm}(x = 0) + \frac{(h - d - x - s_{rm)}}{h - d - x} x\left(0.9 - 1.23\rho_{r,ref}^{0.45}\right) \tag{7}$$

where x is the distance from the main reinforcement axis to the axis of the reinforcement bars located inside the $A_{ct,eff}$, $\rho_{r,eff} = A_s/A_{ct,eff}$ is the effective reinforcement ratio for the main longitudinal reinforcement A_s.

The distance from the main reinforcement axis x_{max} for which the crack reaches the largest width outside the $A_{ct,eff}$ area, can be determined from the relation:

$$\frac{dw_m(x)}{dx} = 0 \tag{8}$$

The presented procedure makes possible to solve the reverse task, that is determination of cross sectional area of the surface reinforcement and selection of longitudinal bars spacing of surface reinforcement at beam depth for the assumed crack width.

4 Working Example

On the example of three reinforced concrete beams examined by Braam (1990) and presented in Fig. 3, the proposed method for assessing the effectiveness of the surface reinforcement is illustrated and verified.

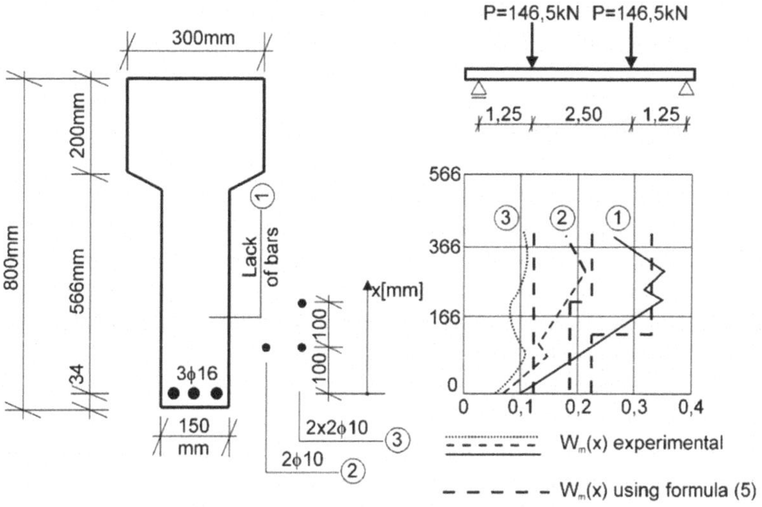

Fig. 3. The width of cracks at the depth of the side faces of RC beams with different surface reinforcement obtained in experiments (Braam 1990) and calculated for effective area of the tensile concrete $A_{ct,eff}$ according to proposed method.

The maximum width of the crack at the depth of the effective tension area g_w = 154 mm, calculated according to the formula (5), does not exceed 0.22 mm for the beam without surface reinforcement, but above 154 mm from the lower edge of the beam's cross-section reaches the maximum value 0.32 mm. For the beam reinforced additionally using 2 bars with the diameter ø10 mm, located 100 mm above the axis of the main reinforcement, the maximum width of crack at the depth of the effective tension area g_w = 244 mm does not exceed 0.18 mm but above 244 mm from the lower edge of the beam's cross-section reaches the maximum value 0.23 mm. Finally, for the beam reinforced additionally using 4 bars with the diameter ø10 mm, located 100 and 200 mm above the axis of the main reinforcement, the maximum width of crack at the depth of the effective tension area g_w = 340 mm does not exceed 0.14 mm while for the depth above 340 mm from the lower edge of the beam's cross-section reaches the maximum value lower than 0.14 mm.

The results of these calculations show quite good agreement with those obtained in experimental studies (Fig. 3). Assuming an acceptable crack width of 0.20 mm, for the considered beam loaded with two concentrated forces 2 × 146.5 kN, it is necessary to use the surface reinforcement in the form of 2 longitudinal bars with a diameter of 10 mm at a distance of 100 mm above the main reinforcement axis.

Surface reinforcement calculated on the basis of the integrity of the beam's cross-section subjected to the distortions is lower by about 25% in the case of the application of formula (1) and about 30% lower with the upper limit of formula (2) than required due to the limitation of crack width at beam's depth.

Cross sectional area of surface reinforcement calculated on the basis of the integrity of the beam's cross-section according to formula (1) is lower by about 25% than required due to limitation of cracks width at the beam's depth, while calculated using the upper limit in formula (2) is lower by about 30%.

5 Concluding Remarks

Presented method of crack's width assessment at the cross-sectional depth of reinforced concrete elements subjected to bending allows to determine the quantity and distribution of surface reinforcement above the axis of main reinforcement necessary for the acceptable width of cracks. Cross sectional area of surface reinforcement needed to limit the width of cracks up to 0.2–0.3 mm is usually much greater than those necessary to limit the width of shrinkage cracks.

Based on the observation of reinforced concrete structures, it seems that the acceptable crack width on the side surfaces of beams can be increased to 0.3–0.4 mm, but also in this case the surface reinforcement of deep beams calculated according to the recommendations of current standards is too small.

References

Basteskar M, Engen M, Kanstad T, Fossa KT (2019) A review of literature and code requirements for the crack width limitations for design of concrete structures in serviceability limit states. Struct Concr J fib 20(2):549–866. https://doi.org/10.1002/suco.201800183

Bazant Z (2002) Concrete fracture models: testing and practice. Eng Fract Mech 69(2002):165–205

Braam CR (1990) Control of crack width in deep reinforced concrete beams. HERON TU Delft 35(3):1–106

Breen J, Frantz G (1978) Control of cracking on side faces of large reinforced concrete beams. Report 198-1F, University of Texas

EN 1992-1-12 (2008) Eurocode 2: design of concrete structures –Part 1-1: general rules and rules for buildings. Brussels, CEN, European Committee for Standardization

fib Model Code for Concrete Structures (2010) Federation Internationale du Beton. Ernst & Sohn

Flaga K (2015) The influence of concrete shrinkage on durability of reinforced concret structural members. Bull Pol Acad Science Tech Sci 63(1). https://doi.org/10.1515/bpasts-2015-0002

Knauff M (2018) Design of concrete structures. Wydawnictwo Naukowe PWN, Warszawa (in Polish)

Rüsch H, Jungwirth D (1979) Shrinkage and creep in concrete structures. Arkady, Warszawa (in Polish)

Subramanian N (2005) Controlling the crack width of flexural RC members. Indian Concr J **79**(11):1–6

Wolinski S (1991) Tensile behavior of concrete and their applications in nonlinear fracture mechanics of concrete. Rzeszow UT, Rzeszow (in Polish)

Wolinski S (2005) New technique of improving the cracking resistance of concrete walls in early age. In: COST conference on improvement of buildings' structural quality by new technologies, Innsbruck, pp 717–724

Ziari A, Kianoush MR (2009) Investigation of flexural cracking and leakage in RC liquid containing structures. Eng Struct 31(5):1056–1067

Selection of the Best Variant of Flood Protection

M. Zeleňáková(✉) and L. Zvijáková

Department of Environmental Engineering, Technical University of Košice, Vysokoškolská 4, 042 00 Košice, Slovakia
martina.zelenakova@tuke.sk

Abstract. In the current period of scientific and technological progress, new concepts, tools and methods of solution must fulfill the principles of a democratic society that dictate us to ensure the quality of the human environment for future generations. Part of this challenging task is the timely assessment of the potential impacts of the proposed activity on the environment and public health with acceptable environmental risk. For this reason, it is necessary to develop the theory and to apply the methods for the systematic investigation of the consequences of projects, constructions, plants, facilities and other interventions on the environment and the population. The aim of the paper is to analyze the current state of the impact assessment process and to point out the existing methods used in the impact assessment process. The work points out the possibility of improving existing methods of impact assessment of proposed activities by applying risk analysis in the process of applying the impact of water structures on the environment. The use of risk analysis methods to assess the impact of activities on the environment and population health is undoubtedly an original and innovative proposal. The paper has a clearly defined original methodology and solution concept.

Keywords: Environmental impact assessment · Flood protection · Proposed activity

1 Introduction

The environmental impact assessment of constructions is seen as a tool that minimizes the implementation of activities that could in any way adversely affect the environment and at the same time allows for the optimal solution, i.e. the one with the least negative environmental impact of the proposed activity.

The entire construction cycle divides the investment construction process into phases, namely pre-investment, investment and implementation. This means that for the proposed construction it is necessary to work out a project, a project for a zoning decision, a project for a building permit, a project for implementation and a project for the final approval decision (Mesaros et al. 2018). EIA is a systematic approach to identifying and evaluating positive and negative impacts on components of the environment that may arise from the implementation of infrastructure projects or policies (Wang et al. 2006). EIA is a mandatory process before approval of infrastructure

Z. Blikharskyy et al. (Eds.): CEE 2019, LNCE 47, pp. 566–574, 2020.
https://doi.org/10.1007/978-3-030-27011-7_72

projects with significant impacts on the environment (Tamura et al. 1994), such as roads (Zhou and Sheate 2011), water supply systems (Al-Agha and Mortaja 2005) and flood protection constructions (Hapciuc et al. 2016). Flood protection structures have been created throughout the centuries to mitigate flood damage (Satrapa et al. 2010).

Nowadays, several authors address this issue of environmental impact assessment worldwide, see Bond and Morrison-Saunders (2011), Bond and Pope (2012), Canter (1996), Cashmore et al. (2009), Demidova and Cherp (2005), Galas et al. (2014), Wang et al. (2006), Zvijáková et al. (2014), Zeleňáková et al. (2014), Zeleňáková and Zvijáková (2017a, 2017b). The paper presents application of the proposed methodology in Zeleňáková and Zvijáková (2017b).

2 Aim

The aim of the paper is to identify and evaluate the expected impacts of selected water construction – with the purpose of flood protection on the environment, estimating their significance and their assessment by proposed method. For the purposes of the paper, the proposed activities are flood protection objects, which are included in the list of proposed activities subject to their environmental impact assessment.

2.1 Study Area

The location of the proposed activity is in the cadastral area of Kružlov, through which the Slatvinec stream flows. The village of Kružlov (Fig. 1) is located in the north-eastern part of Slovakia.

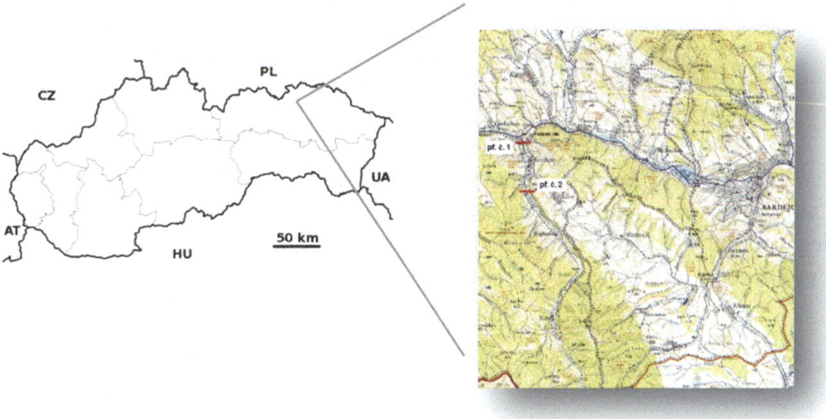

Fig. 1. Localization of the proposed activity within Slovakia

It belongs to the district of Bardejov and to the Prešov region. Figure 1 shows a clearer close-up situation with the location of the proposed activity at a scale of 1:50,000. The area of the proposed activity belongs to the Bodrog river basin and to the Topľa sub-basin. The Slatvinec stream has the character of natural watercourse. On the right bank of the stream are private lands with family houses, on the left bank of the stream there is a local asphalt road. The river bed has an irregular trapezoidal cross-section about 6.0–13.0 m wide. At the site of the planned stream regulation, the flow threatens the existing built-up area with adjacent lands at the flood discharges and at the same time causes its erosion to cause slope instability.

The Slatvinec stream, in the village of Kružlov, is a constant threat of flooding in the village. For this reason, it is necessary to increase the flood protection of the environment and the inhabitants of the municipality. Increased flood protection in the village of Kružlov can be achieved through various measures, from less effective measures to increase retention capacity and anti-erosion protection to highly effective technical measures.

3 Method

The impact assessment process is a systematic approach to assessing the proposed activity. The basic principle of the used methodology is the calculation of risk indexes - an estimate of the level of risk posed by flood protection objects for the environment. The risk analysis is based on the principle that every construction, including flood protection objects, does not only adversely affect hydrological, morphological and geographic features by its technical nature; ecological and archaeological/historical; socio-economic characteristics of the area by equation:

$$R_j = \frac{\sum_{i=1}^{n} P_i w_i}{\sum_{i=1}^{n} w_i} \tag{1}$$

where R is risk (-), P is score of parameter (-), w is weight (-), i is ranking of parameter, j is alternative, n is number of parameters.

Negative impacts can be quantified by calculating the environmental risk index of flood protection objects.

4 Results

The purpose of the proposed activity - flood protection objects is to modify the drainage conditions in order to increase flood protection. The alternative solution to the proposed activity is the basis for the environmental impact assessment. Three variations are proposed:

Variant 0: Current status – the null variant is the current situation in which the bed of the Slatvinec stream and bridge structures in the urban area do not have sufficient capacity to drain the increased flood flows, which often causes recurrent flood situations and consequent damage to the environment and property in the municipality.

Variant I: Water stream regulation – relates to the proposal to modify the Slatvinec stream in the village of Kružlov. Modifications of watercourses have been in the past and are still one of the basic construction activities in the country at present (Šlezingr 2010). The aim of the proposed flood protection measures on the Slatvinec stream in the village of Kružlov is to ensure safe Q_{100} drainage without annual damage to public and private property.

Variant II: Polder construction – considers the design of a dry polder, which would be situated above the village of Kružlov on the Slatvinec stream. Polder will ensure the protection of the urban area of Kružlov under the dam profile immediately and will reduce the peak flow in the lower parts of the Slatvinec stream.

In Table 2 the score (0.2; 0.4; 0.6; 0.8 and 1.0) is assigned for each parameter (A–Z) for variants of proposed activity based on Table 1.

Table 1. Determination of values of individual assessed parameters for individual variants

ID	Parameter	Variant 0	Variant I	Variant II
A	Maximum specific runoff q_{max} ($m^3 \cdot s^{-1} \cdot km^{-2}$)	3.969	3.969	3.969
B	100-year return period discharge Q_{100} ($m^3 \cdot s^{-1}$)	105.0	105.0	95.0
C	Design discharge Q_n ($m^3 \cdot s^{-1}$)	70.0	105.0	95.0
D	Annual precipitation H_z (mm)	600–700	600–700	600–700
E	Forest cover l (%)	60	60	60
F	Basin saturation coefficient S (mm)	3,464	3,464	3,464
G	Nature of the watercourse (-)	0,237	0,237	0,237
H	Average longitudinal slope of flow i_t (%)	2,2	1,3	2,2
I	Basin type (-)	Oblong	Oblong	Oblong
J	River basin area S_p (km^2)	23,085	23,085	23,085
K	Soil type (-)	Sandy-loamy, loamy-sandy	Sandy-loamy, loamy-sandy	Sandy-loamy, loamy-sandy
L	Slope of the basin i_s (%)	4,995	4,995	4,995
M	Ecological significance of the area (-)	Low	Low	Low

(*continued*)

Table 1. (*continued*)

ID	Parameter	Variant 0	Variant I	Variant II
N	Effect on protected species of plants, animals and habitats (number)	0	0	0
O	Change of landscape character (-)	The presence of symptoms, disturbing elements in the country (if non-existent)	The impact of the intention on the future appearance and character of the land is not disturbing (preventive assessment)	The impact of the intention on the future appearance and nature of the country is not disturbing (preventive assessment)
P	Cultural-historical significance of the territory (-)	International and national	International and national	International and national
Q	Affected sites and geological sites (number)	0	0	0
R	Permanent residents in the village (number)	967	967	967
S	Coefficient of built area in municipality (-)	0,0381	0,0381	0,0381
T	Type and importance of transport (point)	1	1	1
U	Infrastructure of municipality (point)	5	5	5
V	Production activity in the village (point)	4	4	4
W	The level of flood protection of the population and the environment (-)	Slight loss of human lives is expected, and environmental damage is likely	Loss of human life is unlikely and environmental damage is insignificant	Loss of human lives and environmental damage are unlikely
X	Total cost of implementation activity (EUR)	0	3 000 000	4 500 000
Y	Distance of the location of the activity from the built-up area (km)	0,010	0,050	0,515
Z	State of flood protection objects (-)	No technical flood protection measures are implemented	Regulation and stabilization of flows in built-up areas of municipalities (fencing of streams and objects on them)	Construction of dry tanks (polders)

Table 2. Allocation of parameter score for assessed flood protection object variants

ID	Variant 0			Weight (w_i)	Variant II		
	0	I	II		0	I	II
	Score of parameter (P_i)				Score of parameter (P_i)		
A	0,2	0,2	0,2	0,07	0,014	0,014	0,014
B	0,6	0,6	0,6	0,05	0,03	0,03	0,03
C	0,4	0,2	0,2	0,1	0,04	0,02	0,02
D	0,6	0,6	0,6	0,08	0,048	0,048	0,048
E	0,4	0,4	0,4	0,04	0,016	0,016	0,016
F	1	1	1	0,03	0,03	0,03	0,03
G	0,4	0,4	0,4	0,01	0,004	0,004	0,004
H	0,4	0,2	0,4	0,04	0,016	0,008	0,016
I	0,4	0,4	0,4	0,01	0,004	0,004	0,004
J	0,4	0,8	0,8	0,02	0,008	0,016	0,016
K	0,4	0,4	0,4	0,02	0,008	0,008	0,008
L	0,4	0,4	0,4	0,02	0,008	0,008	0,008
M	0,4	0,4	0,4	0,01	0,004	0,004	0,004
N	0,2	0,2	0,2	0,02	0,004	0,004	0,004
O	0,8	0,4	0,4	0,01	0,008	0,004	0,004
P	1	1	1	0,01	0,01	0,01	0,01
Q	0,2	0,2	0,2	0,01	0,002	0,002	0,002
R	0,8	0,8	0,8	0,05	0,04	0,04	0,04
S	1	1	1	0,03	0,03	0,03	0,03
T	0,2	0,2	0,2	0,04	0,008	0,008	0,008
U	0,6	0,6	0,6	0,05	0,03	0,03	0,03
V	0,4	0,4	0,4	0,05	0,02	0,02	0,02
W	0,8	0,4	0,6	0,1	0,08	0,04	0,06
X	0,2	1	1	0,03	0,006	0,03	0,03
Y	1	0,8	0,6	0,03	0,03	0,024	0,018
Z	1	0,4	0,2	0,07	0,07	0,028	0,014

From this it is possible to calculate the risk indices (R_j) of the proposed activity in the assessed locality of Kružlov according to Eq. (1) (Fig. 2).

The first place is a variant that is optimal in terms of the level of risk it poses to the environment. The second place is a variant that is less acceptable and the third place is a variant that is least acceptable in terms of the risk to the environment.

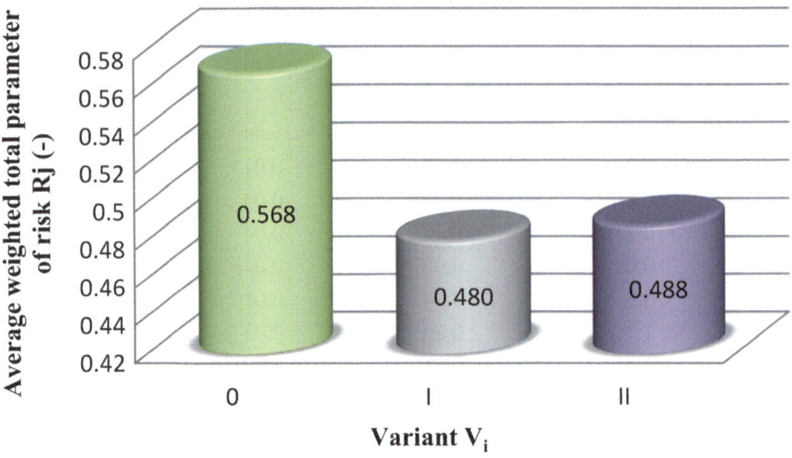

Fig. 2. Assessment of variants of flood protection objects in the municipality of Kružlov based on the average sum of R_j

5 Scientific Novelty and Practical Significance

Within the presented methodological procedure for environmental impact assessment of flood protection objects the scientific novelty and practical significance is presented by:

- proposed A–Z indicators related to the design of flood protection facilities;
- assigned standardized weight for each of the A–Z indicators to determine their importance;
- based on the method of multi-parametric risk expression, the calculated average weighted sum parameter of the risk of *R*j for each assessed variant of the construction - the flood protection object on the basis of which the significance of the risk is determined.

The theoretical but also practical benefits that can be used also in the educational process is the proposed methodology that is part of the assessment of the impact of water structures on the environment. The proposed methodology evaluates the severity of the impact of water structures and activities in the field of water management for the environment, for the purposes of Act no. 24/2006 Coll. On EIA in Slovakia as amended. It can also be used to assess and prioritize risks in the proposed activities.

6 Conclusions

The territory of interest in the realization of the activity - flood protection objects is located in the cadastral area of Kružlov and partly reaches the cadastral area of Krivé.

The aim of the research was to propose, based on scientific results, the optimal variant of flood protection, which would protect the inhabitants of the village Kružlov and the surrounding environment against the flash floods.

The comparison of the proposed activities - flood protection objects in the village of Kružlov on the Slatvinec stream shows that all three assessed variants (Variant 0, Variant I and Variant II) represent different risks based on the calculated average weighted total risk parameter *R*j to the environment. It is on the basis of these calculated risk parameters of *R*j that overall it can be concluded that Variant I is optimal in terms of expected environmental effects, since it achieves the lowest calculated risk indices (*R*j = 0.480); and therefore the construction of the "Watercourse Regulation" on the Slatvinec stream in the village of Kružlov is recommended. This flood protection facility represents an overall low level of environmental risk.

Acknowledgement. This work was supported by projects VEGA 1/0217/19 and APVV-18-0360.

References

Al-Agha MR, Mortaja RS (2005) Desalination in the Gaza Strip: drinking water supply and environmental impact. Desalination 173(2):157–171

Bond AJ, Morrison-Saunders A (2011) Re-evaluating sustainability assessment: aligning the vision and the practice. Environ Impact Assess Rev 31(1):1–7

Bond A, Pope J (2012) The state of the art of impact assessment in 2012. Impact Assess Proj Apprais 30(1):1–4

Cashmore M, Bond A, Sadler B (2009) Introduction: the effectiveness of impact assessment instruments. Impact Assess Proj Apprais 27(2):91–93

Demidova O, Cherp A (2005) Risk assessment for improved treatment of health considerations in EIA. Environ Impact Assess Rev 25(4):411–429

Gałaś S (2014) Assessment of the quality of the environment in the v4 countries. AGH Univ Sci & Technol, Krakow, Poland, p 2014

Hapciuc OE, Romanescu G, Minea I, Iosub M, Enea A, Sandu I (2016) Flood susceptibility analysis of the cultural heritage in the Sucevita catchment (Romania). Int J Conserv Sci 7 (2):501–510

Mesaros P, Mandicak T, Behun M (2018) Applications of knowledge technology in construction industry. In: 16th international conference on emerging eLearning technologies and applications (ICETA). Stary Smokovec, Slovakia

Satrapa L, Fosumpaur P, Horsky M (2010) Experience in assessing the design of land consolidation project focusing on flood protection. In: Conference on landscape engineering 2011. Praha, Czech Republic

Slezingr M, Foltynova L, Zelenakova M (2010) Assessment of the current condition of riparian and accompanying stands. In: Colloquium on Landscape Management, Mendel Univ. Brno, pp 24–27

Tamura H, Fujita S, Hiroshi K (1994) Decision analysis for environmental impact assessment and consensus formation among conflicting multiple agents - including case studies for road traffic. Sci Total Environ 153(3):203–210

Wang YM, Yan JB, Xu DL (2006) Environmental impact assessment using evidential reasoning approach. Eur J Oper Res 174(3):1885–1913

Zeleňáková M, Zvijáková L, Gaňová L (2014) Environmental impact study Bodva river basin Water resources and wetlands. In: 2nd international conference, Tulcea, Romania, 11–13 September 2014. Transversal publishing house, Targoviste, pp 329–335
Zeleňáková M, Zvijáková L (2017a) Risk analysis within environmental impact assessment of proposed construction activity. Environ Impact Assess Rev 62(76–89):2017
Zeleňáková M, Zvijáková L (2017b) Using risk analysis for flood protection assessment. Springer, Cham, 127p
Zhou K, Sheate WR (2011) EIA application in China's expressway infrastructure: clarifying the decision-making hierarchy. J Environ Manage 92(6):1471–1483
Zvijáková L, Zeleňáková M, Purcz P (2014) Evaluation of environmental impact assessment effectiveness in Slovakia. Impact Assess Proj Apprais 32(2):150–161

Author Index

Z. Blikharskyy et al. (Eds.): CEE 2019, LNCE 47, pp. 575–577, 2020.
https://doi.org/10.1007/978-3-030-27011-7

The manufacturer's authorised representative in the EU is Springer Nature Customer Service Centre GmbH, Europaplatz 3, 69115 Heidelberg, Germany. If you have any concerns regarding our products, please contact ProductSafety@springernature.com

Printed and bound by CPI Group (UK) Ltd, Croydon, CR0 4YY
05/10/2026
02228130-0020